T0136056

QUADRATICS

The CRC Press Series on

DISCRETE MATHEMATICS and ITS APPLICATIONS

Series Editor

Kenneth H. Rosen, Ph.D.

AT&T Bell Laboratories

Published Titles

Cryptography: Theory and Practice, *Douglas R. Stinson*

Network Reliability, *Daryl D. Harms, Miroslav Kraetzl, and Charles J. Colbourn*

Forthcoming Handbooks

Handbook of Discrete Mathematics, *Kenneth H. Rosen*

Standard Reference of Discrete Mathematics, *Kenneth H. Rosen*

Handbook of Graph Theory, *Jonathan Gross*

Handbook of Combinatorial Designs, *Charles Colbourn and Jeffrey Dinitz*

Handbook of Cryptography, *Scott Vanstone, Paul Van Oorschot, and Alfred Menezes*

Handbook of Discrete and Computational Geometry, *Jacob E. Goodman and Joseph O'Rourke*

Forthcoming Textbooks and Monographs

Graph Theory with Computer Science Applications, *Jonathan Gross and Jay Yellen*

Error-Correcting Codes and Algebraic Curves, *Serguei Stepanov*

Frames and Resolvable Designs, *Steven Furino*

Introduction to Information Theory and Data Compression, *Darrel R. Hankerson, Greg A. Harris, and Peter D. Johnson*

QUADRATICS

Richard A. Mollin

CRC Press is an imprint of the
Taylor & Francis Group, an **informa** business

CRC Press
Taylor & Francis Group
6000 Broken Sound Parkway NW, Suite 300
Boca Raton, FL 33487-2742

First issued in paperback 2019

© 1996 by Taylor & Francis Group, LLC
CRC Press is an imprint of Taylor & Francis Group, an Informa business

No claim to original U.S. Government works

ISBN-13: 978-0-8493-3983-7 (hbk)
ISBN-13: 978-0-367-40152-8 (pbk)

This book contains information obtained from authentic and highly regarded sources. Reasonable efforts have been made to publish reliable data and information, but the author and publisher cannot assume responsibility for the validity of all materials or the consequences of their use. The authors and publishers have attempted to trace the copyright holders of all material reproduced in this publication and apologize to copyright holders if permission to publish in this form has not been obtained. If any copyright material has not been acknowledged please write and let us know so we may rectify in any future reprint.

Except as permitted under U.S. Copyright Law, no part of this book may be reprinted, reproduced, transmitted, or utilized in any form by any electronic, mechanical, or other means, now known or hereafter invented, including photocopying, microfilming, and recording, or in any information storage or retrieval system, without written permission from the publishers.

For permission to photocopy or use material electronically from this work, please access www.copyright.com (http://www.copyright.com/) or contact the Copyright Clearance Center, Inc. (CCC), 222 Rosewood Drive, Danvers, MA 01923, 978-750-8400. CCC is a not-for-profit organization that provides licenses and registration for a variety of users. For organizations that have been granted a photocopy license by the CCC, a separate system of payment has been arranged.

Trademark Notice: Product or corporate names may be trademarks or registered trademarks, and are used only for identification and explanation without intent to infringe.

Library of Congress Cataloging-in-Publication Data

Mollin, Richard A., 1947–
Quadratics / Richard A. Mollin.
p. cm. -- (Discrete mathematics and its applications)
Includes bibliographical references and index.
ISBN 0-8493-3983-9 (alk. paper)
1. Quadratic fields. I. Title. II. Series.
QA247.M65 1995
512'.7--dc20 95-23311
CIP

Library of Congress Card Number 95-23311

Visit the Taylor & Francis Web site at
http://www.taylorandfrancis.com

and the CRC Press Web site at
http://www.crcpress.com

Dedicated to my wife

Bridget Lynn Mollin

who made it all possible

by always being there.

Contents

List of Symbols

Symbol

Preface

A "preface" is a place to state the motivation for the writing of a book. There can be no stronger motivation in mathematical inquiry than the search for truth and beauty. It is this author's longstanding conviction that number theory has the best of both of these worlds. In particular, algebraic and computational number theory have reached a stage where the current state of affairs richly deserves a proper elucidation. It is this author's goal to attempt to shine the best possible light upon the subject.

In recent years, the development of the theory of continued fractions in conjunction with the theory of reduced ideals has rendered numerous results in algebraic and computational number theory, especially when applied to quadratic orders. It is the purpose of this book to present this material in detail, together with the attendant techniques, methodologies and tools used to achieve them. The current literature on the subject is scattered, and anyone trying to find their way through the maze of results will invariably come upon conflicting notation, terms which are used to mean two different things (such as "primitive"), some obscurities as well as some misinterpretations stemming from the translation from forms to ideals. We will therefore take a different approach from that of most authors who use Gauss' genus theory and composition of binary quadratic forms, and who use class field theory as a developmental tool. We will instead look at the *infrastructure*, introduced by Dan Shanks, of a quadratic order which is the coupling of continued fractions and reduced ideals described in Chapter Two. The infrastructure (*not* known to Gauss) and the entire continued fraction approach have not received the respect and currency which they so richly deserve. The writing of this book then has the additional aim of bringing this approach to the forefront as the principal algorithmic tool. In point of fact, the continued fraction algorithm has the power of modern computational number theory at its disposal in the sense that results ensuing from it may be readily checked and illustrated via such software packages as MAPLE (e.g. see Chapter Seven, section one). This brings the infrastructure to the doorstep of modern computation.

Readers interested in form-theory and the class-field theory approach have a wealth of books from which to choose. They range from such standard texts as Borevich–Shafarevich [30] and Cohn [57] to the more recent, texts by Buell [41], and Cox [67]. There seems to be little point in trying to improve upon such a well-documented approach with so many well-written texts (and more coming from colleagues such as the Buchmann–Williams collaboration and Andrew Granville, both of whom are completing texts from the perspective of form theory at the time of this writing). Rather, this book uses the continued-fraction-ideal-theoretic approach via the infrastructure to concentrate *solely* upon quadratic orders and attendant topics in "quadratics", such as quadratic diophantine equations (Chapter Three), prime-producing quadratic polynomials (Chapter Four), class numbers of quadratic orders (Chapter Five), ambiguous ideals in quadratic orders (Chapter Six), quadratic residue covers (Chapter Seven), and algorithms from cryptography based upon ideals in the class group of a complex quadratic order (Chapter Eight). Also, numerous tables are provided in Appendices A–D. Furthermore, for the sake of balance and completeness we have added Appendix E which gives an overview of the theory of binary quadratic forms, genus theory and composition as it relates to the ideal theory presented herein. Finally, Appendix F contains background from analytic number theory, especially as it relates to Chapter Five, section four.

Given the development in this book on the most recent advances, one of the goals is to have brought the reader to the frontiers of research on the subject. In fact, the proof of any number of the conjectures in this book could lead to a doctorate in itself. Certainly there is enough material to provide a basis for a variety of graduate courses. The reader is invited to view and learn these topics from this new perspective. We trust that no damage has been done to the beauty of this subject by this presentation.

The following people deserve thanks for proofreading parts of the text or for useful conversations. With that being said, this author is solely responsible for any errors, omissions or oversights in this text; and should the reader find any, this author would welcome any correspondence on the matter. Moreover, there has been the most careful attempt to honour priority in the proof of a given fact. If someone was missed, or priority misplaced, it is not for the lack of having searched diligently to ensure otherwise. This author trusts that the extent of the list of references itself is testimony to this assertion. Thanks go to Duncan Buell, John Burke, Harvey Cohn, J.H.E. Cohn, Andrew Glass, Mike Jacobson, Franz Lemmermeyer, Nikos Tzanakis, Gary Walsh, Hugh Williams, Kenneth Williams and Liang-Cheng Zhang. Also, thanks go to Hugh Williams for inspired discussions over a decade ago concerning continued fractions; which ultimately led to a collaboration resulting in over two dozen coauthered papers. Much appreciation is also due to my typist, Joanne Longworth, for her long and arduous hours put into the typing of this text, and for tolerating my numerous revisions. Finally, I am truly indebted to the Killam Foundation for awarding me a resident Killam Fellowship at the University of Calgary during the fall of 1994 for the purpose of writing this book.

Introduction

The preface delineated the motivation for writing the book. The Introduction will now describe the methodology, organization and style used in that writing.

Chapter One is an introductory device intended to set the stage by giving all relevant background needed from algebraic number theory. We do this through discussion of much of this material as statements of fact. However, as is the case with each subsequent chapter, the sections each contain exercises, most of which are accompanied by elaborate "hints" to aid the reader interested in working through the proofs of these facts. Nevertheless, we do provide some proofs which do not exist elsewhere, since it is a fact, to which we alluded in the Preface, that the literature is seriously deficient in terms of ideal-theoretic proofs of keystone concepts. For instance, Theorem 1.4.2 has some truly elegant consequences for complex quadratic orders as evidenced by the whole of section one of Chapter Four and, in particular, Theorem 4.1.10, which describes a criterion for the class group to have exponent at most two in terms of the form of the discriminant. However, the only place where this result is stated in ideal-theoretic terms is in Buchmann–Williams [37] where it appears without proof. Furthermore, there are some other rather enlightening consequences of Theorem 1.4.2, such as, for instance, Corollary 1.4.6 which we apply to class number problems in, for example, Chapter Five. In fact, in section five of Chapter One, we look in detail at arbitrary quadratic orders. We have separated off this general consideration for the sake of clarity of presentation, since it is necessarily the case that once we move up into greater generality, there are more complications, more notation and generally more things that can "go wrong", all of which makes for greater interest, of course. With that being said, the reader who is only interested in the maximal order, i.e. the field case, can skip this section without breaking continuity since all subsequent use of the greater power of the more general case can be interpreted for the maximal order if so desired. Section six of Chapter One is an application of real quadratics to powerful numbers. This is an optional section, which is not part of the background material, but can have some rather deep consequences. In fact, we list and discuss more than twenty outstanding conjectures which have ramifications for such problems as Fermat's last theorem, and even deeper considerations. The proofs of relevant facts provided in Chapter One are thus made as self-contained as practicable. In fact, the decisions as to whether or not to include proofs throughout the text, especially in Chapter One, were based upon whether or not the proofs in question exist in easily accessible places, and whether or not inclusion/exclusion of a proof would detract from central themes in the book. For example, the proof of Theorem 4.1.3 is extremely well-known and is proved at the end of Cox's book [67] using modular functions and complex multiplication, so it makes perfect sense to merely cite the latter. Thus, in the interest of having the book as self-contained as *practicable*, any sources used are included unless doing so would severely alter the scope of the book.

Chapter Two describes the continued fraction algorithm for real quadratic orders and its "analogue" for complex quadratic orders. This infrastructure, as introduced by the work of Dan Shanks for real quadratic orders, is the principal tool used in subsequent chapters. In order to aid the reader in learning this key material, numerous exercises with very detailed hints have been provided in order to illustrate the applicability of the algorithm. Section two contains the complex analogue as introduced by Buchmann and Williams in [37]. We return to this algorithm in Chapter Eight where we describe their cryptographic algorithm for a secure key-exchange system.

Chapter Three explores the relationship between the solutions to diophantine equations and the divisibility of class numbers of quadratic orders. In section one, proofs of some new results are presented together with some consequences and examples based upon *Lucas–Lehmer theory* such as Theorem 3.1.3. All of section one deals with negative discriminants. Section two turns to positive discriminants, especially those of *Extended-Richaud-Degert* (ERD) type. A complete description of the simple continued fraction expansion of the principal class based upon the use of the infrastructure is given in Theorem 3.2.1, and numerous examples and exercises are provided as illustrations. Section three focuses on reduced ideals and solutions of quadratic diophantine equations. Much of this is work of Halter–Koch or generalizations thereof which we have gleaned from the literature and reworked. The power of ideal theory is shown here. Section four deals with solutions of diophantine equations (viewed as positive discriminants) having implications for divisibility of class numbers of real quadratic orders, and is illustrated by numerous examples. Section five presents a new method of Mollin–van der Poorten–Williams [240] for solving quadratic diophantine equations via the infrastructure. It is described in such a way as to be a paradigm with an invitation, via the exercises, for the reader to develop the general theory.

Chapter Four looks at prime-producing quadratic polynomials. Section one deals with the complex side, beginning with Euler's celebrated polynomial x^2+x+41 and culminating with this author's complete classification of all quadratic polynomials of *Euler-Rabinowitsch-type* (Definition 4.1.4) which produce consecutive, distinct, initial prime values up to both a Rabinowitsch bound (Definition 4.1.4) and a Minkowski bound (Definition 1.3.2). We provide a list of all such polynomials, which we know is complete under the assumption of the *generalized Riemann hypothesis* (GRH), discussed in detail in section four of Chapter Five. Section two provides a similar list for polynomials of positive discriminant, and we prove that this list must necessarily contain only ERD-types. This list is known to be complete with one possible exceptional value of the discriminant which we show (in Chapter Five) would have to be a counterexample to the GRH. Some conjectures are left in these two sections for the reader's consideration, albeit the conjectures are known to be true under the assumption of the GRH. Section three of Chapter Four deals with the question of *density* of primes produced by quadratic polynomials via Hardy and Littlewood's well-known conjecture (see Conjecture 4.3.1). In particular, we discuss recent advances of Lukes *et al.* [207].

Chapter Five primarily deals with class number bounds. In section one, we look at factoring the Euler–Rabinowitsch polynomial. The infrastructure is used to describe algorithms for factoring the Euler–Rabinowitsch polynomial based upon a parameterization of the discriminant and the period length of the simple continued fraction expansion of the principal class. Section two lists and compares numerous class number one criteria for real quadratic fields, including some very recent and highly useful ones which were developed using the infrastructure. Section three shows how to use the *divisor function* to bound class numbers of both real and complex fields from below. Some of these recent results yield some very sharp bounds which we illustrate with numerous examples. Section four delineates a history of the *Riemann hypothesis* (RH) and the GRH. Then, using the development of the analytic class number formula, it is shown how to compile lists of various class numbers which are known to be complete with one possible GRH-ruled out exception (via a result of Tatuzawa which we discuss). We demonstrate how arbitrarily large lists of *any* class number desired may be compiled by these methods subject only to computational considerations.

Chapter Six uses the new algorithmic tool, called the *Palindromic Index*, to classify certain ambiguous cycles of reduced ideals without any ambiguous ideals in them. Herein, we discuss equivalence of ideals *without* reference to a class group structure since the reduced ideals in the cycles are not necessarily invertible. Section two solves the exponent two problem for real quadratic fields of ERD-type using the results of Chapter Five. Moreover, we classify certain prime-producing polynomials related to the exponent two problem, then relate all this back to Chapter Four wherein we achieved an exponent two criterion in the complex case.

Chapter Seven describes in detail the new tool called a *Quadratic Residue Cover* or QRC (Definition 7.1.1) introduced in [265] which provided solutions of class number problems such as that of the *Shanks sequence*, i.e. discriminants of the form $\Delta = (2^s+1)^2+2^{s+2} = (2^s+3)^2-8$. This sequence was conjectured to have class number one for a specified small list. The relative ease with which QRC's were used to prove this conjecture led to investigations of more general parameterized discriminants, namely $\Delta = (ba^n + (a-1)/b)^2 + 4a^n$. Results, conjectures, examples and prospects for the future development of QRC's are all discussed at length in section one. In section two, we briefly discuss the solution of the problem of classifying all real quadratic fields which have three or more consecutive norms of reduced principal ideals in a row (as determined by the infrastructure) as powers of a given integer $c > 1$. This was inspired by consideration of such sequences as the Shanks sequence where *all* norms of principal reduced ideals are powers of 2.

In Chapter Eight, section one we give an extensive history of computation of class numbers of real quadratic fields beginning with Gauss and ending with recent advances. We begin the section with definitions of basic concepts and notation in computer science in order to more effectively guide the uninitiated reader through the material. In any case, a general overview should be obtained by every reader. This leads naturally into section two where we describe computational algorithms in cryptography, including some very brief descriptions of the Diffie–Hellman scheme, the discrete log problem and the RSA key exchange cryptosystem, together with an illustration and exercises. This is meant more to pique the reader's interest in the subject than to give any comprehensive overview, and so numerous references are given to steer the reader in the direction of the excellent, in-depth texts and articles on this currently very popular subject. We culminate with a very detailed description of the Buchmann–Williams algorithm for a secure key-exchange system based upon the difficult problem of determining the order of an ideal class in an imaginary quadratic field. This was one of the very first applications of algebraic number theory to computing in quadratic fields and so fits in perfectly as one of the closing statements in the book in that it fits the "key" into the lock of the complex interface between algebraic number theory and modern day computer technology. Section three of Chapter Eight contains a recent article by Andy Lazarus, which gives a human face to the computational efforts by discussing the philosophical implications.

Appendices A–D contain numerical tables. Appendix A has numerous tables which list results proved in text, as described in the Table of Contents. Appendix B has a table of fundamental units of real quadratic fields with radicands less than $2 \cdot 10^3$ compiled using PARI, obtained from colleagues at the University of

Bordeaux I. Appendix C has a table of class numbers and norms of fundamental units for real quadratic fields with radicands less than 10^4, which we gleaned from an unpublished table of Buchmann, Sands, and Williams. Appendix D consists of a table of all class numbers of complex quadratic fields with radicands D having $|D| < 2 \cdot 10^3$, together with the decomposition of the class group into cyclic subgroups. These appendices represent tables which this author has wanted to see in one easily accessible source for some time, and so we trust that the reader will find them to be useful. Appendix E is provided as an overview of the theory of binary quadratic forms, and its relationship to our presentation via the ideal-theoretic methods in text. This gives an historical perspective to the development which completes and balances the presentation. Finally, Appendix F continues in the theme of completeness by providing background facts from analytic number theory.

Throughout the text, for the sake of some enlightened interest, we provide numerous footnotes of either an historical nature or with deeper applications of the concept in mind. In this way, the reader concerned with efficiency may ignore these digressions (possibly returning to them at a more leisurely time for interest sake in order to pursue a given sideline, or perhaps to look more deeply into the historical anecdotes provided). Furthermore, the *List of Symbols* is provided with the "browser" in mind, as are numerous references, in text, to previously defined concepts.

The text is intended for either graduate students with a desire to learn algebraic and computational number theory from this perspective, or enlightened undergraduates with a first course in algebra and number theory under their belts. The assumption therefore is that the reader is knowledgeable in basic algebra such as elementary facts about groups, rings and modules, and some basic elementary number theory such as congruences, polynomials and the Euclidean algorithm. Of course, research mathematicians and computer scientists interested in the interface will be looking at computer runs described in Chapters Five and Seven, and complexity questions described in Chapter Eight as well as the discussion of Cryptology.

This book was written, not only because this author wanted to see this material in one easily accessible place, but also because numerous colleagues have become increasingly insistent over the years that such a text be written and that this author undertake the task. It is hoped that this has been properly accomplished.

Chapter 1

Background from Algebraic Number Theory.

The essential purpose of this introductory chapter is to set the stage by introducing all relevant background required. We present the material in such a way that the uninitiated may be easily convinced of the veracity of the results by working through the exercises, whereas the initiated will be reminded of the relevant facts in an uncluttered fashion from a modern perspective. For the uninitiated we provide, as well, brief parenthetical reminders of the meaning of concepts assumed to be known. In this way, we provide a compact insight into the foundations and applications of this beautiful topic.

We now give a brief historical overview concerning the lives of the pioneers of the subject.[(1.0.1)]

Kummer (1810–1893), Kronecker (1823–1891) and Dedekind (1831–1916) may be considered to be the inventors of modern algebraic number theory. Ernst Eduard Kummer earned his doctorate at the age of 21 from the University of Halle after which he taught in Liegnitz at the Gymnasium (a school which prepares students for university entrance) for over a decade. In 1842, he was appointed Professor of Mathematics at the University of Breslau. Kummer's work on Fermat's last theorem led him to his theory of "ideal numbers" in 1846, an idea which Dedekind later refined. When Gauss died in 1855, Dirichlet succeeded Gauss at Göttingen, and Kummer succeeded Dirichlet at the University of Berlin. Kummer remained at Berlin until his retirement in 1883. Kummer's monumental contribution may be considered to be his proof of Fermat's last theorem (FLT) (i.e. $x^n + y^n = z^n$ is not solvable in nonzero integers (x, y, z) when $n > 2$ is a rational integer) for regular prime exponents.[(1.0.2)] Furthermore, his failed attempt at proving Fermat's last theorem in general, which led to his aforementioned ideal numbers and ultimately

---

(1.0.1) For in-depth historical background, the reader is referred to Bell[21], and Boyer[33].

(1.0.2) A prime p is called *regular* if p does not divide the class number of $Q(\zeta_p)$ where ζ_p is a primitive pth root of unity (see section three of this chapter). In fact, Kummer claimed that there are infinitely many regular primes. This remains an open problem to this day. Furthermore, Kummer was less interested in FLT than he was in the higher reciprocity laws which he saw as his main motivation. (See Edwards [81].) Both Kummer and Gauss saw FLT as being an isolated curiosity. Gauss' attitude toward FLT is best illustrated by the following: In 1816 the Paris Academy set FLT as its prize problem, and on March 7, 1816, H.W.M. Olbers wrote to Gauss from Bremen suggesting that Gauss should get into the competition for the prize, to which Gauss replied: "I am very much obliged for your news concerning the Paris prize. But I confess that Fermat's Theorem as an isolated proposition has very little interest for me, because I could easily lay down a multitude of such propositions, which one could neither prove nor dispose of." (See Bell [21, p. 238]. Copyright ©1937 by E.T. Bell. Copyright renewed ©1965 by Thaine T. Bell. Reprinted by permission of Simon and Schuster, Inc.)

to Dedekind's theory of ideals, is one of those "accidents" of mathematical inquiry which teaches us that the end goal may turn out to be less important than the mathematics which is created out of the search itself.[(1.0.3)]

Leopold Kronecker, inspired by Kummer's work on the theory of numbers, wrote his dissertation (On Complex Units) at the age of 22 at the University of Berlin. In fact, Kummer launched Kronecker's career at the Gymnasium in Liegnitz. Kronecker subsequently became a successful businessman when the estate of a rich uncle fell into his hands after the uncle's death. During the years 1848–1852, he produced no mathematics. Yet in 1853, he published a memoir on the algebraic solvability of equations. In fact, it is fair to say that Kronecker was one of the very few mathematicians to understand the depths of Galois' ideas. Also, Kronecker was very much a "finitist". "God made the integers", he used to say, "and all the rest is the work of man".[(1.0.4)] This attitude brought him into conflict with the likes of George Cantor (1845–1918) who had developed a theory of transfinite numbers. Kronecker went so far as to block a position for Cantor at Berlin. Kronecker's attacks on Cantor and his new discipline may have contributed to Cantor's nervous breakdown in 1884, the first of many which he suffered over the next thirty-three years of his life. Kronecker also came into conflict with Karl Weierstrass (1815–1897). Basically, Kronecker's disagreement with Weierstrass stemmed from the fact that Kronecker demanded that analysis be replaced by finite arithmetic. Unlike Cantor, Weierstrass was able to laugh off the attacks. Kronecker was a member of the Berlin Academy and, as such, without any employer to whom he had to answer, lectured at the University of Berlin from 1861–1883. He finally accepted a professorship in 1883 succeeding Kummer at the University of Berlin. He died of bronchial illness in Berlin on December 29, 1891 at the age of sixty-nine.

Before discussing Dedekind, we digress to point out the impressive faculty which Berlin held during Kronecker's time: Dirichlet, Jacobi and Steiner. Dirichlet came to Berlin in July of 1828, and held his first lecture in the winter semester of 1828/29. However, he was not fully accepted there, because he would/could not give his *Antrittsvorlesung* (or first lecture of a professor) in Latin, as was required in those times. This had some relatively serious implications for the amount of money which he was allowed to earn, so when he received a call to Heidelberg in 1846, he almost left Berlin. However, he stayed after being granted a raise in pay. He finally gave his Latin lecture in 1851, but as mentioned earlier, he was already to leave for Göttingen in 1855, as a successor to Gauss. Jakob Steiner was in Berlin from 1834–1863, and it seems that he too had problems with the Latin language. C.G.J. Jacobi was *Privatdozent* (i.e. allowed to teach but not (yet) a professor) in Berlin from 1824–1827, and went to Könisberg afterwards. He fell ill in 1842, and made a journey to Rome together with Steiner, Dirichlet, C.W. Borchard, and an expert linguist, Schläfli who spoke many languages. Schläfli learned a great deal from his discussions with these mathematicians. He worked in geometry, arithmetic, and function theory. His theory of continuous manifolds was published posthumously in 1901. Upon Jacobi's return from Italy, he came back to Berlin, where he remained until his death in 1851.[(1.0.5)]

[(1.0.3)]There is actually some serious question as to what actually happened with Kummer's "error" of assuming unique factorization. See Edwards [81]–[82].

[(1.0.4)]See Boyer [33, p. 616ff]. Copyright ©1968 by John Wiley and Sons, Inc. Reprinted by permission of John Wiley and Sons, Inc.

[(1.0.5)]For more information on Berlin, see Biermann[28]. Another useful reference by the same author is [29].

Richard Dedekind entered Göttingen at the age of 19 and earned his doctorate in three years with a thesis on the calculus which Gauss praised. Although he stayed at Göttingen for a few years attending Dirichlet's lectures, he ultimately became a secondary school teacher, a position which he maintained for the rest of his life. Dedekind's creation of "ideals", which was inspired by Kummer's "ideal numbers", gives us the modern edifice of algebraic number theory which we enjoy today. In fact, a quote from E.T. Bell's historical book on the "Men of Mathematics"[21, p. 474] is in order: "Kummer, Kronecker and Dedekind in their invention of the modern theory of algebraic numbers, by enlarging the scope of arithmetic ***ad infinitum*** and bringing algebraic equations within the purview of number, did for the higher arithmetic and the theory of algebraic equations what Gauss, Lobatchewsky, Johann Bolyai, and Riemann did for geometry in emancipating it from the slavery in Euclid's too narrow economy."[1.0.6]

We conclude with a discussion of the "Prince of Mathematicians", without which no historical overview of number theory would be complete no matter how brief. The following initial anecdote is part of accepted folklore.

Carl Friedrich Gauss (1777–1855) was a child prodigy and arguably the greatest mathematician of all time. When Gauss was 8 and attending his local school, the teacher asked the students to add up the numbers $1, 2, \ldots, 100$ in order to keep the class occupied, with the stipulation that students write the answer on their slates. Carl promptly did so with 5050 as his answer. The astonished teacher asked how he got such a quick answer, to which the prodigy explained, "I imagined the numbers 1 to 100 written in a row, and then again the same numbers written in a row, but backwards:

$$\begin{array}{ccccccc} 1 & 2 & 3 & \ldots & 98 & 99 & 100 \\ 100 & 99 & 98 & \ldots & 3 & 2 & 1 \end{array}.$$

I noted that the two numbers in each column added to 101. There are 100 columns; this gives a total of 10100 but I have counted each number twice, so the sum asked is one-half of 10100, that is 5050." At age 11, Gauss entered the Gymnasium. At 15, Gauss entered the college of Brunswick with the help of the Duke of Brunswick, and in 1795 he entered Göttingen with the Duke continuing to finance him. Gauss left Göttingen for the University of Helmstädt, where he received his doctorate in 1798. Essentially, his thesis title is what we now call "The Fundamental Theorem of Algebra", and it was published in 1799. Two years later Gauss published, with a dedication to his patron the Duke of Brunswick, "Disquisitiones Arithmeticae". This was volume one of a planned two-volume work, but Gauss never published the second volume since he was interested in too many other subjects; digressions which he *later* regretted. Gauss' contributions to mathematics are too numerous to elucidate here. He once described mathematics as the Queen of the Sciences and number theory as the Queen of Mathematics. His death in 1855 left a void in mathematics which remains to be filled.

[1.0.6]Copyright ©1937 by E.T. Bell. Copyright renewal ©1965 by Taine T. Bell. Reprinted by permission of Simon and Schuster, Inc.

1.1 Quadratic Fields: Integers and Units.

The simplest and most studied of algebraic number fields (finite extensions of the rational number field Q) is the quadratic field, i.e. an extension of Q of degree 2, which we now describe.

Let $D \neq 1$ be a square-free integer (positive or negative) and let $\Delta = 4D/\sigma_0^2$, where $\sigma_0 = 2$ if $D \equiv 1 \pmod 4$ and $\sigma_0 = 1$ otherwise. The value Δ is congruent to either 1 or 0 modulo 4 and is called *a fundamental discriminant* with *fundamental radicand* D. If we set $K = \mathbf{Q}(\sqrt{D})$, i.e. the adjunction of a root of the irreducible polynomial $x^2 - D$ to Q, then K is called *a quadratic field of discriminant* Δ.(1.1.1) If $\Delta > 0$, then K is called a *real quadratic field*, and if $\Delta < 0$ then K is called a *complex quadratic field*. Thus, there is one-to-one correspondence between quadratic fields and square-free rational integers $D \neq 1$. We use the term "rational integers" here as terminology to distinguish these integers, Z of Q, from the integers of K which we now describe.

A complex number is an *algebraic integer* if it is the root of a monic (leading coefficient 1) polynomial with coefficients in Z. Also, if f is a monic polynomial over Z of least degree having α (an algebraic integer) as a root, then f is irreducible over Q.(1.1.2) This is equivalent to saying that the monic irreducible polynomial over Q having α as a root has coefficients in Z. Moreover, the set of all algebraic integers in the complex field C is a *ring* which we denote by A. Therefore, $\mathbf{A} \cap K = \mathcal{O}_\Delta$ is the *ring of integers of the quadratic field* K of discriminant Δ. In order to explicitly describe $\mathcal{O}_\Delta$, we need to introduce some notation which will be useful throughout the remainder of the book as well.

Let $[\alpha, \beta]$ (for $\alpha, \beta \in K$) denote the Z-*module* $\alpha\mathbf{Z} \oplus \beta\mathbf{Z}$, i.e. the additive abelian group, with *basis elements* α and β, consisting of $\{\alpha x + \beta y : x, y \in \mathbf{Z}\}$, then by Exercise 1.1.7 we have

$$\mathcal{O}_\Delta = [1, (1 + \sqrt{D})/\sigma_0].$$

We may also write $\mathcal{O}_\Delta = \mathbf{Z}[(1 + \sqrt{D})/\sigma_0]$ to mean the same thing.

We now introduce some further notation which we will need throughout.

Definition 1.1.1. If Δ be a fundamental discriminant, then

$$w_\Delta = (\sigma_0 - 1 + \sqrt{D})/\sigma_0$$

is called the *principal surd*.

What we have shown is that the ring of integers of a quadratic field of discriminant Δ has the principal surd as its non-trivial basis element. Thus, we may

(1.1.1)We use the term "complex" here to refer to the case $\Delta < 0$ rather than the older, more frequently used term "imaginary", since the complex quadratic fields are no more "imaginary" than the real ones. Furthermore, "complex" is more accurate and descriptive, since it refers to the complex numbers $(a + bi)/\sigma$ (for $b \neq 0$) when $\Delta < 0$. The term "field" itself comes from Dedekind who introduced "Körper" in 1871 (or "corpus" in Latin), although Riemann introduced the *concept* of field in 1857. Note that we will discuss more general discriminants in section five of this chapter.

(1.1.2)A polynomial $f(x)$, which is not identically zero, is called *irreducible over* Q if there is no factorization $f(x) = g(x)h(x)$ into polynomials $g(x)$ and $h(x)$ of positive degree with coefficients in Q.

restate the ring of integers [(1.1.3)] of K as

$$\mathcal{O}_\Delta = [1, w_\Delta] = \mathbb{Z}[w_\Delta].$$

$\mathcal{O}_\Delta$ is an example of an *integral domain* (i.e. no zero divisors, which means that we cannot have $\alpha\beta = 0$ without at least one of α or β being zero).[(1.1.4)] We say that α *divides* β (for $\alpha, \beta \in \mathcal{O}_\Delta$) if there exists a $\gamma \in \mathcal{O}_\Delta$ such that $\beta = \alpha\gamma$. The *invertible* elements of $\mathcal{O}_\Delta$, or *units*, are those divisors of 1, i.e. $\alpha \in \mathcal{O}_\Delta$ is a unit if $\alpha\beta = 1$ for some $\beta \in \mathcal{O}_\Delta$. It follows that $\alpha \in \mathcal{O}_\Delta$ is a unit if and only if $|N(\alpha)| = 1$ where $N(\alpha) = \alpha\alpha'$ is the *norm* of α with α' being the *algebraic conjugate* of α (i.e. if $\alpha = (a + b\sqrt{D})/\sigma_0$ then $\alpha' = (a - b\sqrt{D})/\sigma_0$). If $\alpha, \beta \in \mathcal{O}_\Delta$ with $\alpha = u\beta$ for a unit u, then α and β are called *associates*.

We now demonstrate that the units of $\mathcal{O}_\Delta$ actually form a group, U_Δ, which we call the *unit group* of $\mathcal{O}_\Delta$ (or by abuse of language, the unit group of K). When $\Delta < 0$, then U_Δ is finite cyclic, and when $\Delta > 0$ then the positive units of U_Δ form a multiplicative group isomorphic to $\mathbb{Z}$, and so U_Δ contains *exactly* one generator larger than 1. We describe U_Δ by showing how to pick a canonical generator.

Since units $\alpha = (a + b\sqrt{D})/\sigma_0 \in \mathcal{O}_\Delta$ have $|N(\alpha)| = 1$, then[(1.1.5)]

$$a^2 - b^2 D = \pm\sigma_0^2. \tag{1.1.1}$$

When $D < 0$ equation (1.1.1) has few solutions since $D \geq -4$ when $b \neq 0$. In fact it takes little effort (Exercise 1.1.1) to see that when $\Delta < 0$ then

$$U_\Delta = \begin{cases} \{i^j : j = 0, 1, 2, 3\} & \text{if } \Delta = -4, \\ \{[(1 + \sqrt{3}i)/2]^j : j = 0, 1, 2, 3, 4, 5\} & \text{if } \Delta = -3, \\ \{\pm 1\} & \text{otherwise.} \end{cases}$$

(1.1.3) For example, the integers $\mathbb{Z}[i]$ of $K = Q(i) = \mathbb{Q}(\sqrt{-1})$ are called the *Gaussian* integers. Also, $Z[(1+\sqrt{5})/2]$ is the ring of integers of $K = \mathbb{Q}(\sqrt{5})$ from which the *Fibonacci numbers* may be defined: $F_n = \frac{1}{\sqrt{5}}\left\{\left(\frac{1+\sqrt{5}}{2}\right)^n - \left(\frac{1-\sqrt{5}}{2}\right)^n\right\}$. Fibonacci was actually Leonardo of Pisa (ca. 1180–1250), the son of Bonaccio, an Italian merchant. The term "Gaussian integer" arises from the fact that Gauss considered numbers of the form $a+bi$ to be integers ($a, b \in \mathbb{Z}$). However, Gauss did not consider $(1+\sqrt{3}i)/2$ to be an integer since it did not "fit in" with his theory of quadratic forms given that it does not come from a quadratic form with even middle coefficient. (See Appendix E.) Although Gauss' favourite student, Ferdinand Gotthold Eisenstein (1823–1852), did consider this particular principal surd to be an integer in order to obtain number-theoretic results concerning cubic residues, it was not until Kummer's, and ultimately Dedekind's, work that all elements of $\mathbb{Z}[w_\Delta]$ were perceived as integers. Gauss' admiration of Eisenstein is reflected in this quote: "There have been only three epoch-making mathematicians, Archimedes, Newton and Eisenstein." Unfortunately Eisenstein did not live long enough to fulfill such a glowing testimony. (See Bell [21, pp. 237]. Copyright ©1937 by E.T. Bell. Copyright renewed ©1965 by Taine T. Bell. Reprinted by permission of Simon and Schuter, Inc.)

(1.1.4) Even more than that, as we shall see in the next section, it is a Dedekind domain. See footnote (1.2.3) for the definition.

(1.1.5) The equation $x^2 - y^2 D = 1$ was erroneously dubbed "Pell's equation" by Euler. The honour actually goes to Lord Brouncker. However, as noted by E.E. Whitford [379], "to attempt to rename it would be like trying to give another name to North America because Vespucius was not its discoverer." Amerigo Vespucci (1454–1512) was actually a contemporary of Christopher Columbus, and was an Italian merchant and navigator. In a voyage (1501–1502), he established, under the auspices of Portugal, that the new lands discovered to the west were not part of Asia, but rather a separate land mass. In fact, he was the first to discover *Rio de la Plata* (which today includes Argentina, Uruguay, Paraguay, and Bolivia). In the early sixteenth century, the "new world" became synonymous with Amerigo, the discoverer of the land of "Americus" or "America". Although this reference applied only to South America, the name was later extended to North America.

We have that when $\Delta < 0$, U_Δ is identical to *the group of roots of unity*, R_Δ, of $\mathcal{O}_\Delta$, i.e. when $D = -4$, $R_\Delta = U_\Delta$ is generated by the primitive fourth root of unity $i = \sqrt{-1}$, when $D = -3$, $R_\Delta = U_\Delta$ is generated by the primitive sixth root of unity $(1+\sqrt{-3})/2$, and $R_\Delta = U_\Delta = \{\pm 1\}$ otherwise.

If $\Delta > 0$, then $R_\Delta = \{\pm 1\}$ and the positive units in U_Δ have a generator which is the smallest unit which exceeds 1. This selection is unique and is called *the fundamental unit* of K, denoted ε_Δ.

It is easy to verify (Exercise 1.1.2) that α is a unit in $\mathcal{O}_\Delta$ if and only if $\alpha^2 - \mathrm{Tr}(\alpha)\alpha \pm 1 = 0$ where $\mathrm{Tr}(\alpha) = \alpha + \alpha'$ is the *trace* of α. Moreover ε_Δ is that unit for which $|\mathrm{Tr}(\varepsilon_\Delta)|$ is at a minimum (Exercise 1.1.3 (a)). Hence, we may choose $\varepsilon_\Delta = (T + U\sqrt{D})/\sigma_0$ where T is the smallest positive integer such that

$$T^2 - DU^2 = \pm\sigma_0^2 \tag{1.1.2}$$

holds (with $U > 0$).

To illustrate the notion of a fundamental unit, we have

Example 1.1.1. Exercise 1.1.6 tells us that if $\varepsilon_\Delta = (T_1 + U_1\sqrt{D})/\sigma_0$, then T_1 and U_1 are determined by the sequence $\{DU^2\}$ for $U = 1, 2, 3, \ldots$, i.e. the first value of the sequence which differs from $\pm\sigma_0^2$ by a square is DU_1^2, and that square value is T_1^2. For instance, take $D = 13$ and look at $13U^2$ for $U = 1, 2, \ldots$. We get the value 13 for $U = 1$ immediately, and 13 differs from $\sigma_0^2 = 4$ by the square $T^2 = 9$, so $\varepsilon_\Delta = (3+\sqrt{13})/2$. A more interesting example is $D = 41$. The sequence DU^2 produces values $41, 164, 369, 656, 1025, 1476, 2009, 2624, 3321, 4100, \ldots$. The value 4100 is the first of the sequence to differ from $\pm\sigma_0^2 = \pm 4$ by a square, namely $4100 - 4 = 64^2$. Therefore, $T_1 = 64$ and $U_1 = 10$, i.e. $\varepsilon_{41} = (64 + 10\sqrt{41})/2 = 32 + 5\sqrt{41}$.

Not all fundamental units are so easy to find in practice, even for small values of Δ. For instance, the fundamental unit for $\Delta = 7756$ is

$$\varepsilon_\Delta = 66721654740119159 35 + 151522771012903896 w_\Delta.$$

(See Appendix B for a list of all fundamental units for fundamental radicands $D < 2 \cdot 10^3$.)

We note that it is still an open problem to give a simple criterion for $N(\varepsilon_\Delta) = -1$ in general.(1.1.6) However, we *do* know that if D has a prime divisor $p \equiv 3 \pmod 4$ then $N(\varepsilon_\Delta) = 1$ (Exercise 1.1.4). If D is not divisible by such a prime, then there is no general criterion for $N(\varepsilon_\Delta) = -1$ (albeit certain special cases are known such as $N(\varepsilon_\Delta) = -1$ whenever $\Delta = D$ is prime (Exercise 1.1.5)).

Thus, for $\Delta > 0$ the most general possible unit in U_Δ is of the form $\pm\varepsilon_\Delta^n$ where $n \in \mathbb{Z}$.(1.1.7) Thus, we may write any positive unit α as

$$\alpha = ((T_1 + U_1\sqrt{D})/\sigma_0)^n = (T_n + U_n\sqrt{D})/\sigma_0 \tag{1.1.3}$$

(1.1.6)There is a more general related question involving algebraic integers in an extension of number fields L/K: "When is an integer of K the norm of an *integer* from L?" This was answered for certain cases in [128]. In fact, the open question for a quadratic field is tantamount to asking when $U_\Delta^+ = U_\Delta^2$ where U_Δ^+ is the subgroup of U_Δ consisting of *totally positive units* (i.e. they map into positive elements in every real imbedding of K), and U_Δ^2 consists of the squares of elements of U_Δ. This question was investigated in [79] for more general number fields K. See also Janusz [132, Chapter VI.3, pp. 203–211] and Appendix E, especially footnote (E.5). For further insights, see the classic result by Scholtz [320] from 1935.

(1.1.7)In the next chapter, we will learn how to explicitly determine ε_Δ and $N(\varepsilon_\Delta)$ via continued fractions.

where $\sigma_0 U_{n+1} = T_1 U_n + U_1 T_n$.

Of all real quadratic fields, the *smallest* fundamental unit is $\varepsilon_\Delta = (1+\sqrt{5})/2$ where $\Delta = D = 5$, from which the Fibonacci numbers arise (see footnote (1.1.3)). This particular ε_Δ occurs often in geometry and other areas of mathematics as well, so that it has been dubbed the "golden section". In fact, entire books have been devoted to its study.[(1.1.8)] Dirichlet and Legendre[(1.1.9)] used $\mathbf{Z}[(1+\sqrt{5})/2]$ in 1825 to prove the case of FLT for $n = 5$. Earlier, in 1770, Euler[(1.1.10)] used $\mathbf{Z}[(1+\sqrt{-3})/2]$ to prove FLT for $n = 3$.

We now examine the failure of unique factorization of algebraic integers in $\mathcal{O}_\Delta$ which will motivate the next section on ideals (for which there is a unique factorization). We call a factorization of an algebraic integer $\alpha \in \mathcal{O}_\Delta$ *trivial* if $\alpha = u\beta$ where $u \in U_\Delta$ and $\beta \in \mathcal{O}_\Delta$. We call an algebraic integer *irreducible* if it has no factorizations other than the trivial ones, and is not itself a unit or the zero element. Also, $\gamma \in \mathcal{O}_\Delta$ is called *prime* if γ satisfies the property that γ divides $\alpha\beta$ $(\alpha, \beta \in \mathcal{O}_\Delta)$ implies that γ divides α or γ divides β. (Here "divides" means in the ring $\mathcal{O}_\Delta$, i.e. a divides b in $\mathcal{O}_\Delta$ implies that $b = ac$ for some $c \in \mathcal{O}_\Delta$.) Not all irreducibles are prime. For example, 3 divides $(1+\sqrt{-5})(1-\sqrt{-5})$ without dividing either factor, so 3 is irreducible but not prime in $\mathbf{Z}[\sqrt{-5}]$. Hence, *one may view the failure of unique factorization as the failure of irreducible elements to be prime in general.* Of course, with $\mathbf{Z}$, we have that irreducibility is tantamount to primality and this is the underlying key to the proof of the fundamental theorem of arithmetic (namely that there is unique factorization for rational integers) as proved by Euclid. [(1.1.11)] Nevertheless, unique factorization is restored at the ideal level as we shall see in the next section.

(1.1.8)See Dodd [75].

(1.1.9)Peter Gustav Lejeune Dirichlet (1805–1859) did the most to amplify Gauss' *Disquisitiones* (see the introduction to this chapter). Adrien Marie Legendre (1752–1833), on the other hand, made an enemy of Gauss when he wrote a letter to him almost accusing him of dishonesty when Gauss claimed to have known of the "method of least squares" ten years before Legendre's publication of it in 1806. Legendre felt that, with Gauss' numerous discoveries, he did not need to lay claim to what Legendre felt was purely his own brain-child.

(1.1.10)Leonard Euler (1707–1783) studied under Jean Bernoulli (1667–1748) and became friends with his sons, Nicolaus and David. In 1730, he held the chair of natural philosophy at the St. Petersburg Academy in Russia. Euler was so prolific that the French academician Francois Arago commented that Euler could calculate "just as men breathe, as eagles sustain themselves in the air." (See Boyer [33, Chapter XXI]. Copyright ©1968 by John Wiley and Sons, Inc. Reprinted by permission of John Wiley and Sons, Inc.) In fact, during his lifetime, Euler published over 500 books and papers. He is estimated to have written over 800 pages a year.

(1.1.11)Very little is known about Euclid's life, not even the city in which he was born. However, he is the author of arguably the most successful textbook in mathematics ever written — the *Elements*. Nevertheless, certain editions of the Elements falsely identified the author as "Euclid of Negara". The actual Euclid of Negara was a student of Socrates, but not really a mathematician. The author of the Elements is *Euclid* of *Alexandria*. It is widely held that Ptolemy I (and for that matter, also Alexander the Great) sought an easy introduction to geometry, to which Euclid replied: "There is no royal road to geometry." Euclid was a "purist", as illustrated by the tale that, in response to a student's questioning the practicality of geometry, Euclid had some coins given to the student "since he must needs make gain of what he learns". (For more details, see Boyer [33, p. 111–113]. Copyright ©1968 by John Wiley and Sons, Inc. Reprinted by permission of John Wiley and Sons, Inc.) Perhaps modern-day students could take a lesson from a teacher of 300 B.C.

Exercises 1.1

1. Show that $U_\Delta = R_\Delta$ when $\Delta < 0$ is a fundamental discriminant. (*Hint.* Use (1.1.1)).

2. Verify that if $\Delta > 0$ is a fundamental discriminant, then $\alpha \in U_\Delta$ if and only if
$\alpha^2 - \mathrm{Tr}(\alpha)\alpha \pm 1 = 0$.

3. (a) Prove that if $\Delta > 0$ is a fundamental discriminant, then $|\mathrm{Tr}(\varepsilon_\Delta)| \le |\mathrm{Tr}(\alpha)|$ for all $\alpha \in U_\Delta - R_\Delta$. (*Hint*: Use the quadratic formula to solve for α in the preceding problem.)

   (b) Verify that $\varepsilon_\Delta = (T + U\sqrt{D})/\sigma_0$ where T is the smallest positive integer satisfying (1.1.2). (*Hint*: Show that any $\alpha = (T + U\sqrt{D})/\sigma_0 \in U_\Delta$ satisfies the equation in the preceding problem.)

4. (a) Let $\Delta > 0$ be a fundamental discriminant divisible by a prime $p \equiv 3 \pmod 4$. Show that $N(\varepsilon_\Delta) = 1$. (*Hint*: Assume $N(\varepsilon_\Delta) = -1$ and use (1.1.2) to get a contradiction to $(-1/p) = -1$ where $(*/p)$ is the *Legendre symbol*, i.e. if a is an integer relatively prime to p, then $(a/p) = 1$ if a is a quadratic residue of p, and $(a/p) = -1$ otherwise.)

   (b) Let $\Delta \equiv 0, 1 \pmod 4$ be a non-square integer.[(1.1.12)] The *Kronecker symbol* for Δ is defined as follows: (Δ/p) is just the Legendre symbol if $p > 2$ is a prime not dividing Δ, whereas

$$(\Delta/2) = \begin{cases} 1 & \text{if } \Delta \equiv 1 \pmod 8, \\ -1 & \text{if } \Delta \equiv 5 \pmod 8, \end{cases}$$

   and

$$(\Delta/n) = \prod_{j=1}^{r} (\Delta/p_j)^{a_j},$$

   if $\gcd(\Delta, n) = 1$, where $n = \prod_{j=1}^{r} p_j^{a_j}$ is the canonical prime factorization of n, and $(\Delta/p) = 0$ if p divides Δ.
   Let $n_1, n_2 \in \mathbf{Z}$ be positive. Prove:

   (i) $(\Delta/n_1 n_2) = (\Delta/n_1)(\Delta/n_2)$.

   (ii) If $n_1 \equiv n_2 \pmod{|\Delta|}$ then $(\Delta/n_1) = (\Delta/n_2)$.

   (c) Prove that, if Δ is a sum of two relatively prime squares, then Δ is not divisible by any prime $p \equiv 3 \pmod 4$. (*Hint*: Assume $p \mid \Delta$ with $p \equiv 3 \pmod 4$ and get a contradiction via $(-1/p) = -1$.)

   (d) Let $\gcd(\Delta, n) = 1$. Prove that Δ is a quadratic residue modulo n if and only if $(\Delta/p) = 1$ for all primes p dividing n.

   (e) Use (d) to provide a counterexample to the assertion that $(\Delta/n) = 1$ implies Δ is a quadratic residue modulo n.

   (f) Let $\gcd(\Delta, n) = 1$ where $\Delta \equiv 1 \pmod 4$. Prove that $(\Delta/n) = (n/|\Delta|)$.

[(1.1.12)] This is called the discriminant of an order (see section five of this chapter).

5. (a) If $u \in U_\Delta$ for a fundamental discriminant $\Delta > 0$ with $N(u) = 1$, then there exists an $\alpha \in \mathcal{O}_\Delta$ such that $u = \alpha/\alpha'$. This is known by its generalization as Hilbert's Theorem 90. Prove it. (*Hint*: Formulate α in terms of ε_Δ, observing that $1 + u' = 1 + 1/u$.)

   (b) Use part (a) to prove that $N(\varepsilon_\Delta) = -1$ when $\Delta = D$ is a prime discriminant.

6. Let $D > 0$ be a fundamental radicand and consider the sequence $\{DU^2\}$ where $U = 1, 2, \ldots$. Prove that $\varepsilon_\Delta = (T_1 + U_1\sqrt{D})/\sigma_0$ where T_1 is determined as that number for which DU_1^2 is the first value of the sequence which differs from $\pm\sigma_0^2$ by a square, T_1^2.

7. Let Δ be a fundamental discriminant with radicand D. Prove that the ring of integers $\mathcal{O}_\Delta$ of a quadratic field $K = \mathcal{O}_\Delta$ is given by $\mathcal{O}_\Delta = [1, (1 + \sqrt{D})/\sigma_0]$. (*Hint*: Let $\alpha = a + b\sqrt{D} \in K$, with $a, b \in \mathbf{Q}$. Prove that $\alpha \in \mathcal{O}_\Delta$ if and only if $\alpha\alpha' \in \mathcal{O}_\Delta$ and $\alpha + \alpha' \in \mathcal{O}_\Delta$.)

1.2 The Arithmetic of Ideals in Quadratic Fields.

We will be treating ideals as special kinds of Z-modules. The reader will recall that I is an *ideal* of $\mathcal{O}_\Delta$ if $I \subseteq \mathcal{O}_\Delta$ and I is closed under addition as well as multiplication from $\mathcal{O}_\Delta$, i.e. $\alpha + \beta \in I$ whenever $\alpha, \beta \in I$ and $\alpha\gamma \in I$ whenever $\alpha \in I$ and $\gamma \in \mathcal{O}_\Delta$.

Theorem 1.2.1. (Ideal Criterion). *If $I = [a, b + cw_\Delta]$, then I is a non-zero ideal of $\mathcal{O}_\Delta$ if and only if $c \mid a$, $c \mid b$ and $ac \mid N(b + cw_\Delta)$.*

Proof. See Exercise 1.2.1. □

Remark 1.2.1. Note that I is a zero ideal if and only if $I \subseteq \mathbf{Z}$. Thus, if I is a non-zero ideal in $\mathcal{O}_\Delta$, then $I = [a, b + cw_\Delta]$ where $a, b, c \in \mathbf{Z}$, $a > 0$, $c > 0$, $c \mid b$ and $c \mid a$. In fact, for a given non-zero ideal I in $\mathcal{O}_\Delta$, the integers a and c are unique. Indeed, a is *the least positive rational integer* in I, which we denote by $L(I)$. Also, we denote the value of $cL(I)$ by $N(I)$, called the *norm* of I. We call c a *rational integer factor* of I. *Throughout, we assume our ideals to be non-zero.*

Definition 1.2.1. If I is an $\mathcal{O}_\Delta$-ideal with $L(I) = N(I)$, i.e. $c = 1$, then I is called *primitive* which means that I has no rational integer factors other than ± 1. (When I is primitive then $N(I) = L(I) = |\mathcal{O}_\Delta : I|$, the *index* of I in $\mathcal{O}_\Delta$.)

Since an ideal is a special type of module (i.e. satisfies the ideal criterion given in Theorem 1.2.1) then, in order to understand what it means for two ideals in $\mathcal{O}_\Delta$ to be equal, we must understand what $[\alpha, \beta] = [\gamma, \delta]$ means (for $\alpha, \beta, \gamma, \delta \in K$). This equality holds if and only if

$$\begin{pmatrix}\gamma\\ \delta\end{pmatrix} = X\begin{pmatrix}\alpha\\ \beta\end{pmatrix}$$

where $X \in GL_2(\mathbb{Z})$, the group of all 2×2 matrices with entries in $\mathbb{Z}$ and determinant $|X| = \pm 1$. When we are discussing ideals, this translates into the following, which is the above equivalence operation in practice,

Theorem 1.2.2. (Criterion for Ideal Equality). *If Δ is a fundamental discriminant and $I = [a, \alpha]$ is a primitive $\mathcal{O}_\Delta$-ideal, then $I = [a, na \pm \alpha]$ for any $n \in \mathbb{Z}$.*

Proof. Exercise 1.2.2. □

For instance

Example 1.2.1.

(a) If $D = 15$, then $[7, 1 + \sqrt{15}] = [7, \pm 6 \mp \sqrt{15}] = [7, \pm 13 \mp \sqrt{15}] = \cdots = [7, 7n \pm (1 + \sqrt{15})]$.

(b) If $\Delta = 13$, then $[3, (1 + \sqrt{\Delta})/2] = [3, (-5 + \sqrt{\Delta})/2] = [3, (-11 + \sqrt{\Delta})/2] = \cdots = [3, (-7 - \sqrt{\Delta})/2] = [3, (-13 - \sqrt{\Delta})/2] = \cdots = [3, (5 - \sqrt{\Delta})/2] = [3, (11 - \sqrt{\Delta})/2] = \cdots = [3, (7 + \sqrt{\Delta})/2] = [3, (13 + \sqrt{\Delta})/2] = \cdots = [3, 3n \pm w_\Delta]$.

With so many possible choices for the representation of the ideal, it is a good idea to have a convention for a canonical choice. This is the content of

Remark 1.2.2. If $I = [a, \alpha]$ is a primitive ideal in $\mathcal{O}_\Delta$, then we may always ensure that $|\mathrm{Tr}(\alpha)| \le N(I)$, where $|\mathrm{Tr}(\alpha)|$ is uniquely determined (see Exercise 1.2.7). Thus, if $b \in \mathbb{Z}$ with $\alpha = b + w_\Delta$, then we may assume that

$$-a/2 \le b < a/2.$$

If we find it more convenient to write $\alpha = (b + \sqrt{\Delta})/2$, then we may assume that

$$-a \le b < a.$$

If we wish to ensure that b is nonnegative, then we may assume that

$$0 \le b < a$$

whenever $\alpha = b + w_\Delta$, and

$$0 \le b < 2a$$

when $\alpha = (b + \sqrt{\Delta})/2$. Observe that *at least one of the endpoints*, in either case, must be included (see Exercise 1.2.8). What we are tacitly observing, in the above, is that $\sigma b + \sigma - 1 \equiv \Delta \pmod 2$, necessarily.

Now we explore multiplication of ideals via their module structure.

Multiplication Formulae.

If $I_i = [a_i, (b_i + \sqrt{\Delta})/2]$ for $i = 1, 2$ are primitive ideals in $\mathcal{O}_\Delta$, then

$$I_1 I_2 = (g)[a_3, (b_3 + \sqrt{\Delta})/2], \tag{1.2.1}$$

where

$$a_3 = a_1 a_2 / g^2, \tag{1.2.2}$$

with

$$g = \gcd(a_1, a_2, (b_1 + b_2)/2), \tag{1.2.3}$$

and

$$b_3 \equiv \frac{1}{g}(\lambda a_2 b_1 + \mu a_1 b_2 + \frac{\nu}{2}(b_1 b_2 + \Delta)) \pmod{2a_3}, \tag{1.2.4}$$

where λ, μ and ν are determined by

$$\lambda a_2 + \mu a_1 + \frac{\nu}{2}(b_1 + b_2) = g. \tag{1.2.5}$$

Another formulation for b_3 which is highly useful is *Dan Shanks' formula*: (1.2.1)

$$b_3 \equiv b_2 + \frac{2a_2}{g}(\lambda(\frac{b_1 - b_2}{2} - \nu c_2)) \pmod{2a_3},$$

where

$$c_2 = (b_2^2 - \Delta)/(4a_2),$$

and $\lambda(b_1 - b_2)/2 - \nu c_2$ may be taken modulo a_1/g.

A simple application of the above is for $g = 1$. In this case,

$$I_1 I_2 = [a_1, (b_1 + \sqrt{\Delta})/2][a_2, (b_2 + \sqrt{\Delta})/2] = [a_3, (b_3 + \sqrt{\Delta})/2],$$

where $a_3 = a_1 a_2$ and $b_3 \equiv b_i \pmod{2a_i}$ for $i = 1, 2$.

An illustration of this process is

Example 1.2.2. Let $I_1 = [6, (1+\sqrt{97})/2]$ and $I_2 = [18, (5+\sqrt{97})/2]$, then $g = 3$ so $a_3 = 12$ by (1.2.2)–(1.2.3). Since $a_1 = 6$, $a_2 = 18$, $b_1 = 1$ and $b_2 = 5$, then by (1.2.5) we have $6\lambda + 2\mu + \nu = 1$. Taking $\lambda = 1$, $\mu = -3$ and $\nu = 1$, we have (via (1.2.4)) $b_3 \equiv -7 \pmod{36}$. Thus, $I_1 I_2 = (3)[12, (-7 + \sqrt{97})/2] = (3)[12, (17 + \sqrt{97})/2]$ by Theorem 1.2.2.

Now we need some other important aspects of ideals, i.e.

Definition 1.2.2. If Δ is a fundamental discriminant, then

(1) The *conjugate* of an $\mathcal{O}_\Delta$-ideal $I = [a, b + w_\Delta]$ is the ideal $I' = [a, b + w'_\Delta]$. If I is a primitive ideal and $I = I'$, then I is said to be an *ambiguous ideal.* (1.2.2)

(2) A *principal ideal,* I is one with a single generator, i.e. $I = (\alpha)$ for some $\alpha \in \mathcal{O}_\Delta$, where $(\alpha) = \alpha\mathcal{O}_\Delta = \{\alpha\beta : \beta \in \mathcal{O}_\Delta\}$.

(3) A *non-unit* ideal I, i.e. $I \neq \mathcal{O}_\Delta$ is a *prime* ideal if, whenever I divides a product of ideals $I_1 I_2$ in $\mathcal{O}_\Delta$, then $I \mid I_1$ or $I \mid I_2$.

(1.2.1) For this and the formulae above it, see Lenstra [178] and Shanks [328]. Also, see Appendix E for the connection with Dirichlet's "united forms". The hints to Exercise 1.2.3(a)–(b) show that (1.2.1)–(1.2.5) are simpler than they may appear on the surface.

(1.2.2) The entirety of Chapter Six is devoted to a discussion of ambiguous ideals and their importance.

Remark 1.2.3. Here "divides" means if $I \mid J$, then there exists an $\mathcal{O}_\Delta$-ideal H such that $J = HI$. It follows that $J \subseteq I$. Conversely, if $J \subseteq I$, then there exists an ideal H such that $J = HI$. This means that ideals which "divide" are those which "contain". Therefore, I is a prime ideal if and only if it contains some non-trivial factor of any product it divides. In fact, some authors use "contain" as the definition of divide (e.g. see Cohn [57]). Henceforth, when we say that a prime ideal $\mathcal{P}$ is *above* p, we mean that $\mathcal{P}$ divides (p).

We now describe an essential feature of the arithmetic of quadratic fields, namely *prime decomposition*. (In what follows $(*/p)$ is the Kronecker symbol defined in Exercise 1.1.4.)

Let p be any rational prime and set $(p) = [p, p\omega_\Delta]$ then, by Exercise 1.2.3(c),

$$(p) = \mathcal{P}_1^2 \quad \text{if } p \mid \Delta, \tag{1.2.6}$$

$$(p) = \mathcal{P}_1\mathcal{P}_2 \quad \text{if } (\Delta/p) = 1, \tag{1.2.7}$$

$$(p) = \mathcal{P} \quad \text{if } (\Delta/p) = -1, \tag{1.2.8}$$

where $\mathcal{P}_1$, $\mathcal{P}_2$ and $\mathcal{P}$ are prime $\mathcal{O}_\Delta$-ideals with $N(\mathcal{P}_1) = N(\mathcal{P}_2) = p$ and $N(\mathcal{P}) = p^2$.

When (1.2.6) occurs, we say that p *ramifies* in K. If (1.2.7) occurs, then we say that p *splits* in K (and we often refer to this rational prime p as a *split prime*). Finally, when (1.2.8) occurs, we say that p is *inert* in K.

Now we describe how any ideal of $\mathcal{O}_\Delta$ may be represented as a unique product of certain ideals. As noted in the last section, $\mathcal{O}_\Delta$ is an integral domain. In point of fact, it is a Dedekind domain [(1.2.3)] in which unique factorization of ideals occurs.

Theorem 1.2.3. (Unique Factorization).[(1.2.4)] *Any non-zero ideal in $\mathcal{O}_\Delta$ has a unique decomposition, i.e. $I = \mathcal{P}_1^{a_1}\mathcal{P}_2^{a_2}\dots\mathcal{P}_n^{a_n}$ where the $\mathcal{P}_i$ are the distinct prime ideals of $\mathcal{O}_\Delta$ containing I, and $a_i \geq 1$ are rational integers for $i = 1, 2, \dots, n$.*

Proof. See Marcus [210, Theorem 16, p. 59]. □

Thus, Dedekind's work restored uniqueness of factorization at the ideal level, which had been lost at the level of the algebraic integer, described in the last section. In the next section, we look at the class group and its order, the class number, which somehow measures lack of unique factorization at the integer level.

Exercises 1.2

1. (a) Prove Theorem 1.2.1 (*Hint*: Verify that $a\omega_\Delta = ab/c - a(b - c\omega_\Delta)/c$ and $b\omega_\Delta + c\omega_\Delta^2 = -N(b + c\omega_\Delta)/c + (b + c(\sigma - 1))(b + c\omega_\Delta)/c$.)

(1.2.3) A Dedekind domain is an integral domain in which all fractional ideals are invertible. A *fractional ideal* is a non-zero, finitely generated, $\mathcal{O}_\Delta$-submodule of K (which includes all non-zero ideals of $\mathcal{O}_\Delta$, which are called *integral* ideals to distinguish them from more general fractional ideals). A fractional $\mathcal{O}_\Delta$-ideal I is called *invertible* if $II^{-1} = \mathcal{O}_\Delta$ where $I^{-1} = \{x \in K : xI \subseteq \mathcal{O}_\Delta\}$, and II^{-1} consists of all finite sums of products of α, β with $\alpha \in I$ and $\beta \in I^{-1}$. (See section five of this chapter.)

(1.2.4) "Unique" here is used in the sense that the order of the factors is ignored. Moreover, the result is actually more generally stated in terms of fractional ideals (see footnote (1.2.3)), namely every fractional ideal I is uniquely representable as $I = \mathcal{P}_1^{a_1}\mathcal{P}_2^{a_2}\cdots\mathcal{P}_n^{a_n}$ where the $\mathcal{P}_i$ are distinct prime ideals and $a_i \in \mathbb{Z}$ for $i = 1, 2, \dots, n$.

(b) If Δ is a discriminant and I and J are (non-zero) $\mathcal{O}_\Delta$-ideals, then we define

$$(I, J) = \{\alpha + \beta : \alpha \in I,\ \beta \in J\}.$$

Verify that (I, J) is an ideal. Conclude that if $\gamma, \delta \in \mathcal{O}_\Delta$ and $(\gamma, \delta) = ((\gamma), (\delta))$ is an ideal where $I = (\gamma)$ and $J = (\delta)$, then $(\gamma, \delta) = [\gamma, \delta]$.

(c) Two *ideals* I and J *are called relatively prime* if there does not exist an ideal $H \neq (1) = \mathcal{O}_\Delta$ which divides both I and J (or, in view of Remark 1.2.3, there is no nontrivial ideal containing both I and J). Prove that I and J are relatively prime if and only if $(I, J) = (1)$.

(d) Determine (I, I').

(e) Prove that if $I \subseteq J$ for non-zero $\mathcal{O}_\Delta$-ideals I and J, then there exists an element $\beta \in J$ such that $J = (I, \beta)$. (*Hint*: Use unique factorization, i.e. set $I = \prod_{i=1}^{r} \mathcal{P}_i^{a_i}$, $J = \prod_{i=1}^{r} \mathcal{P}_i^{b_i}$ and deduce the existence of β_i such that $\sum_{i=1}^{r}(\beta_i) = IQ$ where $(I, Q) = (1)$. Then set $\beta = \sum_{i=1}^{r} \beta_i$ and deduce that $J + (\beta) = I$.)

2. Using the definition of module equality given before the statement of Theorem 1.2.2, prove the theorem.

3. (a) Verify the multiplication formulae (1.2.1)–(1.2.5). (*Hint*: Multiply the basis elements of I_1, I_2 and compare coefficients of $\sqrt{\Delta}$. Also (1.2.5) is verified using the Euclidean algorithm. Thus, given the fact that $(b_3 + \sqrt{\Delta})/2$ is a basis element of the ideal $I_1 I_2$ then b_3 is determined modulo $2a_3$.)

   (b) Verify the Shanks formula stated after the multiplication formulae (1.2.1)–(1.2.5) (*Hint*: Eliminate μa_1 from (1.2.4) and (1.2.5).)

   (c) Verify the decomposition formulae (1.2.6)–(1.2.7). (*Hint*: For example, if $p > 2$ and $\Delta \equiv b^2 \pmod{p}$ for some non-zero $b \in \mathbb{Z}$ with $0 < b < p$, then show that $(p) = [p, (b + \sqrt{\Delta})/2][p, (b - \sqrt{\Delta})/2]$.)

4. Let $D = p_1 p_2 \ldots p_n$ be the canonical prime factorization of the fundamental radicand $D > 0$ such that $p_i \not\equiv 3 \pmod{4}$ for any $i = 1, 2, \ldots, n$. Prove that the following are equivalent:

   (a) $N(\varepsilon_\Delta) = -1$.

   (b) $[r, (r + \sqrt{D})/2]$ is not principal in $\mathcal{O}_\Delta$ for any divisor r of D with $r \neq 1, D$.

   (c) $|rx^2 - sy^2| = 4$ has no solutions such that $rs = D$, unless either $r = \pm 1$ or $s = \pm 1$.[(1.2.5)] (*Hint*: Use Exercise 1.1.5 to prove that (b) implies (a). The balance is straightforward.)

5. Prove that if I is an ambiguous primitive ideal of $\mathcal{O}_\Delta$ where Δ is a fundamental discriminant and $N(I) < \sqrt{\Delta}/2$, then $N(I) \mid \Delta$. Conversely, prove that if $N(I) \mid \Delta$ for a primitive $\mathcal{O}_\Delta$-ideal, then I is ambiguous.

6. (a) Let $p > 2$ be a prime and set $p^* = (-1)^{(p-1)/2} p$. Prove that the prime $q \neq p$ splits in $\mathbb{Q}(\sqrt{p^*})$ if and only if $(q/p) = 1$, where $(*/*)$ is the Legendre symbol defined in Exercise 1.1.4. (*Hint*: see (1.2.7)).

[(1.2.5)]This result is due to Gauss [91, Article 187, p. 156], but has been rediscovered by Trotter [364]. A corollary of this result is that if there are no nontrivial factorizations of $D = rs$ such that r and s are quadratic residues of one another, then $N(\varepsilon_\Delta) = -1$. Deduce this fact as an exercise.

(b) Prove that 2 splits in $Q(\sqrt{p^*})$ if and only if $(2/p) = 1$. (*Hint*: see (1.2.7)).

(c) Let (Δ/n) be the Kronecker symbol defined in Exercise 1.1.4 for a discriminant Δ. Is it true that $(\Delta/n) = -1$ implies all primes dividing n are quadratic non-residues modulo $n > 0$? (*Hint*: Let n be a square times a prime p with $(\Delta/p) = -1$.)

(d) If $(\Delta/p_1p_2) = 1$ where p_1 and p_2 are primes not dividing Δ, does this imply that p_1 and p_2 both split in $Q(\sqrt{D})$?

7. Let $\Delta > 0$ be a fundamental discriminant. Prove that, if I is a primitive ideal in $\mathcal{O}_\Delta$, then there exists some $\alpha \in I$ with $I = [N(I), \alpha]$, and $|\mathrm{Tr}(\alpha)| \leq N(I)$. Furthermore, establish that $|\mathrm{Tr}(\alpha)|$ is unique, i.e. if $I = [N(I), \alpha] = [N(I), \beta]$, $|\mathrm{Tr}(\alpha)| \leq N(I)$, and $|\mathrm{Tr}(\beta)| \leq N(I)$, then $|\mathrm{Tr}(\alpha)| = |\mathrm{Tr}(\beta)|$. Also, show that $\mathrm{Tr}(\alpha) = -\mathrm{Tr}(\beta)$ only if $I = I'$.

8. Let Δ be a fundamental discriminant.

(a) Prove that, if $p > 2$ is prime, then $(\Delta/p) \not\equiv -1$ if and only if there exists an integer $b \in Z$ such that $I = [a, (b + \sqrt{\Delta})/2]$ is an $\mathcal{O}_\Delta$-ideal.

(b) Show that, in (a), we may assume, without loss of generality, that either $0 < b \leq 2a$ or $0 \leq b < 2a$ but not that $0 < b < 2a$. (*Hint*: Look at $I = [3, \sqrt{-6}]$ and at $J = [3, (5 + \sqrt{-95})/2]$.)

9. Let $\Delta \equiv 1 \pmod 4$ be a positive fundamental discriminant. Explain why $[r, \sqrt{\Delta}]$ for $r \mid \Delta$ is not an ideal in $\mathcal{O}_\Delta$.

10. Let Δ be a fundamental discriminant, and let $I = [ac, (b + \sqrt{\Delta})/2]$ be a primitive $\mathcal{O}_\Delta$-ideal, for positive $a, c \in Z$. Prove that $\gcd(a, b, c) = 1$.

11. Let $\Delta < 0$ be a fundamental discriminant, and let I be a primitive, principal $\mathcal{O}_\Delta$-ideal. Prove that either $N(I) = 1$, or $N(I) > N(w_\Delta)$.

1.3 The Class Group and Class Number.

First we establish the notion of equivalence between ideals of $\mathcal{O}_\Delta$.

Definition 1.3.1. Let Δ be a fundamental discriminant. If I and J are non-zero $\mathcal{O}_\Delta$-ideals, then we say that I and J are *equivalent* (denoted $I \sim J$) if there exist non-zero $\alpha, \beta \in \mathcal{O}_\Delta$ such that $(\alpha)I = (\beta)J$.

This equivalence relation partitions the $\mathcal{O}_\Delta$-ideals I into disjoint *ideal classes* $\{I\}$ which form an abelian group C_Δ, called the *class group* of $\mathcal{O}_\Delta$ (or simply of K). The order of C_Δ, denoted h_Δ, is called the *class number* of K.[(1.3.1)]

[(1.3.1)] Essentially, the class group C_Δ is the quotient group H/P, where H is the group of fractional ideals under multiplication and P is the subgroup of principal fractional ideals (see footnote (1.2.3)). This means that when C_Δ is trivial (i.e. $h_\Delta = 1$), then every ideal in $\mathcal{O}_\Delta$ is principal (i.e. $\mathcal{O}_\Delta$ is a *principal ideal domain* or (PID)). A *unique factorization domain* (UFD) is an integral domain in which every non-zero, non-unit element factors into a product of irreducible elements (see section one of this chapter), and the factorization is unique up to unit multiples and the order of the factors. It can be shown that a PID is a UFD. Furthermore, for a Dedekind domain, the converse holds. Hence, $h_\Delta = 1$ *is equivalent to unique factorization in* $\mathcal{O}_\Delta$.

The following Minkowski type result points to the finiteness of h_Δ. First we need

Definition 1.3.2. Let Δ be a fundamental discriminant and set

$$M_\Delta = \begin{cases} \sqrt{-\Delta/3} & \text{if } \Delta < 0 \\ \sqrt{\Delta}/2 & \text{if } \Delta > 0, \end{cases}$$

then M_Δ is called the *Minkowski bound.* [(1.3.2)]

Theorem 1.3.1. *If Δ is a fundamental discriminant, then every class of C_Δ contains a primitive ideal I with $N(I) \le M_\Delta$. Furthermore, C_Δ is generated by the primitive non-inert prime ideals $\mathcal{P}$ with $N(\mathcal{P}) < M_\Delta$.*

Proof. The first statement follows from Minkowski's calculated bound which, when applied to quadratic fields, says that there is a primitive ideal I in each class C_Δ with

$$N(I) \le (4/\pi)^\delta/2\sqrt{|\Delta|},$$

where $\delta = 0$ if $\Delta > 0$ and $\delta = 1$ if $\Delta < 0$ (see Theorem 1.4.2 (c) and see footnote (1.3.2)). The second statement is now an easy exercise. □

An instance which illustrates Theorem 1.3.1 is

Example 1.3.1. If $\Delta = 145$, then C_Δ is cyclic of order 4, and $\lfloor M_\Delta \rfloor = 6$.[(1.3.3)] Since the $\mathcal{O}_\Delta$-ideal above 5 is necessarily ambiguous by Exercises 1.2.5, it is of order 2, and the only possibility for a generator of C_Δ is an ideal above 2 or an ideal above 3. In fact, this is a good opportunity to show how to use equations (1.2.1)–(1.2.5) to establish that $I_1 = [2, w_\Delta] \sim I_2 = [3, (5+\sqrt{145})/2]$.

Consider $I_1 I_2' = [2, w_\Delta][3, (5-\sqrt{145})/2] = [2, w_\Delta][3, w_\Delta]$ (by Theorem 1.2.2). Using (1.2.1)–(1.2.5) with $a_1 = 2$, $a_2 = 3$ and $b_1 = b_2 = 1$, we get $I_1 I_2' = [6, (13 + \sqrt{145})/2]$ (having chosen $\lambda = 1 = -\mu$ and $\nu = 0$ in equation (1.2.5)). Since $N[(13+\sqrt{145})/2] = 6$, then $I_1 I_2' = ((13+\sqrt{145})/2)$, i.e. $I_1 I_2' \sim 1$, so $I_1 \sim I_2$ and we have verified our initial assertion about the ideals. We also point out that we must be careful in choosing our ideal of norm 6 since there is in fact another ideal of norm 6 whose class has order 4 in C_Δ. It is $J = [6, (5+\sqrt{145})/2]$. Using the above techniques and equations (1.2.1)–(1.2.5), the reader can verify that $J \sim (I_2')^3$ whose class has order 4 by Exercise 1.3.4(d).

In the next chapter, we will be able to prove the above facts via continued fraction theory without resorting to our multiplication formulae or a knowledge of the class group structure. All of this will come from the infrastructure (see Chapter Two, section one).

It is often useful (as we shall see in section four of this chapter) to know when a given ideal is principal in terms of its basis elements. There is a result which tells us when this happens for $\Delta < 0$. We isolate this fact as

---

(1.3.2) Actually, to be strictly accurate, when $\Delta < 0$, $\sqrt{-\Delta/3}$ is called the *Gauss bound*, and the Minkowski bound is $2\sqrt{-\Delta}/\pi$ (e.g. see Stewart–Tall [352, Corollary 10.3, p. 186]). However, the bound can be improved to $\sqrt{-\Delta/3}$, as Theorem 1.4.2(c) shows. Thus, $\sqrt{-\Delta/3}$ has come to be regarded by some as the Minkowski bound (e.g. see Cohn [57, Theorem 11, p. 141]).

(1.3.3) The notation $\lfloor x \rfloor$ refers to the *floor* of x, i.e. the greatest integer less than or equal to x. Its close cousin is $\lceil x \rceil$ which is the *ceiling* of x, i.e. the least integer greater than or equal to x.

Theorem 1.3.2. [(1.3.4)] *If $\Delta < 0$ is a fundamental discriminant and $I = [a, b+w_\Delta]$ is a primitive ideal of $\mathcal{O}_\Delta$ with $N(b+w_\Delta) < N(w_\Delta)^2$, then I is principal if and only if $a = 1$ or $a = N(b+w_\Delta)$.*

Corollary 1.3.1. *If $\Delta < 0$ is a fundamental discriminant and $I = [a, b+w_\Delta]$ is a primitive ideal in $\mathcal{O}_\Delta$ with $1 < a < N(w_\Delta)$, $N(b+w_\Delta) < N(w_\Delta)^2$, then I is not principal.*

Proof. See Exercise 1.3.2. □

Theorem 1.3.2 is very specific to negative discriminants, i.e. it does not hold for positive discriminants. For example, the ideal $\mathcal{P} = [5, 2+\sqrt{19}] = (9+2\sqrt{19}) \sim 1$ but $N(2+\sqrt{19}) = 15$ and there does not exist a representation of $\mathcal{P}$ of the form $[5, b+\sqrt{19}]$ with $|N(b+\sqrt{19})| = 5$.

We now describe the 2-rank of C_Δ, i.e. the order of *the elementary abelian 2-subgroup*, $C_{\Delta,2}$ of C_Δ. First we need

Definition 1.3.3. Let Δ be a fundamental discriminant divisible by exactly $N+1$ ($N \geq 0$) distinct primes, and set

$$t_\Delta = \begin{cases} N-1 & \text{if } \Delta > 0, \text{ some prime } p \equiv 3 \pmod 4 \text{ divides } \Delta. \\ N & \text{otherwise.} \end{cases}$$

Theorem 1.3.3. *If Δ is a fundamental discriminant, then $C_{\Delta,2}$ has order 2^{t_Δ}.*

Proof. We begin with $\Delta < 0$. Let $h_{\Delta,2}$ denote the order of $C_{\Delta,2}$.

Claim. Every class of $C_{\Delta,2}$ has an ambiguous ideal in it.

By Theorem 1.3.1, it suffices to show that every prime ideal $I = [p, (b+\sqrt{\Delta})/2]$, with $|b| < p$ and $p < M_\Delta$, is equivalent to an ambiguous ideal. We may assume that p splits. Thus, by formulae (1.2.1)–(1.2.5), $I^2 = [p^2, (b_3+\sqrt{\Delta})/2] \sim 1$, where $|b_3| < p^2$ is determined by (1.2.4). An easy check shows that we may assume $\Delta \neq -3, -4$, so $N((b_3+\sqrt{\Delta})/2) < N(w_\Delta)^2$. Thus, we may invoke Theorem 1.3.2 to get $N((b_3+\sqrt{\Delta})/2) = p^2$. The reader may use (1.2.4), or Shanks' formula to verify that this forces $b_3 = b$. Thus, we have $\Delta = b^2 - 4p^2$. We may form $\beta = (b+2p+\sqrt{\Delta})/2 \in \mathcal{O}_\Delta$, and set $J = [N(\beta)/N(I), -\beta']$. By Exercise 1.3.7(a), $J \sim I$. However, $N(J) = b+2p$ which divides $\Delta = (b+2p)(b-2p)$. This establishes the Claim.

By Exercise 1.3.7(b), we may assume that $\Delta \not\equiv 4 \pmod 8$. Thus, if $|\Delta|$ is divisible by exactly $N+1$ distinct primes $q_1 < q_1 < \cdots < q_{N+1}$ for $N \geq 0$, then $-D = q_1 q_2 \ldots q_{N+1}$. Set $q = -D/q_{N+1}$. If r and s are any two distinct divisors of q and $[r, \sqrt{D}] \sim [s, \sqrt{D}]$, then $[rs/g^2, \sqrt{D}] \sim 1$ where $g = \gcd(r, s)$. By Theorem 1.3.2, $rs/g^2 = -D$, contradicting that q_{N+1} divides D. Hence, distinct divisors of q yield pairwise inequivalent ambiguous ideals, i.e. $h_{\Delta,2} \geq 2^N$. By the Claim, $h_{\Delta,2} \leq 2^{N+1}$. However, $\prod_{i=1}^{N+1} \mathcal{Q}_i \sim 1$ where $\mathcal{Q}_i$ is the $\mathcal{O}_\Delta$-prime above q_i, so $h_{\Delta,2} = 2^N$.

[(1.3.4)] This result was proved by R. Sasaki (see [314]), but was later generalized to arbitrary orders in [232]. In any case, the proof is elementary, as the hint to Exercise 1.3.2 shows. (See section five of this chapter.)

If $\Delta > 0$ with $N(\varepsilon_\Delta) = -1$, then, by Exercise 1.2.4(b), $h_{\Delta,2} \geq 2^N$. By Exercise 1.3.7(c), every class of C_Δ has an ambiguous ideal in it. Hence, using Exercise 1.2.4, we deduce that $h_{\Delta,2} = 2^N$.

The case where $\Delta > 0$ and $N(\varepsilon_\Delta) = 1$ is Exercise 1.3.7(d). □

Corollary 1.3.2. *Let Δ be a fundamental discriminant.*

(1) *If $\Delta < 0$, then h_Δ is odd if and only if either $-D = 1, 2$ or $-D \equiv 3 \pmod 4$ is prime.*

(2) *If $\Delta > 0$, then h_Δ is odd if and only if $D = p, 2p_1$, or p_1p_2 where p is prime and $p_1 \equiv p_2 \equiv 3 \pmod 4$ are primes.*

Basically, $C_{\Delta,2}$ represents the subgroup generated by the *ambiguous classes*, i.e. those classes $\{I\}$ for which $I \sim I'$, which means $I^2 \sim 1$ (see Exercise 1.3.3). Hence, the ambiguous classes are the elements of order 1 or 2 in C_Δ. When $\Delta > 0$, however, it is possible to have an ambiguous class without any ambiguous ideals in it (see Definition 1.2.2). Such classes are a major topic in Chapter Six where we give a complete description of these classes, and show how they arise from representations of the discriminant as a sum of two squares. When $\Delta < 0$ this phenomenon cannot occur, but we show how ambiguous classes arise from representations of the discriminant as a difference of two squares.

At the level of the algebraic integer, rather than at the ideal level, we may discuss factorization as follows. If $\alpha \in \mathcal{O}_\Delta$ is any non-zero, non-unit element then $h_\Delta = 1$ if and only if $\mathcal{O}_\Delta$ is a U.F.D. (see footnote (1.3.1). Hence, $h_\Delta = 1$ if and only if every such α satisfies the property that any factorization $\alpha = \alpha_1\alpha_2\ldots\alpha_r = \beta_1\beta_2\ldots\beta_s$ into irreducible elements has

$$r = s \tag{1.3.1}$$

and

$$\alpha_i \text{ is an associate of } \beta_i \tag{1.3.2}$$

for all $i = 1, 2, \ldots, r$ (after possibly renumbering).[(1.3.5)] Carlitz [46] proved that $h_\Delta = 2$ if and only if (1.3.1) always holds but (1.3.2) does not always hold. For instance, in $\mathcal{O}_\Delta = [1, \sqrt{-5}]$ where $h_\Delta = 2$, we have $3 \cdot 3 = (2+\sqrt{-5})(2-\sqrt{-5})$. Carlitz also showed that $h_\Delta > 2$ if and only if (1.3.1) does not always hold. For example, $h_{-23} = 3$ and $3 \cdot 3 \cdot 3 = (2+\sqrt{-23})(2-\sqrt{-23})$.

We conclude this section with the statement of a result, proved by Weber[(1.3.6)] in 1882, concerning the number of primes in a given ideal class.

Theorem 1.3.4. *If Δ is a fundamental discriminant, then every class of C_Δ contains an infinite number of prime ideals.*

Proof. See Cohn [57, Theorem 4, p. 179]. □

[(1.3.5)]See Masley [211] for a nice survey of class number $h_\Delta \leq 2$ for various types of fields, especially quadratic.

[(1.3.6)]Heinrich Weber (1842–1913) achieved his Ph.D. at Heidelberg in 1863, and became Privatdozent there in 1866. He became a Professor in Heidelberg in 1869, went on to Königsberg, and ultimately Berlin in 1883. Weber was also one of Kronecker's official biographers, along with Adolf Kneser. Weber's monumental contribution was his *Lehrbuch der Algebra* in which there is a great wealth of material.

Exercises 1.3

1. Prove that, if $n \in \mathbf{Z}$ is positive, and Δ is a fundamental discriminant with $\gcd(n, h_\Delta) = 1$, then $I^n \sim J^n$ implies $I \sim J$ for classes of ideals I and J in C_Δ.

2. Prove Theorem 1.3.2. (*Hint*: Assume $ac = N(b + w_\Delta)$ so either $a < N(w_\Delta)$ or $c < N(w_\Delta)$. Given $ax + (b + w_\Delta)y = \alpha$ with $N(\alpha) = a$, deduce that $a > N(w_\Delta)$ if $y \neq 0$).

3. Prove that for any class $\{J\}$ in C_Δ, Δ a given discriminant, $\{J\}^{-1} = \{J'\}$ where $\{J\}^{-1}$ denotes the *inverse class* of $\{J\}$. (*Hint*: See Exercise 1.2.1(b).)

4. Let Δ be a fundamental discriminant and I an $\mathcal{O}_\Delta$-ideal. Prove that
   (a) $I^{h_\Delta} \sim 1$.
   (b) If $\gcd(h_\Delta, n) = 1$ and $I^n \sim 1$, then $I \sim 1$.
   (c) If $I^n \sim 1$ $(n \geq 1)$, then $d \mid n$ where d is the order of $\{I\}$ in C_Δ
   (d) If d is the order of $\{I\}$ in C_Δ and $\gcd(h_\Delta, n) = 1$, then d is the order of $\{I\}^n$ in C_Δ. Furthermore, $d \mid h_\Delta$.

5. Let $D > 0$ be a fundamental radicand with discriminant Δ. Prove that if $h_\Delta = 2$, then D has to have one of the following shapes:
   (a) $D = pq$ with $p \equiv q \equiv 1 \pmod 4$ primes.
   (b) $D = 2q$ with $q \equiv 1 \pmod 4$ prime.
   (c) $D = pq$ with $p \not\equiv q \pmod 4$ odd primes, and show that $N(\varepsilon_\Delta) = 1$.
   (d) $D = 2pq$ with one of the primes p or q congruent to three modulo four, and show that $N(\varepsilon_\Delta) = 1$.
   (e) $D = pqr$ where $p \equiv q \equiv 3 \pmod 4$ and $r \equiv 1 \pmod 4$ are all primes, and show that $N(\varepsilon_\Delta) = 1$. (*Hint*: Use Theorem 1.3.3.).

6. (a) Prove that if $\Delta < 0$ is a fundamental discriminant with $\Delta \equiv 0 \pmod 4$, then the unique primitive $\mathcal{O}_\Delta$-ideal above 2 is not principal unless $\Delta = -4$ or -8. (*Hint*: Use Theorem 1.3.2.)
   (b) Use (a) to verify Corollary 1.3.2(1) for $\Delta \equiv 0 \pmod 4$.
   (c) Prove that, for Δ as above, if I is an $\mathcal{O}_\Delta$-ideal with $4 \mid N(I)$, then I is not primitive.

7. (a) Let $I = [N(I), \beta]$ be a primitive $\mathcal{O}_\Delta$-ideal, where Δ is a fundamental discriminant. Prove that if $J = [N(\beta)/N(I), -\beta']$, then $(-\beta')I = (N(I))J$.
   (b) Prove Theorem 1.3.3 for $\Delta < 0$, when $\Delta \equiv 4 \pmod 8$. (*Hint*: Use the ideas of the proof of Theorem 1.3.3 for $\Delta \not\equiv 4 \pmod 8$, and see Exercise 1.3.6(a).)
   (c) Prove that, if J is an ideal such that $J^2 = (\alpha)$ for some $\alpha \in \mathcal{O}_\Delta$ with $N(\alpha) > 0$, then there exists an ambiguous ideal H with $H \sim J$. (*Hint*: Set $\beta = N(J) + \alpha'$. Prove that $(N(J))J' = (\alpha')J$ and $(\beta N(J))J' = (\alpha'\beta)J$. Deduce that $I = I'$ where $I = (\beta)J$.)

(d) Prove that if $\Delta > 0$, and $N(\varepsilon_\Delta) = 1$, then there is an ambiguous class in C_Δ without ambiguous ideals in it if and only if Δ is a sum of two squares. Conclude that, if this happens, then $|C_\Delta : C_{\Delta,2}| = 2$. (*Hint*: If $\Delta = 4a^2 + b^2$, then consider the class of $I = [a, (b + \sqrt{\Delta})/2]$. Conversely, use part (c).)[(1.3.7)]

1.4 Reduced Ideals.

First we define the central topic of this section.

Definition 1.4.1. If Δ is a fundamental discriminant, then an ideal I of $\mathcal{O}_\Delta$ is said to be *reduced* if it is primitive, and does not contain any non-zero element α such that both $|\alpha| < N(I)$ and $|\alpha'| < N(I)$. (Note that when $\Delta < 0$ this means that there is no $\alpha \in I$ such that $|\alpha| < N(I)$ where $|\alpha|^2 = \alpha\alpha' = N(\alpha)$.)

An algebraic way of looking at reduced ideals is

Lemma 1.4.1. *Let $\Delta > 0$ be a fundamental discriminant. I is a reduced ideal in $\mathcal{O}_\Delta$ if and only if there exists some $\beta \in I$ such that $I = [N(I), \beta]$ with $\beta > N(I)$ and $-N(I) < \beta' < 0$.*

Proof. See Exercise 1.4.1(a). □

As an immediate consequence, we have

Corollary 1.4.1. *If $\Delta > 0$, and $I = [a, b + w_\Delta]$ is a primitive ideal in $\mathcal{O}_\Delta$ with $\gamma = (b + w_\Delta)/a > 1$ and $-1 < (b + w'_\Delta)/a < 0$, then I is reduced.*

Theorem 1.4.1. [(1.4.1)] *If $\Delta > 0$ is a fundamental discriminant and $I = [a, b + w_\Delta]$ is a primitive ideal in $\mathcal{O}_\Delta$, then I is reduced if and only if $\lfloor -(b + w'_\Delta)/a \rfloor a > a - b - w_\Delta$.*

Corollary 1.4.2. *If $\Delta > 0$, and $[a, b + w_\Delta]$ is reduced, then $a < \sqrt{\Delta}$.*

Corollary 1.4.3. *If $\Delta > 0$, and $[a, b + w_\Delta]$ is a primitive ideal with $a < \sqrt{\Delta}/2$, then I is reduced.*

Corollary 1.4.4. *If $\Delta > 0$, and $[a, b + w_\Delta]$ is a primitive ideal with $0 \le b < a$ and $\sqrt{\Delta}/2 \le a < \sqrt{\Delta}$, then I is reduced if and only if $a - w_\Delta < b < -w'_\Delta$.*

Proof. See Exercise 1.4.1(b). □

Remark 1.4.1. Notice that by Theorem 1.3.1 and Theorem 1.4.1 we have that every class of C_Δ has a reduced ideal when $\Delta > 0$. This is also true when $\Delta < 0$.[(1.4.2)]

[(1.3.7)] Note that we are *not* saying that there is at most one ambiguous class without an ambiguous ideal, but rather that we may *choose* such classes. See Chapter Six especially Remark 6.1.1.

[(1.4.1)] This was proved in [268].

[(1.4.2)] Both of these facts will ensue from the continued fraction algorithm and its complex analogue

For $\Delta < 0$ we also have the following fundamental result.

Theorem 1.4.2. *If $\Delta < 0$ is a fundamental discriminant, then*

(a) *If I is a primitive ideal of $\mathcal{O}_\Delta$, then there exists some $\alpha \in I$ with $I = [N(I), \alpha]$ and $|\mathrm{Tr}(\alpha)| \le N(I)$. Furthermore, $|\mathrm{Tr}(\alpha)|$ is unique (i.e, if $I = [N(I), \alpha] = [N(I), \beta]$ and $|\mathrm{Tr}(\alpha)| \le N(I)$, $|\mathrm{Tr}(\beta)| \le N(I)$, then $|\mathrm{Tr}(\alpha)| = |\mathrm{Tr}(\beta)|$).*

(b) *If I is a primitive ideal of $\mathcal{O}_\Delta$ and $I = [N(I), \alpha]$ with $|\mathrm{Tr}(\alpha)| \le N(I)$, then I is a reduced ideal if and only if $|\alpha| \ge N(I)$.*

(c) *If I is a reduced ideal of $\mathcal{O}_\Delta$, then $N(I) < \sqrt{|\Delta|/3}$.*

(d) *If I is a primitive ideal of $\mathcal{O}_\Delta$ and $N(I) < \sqrt{|\Delta|/4}$, then I is a reduced ideal.*

(e) *If $I_i = [a_i, (b_i + \sqrt{\Delta})/2]$ (for $i = 1, 2$) are two distinct, equivalent, reduced ideals in $\mathcal{O}_\Delta$, with $|b_i| \le a_i$ and $c_i = (b_i^2 - \Delta)/(4a_i)$, then $a_1 = a_2 = a$, $|b_1| = |b_2| = b$, and $c_1 = c_2 = c$. Also, if $b \ne a$, then $c = a$. (This tells us that there are at most two reduced ideals in any class of C_Δ, and when two distinct such ideals are in a class, then they are conjugates of one another.)*

(f) *If $I = [N(I), \alpha]$ is an ideal of $\mathcal{O}_\Delta$ with $|\mathrm{Tr}(\alpha)| \le N(I) < M_\Delta$, and I is in an ambiguous class of $\mathcal{O}_\Delta$, then either $N(I)$ or $2N(I) + |\mathrm{Tr}(\alpha)|$ is a divisor of Δ.*[(1.4.3)]

Proof. (a) follows quite clearly from Theorem 1.2.2 and Remark 1.2.1. Part (b) is clear from the parenthetical remark in Definition 1.4.1.

To prove (c), we observe that if $I = [N(I), \alpha]$, then $4N(\alpha) - \mathrm{Tr}(\alpha)^2 = -\Delta$. If I is reduced, then $|\alpha| \ge N(I)$, and since $|\mathrm{Tr}(\alpha)| \le N(I)$, then $-\Delta = 4N(\alpha) - \mathrm{Tr}(\alpha)^2 \ge 4N(\alpha) - N(I)^2 = 4|\alpha|^2 - N(I)^2 \ge 4N(I)^2 - N(I)^2 = 3N(I)^2$.

To prove (d), we note that if $N(I) < \sqrt{|\Delta|/4}$, and $|\alpha| < N(I)$, then $N(\alpha) < |\Delta|/4 = N(\alpha) - \mathrm{Tr}(\alpha)^2/4$, a contradiction.

The proof of (e) is more involved. Since the I_i are reduced, then $c_i \ge a_i$ for $i = 1, 2$ by part (b). We may assume, without loss of generality, that $a_1 \ge a_2$. Also, since $I_1 I_2' \sim 1$, then $I_1 I_2' = (\alpha)$ where $\alpha \in I_1$, by Remark 1.2.3. Therefore, there exist $x, y \in \mathbf{Z}$ such that

$$2\alpha = 2xa_1 + b_1 y + y\sqrt{\Delta}. \tag{1.4.1}$$

By taking norms and dividing by $4a_1$ in (1.4.1) we get

$$a_1 \ge a_2 = x^2 a_1 + xyb_1 + y^2 c_1. \tag{1.4.2}$$

Therefore,

$$a_1 \ge x^2 a_1 + xyb_1 + y^2 a_1. \tag{1.4.3}$$

Hence, $xyb_1 \le 0$. We now show that $xy = 0$. If $b_1 > 0$ and $xy < 0$, then $xyb_1 \ge xya_1$. By (1.4.3) we get

$$a_1 \ge x^2 a_1 + xya_1 + y^2 a_1,$$

---

in Chapter Two, where it is shown that every class must have a reduced ideal in it. See Exercise 2.2.2.

(1.4.3) The first mention of Theorem 1.4.2 in the literature, in ideal-theoretic terms, is in Buchmann and Williams [37], but it appears without proof.

i.e.

$$1 \geq x^2 + xy + y^2 = (x+y)^2 - xy,$$

or

$$1 > 1 + xy > (x+y)^2,$$

a contradiction. We leave the case where $b_1 < 0$ and $xy > 0$ for the reader since it is similar. We have $xy = 0$. We observe that $b_1 \neq 0$, since that would force $b_2 = 0$ and $a_1 = a_2$ via (1.4.1)–(1.4.3), contradicting that $I_1 \neq I_2$. If $y = 0$, then $a_1 = a_2 = |b_1| = |b_2|$ by (1.4.1)–(1.4.3). If $y \neq 0$ and $x = 0$, then by (1.4.1)–(1.4.3), $a_1 = a_2 = c_1 = a$, say. A similar argument shows that $c_2 = a$.

Finally, we establish (f). Let $I = [N(I), (b + \sqrt{\Delta})/2] = [N(I), \alpha]$ be in an ambiguous class with $|\mathrm{Tr}(\alpha)| = b \leq N(I) < M_\Delta$. First, we assume that $N(I) > 1$ and $\gcd(N(I), |\Delta|) = 1$. By (1.2.1)–(1.2.3), $I^2 = [N(I)^2, \beta] \sim 1$, where $\beta = (b_3 + \sqrt{\Delta})/2$ and b_3 is determined by (1.2.4). Therefore, $|b_3| < 2N(I)^2 < -2\Delta/3$. Thus, $N(\beta) < N(w_\Delta)^2$, provided that $-\Delta > 8$, which we may assume. Now we invoke Theorem 1.3.2 which yields that $N(\beta) = N(I)^2$, i.e. $\Delta = b_3^2 - 4N(I)^2$. Using (1.2.4), we may verify that $b_3 = b$. Hence, $\Delta = (b - 2N(I))(b + 2N(I))$, i.e. $2N(I) + |\mathrm{Tr}(\alpha)|$ divides $|\Delta|$.

If $d = \gcd(N(I), \Delta) > 1$, then by Exercise 1.2.10, $\gcd(d, N(I)/d) = 1$, so $I = [d, \alpha][N(I)/d, \alpha] = I_1 I_2$, say. Since $d \mid |\Delta|$, then $I_1^2 \sim 1$ so $I_2^2 \sim 1$. Therefore, by the above argument, $2N(I)/d + |\mathrm{Tr}(\alpha)|$ divides $|\Delta|$, if $N(I) > d$. However, $d \mid |\mathrm{Tr}(\alpha)|$ so $d \mid N(I)/d$, a contradiction. Hence, $N(I) = d$ divides $|\Delta|$. □

To help the reader better grasp the concepts in Theorem 1.4.2, we provide the following examples with (a)–(f) corresponding to the parts of the theorem.

Example 1.4.1.

(a) $I = [29, (7 + \sqrt{-3315})/2]$ is a primitive ideal in $\mathcal{O}_\Delta$ where $\Delta = -3315$. Here, $\alpha = (7 + \sqrt{-3315})/2$, with $\mathrm{Tr}(\alpha) = 7$ and $N(I) = 29$. Furthermore, α is unique with respect to $|\mathrm{Tr}(\alpha)| \leq 7$, see Remark 1.2.1.

(b) In part (a), I is reduced since $|\alpha| = \sqrt{841} = 29 = N(I)$.

(c) The ideal $J = [N(J), \beta] = [53, (89 + \sqrt{-3315})/2]$ is *not* reduced, since $53 > \sqrt{|\Delta|/3}$. Observe here that $N(\beta) = N(I) = 53$. However, $\mathrm{Tr}(\beta) = 89 > N(I)$.

(d) The ideal $I = [2, 1 + \sqrt{-5}]$ is reduced with $N(I) < \sqrt{-\Delta}/2$.

(e) Let I be the ideal in part (a). Both I and I' are reduced, and they are the *only* reduced ideals in their class. Similarly, $J = [31, (23 + \sqrt{\Delta})/2]$ is reduced, along with J', and they are the only reduced ideals in their class. We will explore this and similar examples in great detail when we get to Chapter Four (e.g. see Example 4.1.4).

(f) Let I be as in part (a). $N(I) = 29 < M_\Delta$, and I is reduced in an ambiguous class. Also, $2N(I) + |\mathrm{Tr}(\alpha)| = 65$ which divides $|\Delta|$. The only other split prime less than M_Δ in an ambiguous class is 31, with $J = [31, (23 + \sqrt{-3315})/2] = [31, \beta]$. Also, $2N(J) + \mathrm{Tr}(\beta) = 85$ which divides $|\Delta|$. (see Table 4.1.7.)

Theorem 1.4.2 has some interesting consequences which will prove to be valuable in Chapter Four (e.g. see Theorem 4.1.10), such as

Corollary 1.4.5. *If $\Delta < 0$ is a fundamental discriminant, and $I = [N(I), \alpha]$ is an ideal in an ambiguous class of C_Δ, with $N(I) < M_\Delta$ not dividing $|\Delta|$, then*

(a) *There exists a square-free divisor $q > N(I)$ of $|\Delta|$ such that $\Delta = q^2 - 4qN(I)$.*

(b) $\Delta \not\equiv 0 \pmod 8$.

Proof. Continuing from the proof of Theorem 1.4.2(f), $\Delta = b^2 - 4N(I)^2$. If $q = 2N(I) + |b|$, then $q > N(I)$ and q is square-free. This is (a). If $\Delta \equiv 0 \pmod 8$, then q must be even in the above. Thus, $\Delta \equiv q^2 \pmod 8$ forcing $q \equiv 0 \pmod 4$, a contradiction. This is (b). □

Corollary 1.4.5 will have some applications in Chapter Four, where we specialize to a discussion of class groups of exponent 2 in complex quadratic orders. Now we present another consequence of the proof of Theorem 1.4.2.

Corollary 1.4.6. *If $I_i = [a_i, \alpha_i]$ for $i = 1, 2$ are two primitive $\mathcal{O}_\Delta$-ideals, with $\Delta < 0$, $1 \le a_i < \sqrt{-\Delta}/2$, and $I_1 \sim I_2$, then $I_1 = I_2$.*

Proof. By Theorem 1.4.2(d), the I_i are reduced, so by part (e), $a_1 = a_2$ and $I_1' = I_2$. If $I_1 \neq I_2$, then by Theorem 1.4.2(f), $\Delta = b^2 - 4N(I_1)^2$, where $b = |\text{Tr}(\alpha_1)|$. However, $N(I_1) < \sqrt{-\Delta}/2$, so $\Delta = b^2 - 4N(I_1)^2 > b^2 + \Delta$, a contradiction. □

Remark 1.4.2. Theorem 1.4.2(e) says a great deal more than it appears on the surface, namely, if there is more than one reduced ideal in a class of C_Δ for $\Delta < 0$ (and there must be at least one by Theorem 1.3.3 and Remark 1.4.1), then there is at most one other, and that other one is the *conjugate* of the first. Thus, two distinct reduced ideals can exist in the class if and only if the class is *ambiguous* and there are *no ambiguous reduced ideals* in the class. Furthermore, it establishes that, if an ambiguous class has a reduced ambiguous ideal in it, then the only *reduced* ideal is the reduced ambiguous ideal (the extreme example being the principal class which contains only the *trivial* reduced ideal). In all other cases, there is exactly one reduced ideal in the class. Corollary 1.4.6 is more enlightening than it appears on the surface as well, namely, it shows that if there are indeed two distinct reduced ideals in the class, then those ideals $I \neq I'$ must satisfy that $\sqrt{-\Delta}/2 < N(I) = N(I') < \sqrt{-\Delta/3}$. Another interesting fact, which comes out of the proof of Theorem 1.4.2(f), is that we cannot have a reduced ideal $I = [N(I), \alpha]$ where $\gcd(N(I), |\Delta|) > 1$, unless $N(I) \mid |\Delta|$.

Exercises 1.4

1. (a) Establish Lemma 1.4.1. (*Hint*: Show that if $I = [N(I), \alpha]$ is reduced, then we can always find a least element $\beta \in I, \beta > 0$ with $|\beta'| < N(I)$ and $\beta > N(I)$ using Definition 1.4.1. Next, set $\beta = xN(I) + y\alpha$ and prove that this selection of β forces $|y| = 1$. The converse is straightforward.)

(b) Use (a) to prove Theorem 1.4.1 and deduce Corollaries 1.4.2 - 1.4.4. (*Hint*: Once (a) is established, use $\beta = \lfloor(-b + w'_\Delta)/a\rfloor a + b + w_\Delta$.)

2. (a) Prove that, for any fundamental discriminant Δ, if

$$[ca_1, (b_1 + \sqrt{\Delta})/2] \sim [ca_2, (b_2 + \sqrt{\Delta})/2],$$

then

$$[a_1, (b_1 + \sqrt{\Delta})/2] \sim [a_2, (b_2 + \sqrt{\Delta})/2].$$

(*Hint*: First prove the result for $b_1 = b_2$. Then show that $[a_i, (b_i + \sqrt{\Delta})/2] = [a_i, (b_3 + \sqrt{\Delta})/2]$ (for $i = 1, 2$), where $b_3 \equiv b_i \pmod{a}$.)

(b) Prove that, if $b_i \leq a_i$ in part (a), then we can represent b_3 as $b_i + n_i a_i$ for some $n_i \in \mathbb{Z}$ and $b_3 = b_i + n_i a_i \leq a_1 a_2$.

3. Let Δ be a fundamental discriminant with radicand $D > 0$, and let I be a primitive ambiguous ideal of $\mathcal{O}_\Delta$.[(1.4.4)] Prove that

(a) $\sqrt{\Delta} \in I$, i.e. $I \mid (\sqrt{\Delta})$.

(b) If $N(I) < \sqrt{\Delta}$, then I is either reduced or $4 \mid \Delta$ and $\sqrt{\Delta}/2 \in I$.

(c) There exists a reduced ideal $J \sim I$ (see Remark 1.4.1).

(d) If I is reduced and $I \neq (1)$, $I^2 \neq (2)$, then either $N(I)$ or $N(I)/2$ is a non-trivial factor of D.

1.5 Quadratic Orders.

In this section, we give an overview of the more general setting which we will have occasion to access later. The presentation given in sections one through four of this chapter is a special case of the more general case called orders in quadratic fields, of which the ring of integers is called the maximal order. The reader who is interested only in the field case can skip this section and interpret all following results for the maximal order only.[(1.5.1)]

Let Δ_0 be a fundamental discriminant (positive or negative) and set $\Delta = f_\Delta^2 \Delta_0$ for positive $f_\Delta \in \mathbb{Z}$. If $g = \gcd(\sigma_0, f_\Delta)$ and $\sigma = \sigma_0/g$, then $\Delta = 4D/\sigma^2$ is called a *discriminant* with *associated radicand* $D = (f_\Delta/g)^2 D_0$ (and underlying fundamental discriminant Δ_0 having fundamental radicand D_0). Set

$$w_\Delta = f_\Delta w_{\Delta_0} + h,$$

where $h \in \mathbb{Z}$ is called the *principal surd associated with the discriminant* Δ, then $\Delta = (w_\Delta - w'_\Delta)^2$, and

$$\mathcal{O}_\Delta = [1, f_\Delta w_{\Delta_0}] = [1, \omega_\Delta]$$

(1.4.4) These were established by Williams in [383] although, as he points out, the results are well-known.

(1.5.1) On the other hand, the reader who wants to know more about orders than the minimal amount presented here is advised to consult the excellent text by Reiner [303]. The term "order" originated with Gauss' use of the term for quadratic forms (see Appendix E). Dedekind, on the other hand, used the term to mean an integral domain. Hilbert used the term "Zahlring" to mean what we call $\mathcal{O}_\Delta$.

is called an *order* in $K = Q(\sqrt{D_0})$ having *conductor* f_Δ and associated discriminant Δ with radicand D. When $f_\Delta = 1$, $\mathcal{O}_\Delta$ is the ring of integers $\mathcal{O}_{\Delta_0}$ of K called the *maximal order* since, for any f_Δ, $[1, f_\Delta w_{\Delta_0}] \subseteq [1, w_{\Delta_0}]$. In fact, $f_\Delta = |\mathcal{O}_\Delta : \mathcal{O}_{\Delta_0}|$ the index of $\mathcal{O}_\Delta$ in the maximal order $\mathcal{O}_{\Delta_0}$.(1.5.2) Thus, an order in K is uniquely determined by its discriminant Δ, i.e. there is a one-to-one correspondence between nonsquare integers $\Delta \equiv 0, 1 \pmod 4$ and orders in quadratic fields. This generalizes the statement for quadratic fields made in section one of this chapter.

In footnote (1.2.3), we talked about fractional ideals for a maximal order. The same definition holds for arbitrary orders. In point of fact, a *fractional* $\mathcal{O}_\Delta$-ideal is just a set of the form αI $(\alpha \neq 0)$ for some $\alpha \in K = Q(\sqrt{\Delta})$ and some (integral) ideal I of $\mathcal{O}_\Delta$ $(I \neq (0))$. Fractional $\mathcal{O}_\Delta$-ideals are called *invertible* if there is another fractional $\mathcal{O}_\Delta$-ideal J such that $IJ = \mathcal{O}_\Delta$.(1.5.3) As observed in footnote (1.2.3), J may be taken to be

$$I^{-1} = \{\beta \in K : \beta I \subseteq \mathcal{O}_\Delta\}.$$

An $\mathcal{O}_\Delta$-ideal I is called *proper* (1.5.4) if $\mathcal{O}_\Delta = \{\alpha \in K : \alpha I \subseteq I\}$. Thus, if J is a fractional ideal with $J = \alpha I$ as above, then J is called a *proper fractional ideal* if

$$\mathcal{O}_\Delta = \{\beta \in K : \beta I \subseteq I\}.$$

Definition 1.5.1. Let $I(\Delta)$ denote the set of proper fractional ideals in an order $\mathcal{O}_\Delta$. By Exercise 1.5.4, $I(\Delta)$ is a group under multiplication. The principal $\mathcal{O}_\Delta$-ideals form a subgroup $P(\Delta)$ of $I(\Delta)$ and $C_\Delta = I(\Delta)/P(\Delta)$ is the *ideal class group of* $\mathcal{O}_\Delta$. Its order is h_Δ, the *class number* of the order.

The proper ideals are precisely the invertible ones (Exercise 1.5.1). (1.5.5) Note that $\mathcal{O}_\Delta \subseteq \{\beta \in K : \beta I \subseteq I\}$ by the very fact that I is an $\mathcal{O}_\Delta$-ideal. However, equality does not always hold. For instance, we have

Example 1.5.1. Let $\Delta = 1224 = 2^3 \cdot 3^2 \cdot 17$, $D = 306 = 2 \cdot 3^2 \cdot 17$, $D_0 = 34$ and $\Delta_0 = 2^3 \cdot 17$ with $f_\Delta = 3$, and $\sigma = \sigma_0 = g = 1$. Let $\mathcal{O}_\Delta = [1, \sqrt{306}]$ and $I = [9, 15 + \sqrt{306}]$. Here $I^{-1} = (\frac{1}{9})I'$ and $II^{-1} = [3, \sqrt{306}] \neq \mathcal{O}_\Delta$. Hence, I is not invertible. Here $f_\Delta = 3$ since $306 = 9 \cdot 34$.

Notice that, in Example 1.5.1, $\gcd(f_\Delta, N(I)) = \gcd(3, 9) = 3$. In fact, it turns out that $\mathcal{O}_\Delta$-ideals with norms prime to the conductor are proper (Exercise 1.5.2(c)). However, not all proper ideals have norms prime to the conductor as in

Example 1.5.2. Let $\Delta = 725 = 5^2 \cdot 29$ with order $\mathcal{O}_\Delta = [1, (5 + \sqrt{725})/2]$ and conductor $f_\Delta = 5$. Here $w_\Delta = f_\Delta w_{\Delta_0}$ where $w_{\Delta_0} = (1 + \sqrt{29})/2$. Consider the ideal $I = [25, (25 + \sqrt{725})/2]$. This ideal is invertible (as are all principal ideals in a given order (Exercise 1.5.2(a))) since $II^{-1} = \mathcal{O}_\Delta$ with $I^{-1} = (\frac{1}{25})I' = (\frac{1}{25})I$ and in fact $I = (25)$.(1.5.6)

(1.5.2) When $f_\Delta > 1$ then $\mathcal{O}_\Delta$ is *not* a Dedekind domain (see footnote (1.2.3)), i.e. not all ideals in $\mathcal{O}_\Delta$ are invertible. Hence, the definition of C_Δ, the class group, is more involved (see Exercise 1.5.5 and Definition 1.5.2).

(1.5.3) In fact, $\{J\}$ is just the inverse class of $\{I\}$ in C_Δ (see Exercises 1.3.3 and 1.5.1).

(1.5.4) At various places in the literature, this is also called *regular*.

(1.5.5) Thus, these are the ones which comprise the class group (see Exercise 1.5.5).

(1.5.6) Nevertheless, the class group defined for arbitrary orders (see Exercise 1.5.5) avoids those ideals with norms *not* prime to the conductor. The reason for this is that unique factorization can fail for proper ideals whether or not $h_\Delta = 1$ (see Exercise 1.5.6).

There is also the possibility that we can have a primitive ideal of $\mathcal{O}_\Delta$ which is not invertible, as shown in Example 1.5.1.

Remark 1.5.1. Some cautionary comments are in order for the reader. Several places in the literature use the term "primitive" as in Definition 1.2.1, i.e. ideals having no rational integer factors other than ± 1.[(1.5.7)] However, the very same term is used in other places[(1.5.8)] to mean something totally different. In order to clarify the situation, we introduce a new term for the latter.

Definition 1.5.2. Let $\mathcal{O}_\Delta$ be an order and let $I = [a, (b+\sqrt{\Delta})/2]$ be an $\mathcal{O}_\Delta$-ideal. I is called *strictly primitive* if I satisfies $\gcd(a, b, (b^2 - \Delta)/(4a)) = 1$.

Clearly, strictly primitive ideals are primitive, but the converse fails. The ideal I in Example 1.5.2 is an instance of a strictly primitive ideal which has norm which is not prime to the conductor. The ideal in Example 1.5.1 is an instance of an ideal which is primitive, but not strictly primitive.

We now demonstrate that invertibility is equivalent to strict primitivity. We note that from Exercise 1.5.3, the multiplication formulae (1.2.1)–(1.2.5) hold for invertible ideals in arbitrary orders.

Proposition 1.5.1. *Let Δ be a discriminant and let $I = [a, (b+\sqrt{\Delta})/2]$ be a primitive ideal in the order $\mathcal{O}_\Delta$. If $g_0 = \gcd(a, b, (b^2 - \Delta)/(4a))$, then $I^{-1} = (\frac{1}{a})I'$ and $II' = (a)[g_0, w_\Delta]$. Thus, I is invertible if and only if I is strictly primitive.*

Proof. Since $II' = (a)[a, (b+\sqrt{\Delta})/2, (b-\sqrt{\Delta})/2, (b^2-\Delta)/(4a)]$, then an easy exercise shows that $II' = (a)[g_0, w_\Delta]$. Clearly then $(\frac{1}{a})I'I \subseteq \mathcal{O}_\Delta$, and another exercise verifies that there are no other values $x \in \mathbb{Q}(\sqrt{D})$ such that $xI \subseteq \mathcal{O}_\Delta$. Hence, if I is invertible, then $\mathcal{O}_\Delta = II^{-1} = (\frac{1}{a})I'I = [g_0, w_\Delta]$, so $g_0 = 1$. Conversely, if I is strictly primitive, then $g_0 = 1$ and $I^{-1}I = \mathcal{O}_\Delta$, so I is invertible. □

We have shown that invertibility is equivalent to strict primitivity which means being proper. Also, we have shown that unique factorization can fail for the latter.

Because of the problems involved with ideals whose norms are not prime to the conductor, the group C_Δ is often identified with another quotient group which avoids some of these problems which we discussed above (see Exercise 1.5.5).

The *fundamental unit* ε_Δ, in an order $\mathcal{O}_\Delta$, is that unit $\varepsilon_\Delta > 1$ such that any unit u of $\mathcal{O}_\Delta$ is given by $u = \pm\varepsilon_\Delta^m$ where $m \in \mathbb{Z}$. Furthermore, ε_Δ is a unit in the maximal order $\mathcal{O}_{\Delta_0}$. Hence, $\varepsilon_\Delta = \varepsilon_{\Delta_0}^u$ where u is called the *unit index*. This is used in a determination of the class number h_Δ of $\mathcal{O}_\Delta$, i.e. the order of C_Δ.[(1.5.9)]

[(1.5.7)] For example, see H.C. Williams and M. Wunderlich [389], where this general notion of primitive is needed to consider arbitrary *cycles* of reduced ideals *without* reference to any class group structure (see Chapter Six where we discuss this phenomenon in detail for ambiguous cycles). They need this general setting to describe the continued fraction algorithm and its application to factoring.

[(1.5.8)] See P. Kaplan and K.S. Williams [137], for example, where the term "primitive" is used to mean "strictly primitive" which we define now. The authors of [137] required the more specific notion of strictly primitive since they (unlike the authors of [389] in footnote (1.5.7)) were concerned about class group structure.

[(1.5.9)] There are various equivalent formulations of h_Δ (in terms of h_{Δ_0}) when $\mathcal{O}_\Delta$ is an arbitrary order such as in Borevich–Shafarevich [30, Exercise 11, p. 152–153], Cox [67, Corollary 7.28, and Exercise 7.30, pp. 146-158] (for complex quadratic orders), and H. Cohn [57, Theorem 2, p. 217]. We find the latter to be the most useful, so we state it here for the interested reader. If $f_\Delta > 1$ is the conductor of an order $\mathcal{O}_\Delta$ with fundamental discriminant Δ_0 and unit index u, then

We will have substantially more to say about orders in Chapter Six when we discuss cycles of ambiguous ideals. In particular, we need to separate the notion of a reduced ideal under the initial definition of primitive from that of a reduced ideal which is strictly primitive. We do this via

Definition 1.5.3. In an order $\mathcal{O}_\Delta$ we call any reduced ideal, which is strictly primitive, a *strictly reduced* ideal of $\mathcal{O}_\Delta$.

As a result of the above discussion, we see that the equivalence classes of strictly primitive ideals comprise C_Δ. However, if we merely wish to look at *cycles* of reduced ideals (as defined in Chapter Two) without concern for invertibility of ideals, then we need not consider strict primitivity.

Observe that Definitions 1.2.1–1.2.2, 1.3.1–1.3.2, and 1.4.1 hold for arbitrary orders, as well as Theorems 1.2.1–1.2.2 (see Exercise 1.5.3). Also, Theorem 1.2.3 holds for those ideals with norm relatively prime to the conductor (Exercise 1.5.2(d)). Theorem 1.3.2 (Exercise 1.5.7) and Theorem 1.4.1 (Exercise 1.5.9) hold for arbitrary quadratic orders as well when *proper* ideals are considered in the former case, but are not required in the latter case. Theorem 1.3.3 fails for arbitrary orders.

Remark 1.5.2. A note of caution is in order here. The identification of C_Δ with $I_f(\Delta)/P_f(\Delta)$ as given in Exercise 1.5.5 is an artificial construct used by those interested in identifying certain class groups of forms with class groups of ideals via class field theory (see Cox [67]). However, we are not concerned with this translation and its manifold problems here. In fact, such an identification precludes a consideration of certain facts. For instance, we would like Theorem 1.3.1 to hold for arbitrary orders when the word "primitive" is replaced by "strictly primitive". However, the following example illustrates the problems involved.

Example 1.5.3. If $\Delta = -1467$, then $\lfloor M_\Delta \rfloor = 22$ and $(\Delta/p) = -1$ for all primes $p < M_\Delta$, $p \neq 3$. Hence, the only non-inert prime is 3 and there are no ideals of norm 3 which are strictly primitive (an easy exercise). In fact, $h_\Delta = 4$ (see Exercise 1.5.12). Observe (via Exercise 1.5.12) that $I = [9, (3 + \sqrt{-1467})/2]$ and $I' = [9, (15 + \sqrt{-1467})/2]$ are both strictly primitive of order 4 in C_Δ, whereas $J = [9, (9 + \sqrt{-1467})/2] = J'$ *is* an ambiguous ideal.

Remark 1.5.3. [(1.5.10)] What the above shows is that there is no geometric way of translating something akin to "norms prime to the conductor". Thus, the identification in Exercise 1.5.5 (which is the usual class-field theoretic way of treating the group in the literature) loses some information as depicted in Example 1.5.3, where we have strictly primitive ideals with norm less than M_Δ, but none of them have

---

$h_\Delta = h_{\Delta_0}\psi_{\Delta_0}(f_\Delta)/u$ where $\psi_{\Delta_0}(f_\Delta) = f_\Delta \prod(1 - (\Delta_0/p)/p)$ with the product ranging over all the distinct primes dividing f_Δ and $(*/*)$ denotes the Kronecker symbol (see Exercise 1.1.4). Also if $\Delta < 0$, then $u = 1$ unless $\Delta_0 = -4$, in which case $u = 2$, or $\Delta_0 = -3$ for which $u = 3$. It also follows that $h_\Delta \geq h_{\Delta_0}$. Indeed, $h_\Delta / h_{\Delta_0} \in \mathbf{Z}$, but it is an open question as to when $h_\Delta = h_{\Delta_0}$. These facts involve some analytic number theory considerations (see, for example, Zagier [397], Cohn [57]).

[(1.5.10)] An in-depth discussion of ideals with norms *not* prime to the conductor is given in Butts and Pall [44] wherein they comment about such ideals: "Their theory at first seemed chaotic and we tried to bring some order to the chaos." They also mentioned that: "This was found to be far from trivial ... ". However, they deal primarily with invertible ideals, and so they do not deal with instances such as that which occurs in Example 1.5.4. See footnote (6.1.2).

norm prime to the conductor. Hence, *we maintain throughout the definition of C_Δ given in Definition* 1.5.1 (unless specified otherwise for a given special situation). See Exercise 2.2.2.

We notice as well that we cannot get a statement about primes as in the second assertion of Theorem 1.3.1. Example 1.5.3 shows us why. It is not necessarily the case that there is a *prime* ideal, which is a generator of C_Δ, having norm less than M_Δ. However, there will be prime ideals which are equivalent to the ideal with norm less than M_Δ (see Exercise 2.2.2(d)). Of course, the prime ideal may have norm larger than M_Δ, as in Exercise 1.5.12.[(1.5.11)]

The notion of equivalence given in Definition 1.3.1 does not have to be associated with a class group, i.e. even if the ideals are not invertible, we may speak of such equivalence. For instance, we have

Example 1.5.4. Consider the discriminant $\Delta = 1224$ discussed in Example 1.5.1, where we showed that the ideal $I = [9, 15 + \sqrt{306}]$ is not invertible. Yet, the reader may verify as an exercise that

$$I \sim [18, 12+\sqrt{306}] \sim [15, 6+\sqrt{306}] \sim [15, 9+\sqrt{306}] \sim [18, 6+\sqrt{306}] \sim [9, 12+\sqrt{306}].$$

Of course, this is just the radicand $D = 3^2 \cdot 34$ with $f_\Delta = 3$, so the factor of 3 causes the problems. (This is called an ambiguous cycle of ideals which we will discuss in Chapter Six. See also Chapter Two, section one and footnote (6.1.2).)

We conclude this section with important information concerning reduced ideals in complex quadratic orders. The proof of Theorem 1.4.2(e) did not rely upon being in the maximal order, but rather it used the invertibility of the ideals under consideration. Hence, we have actually proved

Theorem 1.5.1. *If $\Delta < 0$ is a discriminant, and $I_i = [a_i, (b_i + \sqrt{\Delta})/2]$ (for $i = 1, 2$) are two distinct, equivalent, strictly reduced ideals in $\mathcal{O}_\Delta$, with $|b_i| \le a_i$ and $c_i = (b_i^2 - \Delta)/(4a_i)$, then $a_1 = a_2 = a$, $|b_1| = |b_2| = b$, and $c_1 = c_2 = c$, where $c = a$ if $b \ne a$.*

Also, Theorem 1.4.2(f) can be shown to hold for arbitrary complex quadratic orders with the proviso that we have invertible ideals.

Theorem 1.5.2. *If $\Delta < 0$ is a discriminant, and $I = [N(I), \alpha]$ is a strictly primitive ideal with $|\mathrm{Tr}(\alpha)| \le N(I) < M_\Delta$, in an ambiguous class of C_Δ, then either $N(I)$ or $2N(I) + |\mathrm{Tr}(\alpha)|$ is a divisor of $|\Delta|$.*

Corollary 1.5.1. *If $\Delta < 0$ is a discriminant and $I = [N(I), \alpha]$ is a strictly primitive ideal in an ambiguous class of C_Δ with $N(I) < M_\Delta$ and $N(I)$ does not divide $|\Delta|$, then:*

(a) *There exists a square-free divisor $q > N(I)$ of $|\Delta|$ such that $\Delta = q^2 - 4qN(I)$.*

(b) $\Delta \not\equiv 0 \pmod 8$.

Proof. Show that Exercise 1.2.10 is not needed to prove Theorem 1.4.2(f). □

[(1.5.11)] In [321], Schoof notes that C_Δ for a complex quadratic order is generated by split prime ideals with norm $p < c_1 \log^2 |\Delta|$ where $c_1 > 0$ is an absolute effectively computable constant. However, this can only be proved under the assumption of GRH (see Chapter Five, section four).

In the following chapters, we will refer to a fundamental discriminant Δ_0 when we deem it necessary to do so. However, when we refer to a discriminant Δ, we will mean the discriminant of an arbirtrary order as elucidated in this section.

Exercises 1.5

1. Prove that a fractional ideal in an order $\mathcal{O}_\Delta$ is invertible if and only if it is proper. (*Hint*: If J is a fractional $\mathcal{O}_\Delta$-ideal, then $J = \alpha I$ for some non-zero $\alpha \in K$ and some non-zero integral $\mathcal{O}_\Delta$-ideal I. Prove that J is proper if and only if I is proper if and only if $II' = \mathcal{O}_\Delta$. Thus, $J^{-1} = (1/N(\alpha))J'$ if J is proper.)

2. (a) Prove that all principal $\mathcal{O}_\Delta$-ideals are proper.

   (b) Prove that in the maximal order all ideals are proper.

   (c) Prove that $\mathcal{O}_\Delta$-ideals with norms prime to the conductor are proper. (*Hint*: Define an $\mathcal{O}_\Delta$-ideal I to be *prime to* f_Δ provided that $I+f_\Delta\mathcal{O}_\Delta = \mathcal{O}_\Delta$ (see Exercise 1.2.1(c)). Show that I is prime to f_Δ if and only if $N(I)$ is relatively prime to f_Δ. Then, given an $\alpha \in K$ with $\alpha I \subseteq I$, show that $\alpha\mathcal{O}_\Delta \subseteq I + f_\Delta\mathcal{O}_\Delta$.)

   (d) Prove that $\mathcal{O}_\Delta$-ideals with norms relatively prime to the conductor can be factored uniquely into prime $\mathcal{O}_\Delta$-ideals (*Hint*: First establish that if I is a prime $\mathcal{O}_\Delta$-ideal with norm relatively prime to the conductor, then $I\mathcal{O}_{\Delta_0}$ is a prime $\mathcal{O}_{\Delta_0}$-ideal, by showing that $\mathcal{O}_\Delta/I$ is isomorphic to $\mathcal{O}_{\Delta_0}/I\mathcal{O}_{\Delta_0}$. Then use uniqueness of factorization in $\mathcal{O}_{\Delta_0}$.)

   (e) Prove that $\mathcal{O}_\Delta \subseteq \mathcal{O}_{\Delta_0}$ for a given discriminant $\Delta \subseteq \Delta_0$, its maximal order, and that $\mathcal{O}_{\Delta_0}$ is the *only* maximal order in the field K with discriminant Δ_0.

3. (a) Verify that the multiplication formulae (1.2.1)–(1.2.5) hold for invertible ideals in any quadratic order.

   (b) Show that the multiplication formulae (1.2.1)–(1.2.5) fail if the ideals in question are not invertible. (*Hint*: Look at the ideal I in Example 1.5.1. Show that (1.2.1)–(1.2.5) yield $II' = (9)$. Then use Proposition 1.5.1 to get a contradiction.)

   (c) Verify that Theorems 1.2.1–1.2.2 hold for arbitrary quadratic orders.

4. Verify that $I(\Delta)$ is a group under multiplication. (*Hint*: See Exercise 1.5.1 and verify that I^{-1} and IJ are invertible fractional ideals whenever I and J are.)

5. Let $I_f(\Delta)$ denote the fractional $\mathcal{O}_\Delta$-ideals which are prime to the conductor f_Δ of the order (see Exercise 1.5.2). Prove that $I_f(\Delta)$ is a subgroup of $I(\Delta)$ and that the set of principal ideals having norm prime to the conductor, $P_f(\Delta)$ is a subgroup of $I_f(\Delta)$. Conclude that $I_f(\Delta)/P_f(\Delta)$ is isomorphic to $C_\Delta = I(\Delta)/P(\Delta)$. (*Hint*: First establish that every class in C_Δ has an integral ideal with norm prime to the conductor. Do this by showing that there is always a strictly primitive integral ideal whose norm *must* be prime to the conductor

using the ideas in Exercise 1.2.1. This guarantees the surjectivity of the map $I_f(\Delta) \to C_\Delta$. Now just verify that $I_f(\Delta) \cap P(\Delta) \subseteq P_f(\Delta)$.)

6. This exercise is designed to illustrate that unique factorization fails for proper ideals in non-maximal orders *whether or not* $h_\Delta = 1$.

   (a) Consider the order $\mathcal{O}_\Delta = [1, (5+\sqrt{725})/2]$ as discussed in Example 1.5.2. Prove that $\varepsilon_\Delta = (27+\sqrt{725})/2$.

   (b) Prove that $h_\Delta = 2$.

   (c) Factor $(25+\sqrt{725})/2$ in two distinct ways in $\mathcal{O}_\Delta$.

   (d) Let $\mathcal{O}_\Delta = [1, (3+\sqrt{-27})/2]$ and prove that $\varepsilon_\Delta = -1$ and $h_\Delta = 1$. Factor 9 in 2 distinct ways in $\mathcal{O}_\Delta$.

   (e) Let $\mathcal{O}_\Delta = [1, (3^{k-1}+\sqrt{-3^{2k-1}})/2]$ for $k \geq 1$. Verify that $h_\Delta = h_{\Delta_0}$, (see footnote (1.5.9)).

7. (a) Prove that Theorem 1.3.2 holds for arbitrary orders if the word *proper* is added to the condition on I.

   (b) Show that Theorem 1.3.2 fails if the work *proper* is deleted. (*Hint*: Look at the ideal $[2, \sqrt{-4}]$ in the order $\mathcal{O}_\Delta = [1, \sqrt{-4}]$ where $\Delta = -16$, $D = -4$, $f_\Delta = 2$ and $\sigma = \sigma_0 = 1$.).[(1.5.12)]

8. Prove that Definition 1.3.1 holds for an arbitrary discriminant Δ and that therefore, the notion of equivalence given in that definition can be made without reference to a class group structure, i.e. that the equivalence relation partitions the $\mathcal{O}_\Delta$-ideals into disjoint equivalence classes, where the ideals in each class need not be proper.

9. Prove that Theorem 1.4.1 holds for arbitrary orders. (*Hint*: First prove that Lemma 1.4.1 holds for arbitrary orders.)

10. (a) Let I be a primitive ambiguous ideal in an order $\mathcal{O}_\Delta$ with $\gcd(N(I), f_\Delta) = 1$ where f_Δ is the conductor of $\mathcal{O}_\Delta$. Prove that, if J is a primitive $\mathcal{O}_\Delta$-ideal with $N(I) = N(J)$, then $I = J$. (*Hint*: Prove that the primes dividing the norm of an ambiguous ideal must divide the discriminant.)

    (b) Show that if we remove the condition $\gcd(N(I), f_\Delta) = 1$, then (a) fails to be true. (*Hint*: Look at the order $\mathcal{O}_\Delta = [1, \sqrt{1305}]$ with $\Delta = D = 1305$, $f_\Delta = 3$, $\sigma_0 = \sigma = 2$, $g = 1$ and investigate ideals of norm 9.)

    (c) Prove that if $I = I'$, then $N(I) \mid \Delta$. (*Hint*: Here we do not assume $\gcd(N(I), f_\Delta) = 1$. Since $I = I'$, then $I \mid (\sqrt{\Delta})$.)

11. Prove that Theorem 1.4.2 holds for arbitrary complex quadratic orders, with appropriate change of words such as "strictly reduced" replacing "reduced" if necessary. (*Hint*: See Theorems 1.5.1–1.5.2.)

12. In Example 1.5.3, set $H = [41, (-3+\sqrt{-1467})/2]$. Prove that $H^2 = [41^2, (243+\sqrt{-1467})/2]$ and that $H^2 J \sim 1$. Conclude that H is an element of order 4 in C_Δ, and that $H \sim I$. (*Hint*: Use multiplication formulae (1.2.1)–(1.2.5).)

13. Prove that Exercise 1.3.4 holds for arbitrary orders.

[(1.5.12)] See also Theorem 4.1.3.

14. Let Δ be a fundamental discriminant, and let I be a fixed $\mathcal{O}_\Delta$-ideal.

    (a) Prove that every class of $\mathcal{O}_\Delta$ has a proper $\mathcal{O}_\Delta$-ideal whose norm is relatively prime to $N(I)$. (*Hint*: Let $J \in \{I'\}$ and use the ideas in Exercise 1.2.1 to establish that $JL \sim 1$ for some $L \in \{I\}$ with $J = J(I, L)$.)

    (b) Use part (a) to prove that every class of C_Δ has a prime $\mathcal{O}_\Delta$-ideal with norm relatively prime to Δ.

    (c) Do parts (a)–(b) hold for arbitrary orders?

15. Let $\mathcal{O}_\Delta = [1, \sqrt{-3}]$ with $f_\Delta = 2$.

    (a) Prove that 4 has two distinct factorizations as a product of irreducible elements, neither of which are prime.

    (b) Prove that the $\mathcal{O}_\Delta$-ideal, $I = [2, 1 + \sqrt{-3}]$ is not principal. (*Hint*: Use Theorem 1.3.2.)

    (c) Prove that any irreducible element $\alpha \in \mathcal{O}_\Delta$, with $|N(\alpha)|$ odd, must be prime.

    (This exercise shows that one may have an order $\mathcal{O}_\Delta$ with $h_\Delta = 1$, yet $\mathcal{O}_\Delta$ is neither a P.I.D. nor a U.F.D.)

16. Prove that, if $\mathcal{O}_\Delta$ is a maximal order and I is a primitive $\mathcal{O}_\Delta$-ideal, then I is strictly primitive.

17. Prove that, if $\Delta > 0$ is a discriminant with associated radicand D, then $\varepsilon_\Delta = (T + U\sqrt{D})/\sigma$ where T and U are the minimal solutions of $T^2 - DU^2 = \pm\sigma^2$. (*Hint*: Show that, in the sequence $\varepsilon_\Delta^i = (T_i + U_i\sqrt{D})/\sigma$, $T_i > T_{i-1}$ and $U_i > U_{i-1}$. This is done for instance, when $N(\varepsilon_\Delta) = 1$, by proving that $2\sqrt{D}(U_{i+1} - U_i)/\sigma = \varepsilon_\Delta^i(\varepsilon_\Delta - 1) - \varepsilon_\Delta^{-i}(\varepsilon_\Delta^{-1} - 1) > 0$.)

1.6 Powerful Numbers: An Application of Real Quadratics.

A *powerful number* is a positive $n \in \mathbb{Z}$ such that no prime appears to the first power in its canonical prime factorization, i.e. if a prime p divides n then p^2 divides n. The relationship between solutions to diophantine equations, especially the Pellian, and questions involving powerful numbers is a rich one, involving some elementary properties of quadratics on the one hand, and some deep and heretofore unsolved problems on the other. The purpose of this section is to describe these relationships in detail.

In the early 1930's, Erdös and Szekeres [83] investigated positive integers n such that p^i divides n whenever the prime p divides n for $i > 1$. In 1970, Golomb [96] dubbed such n, with $i = 2$, powerful numbers. In particular, he asked whether $(25, 27)$ is the only pair of consecutive odd powerful numbers. In 1981, Sentance [324] gave necessary and sufficient conditions for the existence of such pairs. In 1986, the authors of [242] generalized this result by giving necessary and sufficient conditions for the existence of pairs of powerful numbers spaced evenly apart. This leads naturally to consider integers which are representable as a *proper difference* of two powerful numbers, i.e. $n = p_1 - p_2$ where p_1 and p_2 are powerful numbers which are relatively prime. In [96], Golomb conjectured that 6 is not a proper difference[(1.6.1)] of two powerful numbers and that there are infinitely many numbers which cannot be represented as a proper difference of two powerful numbers. The antithesis of Golomb's conjecture was proved in 1982 by McDaniel [215] who gave an existence proof that *every* non-zero integer is a proper difference of two powerful numbers in infinitely many ways. In the aforementioned 1986 paper [242], the authors gave a simple proof of the McDaniel result, together with an effective algorithm for explicitly determining infinitely many such representations. However, in both McDaniel's proof and that of [242], one of the powerful numbers is almost always a perfect square. In [244], the authors finally established that *every* integer is representable in infinitely many ways as a *proper difference of nonsquare powerful numbers*, i.e. a proper difference of powerful numbers neither of which is a perfect square. This is the content of our first result.

Theorem 1.6.1. *Every non-zero rational integer is representable as a proper difference of powerful nonsquare numbers in infinitely many ways.*

Proof. By Exercise 1.6.1, it suffices to find for each non-zero integer n, positive integers r and s with $\gcd(Ar, Bs) = 1$, and $A^2r - B^2s = \pm n$, as well as a solution $T^2 - rsU^2 = \pm 1$ with $\gcd(U, rs) = 1$. The following chart gives such elements for all positive $n \in \mathbf{Z}$ (hence all non-zero $n \in \mathbf{Z}$ since it clearly suffices to prove the result for either n or $-n$).[(1.6.2)] In what follows $A = B = 1$, except when $n = 2t+1$ and $t \equiv 2 \pmod 5$, in which case $A = t+1$ and $B = 1$. Also, $t > 0$.

n	r	s	T	U
$2t+1,\ t \not\equiv 2 \pmod 5$	t^2+2t+2	t^2+1	t^2+t+1	1
$2t+1,\ t \equiv 2 \pmod 5$	2	$2t^2+2t+1$	$2t+1$	1
$4t+2,\ t > 0$	$2t^2+4t+1$	$2t^2-1$	a^2-1 where $a = 2t^2+2t-1$	a
$4t+2,\ t \not\equiv 1 \pmod 3$	$2t^2+4t+3$	$2t^2+1$	b^2+1 where $b = 2t^2+2t+1$	b
$4t+2,\ t \equiv 1 \pmod 3$	$6t^2+8t+3$	$6t^2+4t+1$	c^2+1 where $c = 18t^2+18t+5$	$3c$
$4t,\ t$ odd, $t > 1$	t^2+2t+2	t^2-2t+2	t^2h where $h = (t^4+3)/2$	$h-1$
$4t,\ t$ even	$2t^2+2t+1$	$2t^2-2t+1$	$2t^2$	1
$4t,\ t$ even	$2t^2+3t+1$	$2t^2-t+1$	$4t^3+2t^2+1$	$2t$
$4t,\ t$ even	$2t^2+t+1$	$2t^2-3t+1$	$4t^3-2t^2-1$	$2t$

The algorithm[(1.6.3)] depicted in the proof of Theorem 1.6.1 is illustrated by

Example 1.6.1. Let $n = 8$ and refer to line 7 in the above chart with $t = 2$, and form the product $(\sqrt{13}+\sqrt{5})(8+\sqrt{65})^k$ with $k \equiv 14 \pmod{65}$. For instance, if

[(1.6.1)] However, $6 = 5^4 7^3 - 463^2$ and is infinitely often representable as a proper difference.

[(1.6.2)] We have listed several choices for $n = 4t+2$ and $4t$ in order to ensure that r and s are both nonsquare and relatively prime, since it is not possible for r and s to be nonsquare and *not* relatively prime in all cases simultaneously. The values $n \in \{1, 2, 4\}$ are not listed, since they are covered by the special cases $(A, B, r, s, T, U) = (4, 5, 11, 7, 351, 40), (1, 1, 5, 3, 4, 1), (1, 1, 11, 7, 351, 40)$ respectively. Observe that what is essentially happening here is that $g^2r - h^2s = \pm n$ where $(A\sqrt{r}+B\sqrt{s})(T+U\sqrt{rs})^k = g\sqrt{r}+h\sqrt{s}$, with k chosen such that $r \mid g$ and $s \mid h$.

[(1.6.3)] The r and s chosen in the chart satisfy $rs = \ell^2 + t$ where $t \mid 4\ell$. This we called Extended Richaud-Degert types (see Definition 3.2.2).

$k = 14$, we have

$$A_k\sqrt{r} + B_k\sqrt{s} = (4741115028961333 \cdot 13)\sqrt{13} + (19876527516465469 \cdot 5)\sqrt{5}.$$

Thus,

$$8 = (4741115028961333)^2 \cdot 13^3 - (19876527516465469)^2 \cdot 5^3.$$

The application of real quadratics via powers of units is very well illustrated by the above solution to the problem of representation of integers as differences of powerful numbers. There is another problem involving powerful numbers as an application of real quadratics. In [96] Golomb observed, (as conjectured by Erdös to always hold), that there are no known sequences of integers $(4k-1, 4k, 4k+1)$, $k \in \mathbf{Z}$ such that all three are powerful.(1.6.4)

The translation of this open problem to real quadratics via the Pellian was given by the authors of [244]. The result is

Theorem 1.6.2. *The following are equivalent:*

(a) *There exist three consecutive powerful numbers.*

(b) *There exist powerful numbers P and Q with P even and Q odd such that $P^2 - Q = 1$.*

(c) *There exists a radicand $D \equiv 7 \pmod 8$ with $\epsilon_\Delta = T_1 + U_1\sqrt{D}$ and, for some odd integer k, T_k is an even powerful number and $U_k \equiv 0 \pmod D$, where*

$$(T_1 + U_1\sqrt{D})^k = T_k + U_k\sqrt{D}.$$

Proof. The equivalence of (a) and (b) is Exercise 1.6.2. If (b) holds, then $Q = DU^2$ where D is a square-free integer dividing U. Thus, $N(P + U\sqrt{D}) = 1$ with $D \equiv 7 \pmod 8$, since Q is odd. Let $\epsilon_\Delta = T_1 + U_1\sqrt{D}$ where $\Delta = 4D$. Thus, there exists a positive $k \in \mathbf{Z}$ such that $P + U\sqrt{D} = (T_1 + U_1\sqrt{D})^k = T_k + U_k\sqrt{D}$. It is an easy task to verify that U_1 is odd using the binomial theorem, (see Exercise 3.1.7), so T_1 is even. If k were even, then T_k would be odd, but $T_k = P$ is even so k is odd and (c) holds. The converse is clear. □

Remark 1.6.1. From a computational point of view, the values generated via Theorem 1.6.2 are astronomical. To illustrate this comment, we observe that $T_k^2 - U_k^2 D$ with $U_k \equiv 0 \pmod D$ if and only if $kU_1 \equiv 0 \pmod D$. Hence, for any non-negative $k \in \mathbf{Z}$, $T_{D(2k+1)} \equiv (-1)^k(2k+1)T_D \pmod{T_D^2}$. Therefore, if T_D is *not* powerful, but $T_{D(2k+1)}$ is powerful for some $k > 0$, then all primes properly dividing T_D must divide $2k+1$. Suppose that we take the smallest possible radicand $D = 7$. Since $T_7 = 2^3 \cdot 29 \cdot 197 \cdot 2857$ then, if $T_{7(2k+1)}$ is powerful for some $k > 0$, we must have $29 \cdot 197 \cdot 2857$ dividing $2k+1$. Therefore, the first possibility for the existence of 3 consecutive powerful numbers, coming from a $T_k + U_k\sqrt{7}$, is $(8+3\sqrt{7})^{114254287}$ (and we have checked that this does *not* produce three consecutive powerful numbers). In fact, Erdös (see Guy [101]) posed

(1.6.4)As noted by Golomb, if such triples of consecutive powerfuls exist, then they *must* be of this form.

Conjecture 1.6.1. *There are only finitely many triples of consecutive powerful numbers.*

Not surprisingly, this remains open.

A related question generated by Theorem 1.6.2 is to find those values $\epsilon_\Delta^k = (T_k + U_k\sqrt{D})$ such that $U_k \equiv 0 \pmod{D}$. In fact, Ankeny, Artin and Chowla [6] conjectured that, if $(T_1 + U_1\sqrt{D})/2 = \epsilon_p$ when $p \equiv 1 \pmod 4$ is prime, then p does not divide U_1. Later, Mordell [275] conjectured that if $T_1 + U_1\sqrt{p} = \varepsilon_p$ for $p \equiv 3 \pmod 4$, then p does not divide U_1. Both of these conjectures remain open but have been computationally verified up to large values.(1.6.5)

Another open question concerning powerfuls involves sums. Which positive integers are a sum of two (or three) powerful numbers? Many years ago, Erdös posed

Conjecture 1.6.2. *Every sufficiently large positive integer N is representable as a sum of at most three powerful numbers.*

In [243], Mollin and Walsh more specifically posed

Conjecture 1.6.3. *Every positive integer is a sum of at most 3 powerful numbers with the exception of 7, 15, 23, 87, 111 and 119.*(1.6.6)

Conjecture 1.6.2 was solved by Heath–Brown [113] as follows:

Theorem 1.6.3. *There is an effectively computable constant c such that $n \geq c$ is a sum of at most three powerful numbers.*

To see how this result is proved, we first refer to footnote (1.6.6) which tells us that we may assume $n \equiv 7 \pmod 8$. What Heath–Brown proved is that for sufficiently large n, the equation $pn = x^2 + y^2 + p^4z^2$ is solvable for $x, y, z \in \mathbb{Z}$ and p a prime with $p \equiv 5 \pmod 8$. Hence, an easy exercise using arithmetic in $\mathbb{Z}[\sqrt{-1}]$ yields that $p^{-1}(x^2 + y^2)$ is a sum of two squares, $z^2 + w^2$, so $n = z^2 + w^2 + p^3z^2$ which completes the proof. In fact, Heath–Brown goes a step further. He posed

Conjecture 1.6.4. *Every sufficiently large integer n is expressible as $n = x^2 + y^2 + 5^3z^2$.*

He went yet another step further with

(1.6.5)The Ankeny *et al.* conjecture has been verified by R. Soleng, in an unpublished manuscript, for all primes $p \leq 100028009$, and Mordell's conjecture has been verified by Beach *et al.* in [18] for all primes $p < 7679299$. Mollin and Walsh asked H.C. Williams whether it was possible to computationally find those more general values of D such that $D \mid U_1$ where $(T_1 + U_1\sqrt{D})/\sigma = \epsilon_\Delta$. In [349], Stephens and Williams did this. They developed an algorithm involving continued fractions and found 8 values of D (less than 10^7) with $D \mid U_1$. They are $D \in \{46 = 2 \cdot 23, 430 = 2 \cdot 5 \cdot 43, 1817 = 23 \cdot 79, 58254 = 2 \cdot 3 \cdot 7 \cdot 19 \cdot 73, 209991 = 3 \cdot 69997, 1752299 = 41 \cdot 79 \cdot 541, 3124318 = 2 \cdot 1562159, 4099215 = 3 \cdot 5 \cdot 273281\}$. They also verified the Ankeny *et al.* conjecture for all $p < 10^9$. Finally, Lu [204] has shown that the Ankeny *et al.* conjecture is reduced to the construction of an absolutely nonsingular projective variety X over a finite field such that a certain Dirichlet series is a factor of the zeta function for X.

(1.6.6)We can easily show that every positive $n \in \mathbb{Z}$ with $n \not\equiv 7 \pmod 8$ is a sum of three powerful numbers by the fact (eg. see Mordell [273, Theorem 1, p. 175]) that all positive integers *not* of the form $4^\lambda(8\mu + 7)$, $\lambda \geq 0$, $\mu \geq 0$ are a sum of three integer squares. Also, it is well-known (eg. see Mordell [273, section 3, p. 178]) that all positive integers *not* of the form $2^{2\lambda+1}(8\mu + 7)$ are representable in the form $x^2 + y^2 + 2z^2$. In particular, if $\lambda > 0$, then $4^\lambda(8\mu + 7) = (2^\lambda x)^2 + (2^\lambda y)^2 + 2^{2\lambda+1}z^2$ leaving only $n \equiv 7 \pmod 8$.

Conjecture 1.6.5. *For a given prime $p \equiv 5 \pmod 8$, there is an $n(p) \in \mathbf{Z}$ such that if $n > n(p)$, then $n = x^2 + y^2 + p^3z^2$ for some $x, y, z \in \mathbf{Z}$.*

The proof of Conjecture 1.6.5 would show that the number of representations of n as a sum of three powerful numbers tends to infinity as n gets large. In fact, computational evidence already indicates that as n gets large, the number of ways n is expressible as a sum of 3 powerful numbers gets large.

Now we turn to the question involving sums of two powerful numbers. For example, all primes $p \equiv 1 \pmod 4$ are sums of two squares. More generally, it is well-known that a positive $n \in \mathbf{Z}$ is a sum of two integer squares if and only if n has no prime $p \equiv 3 \pmod 4$ appearing to an odd exponent in its canonical prime factorization. Also Gauss(1.6.7) tells us that, if $p \equiv 1 \pmod 3$ and 2 is a cube modulo p, then $p = x^2 + 27y^2$. However, when 2 is not a cube modulo $p \equiv 1 \pmod 3$, anything can happen. For example, 7 is not a sum of less than 4 powerful numbers, yet $379 = 6^2 + 7^3$.(1.6.8)

Another problem is to determine which positive integers are a sum of two non-square powerful numbers but *not* a sum of a square and non-square powerful number. For example, $16879 = 3^2 \cdot 2^3 + 7^5 = 2^2 + 3^3 \cdot 5^2$ whereas $78157 = 2^5 + 5^7$ is not representable as a square plus a non-square powerful number.

The number of unsolved problems concerning powerful numbers is large indeed. Two more open questions cited by Paulo Ribenboim at the end of his survey article [305] are:

Conjecture 1.6.6. *There exist only finitely many even powerful numbers n, such that $n^2 - 1$ is powerful.*

and,

Conjecture 1.6.7. *There exist only finitely many even integers m, such that $m^4 - 1$ is powerful.*

We observe that if $n - 1, n$ and $n + 1$ are all powerful numbers then $n \equiv 0 \pmod 4$ so $n^2 - 1$ is powerful if and only if $n - 1$ and $n + 1$ are both powerful. Hence, Conjecture 1.6.1 is equivalent to Conjecture 1.6.6. Also, Conjecture 1.6.7 is weaker, since we just take $n = m^2$. In fact, taking $n = A^r$ in Conjecture 1.6.6, we have

Conjecture 1.6.8. *For every integer A, there are infinitely many values of r for which $A^r - 1$ is not powerful.*

Since $A^{2r} - 1 = (A^r - 1)(A^r + 1)$, then Conjecture 1.6.8 also follows from

Conjecture 1.6.9. *For every even integer A, there are infinitely many values of r for which $A^r + 1$ is not powerful.*

Conjectures 1.6.8–1.6.9 were made by Granville in [99], as was

Conjecture 1.6.10. *The largest prime factor of $1 + x^2y^3$ tends to infinity as $x + |y|$ tends to infinity.*(1.6.9)

(1.6.7)See Mordell [273].

(1.6.8)It is interesting to observe that $N(6 + 7\sqrt{-7}) = 379$ with $h_{-7} = 1$. See Chapter Three.

(1.6.9)As noted by Granville, this is analogous to the 1953 conjecture of Mahler [208] that as $x, y \to \infty$ the largest prime factor of $x^2 + y^3 \to \infty$. Observe that all powerful numbers are of the form x^2y^3

In fact in [99], Granville was able to prove Conjecture 1.6.6 and Conjecture 1.6.10 under the assumption of the validity of the following due to Oesterlé and Masser [284] (see also [99]).

Conjecture 1.6.11. [(1.6.10)] *(The abc Conjecture.) Suppose that a, b and c are positive integers satisfying $a + b = c$ with $\gcd(a, b, c) = 1$. Let $G = G(a, b, c)$ be the product of the primes dividing abc, each to the first power. For all $\epsilon > 0$ there exists a constant $k = k(\epsilon)$ such that $c < kG^{1+\epsilon}$.*

If n and $n^2 - 1$ are both powerful, then by taking $a = 1$, $b = n^2 - 1$ and $c = n^2$ in the *abc* conjecture, we get $G \le \sqrt{bn} < n^{3/2}$, so $n^2 < kn^{3/2+\epsilon}$ which bounds n. Thus, Conjecture 1.6.6 holds.

If $x, y \in \mathbb{Z}$ for which the largest prime factor of $1 + x^2y^3$ is no bigger than t, then take $a = 1, b = x^2y^3$ in the *abc* conjecture, so that $G \le xyT$ where T is the product of the primes no bigger than t. Hence, $x^2y^3 \le c(xy)^{1+\epsilon}$, where $c = kT^{1+\epsilon}$ by the *abc* conjecture, which bounds xy and thus $x + |y|$, from which Conjecture 1.6.10 follows.[(1.6.11)]

The *abc* conjecture is a rather deep and seemingly intractable problem at this time. Therefore, it is not a surprise that we can verify other difficult conjectures by using it with such ease as we have done above. We most likely will have a long wait before we see unconditional proofs of these results.

Pure powers x^m for $x > 1$ are very special cases of powerful numbers. One of the earliest results on pure powers was the proof by Euler in 1738 that the only consecutive integers such that one is a square and one is a cube are 8 and 9. In 1844, Catalan wrote a letter to Crelle, which appeared in Volume one of Crelle's journal, (i.e. Journal für die reine und angewandte Mathematik.), wherein he asked for a proof that the only consecutive powers are 8 and 9. Thus, we have

Conjecture 1.6.12. *(Catalan's Conjecture.) $x^u - y^v = 1$ has only the solutions $x = 3$, $u = 2$, $y = 2$, $v = 3$ for $x, y, u, v \in \mathbb{Z}$ all bigger than 1.*

Significant advances toward a solution of Catalan's conjecture have been made. For instance, Tijdeman [363] proved (observing that one need only consider $x^p - y^q = 1$ with p and q primes)

Theorem 1.6.4. *The equation $x^p - y^q = 1$ for rational integers $x, y, p, q > 1$ has only finitely many solutions, and effective bounds for these solutions can be given.*

Furthermore, in [92], Glass *et al.* considered $x^p - y^q = \pm 1$ with $p > q$ primes, and they gave upper bounds $3.42 \cdot 10^{28}$ and $5.6 \cdot 10^{19}$ for p and q respectively.[(1.6.12)] More recently, however, they were able to use their results on linear forms in three logarithms to reduce these bounds to $4.61 \cdot 10^{18}$ and $4.01 \cdot 10^{12}$ respectively, (see Bennett *et al* [22]). [(1.6.13)] Furthermore, in a private communication to this au-

---

(Exercise 1.6.3).

[(1.6.10)] As noted by Granville [99], if $D > 0$ is a fundamental radicand and $(x + y\sqrt{D})^{2D} = e + f\sqrt{D}$ where $x + y\sqrt{D} \in U_\Delta$, then $e^2 - Df^2 = 1$ where D divides f. Hence, $a = 1$, $b = Df^2$, $c = e^2$ satisfies the *abc* conjecture with $G(a, b, c) \le ef \le c/\sqrt{D}$, so the exponent in Conjecture 1.6.11 cannot be improved.

[(1.6.11)] In [99] Granville also links several conjectures related to Fermat's Last Theorem, as does Walsh [374].

[(1.6.12)] They also refer to the improvements of Okada (one of the author's of [92]), in his thesis, where $8.62 \cdot 10^{23}$ and $1.18 \cdot 10^{17}$ were obtained.

[(1.6.13)] Furthermore, as communicated to this author by Andrew Glass, if in addition, $\min\{p, q\} \equiv 3$

thor, Maurice Mignotte indicated that he recently completed and submitted joint work with Yves Roy which shows that $x^p - y^q = 1$ has no nontrivial solution for $\min(p,q) < 10^4$. This could place the problem of solving Catalan's conjecture within the purview of current computational capabilities.[(1.6.14)] However, some would disagree (see Ribenboim [307, p. 216]).

In [374],[(1.6.15)] Walsh generalized Catalan's conjecture as follows.

Conjecture 1.6.13. *The equation $x^n - m^3y^2 = \pm 1$ is solvable in integers $x, y > 1$, $m \geq 1$ and $n > 2$ if and only if $(x, m, y, n) = (2, 1, 3, 3)$ or $(x, m, y, n) = (23, 2, 39, 3)$.*

Thus, Conjecture 1.6.13 upgrades Catalan's conjecture into the purview of powerful numbers. Walsh also considered other problems involving powerful numbers such as: Which Fibonacci numbers are powerful? From results of J.H.E. Cohn [59]–[60], the only perfect squares which are Fibonacci numbers are $F_1 = F_2 = 1$ and $F_{12} = 144$. London and Finkelstein [193], and Lagarias and Weisser [156] independently proved that $F_1 = F_2 = 1$ and $F_6 = 8$ are the only cubes which are Fibonacci numbers. Are there any more powerful Fibonacci numbers? We pose

Conjecture 1.6.14. *The only powerful Fibonacci numbers are* 1, 8 *and* 144.

From results of Williams [381], it follows that Conjecture 1.6.14 holds for any Fibonacci number divisible by a prime $p < 10^9$. Since Fibonacci numbers are examples of second order linear sequences,[(1.6.16)] then it is natural to ask the more general question: Are there only finitely many powerful numbers in an arbitrary second order linear recurrence sequence? Toward this end, Shorey and Stewart [337] have shown that, given a second order linear recurrence sequence, the Diophantine equation $W_n = ed^q$, with $|d| > 1$ and $q \geq 2$, must satisfy $\max\{n, |d|, |q|\} < c$ for some effectively computable constant c.

(mod 4), then, using an idea of Maurice Mignotte, they can show that $\max\{p,q\} < 2.16e^{16}$ and $\min\{p,q\} < 2.73e^{12}$. Tim O'Neil, a doctoral student of Glass', has just submitted this work as of June 1995. Glass *et al* [93], are using a two-pronged approach. The first relies on the prior algebraic work of Inkeri [130], Inkeri and Hyrrö [131], as unified by Mignotte, and the second is to try to improve the Baker–Wüstholz result [16] on linear forms in logarithms in order to lower the bounds.

[(1.6.14)] Suppose that $x^p - y^q = 1$ has a solution for $x, y \in \mathbb{Z}$, and p, q odd primes. Let $L = \mathbb{Q}(\zeta_p)$ where ζ_p is a primitive pth root of unity, and let M be the subfield of L such that $|M : \mathbb{Q}|$ is the 2-part of $p-1$. In this case, $p^{q-1} \equiv 1 \pmod{q^2}$, or q divides the class number of M. Similarly, we repeat the last statement mutatis mutandis with p and q interchanged. If $p \equiv 3 \pmod 4$, then $M = \mathbb{Q}(\sqrt{-p})$ and h_{-p} can be found quickly. If $p \equiv 5 \pmod 8$, then $\mathbb{Q}(\sqrt{p})$ is the maximal real subfield of M, and we can find the class number of M relatively easily as well, since it is related to h_p. For $p \equiv 9 \pmod{16}$, and other cases, the situation becomes harder to tackle. However, graduate students of Andrew Glass, namely D. Robin Clother and Tim O'Neil, are looking at pairs p, q with $\min\{p,q\} < 2000$. For example, O'Neil has shown that if $\min\{p,q\} \equiv 3 \pmod 4$, and $\max\{p,q\} \equiv 3 \pmod 4$, then (unless $\min\{p,q\} < 113223$, or $\max\{p,q\} < 284575469$) a solution to $x^p - y^q = 1$ implies that $p^{q-1} \equiv 1 \pmod{q^2}$ and $q^{p-1} \equiv 1 \pmod{p^2}$ (sometimes called "double Wieferich conditions"). He has also shown that if $M = \max\{p,q\} \equiv 1 \pmod 4$, then, $M^{m-1} \equiv 1 \pmod{m^2}$ where $m = \min\{p,q\}$, unless $m < 1780549$ and $M < 4.16e^{10}$. Mignotte has a technique to show that these conditions cannot hold. However, his method requires checking a large number of congruences with no guarantee, a fortiori, that the process stops (albeit it always has to date, see Mignotte [217]). Furthermore, if the process stops, can all the bounds be checked in "realistic computational time"? See footnote (8.2.3).

[(1.6.15)] This thesis was written at the University of Calgary under this author's supervision.

[(1.6.16)] A second order recurrence sequence $\{W_n\}_{n \geq 1}$ are sequences which satisfy $W_{n+2} = aW_{n+1} + bW_n$ for some $a, b \in \mathbb{Z}$, and there is no $c \in \mathbb{Z}$ for which $W_{n+1} = cW_n$ for all $n \geq 1$. See Exercise 3.1.5 for another example of such sequences.

If $\Delta = \Delta_0 > 0$ is a discriminant, then $\varepsilon_\Delta^i = (T_i + U_i\sqrt{\Delta})/2$ yields the second order linear recurrence sequences $\{T_i\}$ and $\{U_i\}$ (see Exercise 3.1.5). Which of these are powerful numbers? We have already seen the value of this question, for example, in Theorem 1.6.2.

Cohn [60] and Zhenfu [401] have investigated which elements of $\{T_i\}$ or $\{U_i\}$ can be squares. It is an open and, given developments in this section related to such sequences, seemingly difficult problem to determine which elements of these sequences are powerful.

Erdös called rational integers $u_1^{(k)} < u_2^{(k)} < \cdots$, all of whose prime factors have exponents at least k, *k-ful numbers*, and posed the following sequence of conjectures:

Conjecture 1.6.15. *There exist infinitely many triples of $u_i^{(3)}$ in arithmetic progression.*

Conjecture 1.6.16. *There do not exist triples of $u_i^{(4)}$ in arithmetic progression.*

Conjecture 1.6.17. *There are no consecutive $u_i^{(3)}$ numbers, i.e. $u_i^{(3)} - u_j^{(3)} = 1$ is not solvable.*

Conjecture 1.6.18. *The equation $u_i^{(3)} + u_j^{(3)} = u_k^{(3)}$ has infinitely many solutions.*

Conjecture 1.6.19. *The equation $u_i^{(4)} + u_j^{(4)} = u_k^{(4)}$ has only finitely many solutions.*

Conjecture 1.6.20. *For $k \geq 4$ the equation*

$$u_{i_1}^{(k)} + u_{i_2}^{(k)} + \cdots + u_{i_{k-2}}^{(k)} = u_{i_{k-1}}^{(k)}$$

has only finitely many solutions.

Exercises 1.6

1. Let $n \in \mathbf{Z}$ be non-zero and $r, s \in \mathbf{Z}$ be positive and non-square. Suppose that the following three conditions are satisfied:

   (a) The Diophantine equation $rx^2 - sy^2 = \pm n$ has a positive integer solution $(x, y) = (A, B)$ with $\gcd(Ar, Bs) = 1$.

   (b) The Diophantine equation $x^2 - rsy^2 = \pm 1$ has a positive solution $(x, y) = (T, U) \in \mathbf{Z}^2$ with $\gcd(U, rs) = 1$.

   (c) For $k \in \mathbf{Z}$ positive let $(A\sqrt{r} + B\sqrt{s})(T + U\sqrt{rs})^k = A_k\sqrt{r} + B_k\sqrt{s}$ with

   $$k \equiv -TA(UBs)^{-1} \pmod{r}$$

   and

   $$k \equiv -TB(UAr)^{-1} \pmod{s}.$$

   Prove that the following are equivalent:

   (d) $A_k \equiv 0 \pmod{r}$ and $B_k \equiv 0 \pmod{s}$ with $\gcd(rA_k, sB_k) = 1$ for any positive $k \in \mathbf{Z}$.

(e) $A_k^2 r - B_k^2 s = \pm n$ for all positive $k \in \mathbf{Z}$.(1.6.17) (*Hint*: Let $(T + U\sqrt{rs})^k = T_k + U_k\sqrt{rs}$. Use the binomial theorem to verify that $A_k \equiv T^{k-1}[AT + BskU] \pmod r$, and use a similar argument on B_k to get $A_k \equiv 0 \pmod r$ and $B_k \equiv 0 \pmod s$. To get the gcd condition in (d), assume that $p \mid \gcd(rA_k, sB_k)$ and get a contradiction to the gcd condition in (a). To get (e) prove that $(A_k/r)^2 r^3 - (B_k/s)^2 s^3 = (A^2 r - B^2 s)(T_k^2 - rsU_k^2)$.)

2. Prove the equivalence of (a) and (b) in Theorem 1.6.2.(*Hint*: Assume the form $(4k-1, 4k, 4k+1)$.)

3. Prove that all powerful numbers are of the form x^2y^3.

4. In 1657, Frénicle de Bessy [86] proved that if $p > 2$ is prime and $n > 1$, then $p^n + 1$ is not a square and if $n > 3$, then $2^n + 1$ is not a square. Prove this. (*Hint*: $p^n = x^2 - 1 = (x+1)(x-1)$.)

5. Prove that if $3^m - 2^n = \pm 1$ for $m, n > 1$ then $m = 2, n = 3$. (*Hint*: Prove that, if $2^n - 3^m = 1$, then n is even and, deduce that $2^{n/2} - 1 = 3^{m'}$, and $2^{n/2} + 1 = 3^{m-m'}$ with $0 \le m' < m - m'$. Deduce the contradiction $m' = 0$. The other case is similar.)

6. In 1850, Lebesgue [166] used Gaussian integers to show that $x^m - y^2 = 1$ for positive $x \in \mathbf{Z}$ has only trivial solutions. Prove this. (*Hint*: Factor $y^2 + 1$ in $\mathbf{Z}[\sqrt{-1}] = \mathbf{Z}[i]$, and conclude that $y + i = (u + iv)^m i^s$ with $0 \le s \le 3$, and $y - i = (u - iv)^m(-i)^s$. By subtracting these two equations and using the binomial theorem, deduce that $\sum_{i=0}^{\frac{m-1}{2}} \binom{m}{2i}(-1)^i w^{2i} = \pm 1$ where $w = u$,and $v = \pm 1$ (for even s), or $w = v$ and $u = \pm 1$ (for odd s). Analyze this equation to get the result.)

7. In 1885 Catalan [48] asserted that if $x^y - y^x = 1$, then $x = 2$ and $y = 3$. Prove this.

8. In 1897–1899, Stormer [353]–[354] proved that the only solutions of $1 + x^2 = 2y^n$ (where $n \in \mathbf{Z}, n > 1$, not a power of 2) are $x = \pm 1$. Use this fact to prove Selberg's result, in 1932, that for $n \ge 2$ the equation $x^4 - y^n = 1$ has only trivial solutions. (*Hint*: For n odd and y odd or for n even, this is easy. If n is odd and y is even, look at expressions for $x^2 + 1$ and $x^2 - 1$ using Stormer's result.)

9. In 1961, Cassels [47] proved that when $p, q > 2$ are primes $x, y \ge 2$ and $x^p - y^q = 1$, then $p \mid y$ and $q \mid x$. Use this result to prove that three consecutive integers cannot be pure powers. (*Hint*: Verify that the exponents may be assumed to be primes without loss of generality. Conclude $x^l - y^p = 1 = y^p - z^q$ for primes l, p and q. Now use Cassels' result to get a contradiction to Lebesgue's result, Exercise 1.6.6.)

10. (a) Prove that if $p > 2$ is prime and $m > 1$, then

$$\gcd(m \pm 1, (m^p \pm 1)/(m \pm 1)) = 1 \text{ or } p.^{(1.6.18)}$$

(1.6.17) In other words, (d) and (e) say that n is a proper difference of two non-square powerful numbers in infinitely many ways.

(1.6.18) In 1942–43, Ljunggren [191]–[192] looked at $(x^n - 1)/(x - 1) = y^m$ and completed a result of Nagell by showing that $x^2 + x + 1 = y^m$ only has solutions when either m is odd and $x = -1 = -y$, or m is even and $x = -1 = \pm y$. See Ribenboim [307], for more detail on this topic, as well as the topics in Exercises 1.6.4–1.6.9.

(b) If $p > 2$ is prime and p does not divide $a \in \mathbf{Z}$ with $a > 1$, then $a^{p-1} \equiv 1 \pmod{p}$[1.6.19] and the integer $q_p(a) = (a^{p-1} - 1)/p$ is called the ***Fermat quotient of p with base a***. It is an open problem to prove

Conjecture 1.6.21.[1.6.20] There are infinitely many primes p for which p does not divide $q_p(a)$.

Prove that Conjecture 1.6.8 implies Conjecture 1.6.21. (*Hint*: Assume that Conjecture 1.6.21 is false, then $p \mid q_p(a)$ for all $p > p_0$. Set $t = \prod_{p \le p_0} \phi(p^2)$ and $A = a^t$.) [1.6.21]

11. Prove that Conjecture 1.6.13 follows from Conjecture 1.6.11.

12. Let p and q be odd primes such that $q - 1$ is not a square. Prove that, if $\dfrac{q^p - 1}{q - 1} = n^2$, then $q = 3$, $p = 5$ and $n = 11$. (*Hint*: Prove that either $\gamma = q^{(p-1)/2}\sqrt{q} + n\sqrt{q-1}$ or γ^3 is a solution to $x^2 q - y^2(q-1) = 1$, and observe that $\beta = \sqrt{q} + \sqrt{q-1}$ is also a solution.)

[1.6.19] Called "Fermat's little theorem", a generalization of which is *Euler's theorem*: $a^{\phi(m)} \equiv 1 \pmod{m}$ for $\gcd(a, m) = 1$ (see footnote (1.6.21)).

[1.6.20] See Granville [99]. Some people do not believe that this holds, and so they would consider it to be an "open question" rather than a "conjecture".

[1.6.21] $\phi(n)$ in Euler's *totient* function, which equals the number of positive integers less than or equal to $n \in \mathbf{Z}$ which are relatively prime to n.

Chapter 2

Continued Fractions Applied to Quadratic Fields.

This chapter is concerned with the introduction of continued fractions, developing the theory in ideal-theoretic terms and applications of this interrelationship to class number problems in quadratic fields.

2.1 Continued Fractions and Real Quadratics: The Infrastructure.

A continued fraction is an expression of the form

$$a_0 + \cfrac{1}{a_1 + \cfrac{1}{a_2 + \ddots + \cfrac{1}{a_{i-1} + \cfrac{1}{a_i + \ddots}}}}$$

where $a_i \in \mathbf{R}$ are called the *partial quotients* of the continued fraction expansion. If $a_i \in \mathbf{Z}$ and $a_i > 0$ for $i > 0$ we call the expression an *infinite simple continued fraction* (which is equivalent to being an irrational number (Exercise 2.1.2)) and use the notational convenience $\langle a_0; a_1, a_2, \ldots, a_i, \ldots \rangle$ to denote it; whereas, if the expression terminates at a_N say, then we call the expression a *finite simple continued fraction* which is equivalent to being a rational number (Exercise 2.1.1).

Furthermore, we will be concerned with continued fraction expressions involving quadratics, so we need

Definition 2.1.1. A real number γ is called a *quadratic irrational*, associated with the radicand D, if γ can be written as

$$\gamma = (P + \sqrt{D})/Q$$

where $P, Q, D \in \mathbf{Z}$, $D > 0$, $Q \neq 0$, and $P^2 \equiv D \pmod{Q}$.

We therefore see that a quadratic irrational γ satisfies the equation

$$x^2 - \operatorname{Tr}(\gamma)x + N(\gamma) = 0$$

with γ' as its other root.

We will denote the *continued fraction expansion* of γ by

$$\gamma = \langle a_0; a_1, a_2, \ldots, \gamma_i \rangle, \tag{2.1.1}$$

where (for $i \geq 0$ and $\gamma = \gamma_0$, $P_0 = P$, $Q_0 = Q$) we recursively define

$$\gamma_i = (P_i + \sqrt{D})/Q_i, \tag{2.1.2}$$

$$a_i = \lfloor \gamma_i \rfloor, \tag{2.1.3}$$

$$P_{i+1} = a_i Q_i - P_i, \tag{2.1.4}$$

and

$$Q_{i+1} = (D - P_{i+1}^2)/Q_i. \tag{2.1.5}$$

Quadratic irrationals are also special in terms of periodicity, so we need

Definition 2.1.2. An infinite simple continued fraction γ is called *periodic* if $\gamma = \langle a_0; a_1, a_2, \ldots \rangle$, where $a_n = a_{n+\ell}$ for all $n \geq k$ with $k, \ell \in \mathbb{N}$. We use the notation,

$$\langle a_0, a_1, a_2, \ldots a_{k-1}; \overline{a_k, a_{k+1}, \ldots, a_{\ell+k-1}} \rangle,$$

as a convenient abbreviation for $\langle a_0, a_1, a_2, \ldots, a_{k-1}; a_k, a_{k+1}, \ldots, a_{\ell+k-1}, a_k, a_{k+1} \ldots \rangle$. The sequence $a_0, a_1, \ldots, a_{k-1}$ is called the *preperiod* of γ.

Every quadratic irrational is periodic (Exercise 2.1.3). Furthermore, there is a special kind of periodicity, which will provide our link to the theory of reduced ideals. We therefore need to define this special type.

Definition 2.1.3. An infinite simple continued fraction γ is called *purely periodic* if $\gamma = \langle \overline{a_0; a_1, a_2, \ldots, a_{\ell-1}} \rangle$, said to have *period length* equal to $\ell = \ell(\gamma)$. (We may also write $\gamma = \langle a_0; \overline{a_1, a_2, \ldots, a_\ell} \rangle$.)

We illustrate this concept by

Example 2.1.1. If $D = 385$ and $\gamma = (7+\sqrt{385})/14$, then $\gamma = \langle \overline{1; 1, 9, 6, 2, 3, 2, 6, 9, 1} \rangle$ with $\ell(\gamma) = 10$.

Example 2.1.2. If $D = 145$ and $\gamma = (9 + \sqrt{145})/8$, then $\gamma = \langle \overline{2; 1, 1, 1, 2} \rangle$ with $\ell(\gamma) = 5$.

Now we look at the quadratic irrationals which satisfy this property.

Definition 2.1.4. A quadratic irrational γ is called *reduced*, provided $\gamma > 1$ and $-1 < \gamma' < 0$.

Theorem 2.1.1. *The continued fraction expansion of a quadratic irrational is purely periodic if and only if it is reduced.*

Proof. First, we establish that (for $m \geq 1$) $\gamma_m = 1/(\gamma_{m-1} - \lfloor \gamma_{m-1} \rfloor)$. Since $\lfloor \gamma_{m-1} \rfloor = a_{m-1}$, then this is equivalent to saying that $\gamma_{m-1}\gamma_m - a_{m-1}\gamma_m = 1$. To see why this holds, we use (2.1.5) to get that

$$\gamma_{m-1}\gamma_m = (P_{m-1} + \sqrt{D})(P_m + \sqrt{D})/(Q_{m-1}Q_m) = (P_{m-1} + \sqrt{D})/(\sqrt{D} - P_m),$$

and

$$a_{m-1}\gamma_m = a_{m-1}(P_m + \sqrt{D})/Q_m = a_{m-1}Q_{m-1}/(\sqrt{D} - P_m).$$

Therefore,

$$\gamma_{m-1}\gamma_m - a_{m-1}\gamma_m = (P_{m-1} - a_{m-1}Q_{m-1} + \sqrt{D})/(\sqrt{D} - P_m).$$

However, $P_{m-1} - a_{m-1}Q_{m-1} = -P_m$ by (2.1.4), which secures the result.

Thus, $1/\gamma_m = \gamma_{m-1} - a_{m-1}$, and so it follows from Exercise 2.1.4 that $1/\gamma'_m = \gamma'_{m-1} - a_{m-1}$. The reader may verify, using a simple induction argument, that $-1 < \gamma'_{m-1} < 0$ for $m \geq 1$. Furthermore, by using (2.1.4)–(2.1.5), we have $\lfloor -1/\gamma'_m \rfloor = \lfloor -Q_m/(P_m - \sqrt{D}) \rfloor = \lfloor -Q_m(P_m + \sqrt{D})/(P_m^2 - D) \rfloor = \lfloor (P_m + \sqrt{D})/Q_{m-1} \rfloor = \lfloor (a_{m-1}Q_{m-1} - P_{m-1} + \sqrt{D})/Q_{m-1} \rfloor = \lfloor a_{m-1} - \gamma'_{m-1} \rfloor = a_{m-1}$, since $0 < -\gamma'_{m-1} < 1$.

By Exercise 2.1.3, $\gamma_i = \gamma_j$ for some $i < j$ so $-1/\gamma'_i = -1/\gamma'_j$, i.e. $a_{i-1} = a_{j-1}$. Also, $\gamma_{i-1} = a_{i-1} + 1/\gamma_i = a_{j-1} + 1/\gamma_j = \gamma_{j-1}$. We see that this may be continued until $\gamma_{i-k} = \gamma_{j-k}$ for $k = 0, 1, \ldots, i$, i.e. $\gamma = \gamma_0 = \langle \overline{a_0; a_1, \ldots, a_{j-i-1}} \rangle$.

Conversely, suppose that $\gamma = \langle \overline{a_0; a_1, \ldots, a_\ell} \rangle$. By Exercise 2.1.2(a),

$$\gamma = \frac{\gamma A_\ell + A_{\ell-1}}{\gamma B_\ell + B_{\ell-1}}.$$

Hence,

$$B_\ell \gamma^2 + (B_{\ell-1} - A_\ell)\gamma - A_{\ell-1} = 0.$$

Similarly, if $\beta = \langle \overline{a_\ell; a_{\ell-1}, \ldots, a_0} \rangle$, then

$$\beta = \frac{\beta A'_\ell + A'_{\ell-1}}{\beta B'_\ell + B'_{\ell-1}},$$

where $C'_i = A'_i/B'_i$ is the ith convergent of β. By Exercise 2.1.1(c),

$$A_\ell/A_{\ell-1} = \langle a_\ell; a_{\ell-1}, \ldots, a_1, a_0 \rangle = A'_\ell/B'_\ell,$$

and

$$B_\ell/B_{\ell-1} = \langle a_\ell; a_{\ell-1}, \ldots, a_2, a_1 \rangle = A'_{\ell-1}/B'_{\ell-1}.$$

By Exercise 2.1.2(c), $A_\ell B_{\ell-1} - A_{\ell-1}B_\ell = (-1)^{\ell-1}$. Hence, $A'_\ell = A_\ell$, $B'_\ell = A_{\ell-1}$, $A'_{\ell-1} = B_\ell$, and $B'_{\ell-1} = B_{\ell-1}$. Therefore,

$$\beta = \frac{\beta A'_\ell + A'_{\ell-1}}{\beta B'_\ell + B'_{\ell-1}} = \frac{\beta A_\ell + B_\ell}{\beta A_{\ell-1} + B_{\ell-1}}.$$

Thus,

$$A_{\ell-1}\beta^2 + (B_{\ell-1} - A_\ell)\beta - B_\ell = 0,$$

from which it follows that

$$B_\ell(-1/\beta)^2 + (B_{\ell-1} - A_\ell)(-1/\beta) - A_{\ell-1} = 0.$$

We have shown that the two roots of $A_\ell x^2 + (B_{\ell-1} - A_\ell)x - A_{\ell-1} = 0$ are γ and $-1/\beta$. Hence, $\gamma' = -1/\beta$, and it follows that $-1 < \gamma' < 0$, i.e. γ is a reduced quadratic irrational. □

Corollary 2.1.1. *If* $\gamma = \langle \overline{a_0; a_1, a_2, \ldots, a_{\ell-1}} \rangle$, *then* $-1/\gamma' = \langle \overline{a_{\ell-1}; a_{\ell-2}, \ldots, a_0} \rangle$. *If* γ *is reduced, then so is* $-1/\gamma'$.

Proof. Continuing with the proof of Theorem 2.1.1, we have $\beta = -1/\gamma' = \langle \overline{a_\ell; a_{\ell-1}, \ldots, a_1, a_0} \rangle$. □

The quadratic irrationals in Examples 2.1.1–2.1.2 are reduced for example.

We are finally in a position to link continued fractions with reduced ideals. We will show that if $\Delta > 0$ is a discriminant, then there exists a one-to-one correspondence between quadratic irrationals, associated with the radicand D, and primitive ideals in the order $\mathcal{O}_\Delta$.

Let $I = [a, b + w_\Delta]$ be a primitive ideal in $\mathcal{O}_\Delta$ and set $\gamma = (b + w_\Delta)/a$. If we let

$$P = (\sigma_0 b + f_\Delta(\sigma_0 - 1) + h\sigma_0)/g \in \mathbf{Z}, \tag{2.1.6}$$

$$Q = a\sigma_0/g \in \mathbf{Z}, \tag{2.1.7}$$

where $g = \gcd(f, \sigma_0)$ and $w_\Delta = f_\Delta w_{\Delta_0} + h$ (defined in Chapter One, section five), then $\gamma = (P + \sqrt{D})/Q$. Since $\sigma = \sigma_0/g$, then an easy exercise (via Exercise 1.5.3(c)) shows that $I = [Q/\sigma, (P + \sqrt{D})/\sigma]$, i.e. to each primitive $\mathcal{O}_\Delta$-ideal there corresponds a quadratic irrational.

Conversely, by Definition 2.1.1 a quadratic irrational can always be written as

$$\gamma = (P + \sqrt{D})/Q = (P + \sqrt{D})/(\sigma Q'),$$

with $P \equiv 1 \pmod{\sigma}$ and $\sigma Q'$ dividing $P^2 - D$. If we now set $a = |Q|/\sigma$ and $b = (P - 1)/\sigma - f_\Delta - h + (f_\Delta + g)/\sigma_0 \in \mathbf{Z}$, then by Exercise 1.5.3(c) we have that $[a, b + w_\Delta]$ is an $\mathcal{O}_\Delta$-ideal, i.e. to each quadratic irrational there corresponds a primitive $\mathcal{O}_\Delta$-ideal. Thus, we have established a one-to-one correspondence between quadratic irrationals and primitive $\mathcal{O}_\Delta$-ideals. This motivates

Definition 2.1.5. To each quadratic irrational $\gamma = (P + \sqrt{D})/Q$, there corresponds the $\mathcal{O}_\Delta$-ideal $I = [|Q|/\sigma, (P + \sqrt{D})/\sigma]$ (via (2.1.6)–(2.1.7)) for any discriminant $\Delta > 0$. We denote this ideal by $[\gamma] = I$, and write $\ell(I)$ for $\ell(\gamma)$.

This association allows us to more conveniently present the following crucial algorithm which produces all of the reduced ideals equivalent to a given $\mathcal{O}_\Delta$-ideal. We will see that, whatever ideal we choose, the application of this algorithm leads to a "periodic part" which contains all of the reduced ideals.

Theorem 2.1.2. (The Continued Fraction Algorithm.) *Let* $\Delta > 0$ *be a discriminant, and let* $I = I_1 = [a, b + w_\Delta]$ *be a primitive ideal in the order* $\mathcal{O}_\Delta$. *Set* $P = P_0$ *and* $Q = Q_0$, *as defined in equations* (2.1.6)-(2.1.7), *and let* P_i *and* Q_i *for* $i > 0$ *be defined by equations* (2.1.2)–(2.1.5) *in the continued fraction expansion of* $\gamma = \gamma_0 = (P + \sqrt{D})/Q$. *If* $I_k = [Q_{k-1}/\sigma, (P_{k-1} + \sqrt{D})/\sigma]$, *then* $I_1 \sim I_k$ *for all* $k \geq 1$. *There exists a least value* $m \geq 1$ *such that* I_m *is reduced, and* I_{m+i} *is reduced for all* $i \geq 0$.

Proof. Suppose that $\gamma_i = (P_i + \sqrt{D})/Q_i$ (for $i \geq 0$), and set $\theta_k = \prod_{i=1}^{k-1} 1/\gamma_i$ for $k \geq 2$, and let $\theta_1 = 1$.

Claim 1. $\theta_k = (-1)^{k-1}(A_{k-2} - \gamma B_{k-2})$, where A_{k-2} and B_{k-2} are defined in Exercise 2.1.2(a).

By the proof of Theorem 2.1.1, we have $\gamma_m = 1/(\gamma_{m-1} - \lfloor \gamma_{m-1} \rfloor)$. Therefore, we get that $\theta_k = \prod_{i=1}^{k-1}(\gamma_{i-1} - a_{i-1})$. Moreover, by Exercise 2.1.2(a),

$$\gamma = \frac{\gamma_i A_{i-1} + A_{i-2}}{\gamma_i B_{i-1} + B_{i-2}}, \tag{2.1.8}$$

so,

$$\gamma_i = \frac{A_{i-2} - \gamma B_{i-2}}{\gamma B_{i-1} - A_{i-1}}. \tag{2.1.9}$$

An induction on (2.1.9) establishes Claim 1.

Claim 2. $(Q_0\theta_k)I_k = (Q_{k-1})I$.

From (2.1.9), we have

$$\begin{pmatrix} \theta_k \\ \theta_{k+1} \end{pmatrix} = X \begin{pmatrix} 1 \\ \gamma \end{pmatrix},$$

where

$$X = (-1)^k \begin{pmatrix} -A_{k-2} & B_{k-2} \\ A_{k-1} & -B_{k-1} \end{pmatrix},$$

and by Exercise 2.1.2(c), we know that the determinant of X is ± 1. Thus, $[\theta_k, \theta_{k+1}] = [1, \gamma]$. Therefore,

$$(Q_0\theta_k)I_k = (Q_{k-1})I.$$

Thus, we have shown that for $k \geq 1$, $I_1 \sim I_k$ (see Definition 1.3.1 and comments in Chapter One, section five).

Next, we must verify that there exists an integer m such that I_m is reduced, and I_n is reduced for all $n \geq m$. We achieve this through the following sequence of claims.

Claim 3. If $\gamma_m' < 0$ for some $m \geq 1$, then $I_{m+1} = [\gamma_m]$ is a reduced $\mathcal{O}_\Delta$-ideal.

First, we observe that, (via Exercise 1.5.3(c)),

$$[\gamma_m] = [Q_m/\sigma, (P_m + \sqrt{D})/\sigma] = [Q_m/\sigma, (Q_m\lfloor -\gamma_m' \rfloor + P_m + \sqrt{D})/\sigma] = [\lfloor -\gamma_m' \rfloor + \gamma_m].$$

Also, since $\gamma_m = 1/(\gamma_{m-1} - \lfloor \gamma_{m-1} \rfloor)$ from the proof of Claim 1, then $\gamma_m > 1$, for $m \geq 1$. Hence, $-1 < \lfloor -\gamma_m' \rfloor + \gamma_m' < 0$.

Therefore, by Corollary 1.4.1, via Exercise 1.5.9, $[\gamma_m]$ is a reduced $\mathcal{O}_\Delta$-ideal, which secures Claim 3.

Claim 4. If $[\gamma_m]$ is a reduced ideal for some $m \geq 1$, then $-1 < \gamma_m' < 0$ for all $n > m$, i.e. $[\gamma_n]$ is reduced for all $n > m$.

First we look at $n = m + 1$. By Exercise 1.5.3(c), we have that $[\gamma_m] = [Q_m/\sigma, (P_m + \sqrt{D})/\sigma] = [Q_m/\sigma, (P_m - a_mQ_m + \sqrt{D})/\sigma] = [Q_m/\sigma, (-P_{m+1} + \sqrt{D})/\sigma]$, where the last equality holds via (2.1.4). Furthermore, $Q_m > 0$, by Remark 1.2.1, since reduced ideals are, a fortiori, primitive.

If $\delta = (-P_{m+1} + \sqrt{D})/\sigma$, then $\delta = Q_m/(\sigma\gamma_{m+1})$, by (2.1.5). The proof of Claim 3 shows us that $\gamma_{m+1} > 1$, so $0 < \delta < Q_m/\sigma$. However, $[\gamma_m]$ is reduced, so $|\delta'| > Q_m/\sigma$, by Definition 1.4.1. Furthermore, $|\delta'| = P_{m+1} + \sqrt{D} = -\delta'$. To see this we observe that $Q_m/\sigma < \sqrt{\Delta} = 2\sqrt{D}/\sigma$, by Corollary 1.4.2, via Exercise 1.5.3(c), so

$$\delta = (-P_{m+1} + \sqrt{D})/\sigma < Q_m/\sigma < 2\sqrt{D}/\sigma,$$

i.e. $\delta' = (-P_{m+1} - \sqrt{D})/\sigma < 0$. Finally, $\gamma'_{m+1} = Q_m/(\sigma\delta')$ implies $-1 < \gamma'_{m+1} < 0$. By Claim 3, $[\gamma_{m+1}]$ is a reduced ideal. Thus, by induction $[\gamma_n]$ is reduced for all $n > m$, with $-1 < \gamma'_n < 0$, which secures Claim 4.

Claim 5. For any $m \geq 1$, we have that $\gamma'_m < 0$ if and only if $P_m < \sqrt{D}$ and $Q_m > 0$. Also, if $\gamma'_m < 0$, then $\gamma'_j < 0$ for all $j \geq m$.

If $\gamma'_m < 0$, then $2\sqrt{D}/Q_m = \gamma_m - \gamma'_m > 0$, so $Q_m > 0$ and $P_m < \sqrt{D}$. Conversely, it is clear that $\gamma'_m < 0$ whenever $Q_m > 0$ and $P_m < \sqrt{D}$. By Claims 3–4, $\gamma'_j < 0$ for all $j \geq m$, whenever $\gamma'_m < 0$. This establishes Claim 5.

The following establishes that γ'_m is "eventually" negative.

Claim 6. If $|\gamma_0 - \gamma'_0| > 1/(B_{m-1}B_{m-2})$ then $\gamma'_m < 0$, where $m \geq 2$ and the B_i are defined in Exercise 2.1.2(a).

From (2.1.8) we have,

$$\gamma_0 = \frac{\gamma_m A_{m-1} + A_{m-2}}{\gamma_m B_{m-1} + B_{m-2}} = \frac{\gamma_m A_{m-1}B_{m-1} + A_{m-2}B_{m-1}}{\gamma_m B^2_{m-1} + B_{m-2}B_{m-1}}$$
$$= \frac{A_{m-1}}{B_{m-1}} + \frac{(-1)^{m-1}}{B^2_{m-1}\gamma_m + B_{m-2}B_{m-1}},$$

where the last equality follows from (2.1.12) in Exercise 2.1.2(c). From this we deduce that

$$(-1)^m(\gamma_0 - \gamma'_0) = \frac{1}{B^2_{m-1}\gamma'_m + B_{m-2}B_{m-1}} - \frac{1}{B^2_{m-1}\gamma_m + B_{m-2}B_{m-1}}.$$

Therefore, if $\gamma'_m > 0$, then

$$|\gamma_0 - \gamma'_0| < \max\left\{\frac{1}{B^2_{m-1}\gamma'_m + B_{m-2}B_{m-1}}, \frac{1}{B^2_{m-1}\gamma_m + B_{m-2}B_{m-1}}\right\} < \frac{1}{B_{m-1}B_{m-2}}.$$

This secures Claim 6.

Claim 7. Let $M_0 = \max\left\{2, \frac{5}{2} + \frac{\log(|Q_0|/(2\sqrt{D}))}{2\log\tau}\right\}$, where $\tau = (1+\sqrt{5})/2$. If $m \geq M_0$, then $\gamma'_m < 0$.

First we establish that $B_n \geq \tau^{n-1}$ for all $n \geq 0$. From the definition of B_n, given in Exercise 2.1.2(a), and the fact that $a_n \geq 1$ for $n \geq 1$, we have $B_n \geq F_{n+1}$, where F_i is the ith Fibonacci number (since $F_{n+1} = F_n + F_{n-1}$ with $F_0 = 0$ and $F_1 = 1$). Since $F_{n+2} \geq \tau^n$, then $B_n \geq \tau^{n-1}$ for $n \geq 1$. Hence, $B_{m-1}B_{m-2} \geq \tau^{2m-5} \geq |Q_0|/(2\sqrt{D})$, if $m \geq M_0$. However, $|Q_0|/(2\sqrt{D}) = 1/|\gamma_0 - \gamma'_0|$. By Claim 6, $\gamma'_m < 0$, and we have Claim 7.

We have established the existence of an $M_0 \in \mathbf{Z}$ such that I_m is reduced for all $m \geq M_0$. This establishes the existence of reduced ideals equivalent to I in $\mathcal{O}_\Delta$. We must now complete the task by establishing uniqueness, i.e. that these are the *only* reduced ideals equivalent to I. We do this via the following final sequence of claims.

Claim 8. If I and J are equivalent, primitive $\mathcal{O}_\Delta$-ideals, then there exists a $\delta \in I$ such that $(\delta)J = (N(J))I$ and $0 < \delta < N(I)$.

Since $I \sim J$, then there exist non-zero $\alpha, \beta \in \mathcal{O}_\Delta$ with $(\alpha)I = (\beta)J$. Therefore, $|\beta|N(J) = |\alpha|\lambda$ for some $\lambda \in I$. Let $u \in U_\Delta^+$, (see footnote (1.1.6)). Thus, there is a $j \in \mathbf{Z}$ with $u^j\lambda < N(I)$. Set $\delta = u^j\lambda$, and $L = (\delta)J = (\lambda)J$. Therefore, $(N(J)\beta)J = (\alpha)L$, and $(N(J)\alpha)I = (\alpha)L$, so $L = (N(J))I$ and $(\delta)J = (N(J))I$. This secures Claim 8.

Claim 9. If, in Claim 8, $I = I_1$ and J are reduced $\mathcal{O}_\Delta$-ideals, then there exists an $m \in \mathbf{Z}$, $m \geq 1$ such that $\delta = \theta_m N(I) = \theta_m Q_0/\sigma$.

The reader may verify, by an induction argument, that $\theta_{k+1}^{-1} = B_{k-1}\gamma_k + B_{k-2}$ for $k \geq 1$. Furthermore, from the proof of Claim 7, $B_k \geq \tau^{k-1}$ for all $k \geq 1$. Thus, we may conclude that $f(k) = \theta_{k+1}^{-1} = \prod_{i=1}^k \gamma_i$ is a monotonically increasing, and unbounded function of k. Hence, there exists an $m \in \mathbf{Z}$, $m \geq 1$, such that

$$\theta_m^{-1} \leq N(I)/\delta < \theta_{m+1}^{-1}.$$

Hence,

$$\theta_{m+1} < \delta/N(I) \leq \theta_m.$$

If $\delta \neq N(I)\theta_m$ then, since $\delta \in I = [Q_0\theta_{m+1}/\sigma, Q_0\theta_m/\sigma]$, by Claim 2, $\delta/N(I) = x\theta_m + y\theta_{m+1}$ for some $x, y \in \mathbf{Z}$. We now establish that $|\delta'| < |\theta'_{m+1}|N(I)$.

Suppose $|\delta'| > |\theta'_{m+1}|N(I)$. Since $\lambda = N(I)\theta_{m+1} \in I$, then $N(J)\lambda = \delta\rho$ for some non-zero $\rho \in J$. Also, $|\rho| = N(J)|\lambda/\delta| < N(J)$, and $|\rho'| = N(J)|\lambda'/\delta'| < N(J)$. By Definition 1.4.1, this contradicts the fact that J is reduced, and thereby establishes that $|\delta'| < |\theta'_{m+1}|N(I)$. Since $\delta/N(I) < \theta_m$, then $|x\theta_m + y\theta_{m+1}| < |\theta_m|$, and $|x\theta'_m + y\theta'_{m+1}| < |\theta'_{m+1}|$. However, I is reduced, so $\gamma'_i < 0$ for all $i \geq 1$, by Claim 4. Thus, θ'_{m+1} and θ'_m have different signs, contradicting that the last two inequalities both hold. Hence, $\delta = N(I)\theta_m$ for some $m \geq 1$. Since $(\delta)J = (N(J))I$, and $(N(I)\theta_m)I_m = (N(I_m))I$, from Claim 2, then $|N(\delta)|N(J) = N(J)^2N(I)$, and $N(I)^2|N(\theta_m)|N(I_m) = N(I_m)^2N(I)$. Since $N(\delta) = N(\theta_m)N(I)^2$, we have $N(J) = N(I_m)$. Since $(\delta)J = (N(I_m))I = (N(I)\theta_m)I_m$, we have $J = I_m$.

Claim 9 establishes that any reduced ideal equivalent to I, must be an I_m for some $m \geq 1$. This completes the proof of the continued fraction algorithm. □

Theorem 2.1.2 shows that we may begin with any primitive $\mathcal{O}_\Delta$-ideal $I = I_1 = [\gamma_0]$, and after applying the continued fraction algorithm to $\gamma = \gamma_0$, we must ultimately reach a reduced ideal $I_m \sim I_1$ for some $m \geq 1$. Furthermore, once we have produced this ideal I_m, we enter into a periodic *cycle of reduced ideals*, and this periodic cycle contains *all* of the reduced ideals equivalent to I.

We have much more to say about this important result. First, we look at valuable and informative consequence of Theorem 2.1.2, namely *a complete solution of Pell's equation.*

Corollary 2.1.2. *Let $D > 0$ be a non-square rational integer with $\gamma = \sqrt{D} = \langle a_0; a_1, a_2, \ldots \rangle$ and $\ell = \ell(\sqrt{D})$. Also, let the sequences $\{A_k\}$ and $\{B_k\}$ be defined as in Exercises 2.1.2(a). If ℓ is even, then all positive solutions of $x^2 - Dy^2 = 1$ are given by $x = A_{k\ell-1}$ and $y = B_{k\ell-1}$ for $k \geq 1$, whereas there are no solutions to $x^2 - Dy^2 = -1$. If ℓ is odd, then all positive solutions of $x^2 - Dy^2 = 1$ are given by $x = A_{2k\ell-1}$ and $y = B_{2k\ell-1}$ for $k \geq 1$, whereas all positive solutions of $x^2 - Dy^2 = -1$ are given by $x = A_{(2k-1)\ell-1}$, $y = B_{(2k-1)\ell-1}$ for $k \geq 1$.*

Proof. By Claim 1 of Theorem 2.1.2 and Exercise 2.1.2(g)(iv), we have that

$$N(\theta_{k\ell+1}) = A_{k\ell-1}^2 - DB_{k\ell-1}^2 = (-1)^{k\ell}Q_{k\ell} = (-1)^{k\ell}.$$

This establishes existence. For uniqueness, we invoke Exercise 2.1.10(b) which tells us that any solution to Pell's equation $x^2 - Dy^2 = \pm 1$ must satisfy $x = A_n$ and $y = B_n$ for some non-negative $n \in \mathbf{Z}$. However, by Claim 1 and Exercise 2.1.2(g)(iv) again, $N(\theta_{n+1}) = A_{n-1}^2 - DB_{n-1}^2 = (-1)^n Q_n$ so $Q_n = \pm 1$. However, Theorem 2.1.2 tells us that $Q_i > 0$ so $Q_n = 1$. Since $Q_n = 1$ if and only if $n \equiv 0 \pmod{\ell}$ by Theorem 2.1.2, we have uniqueness. □

A depiction of Corollary 2.1.2 is contained in

Example 2.1.3. If $D = 145$, then $\sqrt{D} = \langle 12; \overline{24} \rangle$ (see Exercise 2.1.13(c)), so $\ell(\sqrt{D}) = \ell = 1$. Also, $A_0 = 12, A_1 = 289, A_2 = 6948, \ldots$ and $B_0 = 1, B_1 = 24, B_2 = 577, \ldots$ with $A_0^2 - B_0 D^2 = -1$, $A_1^2 - B_1^2 D = 1$, $A_2^2 - B_2 D^2 = -1, \ldots$ Hence, (A_{2k+1}, B_{2k+1}) are all solutions to $x^2 - Dy^2 = 1$ for $k \geq 0$ and (A_{2k}, B_{2k}) are all solutions to $x^2 - Dy^2 = -1$ for $k \geq 0$.

Given the importance of Theorem 2.1.2, a discussion of its essential components is in order. Periodicity is of paramount importance, so we discuss it first. If $\gamma = \gamma_0$ is a quadratic irrational, as given in Theorem 2.1.2, then $I = [\gamma] = I_1$ is the first ideal in the cycle. If $m \in \mathbf{Z}$ is the least non-negative integer such that I_m is reduced, then there exists a least positive integer ℓ, such that $I_{m+\ell} = I_m$. This "period length" (see Definition 2.1.6 below) is thus encountered once we enter into a cycle of reduced ideals. In particular, if $m = 0$, then $P_0 = P_\ell$ and $Q_0 = Q_\ell$, and ℓ is the first integer for which these two events occur together. For instance, if $\gamma = w_\Delta$, then by Exercise 2.1.13(g), we see symmetry properties also enjoyed by ambiguous cycles of ideals, namely

$$Q_i = Q_{\ell-i}$$

and,

$$P_{\ell-i} = P_{i+1}$$

for $0 \leq i \leq \ell - 1$. In particular,

$$Q_{(\ell+1)/2} = Q_{(\ell-1)/2}$$

if ℓ is odd, and

$$P_{\ell/2} = P_{\ell/2+1}$$

if ℓ is even. In the latter case, $Q_{\ell/2}$ divides Δ (Exercise 2.1.13(f)). Also, by Exercise 2.1.13(h),

$$\varepsilon_\Delta = |\theta'_{\ell/2+1}/\theta_{\ell/2+1}|$$

if ℓ is even, and

$$\varepsilon_\Delta = |\theta'_{(\ell+3)/2}/\theta_{(\ell+1)/2}|$$

if ℓ is odd.

Moreover, for any reduced quadratic irrational γ we have

$$0 < Q_i < 2\sqrt{D},$$

$$0 < P_i < \sqrt{D},$$

and

$$a_i < a_0,$$

(see Exercise 2.1.13(e)).

Example 2.1.4. Let $\Delta = D = 85$ and consider the reduced ideal $I = [3, (5 + \sqrt{D})/2]$ (known to be reduced by Corollary 1.4.3). The continued fraction expansion of $\gamma = (5 + \sqrt{85})/6$ is given by

i	0	1	2	3
P_i	5	7	5	5
Q_i	6	6	10	6
a_i	2	2	1	2

We have $I = I_1 = [3, (5 + \sqrt{D})/2] \sim I_2 = [3, (7 + \sqrt{D})/2] \sim I_3 = [5, (5 + \sqrt{D})/2] \sim I_4 = I_1$ and these are the *only* reduced ideals in $\{I\}$. Here $\ell(\gamma) = 3$.

Once we have achieved a reduced ideal I_m via the continued fraction algorithm depicted in Theorem 2.1.2, then the cycle becomes periodic. Hence, it makes sense to have a name for the period length. This is the content of

Definition 2.1.6. If $I = I_1 = [Q/\sigma, (P + \sqrt{D})/\sigma]$ is a reduced ideal in a real quadratic order $\mathcal{O}_\Delta$, and if ℓ is the least positive integer such that $I_1 = I_{\ell+1} = [Q_\ell/\sigma, (P_\ell + \sqrt{D})/\sigma]$, then $\gamma_i = (P_i + \sqrt{D})/Q_i$ for $i \geq 0$ all have the same period length $\ell(\gamma_i) = \ell(\gamma_0) = \ell(\gamma)$, via $[\gamma_i] = I_{i+1} = [Q_i/\sigma, (P_i + \sqrt{D})/\sigma]$. We denote this common value by $\ell = \ell(C)$, where C is the equivalence class of I, and call this value the *period length of the cycle of reduced ideals equivalent to I* (see Exercise 2.1.8). If we wish to keep track of the specific ideal, then we will write $\ell(I)$ for ℓ.

Remark 2.1.1. If $I = [Q/\sigma, (P + \sqrt{D})/\sigma]$ is a reduced ideal in the real quadratic order $\mathcal{O}_\Delta$, then the set $\{Q_i/\sigma\}_{i=1}^{\ell}$ represents the *norms of all reduced ideals equivalent to I* (via the simple continued fraction expansion of $\gamma = (P + \sqrt{D})/Q$).

Remark 2.1.2. It is also worth reminding the reader at this juncture of something we mentioned in Chapter One, section five. As Example 1.5.1 showed us, we can discuss equivalence classes of ideals without reference to a class group structure. In this way, we are not confined to looking at only invertible ideals, and this greater freedom is sometimes necessary as we will see in Chapter Six where we discuss, at great length, ambiguous cycles of reduced ideals which are not necessarily strictly reduced (as in Example 1.5.1). We need this degree of freedom in Chapter Six, for example, to show that we can classify the number of ambiguous *cycles* without ambiguous ideals in them, whereas there may *not* exist any ambiguous *classes* in C_Δ without ambiguous ideals. When necessary, of course, we will restrict ourselves to the strictly primitive ideals, and hence the class group.

To illustrate the process, we have

Example 2.1.5. Let's consider Example 2.1.1 again. First we depict the tableaux from which we will obtain the whole story on the complete cycle C of reduced ideals beginning with $\gamma = (7 + \sqrt{385})/14$, namely

i	0	1	2	3	4	5	6	7	8	9	10
P_i	7	7	17	19	17	15	15	17	19	17	7
Q_i	14	24	4	6	16	10	16	6	4	24	14
a_i	1	1	9	6	2	3	2	6	9	1	1

We read off the ideals $[\gamma_i] = [Q_i/\sigma, (P_i + \sqrt{D})/\sigma]$ as follows: $[\gamma_0] = [7, (7 + \sqrt{385})/2]$, $[\gamma_1] = [12, (7 + \sqrt{385})/2]$, $[\gamma_2] = [2, (17 + \sqrt{385})/2]$, $[\gamma_3] = [3, (19 + \sqrt{385})/2]$, $[\gamma_4] = [8, (17 + \sqrt{385})/2]$, $[\gamma_5] = [5, (15 + \sqrt{385})/2]$, $[\gamma_6] = [8, (15 + \sqrt{385})/2]$, $[\gamma_7] = [3, (17 + \sqrt{385})/2]$, $[\gamma_8] = [2, (19 + \sqrt{385})/2]$, $[\gamma_9] = [12, (17 + \sqrt{385})/2]$, and $[\gamma_{\ell(C)}] = [\gamma_{10}] = [\gamma_0]$ for a complete cycle of reduced ideals in $C = \{[\gamma_0]\}$.

This value $\ell(C)$, which is 10 in this example, is a common value since we may enter the cycle of reduced ideals at any point and consider that ideal to be the beginning ideal. For instance, in this example, there is nothing special about starting with $\gamma = (7 + \sqrt{385})/14$. We could just as easily have started with $\alpha = \gamma_3 = [2, (17 + \sqrt{385})/2]$, say. If we did that and let $\alpha = \alpha_0$, then $[\alpha_1] = [\gamma_3]$, $[\alpha_2] = [\gamma_4], \ldots, [\alpha_7] = [\gamma_9]$, $[\alpha_8] = [\gamma_0]$, $[\alpha_9] = [\gamma_1]$ and $[\alpha_{\ell(C)}] = [\alpha_{10}] = [\gamma_2] = [\alpha_0]$. Basically, by choosing α, we merely permuted the columns of the above tableaux so that column 3 became column 1, i.e. a "shift" of 2 to the left.

There are many more properties involving permutations and symmetry which we will discuss when we devote our attention strictly to ambiguous ideals in Chapter Six.

We now show how this continued fraction algorithm gives the structures of the group of units U_Δ as well as the structure within each class (called the "infrastructure").[2.1.1]

First, we show how each primitive ideal I in $\mathcal{O}_\Delta$ has a well-defined "neighbour".

Definition 2.1.7. Let $\Delta > 0$ be a discriminant and let $I = [Q/\sigma, (P + \sqrt{D})/\sigma]$ be a primitive ideal of $\mathcal{O}_\Delta$, then the *Lagrange neighbour*[2.1.2] of I is the ideal

$$I^+ = [Q^+/\sigma, (P^+ + \sqrt{D})/\sigma],$$

where

$$P^+ = -P + \sigma Q\lfloor (P + \sqrt{D})/(\sigma Q)\rfloor \text{ and } Q^+ = -N(P^+ + \sqrt{D})/(\sigma Q).$$

[2.1.1] See [331] where Dan Shanks introduced the term infrastructure. Therein, he spoke largely in terms of the theory of forms, whereas we speak only in the language of ideals and continued fractions.

[2.1.2] Lagrange first introduced the process, which we now call "Lagrange reduction", in 1766 (see [157]). Joseph-Louis Lagrange (1736–1813) had honours bestowed upon him by Napolean Bonaparte, including being made a grand officer of the Legion of Honour and a Count of the Empire. Euler got Lagrange elected as a foreign member of the Berlin Academy in 1759 at the age of 23. Lagrange was quite modest as he wrote to Laplace in 1777: "I have always regarded mathematics as an object of amusement rather than ambition and I can assure you that I enjoy the work of others much more than my own, with which I am always dissatisfied." See Bell [21]. Copyright ©1937 by E.T. Bell. Copyright renewed ©1965 by Taine T. Bell. Reprinted by permission of Simon and Schuster, Inc. In 1786 he sought release from the Academy in Berlin after the death of Frederick the Great caused the Academy to become a less than desirable place to be. He moved to Paris in 1787 where he lived in the Louvre until the revolution. At the age of fifty-one he felt that he was through as a mathematician and turned to other pursuits, although the revolution changed him and he later returned to mathematics. In 1795, Lagrange was appointed Professor of Mathematics at Ècole Normale which closed in 1797 and the now famous École Polytechnique was founded, at whch Lagrange became its first professor. Despite his interest in his work he was despondent by his mid-fifties. However at the age of fifty-six, he married a young girl almost forty years younger than he. He died in the morning of April 10, 1813 in his seventy-sixth year.

Alternatively (see Remark 1.2.1) we can write $I = [Q, (P + \sqrt{\Delta})/2]$ as a primitive ideal of $\mathcal{O}_\Delta$. The Lagrange neighbour of I is then

$$I^+ = [Q^+, (P^+ + \sqrt{\Delta})/2],$$

where

$$P^+ = -P + 2Q\lfloor (P + \sqrt{\Delta})/(2Q) \rfloor,$$

and

$$Q^+ = (\Delta - P^{+2})/(4Q).$$

Some readers may prefer this notation, since it avoids the cumbersome use of the σ. However, the former does have its bright side, as we shall see.

If I is reduced, then I^+ is reduced (Exercise (2.1.7)). Furthermore I^+ is just the "neighbour to the right", i.e. in the continued fraction expansion of $\gamma = (P + \sqrt{D})/Q$, $I^+ = I_1^+ = I_2$, where $I_1 = [Q/\sigma, (P + \sqrt{D})/\sigma]$. In fact, if $I = I_j$, then $I^+ = I_{j+1}$ for any integer $j \geq 1$ (Exercise 2.1.8), i.e. $I_1^{(+)^n} = I_{n+1}$ for any $n \geq 1$ where "$(+)^n$" means taking the Lagrange neighbour n times. This process of finding successive Lagrange neighbours is called *Lagrange reduction* (see footnote (2.1.2)). Basically, what this means is that if we begin with a primitive $\mathcal{O}_\Delta$-ideal $I = I_1$ and form successive Lagrange neighbours, then for some integer $n \geq 1$, I_n will be reduced. Thus, the continued fraction algorithm described above *eventually* produces *all* of the reduced ideals in the class, and they are linked in this well-ordered fashion via the Langrange neighbours.[(2.1.3)] As an illustration of this process, we have

Example 2.1.6. Consider Example 2.1.4 again. Therein, we have that $I_1^+ = I_2 = [3, (7 + \sqrt{D})/2]$, $I_2^+ = I_3 = [5, (5 + \sqrt{D})/2]$ and $I_3^+ = I_4 = I_1 = [3, (5 + \sqrt{D})/2]$ for a complete cycle of reduced ideals in the class. As observed above, we may begin our reduction process with a *non*-reduced primitive ideal and achieve a reduced one. For example, if $J = [17, (17 + \sqrt{85})/2]$, then J is *not* reduced by Corollary 1.4.1. However, $J^+ = [-3, (-17 + \sqrt{85})/2] = J_1$, $J_1^+ = [-5, (-5 + \sqrt{85})/2] = J_2$, $J_2^+ = [7, (15 + \sqrt{85})/2] = J_3$, and $J_3^+ = [5, (5 + \sqrt{85})/2] = I_3$ above which is reduced! Thus, we enter into the cycle of reduced ideals via I_3 in this case.

Remark 2.1.3. What is implicit in the above is that $I \sim I^+$, and this simply follows from the continued fraction algorithm.

This infrastructure also has relevance to the unit group which we now describe.

Theorem 2.1.3. [(2.1.4)] *Let $\Delta > 0$ be a discriminant, $I = [Q/\sigma, (P + \sqrt{D})/\sigma]$ a reduced ideal in $\mathcal{O}_\Delta$, and $\gamma = (P + \sqrt{D})/Q$. If P_i and Q_i for $i = 1, 2, \ldots, \ell(\gamma) = \ell$ appear in the continued fraction expansion of γ, then*

$$\varepsilon_\Delta = \prod_{i=1}^{\ell} (P_i + \sqrt{D})/Q_i$$

(2.1.3) We will not be concerned herein with the specifics of the continued fraction algorithm such as the determination of bounds on the length of the preperiod, or the actual computational implementation of the algorithm. The interested reader will find all of the details in [389]. For our needs these details are not germane. However, we will look at applications of the algorithm to factoring and cryptology in Chapter Eight.

(2.1.4) This result was proved for the *principal class* by Lagrange in 1769 (see [158]), but easily generalizes to any class, (see also Smith [344]).

and

$$N(\varepsilon_\Delta) = (-1)^\ell.$$

Proof. If $\alpha = \prod_{i=1}^{\ell} \gamma_i$ where $\gamma_i = (P_i + \sqrt{D})/Q_i$, then

$$N(\alpha) = \prod_{i=1}^{\ell} N(\gamma_i) = \prod_{i=1}^{\ell}(P_i^2 - D)/Q_i^2 = (-1)^\ell \prod_{i=1}^{\ell}(Q_iQ_{i-1})/Q_i^2$$

by (2.1.4). Therefore,

$$N(\alpha) = (-1)^\ell \prod_{i=1}^{\ell} Q_{i-1}/Q_i = (-1)^\ell Q_0/Q_\ell = (-1)^\ell.$$

It remains only to show that any unit $u \in U_\Delta$ is of the form $\pm\alpha^q$ for some integer q. The reader may verify that we may assume $u > 1$, since the other cases follow easily from this.

Since $(u)I = I$ for any $u \in U_\Delta$, then $u = \varepsilon_\Delta^q$. However, $(\alpha)I = I$ by Claim 2 of the proof of Theorem 2.1.2. Thus, by periodicity, $k = q\ell$ for some $q \in \mathbf{Z}$, i.e. $u = \alpha^q$. □

We illustrate Theorem 2.1.3 in

Example 2.1.7. Let $\Delta = D = 145$ and $I = [6, 2 + w_\Delta] = [6, (5 + \sqrt{145})/2]$. The continued fraction expansion of $\alpha = (5 + \sqrt{145})/12$ is given by the tableaux

i	0	1	2	3	4	5
P_i	5	7	9	7	5	5
Q_i	12	8	8	12	10	12
a_i	1	2	2	1	1	1

Thus,

$$\varepsilon_\Delta = [(7 + \sqrt{145})/8][(9 + \sqrt{145})/8][(7 + \sqrt{145})/12][(5 + \sqrt{145})/10][(5 + \sqrt{145})/12]$$

$= 12 + \sqrt{145}$, and $N(\varepsilon_\Delta) = (-1)^\ell = (-1)^5 = -1$ as predicted by Theorem 2.1.3.

For an example involving a *non*-invertible ideal in a *non*-maximal order, we have

Example 2.1.8. Consider the discriminant $\Delta = 1224$ discussed in Example 1.5.1. If $\gamma = (15 + \sqrt{306})/9$, then the simple continued fraction expansion of γ is given by

i	0	1	2	3	4	5	6
P_i	15	12	6	9	6	12	15
Q_i	9	18	15	15	18	9	9
a_i	3	1	1	1	1	3	3

Hence, $\varepsilon_\Delta = 35+2\sqrt{306} = \prod_{i=1}^{5}(P_i+\sqrt{D})/Q_i = [(12+\sqrt{306})/18][(6+\sqrt{306})/15][(9+\sqrt{306})/15][(6 + \sqrt{306})/18][(12 + \sqrt{306})/9][(15 + \sqrt{306})/9]$.

Even more than the above, we can give a more specific representation for ε_Δ in terms of continued fractions as promised in the first chapter.

Theorem 2.1.4. *If $\Delta > 0$ is a discriminant, ℓ is the period length of the continued fraction expansion of $\gamma = \sqrt{D}$, and A/B is the $(\ell - 1)$-th convergent of it (i.e. $A = A_{\ell-1}$ and $B = B_{\ell-1}$ in Exercise 2.1.2), then either*

$$\varepsilon_\Delta = A + B\sqrt{D},$$

or

$$\varepsilon_\Delta^3 = A + B\sqrt{D},$$

and the latter can only occur if $D \equiv 5 \pmod 8$.[(2.1.5)]

Proof. An easy verification shows that if $\Delta \not\equiv 5 \pmod 8$, then ε_Δ is in $[1, \sqrt{D}]$. Hence, the result follows from Corollary 2.1.2 and Exercise 1.5.17. If $\Delta \equiv 5 \pmod 8$, then set $\alpha = A_{\ell-1} + B_{\ell-1}\sqrt{D}$ which is a unit by Exercise 2.1.11. Thus, $\alpha = \varepsilon_\Delta^n$. If $n = 1$, then we are done. If $n > 1$, then $\varepsilon_\Delta = (T + U\sqrt{D})/2$, with T and U being odd. Thus, the least n for which ε_Δ^n is in $[1, \sqrt{D}]$ is easily seen to be $n = 3$, which secures the result. □

An illustration of Theorem 2.1.4, as well as a cautionary note to the reader concerning the possibility of confusing the simple continued fraction expansion of $\sqrt{D}$ with that of w_Δ when $\Delta = D$, is

Example 2.1.9. In Example 2.1.7 we saw that $\varepsilon_\Delta = 12 + \sqrt{145}$. However, that came from Theorem 2.1.3 via the continued fraction expansion of $w_\Delta \neq \sqrt{\Delta}$. As seen in Example 2.1.3, we have that $A_0 = A_{\ell-1} = 12$ and $B_0 = B_{\ell-1} = 1$, i.e. $\varepsilon_\Delta = A_{\ell-1} + B_{\ell-1}\sqrt{D}$ as predicted by Theorem 2.1.4.

Example 2.1.10. For an illustration of the case where $\Delta \equiv 5 \pmod 8$, look again at Example 2.1.4 where $\Delta = 85 = D$. From Exercise 2.1.13, we have that $\ell(\sqrt{D}) = 5$ and $\sqrt{D} = \langle 9; \overline{4, 1, 1, 4, 18} \rangle$. Hence, from Exercise 2.1.2(a) we get that $A_4 = A_{\ell-1} = 378$ and $B_4 = B_{\ell-1} = 41$. Moreover, by applying Theorem 2.1.3 to $D = 85$, we get that $\varepsilon_\Delta = (9 + \sqrt{85})/2$, and so $\varepsilon_\Delta^3 = 378 + 41\sqrt{85} = A_{\ell-1} + B_{\ell-1}\sqrt{D}$ illustrating Theorem 2.1.4 in this case. However, it is also possible that $\varepsilon_\Delta = A_{\ell-1} + B_{\ell-1}\sqrt{D}$ when $D \equiv 5 \pmod 8$, since Theorem 2.1.4 does *not* say that ε_Δ^3 *must* be $A_{\ell-1} + B_{\ell-1}\sqrt{D}$ when $D \equiv 5 \pmod 8$, but rather that it can only happen in that particular case. For instance, if $D = \Delta = 37$, then (from an application of Theorem 2.1.3 for example) we have that $\varepsilon_\Delta = 6 + \sqrt{37}$. From Exercise 2.1.13, we have $\sqrt{37} = \langle 6; \overline{12} \rangle$, so from Exercise 2.1.2 we calculate $A_{\ell-1} = A_0 = 6$, and $B_{\ell-1} = B_0 = 1$, i.e. $\varepsilon_\Delta = A_{\ell-1} + B_{\ell-1}\sqrt{D}$.

Remark 2.1.4. In Corollary 2.1.2, we saw how to solve Pell's equation via the convergents of the simple continued fraction expansion. Also, Theorem 2.1.4 gave us the fundamental unit in terms of those convergents. What is revealed by these two facts is that if we let $\varepsilon_\Delta^n = (A+B\sqrt{D})^n = A_n + B_n\sqrt{D}$, then we can determine these A_n and B_n via what is called Lucas-Lehmer theory. This is depicted in Chapter Three, section one (see Exercise 3.1.5). In fact, Chapter Three will be concerned

[(2.1.5)]This result underlies a deeper phenomenon. To understand what that is the reader can work through Exercise 2.1.10(b). Furthermore, the interested reader will note that if $\Delta = \Delta_0$ is fundamental, then we can use footnote (1.5.9) to get the result. The reason is that, when $\Delta_0 \equiv 5 \pmod 8$, we have $\psi_{\Delta_0}(2) = 3$ where $f_{4\Delta_0} = 2 = f_\Delta$, so $u \mid 3$.

with solutions of arbitrary Diophantine equations via the methods developed in the first two chapters.

Exercises 2.1

1. (a) Prove that a *finite* simple continued fraction is a rational number. (*Hint*: Verify that $\langle a_0; a_1, a_2, \ldots, a_n\rangle = a_0 + 1/\langle a_1; a_2, \ldots, a_n\rangle$, and use induction on n.)

   (b) Prove that a rational number has a finite simple continued fraction expansion. (*Hint*: Use the Euclidean algorithm.)

   (c) Let $\{A_k\}$ and $\{B_k\}$ be sequences of integers defined as in Exercise 2(a) below for a *finite* such sequence of integers $\{a_i\}$. Prove that

$$A_n/A_{n-1} = \langle a_n; a_{n-1}, \ldots, a_1, a_0\rangle$$

   and,

$$B_n/B_{n-1} = \langle a_n; a_{n-1}, \ldots, a_2, a_1\rangle.$$

   (*Hint*: Use $A_k = a_k A_{k-1} + A_{k-2}$ to prove that $A_n/A_{n-1} = a_n + 1/(A_{n-1}/A_{n-2})$.)

2. (a) Define two sequences of integers $\{A_k\}$ and $\{B_k\}$ inductively by:

$$A_{-2} = 0, \quad A_{-1} = 1, \quad A_k = a_k A_{k-1} + A_{k-2} \quad (\text{for } k \geq 0),$$

$$B_{-2} = 1, \quad B_{-1} = 0, \quad B_k = a_k B_{k-1} + B_{k-2} \quad (\text{for } k \geq 0),$$

   where $\{a_i\}$ is an infinite sequence of integers with $a_i > 0$ for $i > 0$. Prove that for any $r \in \mathbb{R}$ we have

$$\langle a_0; a_1, a_2, \ldots, a_{n-1}, r\rangle = \frac{rA_{n-1} + A_{n-2}}{rB_{n-1} + B_{n-2}} \quad (n \geq 0).$$

   (*Hint*: Use Exercise 2.1.1 and induction on n.)

   (b) Prove that, if $C_k = \langle a_0; a_1, a_2, \ldots, a_k\rangle$, then $C_k = A_k/B_k$ for all $k \geq 0$. (*Hint*: Apply part (a).)

   (c) Prove the identities

$$C_k - C_{k-1} = (-1)^{k-1}/(B_k B_{k-1}), \quad (k \geq 1), \tag{2.1.10}$$

$$C_k - C_{k-2} = (-1)^k a_k/(B_k B_{k-2}), \quad (k \geq 2), \tag{2.1.11}$$

$$A_k B_{k-1} - A_{k-1} B_k = (-1)^{k-1}, \quad (k \geq 1), \tag{2.1.12}$$

   and

$$A_k B_{k-2} - A_{k-2} B_k = (-1)^{k-1} a_k, \quad (k \geq 1). \tag{2.1.13}$$

   (*Hint*: Use part (a) and induction on k).

(d) Prove that $\lim_{k\to\infty} C_k = r \in \mathbb{R}$. (*Hint*: Use part (c) to prove that $\lim_{k\to\infty} C_{2k} = \lim_{k\to\infty} C_{2k+1} = r \in \mathbb{R}$.)
Note that $r = \lim_{k\to\infty}\langle a_0; a_1, a_2, \ldots, a_k\rangle$ and may be considered to be *the definition of the infinite simple continued fraction* $\gamma = \langle a_0; a_1, a_2, \ldots\rangle$. Moreover, C_k is called the ***kth convergent*** of γ.

(e) Prove that an infinite simple continued fraction $\gamma = \langle a_0; a_1, a_2, \ldots\rangle$ is irrational. (*Hint*: Use part (d) to prove that $C_{2k} < \gamma < C_{2k+1}$, then use part (c) to get that $0 < |B_k\gamma - A_k| < 1/B_{k+1}$. Finally, assume γ is rational and get a contradiction.)

(f) Prove that every irrational number can be uniquely expressed by an infinite simple continued fraction. (*Hint*: Let γ be an irrational number and define the recursion

$$a_k = \lfloor \gamma_k \rfloor \quad \text{and} \quad \gamma_{k+1} = 1/(\gamma_k - a_k)$$

for $k \geq 0$ where $\gamma = \gamma_0$. Conclude that $\gamma = \langle a_0; a_1, a_2, \ldots,\rangle$ using parts (a)–(d). To prove uniqueness, assume equality of two infinite simple continued fractions. Then use part (d) and induction.)

(g) Let $\gamma = (P + \sqrt{D})/Q = \langle a_0; a_1, a_2, \ldots,\rangle$ be a quadratic irrational where $D > 0$ is not a perfect square, and let θ_k be defined as in the proof of Theorem 2.1.2. Define $G_{k-1} = Q_0A_{k-1} - P_0B_{k-1}$ for $k \geq -1$. Prove that

(i)
$$\theta_{k+1} = (-1)^k(G_{k-1} - B_{k-1}\sqrt{D})/Q_0, \tag{2.1.14}$$
for $k \geq -1$,

(ii)
$$G_{k-1} = P_kB_{k-1} + Q_kB_{k-2} \tag{2.1.15}$$
for $k \geq 0$, and

(iii)
$$DB_{k-1} = P_kG_{k-1} + Q_kG_{k-2} \tag{2.1.16}$$
for $k \geq 0$.

(iv) $N(\theta_{k+1}) = (-1)^kQ_k/Q_0$.

(v) Let γ_i be as in (2.1.2) with $\gamma_0 = \gamma$ as above, and let $t \geq 0$ be the least $t \in \mathbb{Z}$ such that $\gamma_t' < 0$. Prove that, if s is the least non-negative $s \in \mathbb{Z}$ such that $0 < Q_s < \sqrt{D}$, then $t = s$ or $t = s+1$ unless $(t, s) = (0, 1)$. (*Hint*: Prove that $a_s = (P_s + \sqrt{D})/Q_s - \varepsilon$, where $0 < \varepsilon < 1$ and $2\sqrt{D}/Q_s > 2 > 1+\varepsilon$, implies $\gamma_{s+1}' < 0$ and $t \leq s+1$.)

(vi) With reference to part (v), let $m \in \mathbb{Z}$ be the least positive integer such that $0 < Q_{m-1} < \sqrt{D}$. Prove that $\theta_m^{-1} < 2Q_0/Q_{m-1}$. (*Hint*: Verify that I_m is reduced via Corollary 1.4.3, and Exercise 1.5.9. For $m \leq 2$ the result is an easy check. Assume that $m \geq 3$, and let $k \in \mathbb{Z}$ be the least non-negative integer such that $\gamma_k' < 0$. Show that $k = 2$ is not possible so $k = m$ or $m-1$ by part (v). For $k = m \geq 3$, $\gamma_{m-1}' > 0$, show that $\theta_m^{-1} < Q_0/(Q_{m-1}B_{m-3}) \leq Q_0/Q_{m-1}$. If $k = m-1 \geq 3$, then $\gamma_{m-2}' > 0$. Verify that $\theta_{m-1}^{-1} < Q_0/(Q_{m-2}B_{m-4})$, so $\theta_m^{-1} = \theta_{m-1}^{-1}\gamma_{m-1} < (Q_0/(Q_{m-2}B_{m-4}))(2\sqrt{D}/Q_{m-1}) < 2Q_0/Q_{m-1}$.)

(vii) Prove that, if γ is reduced ($Q_0 > 0$), then $|\theta'_k| > \tau^{k-2}$ where $\tau = (1+\sqrt{5})/2$ and $k \geq 3$. (*Hint*: Use part(vi) to deduce that both $-1 < (P_1 - \sqrt{D})/Q_1 < 0$ and $(P_1 + \sqrt{D})/Q_1 > 1$, with $\gamma_1 > 0$ and $P_1 > 0$. Deduce that $|\theta'_k| = a_{k-2}|\theta'_{k-1}| + |\theta'_{k-2}|$ for $k \geq 2$. Since $a_{k-2} \geq 1$ for $k \geq 3$ then, if F_k is the kth Fibonacci number, $|\theta'_k| > F_k > \tau^{k-2}$.)

(viii) From part (vii), conclude that $|\theta'_m/\theta'_k| > \tau^{m-k-1}$ when $m \geq k$.

3. (a) Prove that every quadratic irrational γ is periodic. (*Hint*: Apply Exercise 2.1.2(a) to get an expression for γ_i in terms of its convergents, where γ_i is determined by (2.1.2). Use this expression and (2.1.2)–(2.1.5) to bound the number of values for P_i and Q_i when i is sufficiently large, (so that $P_i = P_j$ and $Q_i = Q_j$ for some $i < j$). This forces $\gamma_i = \gamma_j$ from which periodicity may be deduced.)

   (b) Prove that, if the infinite simple continued fraction expansion of an irrational number is periodic, then the number is a quadratic irrational. (*Hint*: Use Exercise 2.1.2(a) to deduce that the periodic part of the irrational is quadratic.)

Note that 3(a)–(b) is called *Lagrange's Theorem*.

4. Let α and β be two quadratic irrationals. Prove that

   (a) $(\alpha + \beta)' = \alpha' + \beta'$.

   (b) $(\alpha - \beta)' = \alpha' - \beta'$.

   (c) $(\alpha\beta)' = \alpha'\beta'$.

   (d) $(\alpha/\beta)' = \alpha'/\beta'$.

5. Let $\Delta > 0$ be a discriminant with radicand D, and let γ be a quadratic irrational associated with D. Prove that, if γ is reduced, then $[\gamma]$ is reduced. Does the converse hold? (*Hint*: Look at $I = [1, w_\Delta]$ with $\gamma = w_\Delta$.) Compare this with Claims 3–4 in the proof of Theorem 2.1.2.

6. Let $\Delta > 0$ be a discriminant and let $I = [Q/\sigma, (P + \sqrt{D})/\sigma]$ be a reduced ideal in $\mathcal{O}_\Delta$. If $\gamma = (P + \sqrt{D})/Q$, then show that $\ell(\gamma_i) = \ell(\gamma_j)$ for all $i, j \in \mathbf{Z}$ with $1 \leq i, j \leq \ell(\gamma) = \ell$. (*Hint*: Show that the continued fraction expansion of $[\gamma_i]$ is just a "permutation" of the continued fraction expansion of $[\gamma_j]$.)

7. If $\Delta > 0$ is a discriminant and I is a reduced ideal in $\mathcal{O}_\Delta$, show that its Lagrange neighbour I^+ is also reduced.

8. If $\Delta > 0$ is a discriminant and we assume that the primitive ideal $I_j = [Q_{j-1}/\sigma, (P_{j-1} + \sqrt{D})/\sigma]$ comes from the continued fraction expansion of $\gamma = (P + \sqrt{D})/Q$, i.e. $I_{j+1} = [\gamma_j]$, then show that $I_j^+ = I_{j+1}$. This shows that the Lagrange neighbour is actually the neighbour on the right.

9. By the continued fraction algorithm, if we are given a reduced ideal I in a real quadratic order $\mathcal{O}_\Delta$, and a reduced ideal $J \sim I = I_1$, then $J = I_k$ for some $k \geq 1$ (see Theorem 2.1.2). By Claim 2 of the proof of that theorem, $(N(I), \theta_k)I_k = (N(I_k))I_1$. Define the *distance* from I to J to be

$$\delta_k = \delta(I_k, I_1) = \log \Psi_k$$

where

$$\Psi_k = \prod_{i=1}^{k-1} \psi_i = |\theta'_k|$$

with $\psi_i = |\gamma'_i|^{-1} = |(P_i + \sqrt{D})/Q_{i-1}|$, and $\psi_1 = 1$. Here γ_i, θ_i are as in the proof of Theorem 2.1.2. (2.1.6)

(i) Prove that $\Psi_i = |(G_{i-2} + B_{i-2}\sqrt{D})/Q_0|$ where G_j and B_j are defined in Exercise 2.1.2(g).

(ii) Show that $\delta_i > \delta_{i-1} \geq 0$ for $i > 1$ and $\delta_i = 0$ if and only if $i = 1$.

10. (a) If γ is an irrational number and r/s is a rational number with gcd $(r, s) = 1$ and $s > 0$, then r/s is a convergent of the continued fraction expansion of γ whenever

$$|\alpha - r/s| < 1/(2s^2)$$

(see Exercise 2.1.2(a)). Prove this. (*Hint*: If r/s is *not* a convergent of γ, then there exists an integer $k \geq 1$ such that $B_k < s < B_{k+1}$. Prove that $|B_k\gamma - A_k| \leq |s\gamma - r|$. Thus, $|\gamma - C_k| < 1/(2sB_k)$, where C_k is defined in Exercise 2.1.2(b). Finally, verify that $1/(sB_k) < |\gamma - C_k| + |\gamma - r/s|$. Deduce the contradiction that $B_k > s$.)

(b) Let $n, D \in \mathbf{Z}$ with $D > 0$, not a perfect square. If $x^2 - Dy^2 = n$ with $|n| < \sqrt{D}$, then x/y is a convergent of the simple continued fraction expansion of $\sqrt{D}$. Prove this using part (a). (*Hint*: Factor $x^2 - Dy^2$ and assume $n > 0$. Deduce that $0 < x/y - \sqrt{D} < 1/(2y^2)$. The case $n < 0$ is then based on the solution of the case for $n > 0$.)

11. Let $\gamma = (P + \sqrt{D})/Q$ be a quadratic irrational and set $\alpha = \prod_{i=0}^{k-1} \gamma_i$ where γ_i is given by (2.1.2) in the simple continued fraction expansion of γ. Prove that $N(\alpha) = (-1)^k Q_k/Q_0$. In particular, conclude that if $k = \ell(\gamma) = \ell$, then $A_{\ell-1}^2 - B_{\ell-1}^2 D = (-1)^\ell$. (*Hint*: See Exercise 2.1.2(g)(iv).)

12. (2.1.7) Prove that there are infinitely many positive radicands $D \equiv 1 \pmod 4$ such that:

(a) $\ell((1 + \sqrt{D})/2) + 4 = \ell(\sqrt{D})$.

(b) $X^2 - DY^2 = 4$ is solvable for odd $X, Y \in \mathbf{Z}$.

(c) $\ell(\sqrt{D})$ is unbounded, i.e. $\ell(\sqrt{D}) = f(n)$ where $f(n) \to \infty$ as $n \to \infty$.

(*Hint*: Prove that $F_{r+t}F_s - F_rF_{s+t} = (-1)^{s-1}F_tF_{r-s}$ $(r \geq s \geq 0,\ t \geq 0)$ where F_i is the ith Fibonacci number. Establish that, for $D = (2F_{6n+1} + 1)^2 + (8F_{6n} + 4)$ $(n \geq 1)$, $\ell(\sqrt{D}) = 6n + 5$ and $\ell((1 + \sqrt{D})/2) = 6n + 1$.)

(2.1.6)This concept was introduced in 1972 by Shanks [331] wherein he used the language of binary quadratic forms, and this was refined in 1982 by Lenstra [178] and in 1983 by Schoof [321]. The first treatment in terms of ideals was given by Williams-Wunderlich in [389] where strict primitivity is not required since they do not mention C_Δ (albeit they still define equivalence which is possible without concern for invertibility). A version which does require strict primitivity is given by Kaplan–Williams in [137].

(2.1.7)This and similar results are due to K.S. Williams and N. Buck (see [390]). Therein, the authors comment: "As far as the authors are aware this is the first example of a continued fraction expansion of $\sqrt{D}$ or $(1 + \sqrt{D})/2$, where D involves the Fibonacci numbers F_n and their squares. Presumably the D's used here are a special case of an infinite family of D's involving Fibonacci numbers for which the continued fractions of $\sqrt{D}$ and $(1 + \sqrt{D})/2$ can be given explicitly."

13. Let $D \in \mathbf{Z}$ be positive and non-square.

(a) Prove that the simple continued fraction expansion of $\sqrt{D}$ is of the form $\langle a_0; \overline{a_1, a_2, \ldots, a_{\ell-1}, 2a_0}\rangle$, where $a_i = a_{\ell-i}$ for $1 \leq i \leq \ell - 1$. (*Hint*: Use Corollary 2.1.1 on the reduced quadratic irrational $\gamma = \lfloor\sqrt{D}\rfloor + \sqrt{D}$. Observe that the first partial quotient of the simple continued fraction expansion of γ is $2a_0$ where $a_0 = \lfloor\sqrt{D}\rfloor$. To prove symmetry, look at $-1/\gamma' = -1/(\lfloor\sqrt{D}\rfloor - \sqrt{D})$.)

(b) Let $\Delta = D \equiv 1 \pmod 4$ be a radicand with $D > 0$. Prove that the simple continued fraction expansion of w_Δ is of the form

$$\langle a_0; \overline{a_1, a_2, \ldots, a_{\ell-1}, 2a_0 - 1}\rangle,$$

where $a_i = a_{\ell-i}$ for $1 \leq i \leq \ell - 1$. Verify as well that all the Q_i which emerge (via 2.1.5) are even. (*Hint*: Use methodology similar to that given in the hint for part (a). Also compare with Exercise 2.1.14.)

(c) Verify each of the following facts concerning the simple continued fraction expansions of the given narrow ERD-types (see Definition 3.2.2) of radicands $D > 0$.

(i) If $D = s^2 + 1$, then $\sqrt{D} = \langle s; \overline{2s}\rangle$, and if s is even, then $w_\Delta = \langle s/2; \overline{1, 1, s-1}\rangle$. Thus, $\ell(\sqrt{D}) = 1$ and, when s is even, $\ell(w_\Delta) = 3$.

(ii) If $D = s^2 - 1$, then $\sqrt{D} = \langle s - 1; \overline{1, 2s-2}\rangle$.

(iii) If $D = s^2 + 4$ for odd s, then

$$\sqrt{D} = \langle s; \overline{(s-1)/2, 1, 1, (s-1)/2, 2s}\rangle$$

and $w_\Delta = \langle (s+1)/2; \overline{s}\rangle$. Thus, $\ell(\sqrt{D}) = 5$ and $\ell(w_\Delta) = 1$ (see Exercise 2.1.14).

(iv) If $D = s^2 - 4$ and $s \geq 4$, then

$$\sqrt{D} = \langle s - 1; \overline{1, (s-3)/2, 2, (s-3)/2, 1, 2s-2}\rangle$$

and $w_\Delta = \langle (s-1)/2; \overline{1, s-2}\rangle$. Hence, $\ell(\sqrt{D}) = 6$ and $\ell(w_\Delta) = 2$.

(d) Prove that, in the continued fraction expansion of $\sqrt{D}$ (or $(1+\sqrt{D})/2 = w_\Delta$ if $\Delta \equiv 1 \pmod 4$), either $P_k = P_{k+1}$ or $Q_k = Q_{k+1}$ where $\ell = 2k$, respectively $\ell = 2k+1$. (*Hint*: Use the symmetry established in parts (a)-(b).)

(e) Prove that, in the simple continued fraction expansion of a reduced quadratic irrational $\gamma = (P_0 + \sqrt{D})/Q_0$ where $D > 0$ is a radicand, we have for $0 \leq i \leq \ell(\gamma) = \ell$

$$0 < Q_i < 2\sqrt{D},$$

$$0 < P_i < \sqrt{D},$$

and $a_i < a_0$ for $1 \leq i < \ell$. (In particular, this holds for $\gamma = w_\Delta$.)

(f) With reference to (e), prove that if $\Delta > 0$ is a fundamental discriminant, and if $[\gamma]$ is reduced and ambiguous with $Q_i \neq Q_0$, then Q_i/σ is a squarefree divisor of Δ if and only if $\ell = 2i$. (*Hint*: Use part (e) of this problem and (2.1.4)–(2.1.5) to establish symmetry in the period.)

(g) Establish that, in the simple continued fraction expansion of w_Δ, $Q_i = Q_{\ell-i}$ and $P_{\ell-i} = P_{i+1}$ where $\ell = \ell(w_\Delta)$ and $0 \le i \le \ell - 1$. (See Lemma 6.1.1.)

(h) Let $I = [Q/\sigma, (P + \sqrt{D})/\sigma]$ be a reduced ambiguous ideal of $\mathcal{O}_\Delta$. Show that, if s is the least positive integer such that $P_s = P_{s+1}$, then $\ell(\gamma) = 2s$ and $\varepsilon_\Delta = |\theta'_{s+1}/\theta_{s+1}|$. If $D > 5$ show that, if t is the least positive integer such that $Q_t = Q_{t+1}$, then $\ell(\gamma) = 2t + 1$ and $\varepsilon_\Delta = |\theta'_{t+2}/\theta_{t+1}|$. (*Hint*: See Exercise 2.1.2(g) for the statements on ε_Δ, and use an induction argument on the P_i and Q_i by establishing first that $a_i = \lfloor (P_{i+1} + \sqrt{D})/Q_i \rfloor$.)

14. Let $d > 0$ be a nonsquare rational integer.

(a) Prove that, if $d \equiv 1 \pmod 4$, then $\ell(\sqrt{d}) \equiv \ell((1 + \sqrt{d})/2) \pmod 2$. (*Hint*: See Theorems 2.1.3–2.1.4. [2.1.8])

(b) Consider the equation

$$x^2 - dy^2 = -4 \text{ with } \gcd(x, y) = 1. \tag{2.1.17}$$

(i) Prove that, if (2.1.17) is solvable and d is even, then $d \equiv 4$ or $8 \pmod{16}$. Then verify that if $d \equiv 4$ or $8 \pmod{16}$, (2.1.17) is solvable if and only if $\ell(\sqrt{d/4})$ is odd. Also, show that if d is even and (2.1.17) is solvable, then $x^2 - dy^2 = -1$ is not solvable.

(c) Let $d \equiv 1 \pmod 4$ and let P_i and Q_i arise from (2.1.2)-(2.1.5) via the simple continued fraction expansion of $\sqrt{d}$. Set $\ell = \ell(\sqrt{d})$ and $\overline{\ell} = \ell((1 + \sqrt{d})/2)$ with $\overline{P}_i$ and $\overline{Q}_i$ arising from the simple continued fraction expansion of $(1 + \sqrt{d})/2$. Assume ℓ is odd. Prove that $Q_{\frac{\ell+1}{2}} = \overline{P}_{\frac{\overline{\ell}+1}{2}}$ and $\overline{Q}_{\frac{\overline{\ell}+1}{2}} = P_{\frac{\ell+1}{2}}$. (*Hint*: If $\sqrt{d} = \langle a_0; \overline{a_1, a_2, ..., 2a_0}\rangle$ and $(1 + \sqrt{d})/2 = \langle b_0; \overline{b_1, b_2, ..., 2b_0 - 1}\rangle$ (via Exercise 2.1.13), then prove that

$$(P_{\frac{\ell+1}{2}} + \sqrt{d})/Q_{\frac{\ell+1}{2}} = \langle \overline{a_{\frac{\ell-1}{2}}, a_{\frac{\ell-3}{2}}, ..., 2a_0, a_1, ... a_{\frac{\ell-1}{2}}} \rangle$$

and,

$$(Q_{\frac{\ell+1}{2}} + \sqrt{d})/P_{\frac{\ell+1}{2}} = \langle \overline{b_{\frac{\overline{\ell}-1}{2}}, b_{\frac{\overline{\ell}-3}{2}}, ..., 2b_0 - 1, b_1, ..., b_{\frac{\overline{\ell}-1}{2}}} \rangle.)$$

(d) Let $x^2 - dy^2 = -1$ be solvable and retain the notation from part (c), with ℓ odd. Prove that there are positive $x, y \in \mathbb{Z}$ such that

$$B_{\ell-1} = x^2 + y^2,$$

and

$$P_{\frac{\ell+1}{2}} A_{\ell-1} = (P_{\frac{\ell+1}{2}} x + Q_{\frac{\ell+1}{2}} y)^2 - y^2 d,$$

where $A_{\ell-1}$ and $B_{\ell-1}$ arise via Exercise 2.1.2 in the simple continued fraction expansion of $\sqrt{d}$. Deduce that

$$A_{\ell-1} P_{\frac{\ell+1}{2}} + (-1)^{\frac{\ell-1}{2}} Q_{\frac{\ell+1}{2}} = d(x^2 - y^2).$$

(*Hint*: Use part (c) and Exercises 2.1.10, 2.1.13.)

[2.1.8] Lagrange was the first to give a proof that Pell's equation $x^2 - dy^2 = 1$ is always solvable for non-zero $y \in \mathbb{Z}$ and $x \in \mathbb{Z}$ using the simple continued fraction expansion of $\sqrt{d}$. Later, the solvability of $x^2 - dy^2 = -1$ was determined via the parity of $\ell(\sqrt{d})$, i.e. it is solvable if and only if $\ell(\sqrt{d})$ is odd.

(e) Let $d \equiv 1 \pmod 4$ and t be the smallest positive integer such that $t^2 - du^2 = -4$, with $u > 0$. Assume ℓ is odd. Prove that there exist integers x_0 and y_0 such that

$$u = x_0^2 + y_0^2$$

and

$$tQ_{\frac{\ell+1}{2}} = (Q_{\frac{\ell+1}{2}}x_0 + P_{\frac{\ell+1}{2}}y_0)^2 - y_0^2 d.$$

Deduce that

$$tQ_{\frac{\ell+1}{2}} + 2(-1)^{\frac{\bar{\ell}-1}{2}} P_{\frac{\ell+1}{2}} = d(x_0^2 - y_0^2),$$

where $\bar{\ell}$ is defined in part (c). (*Hint*: If (2.1.17) is not solvable, then $t = A_{\ell-1}$, $u = B_{\ell-1}$, $x = x_0$ and $y = y_0$, so use part (d). If (2.1.17) is solvable, then prove that $A_{\ell-1} = (t^3 + 3t)/2$, and $B_{\ell-1} = u(t^2+1)/2$. If $t = 2s+1$ and $u = x_0^2 + y_0^2$, then $B_{\ell-1} = (2x^2+2s+1)(x_0^2+y_0^2) = x^2+y^2$ where $x = x_0(s+1) + y_0 s$ and $y = y_0(s+1) - x_0 s$.)

(f) Let $d \equiv 1 \pmod 8$. Prove that if $\ell = \ell(\sqrt{d})$ is odd, then $\ell \equiv \ell((1+\sqrt{d})/2) + 2 \equiv \bar{\ell} + 2 \pmod 4$, and conclude that (2.1.17) is not solvable. (*Hint*: Use (c)–(e).)

(g) Let $d \equiv 5 \pmod 8$, and assume that ℓ is odd. Prove that (2.1.17) is solvable if and only if $\ell \equiv \bar{\ell} \pmod 4$. (*Hint*: Use (d)–(f).) [(2.1.9)]

15. Let $\Delta = \Delta_0 \equiv 5 \pmod 8$ be a fundamental discriminant and consider the equation[(2.1.10)]

$$|x^2 - \Delta y^2| = 4 \text{ with } \gcd(x, y) = 1. \qquad (2.1.18)$$

(a) Prove that (2.1.18) is solvable if and only if there exists an integer $m \geq 1$ such that one of the following occurs. In what follows (as in Exercise 2.1.14), all "barred " objects $\bar{\ell}, \bar{P}_1$ etc. refer to the continued fraction expansion of $(1+\sqrt{\Delta})/2$ whereas all "unbarred objects", ℓ, P_i, etc. refer to the continued fraction expansion of $\sqrt{\Delta}$.

If $\bar{\ell}$ is even, then for some $m \geq 2$, exactly one of the folowing holds:

(i) $Q_{m-1} = 2P_m - 2P_{m-1}$.
(ii) $Q_{m-1} = 2P_{m-1} - 2P_m$.
(iii) $Q_m = 2P_m - 3Q_{m-1}$.
(iv) $Q_{m-2} = 2P_{m-1} - 3Q_{m-1}$.

If $\bar{\ell}$ is odd, then for some $m \geq 1$, exactly one of the following holds:

(v) $Q_m = 4Q_{m-1}$.
(vi) $Q_m = 5Q_{m-1} - 2P_m$.
(vii) $Q_{m-2} = 4Q_{m-1}$.

(2.1.9) This was established by Kaplan-Williams in [136] where they used composition of binary quadratic forms. However, as elucidated in the hints above, what underlies this phenomenon is the continued fraction algorithm.

(2.1.10) Although Eisenstein looked for a criterion for the solvability of this equation, the question was first asked by Gauss. Exercise 2.1.14 gives such a criteria when $N(\varepsilon_\Delta) = -1$. There has been no such criteria given when $N(\varepsilon_\Delta) = 1$. The criterion given in this exercise was discovered by H.C. Williams [384]. Note that if (2.1.17) is solvable, then ε_Δ is not in the order $[1, \sqrt{\Delta}]$, where ε_Δ is the fundamental unit of the order $[1, (1+\sqrt{\Delta})/2]$, $\Delta \equiv 1 \pmod 4$. The converse fails (take $\Delta = 21$, for example). However, the solvability of (2.1.18) is equivalent to ε_Δ *not* being in $[1, \sqrt{\Delta}]$, when $\Delta \equiv 5 \pmod 8$.

(viii) $Q_{m-2} = 5Q_{m-1} - 2P_{m-1}$.

(*Hint*: We look only at the odd case since both cases are similar. Let $\ell = 2k+1$. First establish that exactly one of $\overline{B}_k$ or $\overline{B}_{k-1}$ is even (see Exercise 2.1.2(a)). Do this by setting

$$\gamma = (X - \sqrt{\Delta}Y)/2 = |\overline{\theta}'_{k+2}/\overline{\theta}_{k+1}|,$$

and using Exercise 2.1.2(g) to expand the right hand side, comparing the coefficients of $\sqrt{\Delta}$. Observe that γ is a unit by Exercise 2.1.2(g)(iv) and 2.1.13(h). Also, see footnote (2.1.10) to establish that (2.1.18) is solvable if and only if exactly one of $\overline{\theta}_{k+1}$ or $\overline{\theta}_{k+2}$ is in $[1, \sqrt{D}]$, from the above. Suppose that $\overline{B}_k$ is odd. Show that, if (2.1.18) is solvable, then there exists an integer $m \geq 1$ such that $\theta_m = \overline{\theta}_{k+1}$ and either (v) or (vi) above holds. Do this by observing, from the above, that $\overline{\theta}_{k+1} \in [1, \sqrt{D}]$ and $|N(\overline{\theta}_{k+1})| = \overline{Q}_k/2 < \sqrt{D}$, where $\overline{Q}_k = \overline{Q}_{k+1}$. Thus, $\theta_m = \overline{\theta}_{k+1}$ by Exercise 2.1.9, and $Q_{m-1} = \overline{Q}_k/2$ by Exercise 2.1.2(g)(iv). Then establish that $\overline{P}_k \equiv P_{m-1} \pmod{Q_{m-1}}$ by using the identities in Exercise 2.1.2(g). Show that $\overline{P}_{k+1} = P_m + tQ_{m-1}$ for only $t = 0$ or $t = -1$ which establishes (v) and (vi) (respectively). Other cases are similar.)

(b) Does the above hold for arbitrary orders?

16. Let $\mathcal{O}_\Delta = [1, \sqrt{\Delta}]$ with $\Delta \equiv 1 \pmod 4$ and set $I = [4, 1 + \sqrt{\Delta}]$. Prove that (2.1.18) is solvable if and only if I is a principal ideal of $\mathcal{O}_\Delta$.[2.1.11]

17. In this question, all reference to symbols in a simple continued fraction expansion refer to w_Δ where $\Delta = \Delta_0 > 0$ is a fundamental discriminant with radicand D.

 (a) Prove that if $D = (2a-3)^2 + 8r$ (with $a = a_0$ where $w_\Delta = \langle a_0; a_1, \ldots \rangle$), an odd prime p divides r, and the $\mathcal{O}_\Delta$-primes above 2 are principal, then Q_i has an odd prime factor for some i with $0 < i < \ell(w_\Delta) = \ell$. (*Hint*: Show that some $Q_i = 4$ and look at Q_{i-1} and Q_{i+1}.)

 (b) Show that (a) fails if the $\mathcal{O}_\Delta$-primes above 2 are not principal. (*Hint*: Consider $D = 257$.)

 (c) Suppose that $D - (2a-1)^2$ and $D - (2a-3)^2$ are powers of 2 and that $\lfloor\sqrt{D}\rfloor = 2a - 1$. Prove that $D = (2^m - 2^n + 1)^2 + 2^{n+2}$ with $m > n \geq 1$, and $\lfloor\sqrt{D}\rfloor = 2^m - 2^n + 1$. Furthermore, establish that all Q_i's are powers of 2 if and only if $\ell = 2j + 1 = (m+n)/(m-n)$; $j \geq 1$ and $(j+1)n = jm$ with j properly dividing n. (*Hint*: The first part is easy given the hypothesis. For the second part, just examine the simple continued fraction expansion of w_Δ in light of (2.1.3)–(2.1.5).)

 (d) Use (c) to prove that if $D - (2a-1)^2$, and $D - (2a-3)^2$ are powers of 2 with $\lfloor\sqrt{D}\rfloor = 2a - 1$ and the $\mathcal{O}_\Delta$-primes above 2 are principal, then *not* all Q_i are powers of 2.

[2.1.11]This was established as an initial result by Stephens–Williams in [350] who sought therein to obtain a computationally effective algorithm for determining when (2.1.18) is solvable. They accomplished this using the infrastructure. This was in response to Eisenstein's problem of finding an *a priori* criterion for determining when (2.1.17) is solvable (see footnote (2.1.10)). More recent results have been found by Kaplan [135] using homomorphisms between orders and a "grading" which he defines for reduced ideals.

(e) Prove that if $\lfloor\sqrt{D}\rfloor = 2a$ and both $D-(2a-1)^2$ and $D-(2a-3)^2$ are powers of 2, then $D=(2^m+1)^2+2^{m+2}$, $m\geq 1$, $\ell = 2m+1$ and all Q_i's are powers of 2. (*Hint*: Use (2.1.3)-(2.1.5) and an induction argument on the Q_i.)

(f) Let $D\equiv 1 \pmod 8$ with both $D-(2a-1)^2$ and $D-(2a-3)^2$ being powers of 2. Prove that all Q_i are powers of 2 if and only if either:

(i) $D=(2^m+1)^2+2^{m+2}$; $m\geq 1$ and $\ell=2m+1$, or

(ii) $D=(2^m-2^n+1)^2+2^{n+2}$ with $m>n\geq 1$; $\ell=2j+1=(m+n)/(m-n)$ for some $j\geq 1$, and $(j+1)n=jm$ with j properly dividing n. (*Hint*: Use (a)-(d).)

(g) Assume that $D\equiv 1 \pmod 8$ with the $\mathcal{O}_\Delta$-primes above 2 *not* being principal, and both $D-(2a-1)^2$, $D-(2a-3)^2$ are powers of 2. Prove that all Q_i are powers of 2 if and only if $D=(2a-1)^2+2^{f+2}$, where $2a-1=2^fb+s$ (with $a_1=b$), $0<s<2^f$, and $bs+1=2^{2f/(\ell-1)}$ with $f\equiv 0 \pmod{(\ell-1)/2}$, $\ell>1$ odd and $2f\not\equiv \ell-1$. (*Hint*: Use an induction argument to verify that $P_{2i+1}=2^fb+s$ $(0\leq i\leq(\ell-1)/2)$, $P_{2i}=2^fb-s$ $(1\leq i\leq(\ell-1)/2)$, $Q_{2i}=2^{it+1}$ $(1\leq i\leq(\ell-1)/2)$, $Q_{2i+1}=2^{f+1-it}$ $(0\leq i\leq(\ell-1)/2)$, $a_{2i}=b\cdot 2^{f-it}$ $(1\leq i\leq(\ell-1)/2)$, and $a_{2i+1}=b\cdot 2^{it}$ $(0\leq i\leq(\ell-1)/2)$.)[(2.1.12)]

(h) Suppose that not all Q_i are powers of 2. Prove that there exists an odd prime p dividing Q_i, for some i, such that the $\mathcal{O}_\Delta$-primes above p are reduced. (*Hint*: Assume $p>\sqrt{D}/2$ in view of Corollary 1.4.3, then use part (a) and (2.1.3)-(2.1.5).)

(i) Let $D\equiv 1 \pmod 8$ such that not all Q_i are powers of 2. Prove that, if $D>2^{\ell+1}$, then $h_\Delta>1$. (*Hint*: Look at the argument in the proof of Theorem 3.4.1.)

18. Let $\Delta=\Delta_0>0$ be a fundamental discriminant with radicand D. Use the techniques developed in Exercise 2.1.17 to prove the following general criterion: All Q_i/σ in the simple continued fraction expansion of w_Δ are powers of a given integer $c>1$ if and only if one of the following holds:

(i) $\ell=1$ and $D=(\sigma a-\sigma+1)^2+\sigma^2$.

(ii) $\ell=2$ and $D=(\sigma a-\sigma+1)^2+\sigma^2c^2$, where $ca_1=2a-\sigma+1$.

(iii) $\ell>2$ and $D=(\sigma a-\sigma+1)^2+\sigma^2c^n$, with $n>0$ and

(a) $2a-\sigma+1=c^na_1+g$ with $0<g<c^n$.

(b) $a_1g+1=c^k$ with $k>0$ and $n\equiv 0 \pmod k$.

(c) $\ell=1+2n/k$.

(d) $Q_{2i}/\sigma=c^{ik}$ and $Q_{2i+1}/\sigma=c^{n-ik}$ for $0\leq i\leq n/k$.

19. Derive from Exercise 2.1.18 the following criterion. Let $D\equiv x^2 \pmod p$ for $0<x<p$ a prime with $p\equiv 1 \pmod x$ and suppose that the $\mathcal{O}_\Delta$-primes above p are principal. Prove that all Q_i/σ are powers of p if and only if, either

(2.1.12) All the results of this exercise appeared in [255] motivated by the work of Bernstein [24]-[25] who was interested in the forms in this question from the perspective of the simple continued fraction expansion of $\sqrt{D}$. Lévesque-Rhin [184] followed up on Bernstein's work. What (a)-(f) establishes is a classification of $D\equiv 1 \pmod 8$ having all Q_i as powers of 2. We will have much more to say about Q_i as powers of a single integer in Chapter Seven.

(1) $\ell = 1$ and $D = (\sigma a - \sigma + 1)^2 + \sigma^2$, or

(2) $\ell = 2$ and $D = (\sigma a - \sigma + 1)^2 + \sigma^2 p$ where $pa_1 = 2a - \sigma + 1$, or

(3) $\ell = 1 + 2n \geq 3$, and:

(a) If $D \equiv 1 \pmod 4$, then $D = (p^n(p-1)/x + x)^2 + 4p^n$.

(b) If $D \not\equiv 1 \pmod 4$, then $D \equiv 2 \pmod 4$, $p \equiv 1 \pmod 8$ and $D = (p^n(p-1)/(4x) + 2x)^2 + p^n$.

Finally, $Q_{2i}/\sigma = p^i$ and $Q_{2i+1}/\sigma = p^{n-i}$ for $0 < i \leq n$.

20. Assume that $\Delta = \Delta_0 > 0$ is a discriminant with radicand D, and that in the simple continued fraction expansion of w_Δ, all Q_i/σ are powers of a prime p. Prove that $h_\Delta = 1$ if and only if all primes $q < \sqrt{\Delta}/2$ with $q \neq p$ are inert. (*Hint*: Use Theorem 2.1.2.)

21. Let $\Delta \not\equiv 5 \pmod 8$ be a fundamental discriminant with radicand D such that in the simple continued fraction expansion of w_Δ, all Q_i/σ are powers of a prime $p > 2$. Prove that if $D > 2$, then $h_\Delta > 1$. (*Hint*: Use the fact that $(\Delta/2) \neq -1$.)[(2.1.13)]

2.2 The Continued Fraction Analogue for Complex Quadratics.

Given the result in Theorem 1.4.2(e), i.e. that there are at most two reduced ideals in any class of C_Δ when $\Delta < 0$, we cannot expect a structure theory as rich as that given by the infrastructure for $\Delta > 0$ described in the last section. However, there is a useful analogue for $\Delta < 0$ which allows us to mimic, in a sense, the notion of a continued fraction expansion. First, we present an analogue of the Lagrange neighbour. In order to do this, we require

Definition 2.2.1. Let $Ne(x)$ denote the *nearest integer* to x, i.e.

$$Ne(x) = \lfloor x + 1/2 \rfloor.$$

Now we are in a position to define a "Lagrange neighbour" for $\Delta < 0$.

Definition 2.2.2. Let $I = [Q/\sigma, (P + \sqrt{D})/\sigma]$ be a primitive ideal of $\mathcal{O}_\Delta$ for a discriminant $\Delta < 0$. The *Lagrange neighbour* of I is

$$I^+ = [Q^+/\sigma, (P^+ + \sqrt{D})/\sigma],$$

where

$$P^+ = -P + Ne(P/Q)Q,$$

and

$$Q^+ = N(P^+ + \sqrt{D})/Q.$$

[(2.1.13)] Exercises 2.1.18–2.1.21 all appear in [235]. We will have occasion to reference the above in Chapter Seven, when we look at some deeper considerations involving consecutive powers of the Q_i.

Alternatively, as with the real case in the preceding section, we can write

$$I = [Q, (P + \sqrt{\Delta})/2],$$

in which case

$$I^+ = [Q^+, (P^+ + \sqrt{\Delta})/2],$$

where

$$P^+ = -P + 2QNe(P/(2Q))$$

and,

$$Q^+ = N(P^+ + \sqrt{\Delta})/(4Q).$$

We now show that $I \sim I^+$. By Theorem 1.2.2, we have that $I = [Q/\sigma, (-P^+ + \sqrt{D})/\sigma]$. Now the reader may verify easily that $(\alpha')I = (Q)I^+$ where $\alpha = (-P^+ + \sqrt{D})/\sigma$, i.e. $I \sim I^+$, (see Exercise 1.3.7(a)).

Now we are in a position to give the promised analogue of the continued fraction algorithm, for complex quadratic orders, namely

Algorithm 2.2.1.

(a) *Let $\Delta_0 < 0$ be a fundamental discriminant with associated radicand D, and let $I = [Q/\sigma, (P + \sqrt{D})/\sigma]$ be a primitive ideal of $\mathcal{O}_\Delta$ (with $Q > 0$).*

(b) *Let $P = P_0$, $Q = Q_0$ and compute the following for $i \geq 0$:*

$$a_i = Ne(P_i/Q_i),$$

$$P_{i+1} = a_i Q_i - P_i,$$

and

$$Q_{i+1} = (P_{i+1}^2 - D)/Q_i.$$

(c) *If $I_{i+1} = [Q_i/\sigma, (P_i + \sqrt{D})/\sigma]$ and $Q_{i+1} \geq Q_i$, then I_{i+1} is a reduced ideal of $\mathcal{O}_\Delta$.* [(2.2.1)]

Proof. From (b) the I_i $(i \geq 0)$ are ideals, and $I_{i+1} = I_i^+$ implies, by the preceding discussion, that $I_i \sim I_{i+1}$. Also, since $I_{i+1} = [Q_i/\sigma, \beta_i]$ where $\beta_i = (-P_{i+1} + \sqrt{D})/\sigma$ (by the same discussion), then $|\text{Tr}(\beta_i)| = 2|P_{i+1}|/\sigma = 2|a_i Q_i - P_i|/\sigma = 2Q_i|a_i - P_i/Q_i|/\sigma \leq Q_i/\sigma = N(I_{i+1})$. By Theorem 1.4.2(b), I_{i+1} is reduced if $N(\beta_i) \geq N(I_{i+1})^2$, i.e. if $Q_{i+1} \geq Q_i$. □

The question now arises: Do we necessarily get an i (eventually) such that I_{i+1} is reduced, i.e. such that $Q_{i+1} \geq Q_i$? The answer is provided by Exercise 2.2.1, namely there always is such an i. Therefore, as we have demonstrated, this is a clear analogue of continued fraction algorithm for $\Delta > 0$ described in the last section.

One final important fact is

Theorem 2.2.1. *If $\Delta < 0$ is a discriminant, then every ambiguous class of C_Δ has an ambiguous ideal in it.*

[(2.2.1)] The intention of the authors of [37] was to use Algorithm 2.2.1 to describe a secure key-exchange system (see Chapter Eight, section two).

Proof. By Exercise 2.2.2, every class of C_Δ has a strictly primitive ideal with $N(I) < M_\Delta$. Let $I = [N(I), (b + \sqrt{\Delta})/2]$ be in an ambiguous class of C_Δ with $N(I) < M_\Delta$. If I is not ambiguous, then by Theorem 1.5.2, $2N(I) + b$ divides Δ. Also, by Theorem 1.2.2, $I = [N(I), (b + 2N(I) + \sqrt{\Delta})/2]$, and by Exercise 1.3.7(a), $I \sim [b + 2N(I), (-b - 2N(I) + \sqrt{\Delta})/2] = J = J'$. Hence, $\{I\}$ has an ambiguous ideal in it. □

Exercises 2.2

1. If I is as given in Algorithm 2.2.1, prove that $Q_{i+1} \geq Q_i$ for some $i \leq 2 + \frac{1}{2}\log_2(3Q_0/(5\sqrt{|D|}))$. (*Hint*: Define $\rho_j = Q_j/\sqrt{|D|}$ and $K_j = (5 \cdot 4^j + 1)/3$. Prove that for $m = \lfloor \frac{1}{2}\log_2(3\rho_0/5)\rfloor + 1$, we have $K_m > \rho_0$ and conclude that, for some $i \leq 1 + \frac{1}{2}\log_2(3Q_0/(5\sqrt{|D|}))$, we have $Q_i < 2\sqrt{|D|}$. Now assume $Q_{i+1} < Q_i$ and manufacture an ideal $I_{i+1} = [Q_i/\sigma, \gamma]$ whose Lagrange neighbour is reduced.)

2. (a) Prove that Algorithm 2.2.1 is valid for an arbitrary complex quadratic order if the word *proper* is added to the condition on the ideal.

   (b) Use the new generalized algorithm to verify that in Example 1.5.3, the ideal I has an ideal of norm 41 in its class and the ideal J has an ideal of norm 43 in its class but $I \not\sim J$, (see Exercise 1.5.12).

   (c) Use the results of this section to prove that in an arbitrary quadratic order there always exists a reduced ideal in each class.

   (d) Prove that under Definition 1.5.1, for $\Delta < 0$ a discriminant, C_Δ is generated by those invertible ideals I with $N(I) < M_\Delta$ such that $(\Delta/p) \neq -1$ for all primes $p \mid N(I)$. (*Hint*: Use part (c) and Theorem 1.4.2(c) proved for arbitrary orders. Then let $I_j = [N(I_j), (b_j + \sqrt{\Delta})/2]$ for $j = 1, \ldots, h_\Delta$ be ideals of $\mathcal{O}_\Delta$ such that $N(I_j) < M_\Delta$, and which form distinct classes. Then use the fact that $N(I_j) \mid N((b_j + \sqrt{\Delta})/2)$.)

   (e) Find an analogue of (d) for $\Delta > 0$.

Chapter 3

Diophantine Equations and Class Numbers.

The purpose of this chapter is to use the material in the previous two chapters to describe how solutions to Diophantine equations can have some remarkable consequences for class numbers of quadratic fields. Conversely, divisibility properties for class numbers of quadratic fields have some consequences for the solutions of Diophantine equations.

3.1 Class Numbers and Complex Quadratics.

Solutions of Diophantine equations and applications to class numbers have enjoyed long-standing interest. We will cite numerous authors who were looking, for example, at conditions for the existence of cyclic subgroups of C_Δ for fundamental discriminants Δ. This involved solutions to certain Diophantine equations. For instance, to illustrate the power of the process, we have (see Chapter One, section five)

Theorem 3.1.1. (3.1.1) *Let Δ be a discriminant with radicand $D = b^2 - \sigma^2 m^t < 0$ where $t, m, b \in \mathbb{Z}$, with $t > 1$, $m > 1$, $b > 0$ and $\gcd(m, f_\Delta) = 1$. If t is even, then $t/2$ divides h_Δ, and if $b \neq 2m^{t/2} - 1$, then t divides h_Δ. If t is odd, and $b \neq \lfloor \sigma m^{t/2} \rfloor$, then t divides h_Δ.*

Proof. First we establish two claims.

Claim 1. Either $N(w_\Delta)^2 > N((b + \sqrt{D})/\sigma)$, or else $b = 2m^{t/2} - 1$.

If
$$N(w_\Delta)^2 \leq N((b + \sqrt{D})/\sigma),$$
then
$$b^2 \geq (D + (\sigma - 1))^2/\sigma^2,$$
i.e. $b \geq \sigma m^{t/2} - 1$. However, $b < \sigma m^{t/2}$, so $b = \lfloor \sigma m^{t/2} \rfloor$. If t is odd, then this contradicts the hypothesis, so $b = \sigma m^{t/2} - 1$. If $\sigma = 1$, then
$$N(w_\Delta)^2 = D^2 = 4m^t - 4m^{t/2} + 1 > N(b + \sqrt{D}) = m^t.$$

(3.1.1) This is [236, Theorem 3.1].

This establishes Claim 1.

Form the ideals $I^c = [m^c, (b+\sqrt{D})/\sigma]$ where $1 \le c \le t$.

Claim 2. $I^g \sim 1$ where $g = \gcd(t, h_\Delta)$.

There exist integers u and v such that $g = tu + h_\Delta v$. Therefore, $I^g = I^{tu+h_\Delta v} \sim (I^t)^u(I^{h_\Delta})^v \sim 1$ (since $N(I^t) = m^t = N((b+\sqrt{D})/\sigma)$ and $I^{h_\Delta} \sim 1$ by Exercise 1.5.13). This secures Claim 2.

By Exercise 1.5.7 and Claims 1–2, we conclude that $g = t$, i.e. $t \mid h_\Delta$ unless $b = 2m^{t/2} - 1$. In the latter case, $D = 1 - 4m^{t/2}$. In this instance, we form the ideal $J = [m^{g_1}, (1+\sqrt{D})/2]$ where $g_1 = \gcd(t/2, h_\Delta)$. We then use the same reasoning as above to conclude that $J \sim 1$. Moreover, since $m^{t/2} = N((1+\sqrt{D})/2) < N(w_\Delta)^2 = m^t$, then by Exercise 1.5.7(a), $g_1 = t/2$, i.e. $t/2$ divides h_Δ. □

Theorem 3.1.1 has numerous consequences. For example,

Corollary 3.1.1. [(3.1.2)] *Let $\Delta = D = 1 - 4m^s$ be a discriminant with $m > 1$ and s prime. If* $\gcd(m, f_\Delta) = 1$, *then $s \mid h_\Delta$.*

Corollary 3.1.2. [(3.1.3)] *Let $\Delta = b^2 - 4m^t \equiv 1 \pmod 4$ be a negative discriminant where m and t are odd primes. If one of the primes over m is not principal in $\mathcal{O}_\Delta$, and* $\gcd(m, f_\Delta) = 1$, *then $t \mid h_\Delta$.*

Proof. By Theorem 3.1.1 we need only consider the case where $b = \lfloor 2m^{t/2} \rfloor$. Thus, $I^t = [m^t, (b+\sqrt{D})/2] \sim 1$, by Exercise 1.5.7(a), since $N((b+\sqrt{D})/2) = m^t < N(w_\Delta)^2$. If t does not divide h_Δ then, by Exercise 1.5.13, $I \sim 1$ contradicting the hypothesis. □

Corollary 3.1.1 is actually a special case of a more general phenomenon, namely

Corollary 3.1.3. [(3.1.4)] *Let $\Delta = b^2 - 4m^t \equiv 1 \pmod 4$ be a negative discriminant with* $\gcd(m, f_\Delta) = 1$, *$m > 1$ and $t > 1$. If m^c is not the norm of a primitive, principal $\mathcal{O}_\Delta$-ideal for any c properly dividing t, then $t \mid h_\Delta$.*

Proof. By Theorem 3.1.1 we need only handle the cases where $b = \lfloor 2m^{t/2} \rfloor$ when t is odd and $b = 2m^{t/2} - 1$ when t is even. If the latter case holds, then $J = [m^{t/2}, (1+\sqrt{D})/2] \sim 1$, by Exercise 1.5.7(a), (since $N((1+\sqrt{D})/2) = m^{t/2} < N(w_\Delta)^2 = m^t$), which contradicts the hypothesis. If $b = \lfloor 2m^{t/2} \rfloor$ then, as in the proof of Corollary 3.1.2, we have $I^t = [m^t, (b+\sqrt{D})/2] \sim 1$. If $g = \gcd(h_\Delta, t)$ then, as in the proof of Theorem 3.1.1, $I^g \sim 1$. Thus, $g = t$. □

Corollaries 3.1.2–3.1.3 have an unnecessarily strong hypothesis. What the proof of Theorem 3.1.1 shows is that, with the exception of the cases where $b = \lfloor \sigma m^{t/2} \rfloor$ (t odd) or $b = 2m^{t/2} - 1$ (t even), I^c *cannot* be principal for any integer c with $1 \le c < t$.

Another consequence of Theorem 3.1.1 is

Corollary 3.1.4. [(3.1.5)] *Let Δ be a discriminant with radicand $D = b^2 - \sigma^2 m^t < 0$*

[(3.1.2)] This is proved by Gross and Rohrlich in [100] for the maximal order.

[(3.1.3)] This is proved by Cowles in [66] for the maximal order.

[(3.1.4)] This was proved in [226, Theorem 2.1, p. 15] for the maximal order.

[(3.1.5)] This is proved in [236] as a generalization of the weaker result [226, Corollary 2.4, p. 15].

where $t > 1$, $m > 1$, $b > 0$, *and* $\gcd(m, f_\Delta) = 1$. *If* $b^2 \leq \sigma^2 m^t - 2\sigma^2 m^{t/2} + \sigma - 1$, *then* $t \mid h_\Delta$.

Proof. If $b = \lfloor \sigma m^{t/2} \rfloor$, then

$$(\sigma m^{t/2} - 1)^2 \leq b^2 \leq \sigma^2 m^t - 2\sigma^2 m^{t/2} + \sigma - 1,$$

so

$$2\sigma^2 m^{t/2} \leq 2\sigma m^{t/2} + \sigma - 2,$$

a contradiction. If $b = 2m^{t/2} - 1$, then

$$(2m^{t/2} - 1)^2 = b^2 \leq 4m^t - 8m^{t/2} + 1$$

which leads to $-4m^{t/2} \leq -8m^{t/2}$, a contradiction. □

Corollary 3.1.4 has the nice property that we need only check an inequality in order to give divisibility properties on the class number. Therefore, even the knowledge of the definition of a class number is not needed to perform the test. For example,

Example 3.1.1. Let $D = 34933 - 4 \cdot 15^{13} = -3662498011 = -61 \cdot 60040951$ with $h_\Delta = 12714 = 13 \cdot 978$. Here $t = 13$, $b = 34933$, and $m = 15$ with $b^2 \leq 4m^t - 8m^{t/2} + 1$.

An illustration of Theorem 3.1.1 for the case of $b = 2m^{t/2} - 1$ is given by

Example 3.1.2. Let $D = 249^2 - 4 \cdot 5^6 = -499$ where $b = 249 = 2m^{t/2} - 1 = 2 \cdot 5^3 - 1$. Here $h_\Delta = 3 = t/2$.

An illustration of Theorem 3.1.1 for even t, with $b \neq 2m^{t/2} - 1$ is

Example 3.1.3. If $D = 247^2 - 4 \cdot 5^6$, then $b = 247 = 2m^{t/2} - 3 = 2 \cdot 5^3 - 3$, and $h_\Delta = 12$.

To show that Theorem 3.1.1 fails when $\gcd(m, f_\Delta) > 1$, we have

Example 3.1.4. If $D = 245^2 - 4 \cdot 5^6$, then $m = 5$, $t = 6$, and $f_\Delta = 15$. However, $h_\Delta = 8$ so $t/2 = 3$ does not divide h_Δ. (See Exercise 1.5.7.)

Remark 3.1.1. It should be remarked that Theorem 3.1.1 speaks about divisiblity of class numbers but says little (that cannot be determined easily by other methods) about the actual solutions of the Diophantine equations. The reasons for this are as follows. Consider a radicand $D < 0$, and the equation

$$D = b^2 - \sigma^2 m^t \tag{3.1.1}$$

where b and t are unknowns.

Suppose that

$$t \geq 2\log((1 - D)/(2\sigma))/\log m, \tag{3.1.2}$$

then it is not difficult to see that $-1 \leq b - \sigma m^{t/2} < 0$, i.e. either $b = \lfloor \sigma m^{t/2} \rfloor$, or $b = \sigma m^{t/2} - 1$. Thus, if (3.1.2) holds, and t is odd for example, then the solvability of (3.1.1) is equivalent to the solvability of $D = \lfloor \sigma m^{t/2} \rfloor^2 - \sigma^2 m^t$.

Let us look, for instance, at the classical Diophantine equation $x^2 + 7 = 2^s$, called the Ramanujan–Nagell equation[(3.1.6)] which is known to have solutions for only $s = 3, 4, 5, 7, 15$. If we write $D = -7 = b^2 - 4 \cdot 2^{s-2}$ with $t = s - 2$, then Theorem 3.1.1 tells us that t cannot be even unless $t = 2$, since $h_{-7} = 1$. However, we don't need Theorem 3.1.1 to tell us that $s = 4$ is the only even solution since $-7 = (x - 2^{s/2})(x + 2^{s/2})$, forcing $x - 2^{s/2} = -1$, and $x + 2^{s/2} = 7$, i.e. $s = 4$ and $x = 3$. In fact, we see that (3.1.2) is satisfied if $t \geq 2$. Thus, for $s > 4$, the solutions are given by $-7 = \lfloor 2^{s/2} \rfloor^2 - 2^s$, and Theorem 3.1.1 gives us no information about these solutions.

In view of Remark 3.1.1, it would be desireable to have a result that tells us something about the solutions of such Diophantine equations[(3.1.7)] as $x^2 - D = p^n$ for a prime p and a radicand $D < 0$, (see footnote 3.1.6). The existence of solutions to these generalized Ramanujan–Nagell equations often dictates the form of n and D as in the following result.

Theorem 3.1.2. [(3.1.8)] *Let $q > 2$ be a prime not dividing a given $A \in \mathbb{Z}$ with $A \equiv 3 \pmod 8$ positive, and let d be the least positive integer such that $a^2 + Ab^2 = 4q^d$, for some odd positive $a, b \in \mathbb{Z}$. The Diophantine equation $x^2 + A = q^n$ has a positive solution $(x, n) \in \mathbb{Z}^2$ if and only if $b = 1$ and $A = 3a^2 \pm 8$. The unique solution is given by $x = |a(a^2 - 3A)/8|$ and $n = 3d$.*

Proof. Since d is the least positive integer such that $a^2 + Ab^2 = 4q^d$ for some odd positive $a, b \in \mathbb{Z}$, then d is the order of the $\mathcal{O}_\Delta$-ideal $I = [q, (a + \sqrt{\Delta})/2]$

(3.1.6) Ramanujan [300] knew of these solutions, and Nagell [277] in 1948 proved that these are the *only* solutions. Nagell used the fact that $h_\Delta = 1$ for $\Delta = -7$ to prove Ramanujan's conjecture by explicitly using the unique factorization of algebraic integers in $\mathbb{Q}(\sqrt{-7})$. Skolem, Chowla and Lewis proved it in [341] over ten years later using Skolem's p-adic methods. See also Ramasamy [301] for a survey together with an extensive list of references. See also the related work on $y^2 - D = 2^k$ by Tzanakis [368]. The more general equation is $x^2 - D = 2^n$, called the *generalized Ramanujan–Nagell equation*, which Apéry [7] proved (for $D < 0$) has at most two solutions. Browkin and Schinzel [34] conjectured that it has two solutions if and only if $D = -23$ or $1 - 2^k$ for some $k > 3$. This was proved by F. Beukers in [26]–[27]. He also dealt with the case where $D > 0$ therein, proving that this equation has at most four solutions. For the case where $x^2 - D = p^n$ for $p > 2$ a prime, F. Beukers solved the problem in [26]–[27]. Beukers' methods involved hypergeometric polynomials. There is also a very good survey article by E.L. Cohen [55].

(3.1.7) This was solved by Alter and Kubota for $D = -11$ and $p = 3$ in [5]. An alternative proof of this fact was provided by E.L. Cohen in [54]. There is also the generalization by J.H.E. Cohn [62]. For the case where $p = 3$, see recent work in 1993 of J.H.E. Cohn [63]. In a private communication to this author, Dr. Cohn wrote: "I wrote the paper about $x^2 + 3 = y^n$ entirely in good faith, thinking that it was original work. Presumably the referee thought so too, but I was wrong. Ezra Brown proved the same result in 1975; he too had been preceded by T. Nagell as long ago as 1923!" This shows that in mathematical research it is easy and, as it turns out, quite common to "rediscover" facts overlooked in the literature. This, however, should not detract from the enjoyment of the pursuit since, after all, the publishing of a discoverd fact is merely our attempt to share that discovery with others. Thus, for example, Dr. Cohn's efforts have not been in vain, for his work has made us more aware of the earlier efforts and the inherent attractiveness of this subject matter.

(3.1.8) This was proved by Alter and Kubota in [4] for the case where A is square-free, and it was proved for the case where $d = 1$ and A is not necessarily square-free by Beukers [26]. The case where $A \equiv 7 \pmod 8$ is square-free was also covered by Alter and Kubota (see Exercise 3.1.6). Alter and Kubota employed ideal-theoretic arguments as well as methods of Skolem *et al.* in [4], wherein they comment at one point that they assume A to be square-free in order to represent a certain ideal power in $\mathcal{O}_\Delta$. Our methods (Using Lucas–Lehmer theory, see Exercise 3.1.5.) show that this is not necessary. In [188]–[190], W. Ljunggren considered the equation $Cx^2 + D = y^n$ where C, D and n are fixed positive rational integers and the product CD is square-free, and related such equations. See also the recent work of Le Maohua [209].

in C_Δ, where $\Delta = -Ab^2$. If $x^2 + A = q^n$ for some positive $x, n \in \mathbf{Z}$, then $q = (x+\sqrt{-A})(x-\sqrt{-A})$, so $(q)^n = I^n I'^n \sim (bx+\sqrt{\Delta})(bx-\sqrt{\Delta})$ in $\mathcal{O}_\Delta$. Thus, $I \sim (bx+\sqrt{\Delta}) \sim (bx-\sqrt{\Delta}) \sim I'$, i.e. $I \sim 1$. By Exercise 1.5.13, we have $d \mid n$.

We now invoke Exercise 3.1.5 with $\Delta = -Ab^2$, $R = a^2$, $Q = q^d$, $V_1 = a$ and $U_1 = 1$. In this situation, we have $q^d = N((V_1+\sqrt{\Delta})/2)$, so $q^n = N([(V_1+\sqrt{\Delta})/2]^{n/d})$, i.e. $(V_{n/d} + U_{n/d}\sqrt{\Delta})/2 = \pm(x+\sqrt{-A})$. Hence, $bU_{n/d} = \pm 2$, and $V_{n/d} = \pm 2x$. By Exercise 3.1.5(e), $n/d \equiv 0 \pmod 3$, so $U_3 \mid U_{n/d}$, by Exercise 3.1.5(g). However, $U_3 = (3a^2 - Ab^2)/4$. If $b = 2$, then $(3a^2-4A)/4 = \pm 1$, i.e. $3a^2 = 4(A \pm 1)$, contradicting that a is odd. Therefore, $b = 1$, and $(3a^2 - A)/4$ divides ± 2. If $(3a^2-A)/4 = \pm 1$, then $3a^2 \equiv 7 \pmod 8$, i.e. $a^2 \equiv 5 \pmod 8$, a contradiction. It follows that

$$A = 3a^2 \pm 8. \tag{3.1.3}$$

Now we consider $(V_{n/d} + U_{n/d}\sqrt{\Delta})/2 = ((V_{n/3d} + U_{n/3d}\sqrt{\Delta})/2)^3 = \frac{1}{8}[V_{n/3d}^3 - 3U_{n/3d}^2 V_{n/3d} A + (3V_{n/3d}^2 U_{n/3d} - U_{n/3d}^3 A)\sqrt{\Delta}]$. Hence, $(3V_{n/3d}^2 U_{n/3d} - U_{n/3d}^3 A)/4 = U_{n/d}$, and $V_{n/d} = (V_{n/3d}^3 - 3U_{n/3d} V_{n/3d} A)/4$, i.e.

$$3V_{n/3d}^2 U_{n/3d} - U_{n/3d}^3 A = \pm 8, \tag{3.1.4}$$

and

$$V_{n/3d}^3 - 3U_{n/3d}^2 V_{n/3d} A = \pm 8x. \tag{3.1.5}$$

From (3.1.4), $U_{n/3d} = \pm 1$ or ± 2. If $U_{n/3d} = \pm 1$, then (3.1.4) becomes

$$3V_{n/3d}^2 - A = \pm 8. \tag{3.1.6}$$

In view of (3.1.3), we must have $V_{n/3d} = a = V_1$. Also, since $U_{n/3d} = -1$ is not possible, then $U_{n/3d} = 1 = U_1$. Therefore, $n = 3d$. Furthermore, from (3.1.5), $x = |a(a^2-3A)/8|$, as required.

If $U_{n/3d} = \pm 2$, then (3.1.3) forces (3.1.4) to become $(V_{n/3d}/2)^2 - a^2 = \pm 3$, for which only $a = 1$ is possible. From (3.1.3), we get $A = 11$, $q^d = 3$, $n = 3$, and $x = 4$, as required.(3.1.9) □

Some illustrations of Theorem 3.1.2 are

Example 3.1.5. Let $A = 315 = 3^2 \cdot 5 \cdot 7$ and look at $x^2 + 315 = 11^n$. Since the smallest solution to $a^2 + 315 = 4 \cdot 11^d$ is for $a = 13$ and $d = 2$, then Theorem 3.1.2 tells us that the Diophantine equation has no solutions.

Example 3.1.6. If $A = 2970067$, and $x^2 + A = q^n$ for $q = 990023$, then the smallest d for which $a^2 + A = 4q^d$ is $d = 1$ where $a = 995$. Therefore, $n = 3$ and $x = |a(a^2-3A)/8| = 985071890$ with $q^3 = 970366628471142167$.

Example 3.1.7. If $A = 41075 = 5^2 \cdot 31 \cdot 53$ and we consider $x^2 + A = 13691^n$ for $q = 13691$, then the smallest d for which $a^2 + A = q^d$ is $d = 1$ where $a = 117$. Since $A = 3 \cdot 117^2 + 8$, the unique solution occurs for $n = 3$ and $x = |a(a^2-3A)/8| = 1601964$.

(3.1.9) What we have implicitly established in particular is the solution of the Diophantine equation, $x^2 + 11 = 3^n$, discussed in footnote (3.1.7).

Special cases of the Ramanujan–Nagell equation are also interesting and have other applications. For example, consider the equation $x^2 = 4q^n - 4q + 1$ where q is any odd prime (and if $q = 2$, we have the Ramanujan–Nagell equation).

Theorem 3.1.3. (3.1.10) *Let $D = 1 - 4q$ where q is an odd prime. If $q > 5$, then the only solutions to $x^2 - D = 4q^n$ occur when $n = 1$ or 2. If $q \le 5$, then:*

(a) *If $q = 3$, then $(x, n) \in \{(1,1), (5,2), (31,5)\}$.*

(b) *If $q = 5$, then $(x, n) \in \{(1,1), (9,2), (559,7)\}$.*

Proof. If n is even, then $n = 2m$. Thus, $4q - 1 = (2q^m - x)(2q^m + x) > 2q^m$, forcing $n = 2$. Now assume that n is odd and invoke Exercise 3.1.5, with $\Delta = 1-4q$, $R = 1$, $Q = q$, $U_1 = V_1 = 1$, then

$$w_\Delta^n = [(1 + \sqrt{\Delta})/2]^n = (V_n + U_n\sqrt{\Delta})/2 = \pm(x + \sqrt{\Delta})/2.$$

Hence, $U_n = \pm 1$ and $V_n = \pm x$. However, $U_n \ne -1$ for any $n \ge 1$ (since $U_n \equiv 1 \pmod q$ by a simple induction), so we assume that $U_n = 1$ and $V_n = x$. A simple induction argument shows that $U_i \equiv 1 \pmod{q-1}$ if $i \equiv 1, 2 \pmod 6$, $U_i \equiv 0 \pmod{q-1}$ if $i \equiv 3 \pmod 6$, and $U_i \equiv -1 \pmod{q-1}$ if $i \equiv 4, 5 \pmod 6$. Thus, if $q > 3$, then the fact that $U_n = 1$ forces $n \equiv 1, 2 \pmod 6$. We have already dealt with the case where n is even, so $n \equiv 1 \pmod 6$. Set $n = 3r + 1$, so we have that $1 = U_n = (w_\Delta^n - w_\Delta'^n)/(w_\Delta - w'_\Delta)$, so $\sqrt{\Delta} = w_\Delta w_\Delta^{3r} - w'_\Delta w_\Delta'^{3r}$. However, an easy check shows that $w_\Delta^3 = U_3(w_\Delta + 1) - 1 = (1-q)(w_\Delta + 1) - 1$. Since r is even, we may write

$$\sqrt{\Delta} = w_\Delta((q-1)(w_\Delta + 1) + 1)^r - w'_\Delta((q-1)(w'_\Delta + 1) + 1)^r$$
$$= \sqrt{\Delta} + \sum_{k=1}^{r} \binom{r}{k} (q-1)^k [w_\Delta(w_\Delta + 1)^k - w'_\Delta(w'_\Delta + 1)^k].$$

Thus, since Exercise 3.1.7 tells us that $\binom{r}{k} = \frac{r}{k}\binom{r-1}{k-1}$ (when $r > 0$), and

$$w_\Delta(w_\Delta + 1)^k - w'_\Delta(w'_\Delta + 1)^k = \sqrt{\Delta}\sum_{j=0}^{k}\binom{k}{j}U_{j+1},$$

then

$$r\sum_{k=1}^{r}\binom{r-1}{k-1}/k(q-1)^{k-1}(\sum_{j=0}^{k}\binom{k}{j}U_{j+1}) = 0.$$

By Exercise 3.1.7(d), $(q-1)/2$ divides all terms if $r > 0$. However, the first term is $\sum_{j=0}^{1}\binom{k}{j}U_{j+1} = U_1 + U_2 = 2$, a contradiction, unless $q \le 5$. Now we verify the last part of the theorem.

The result will ensue from the following claims.

Claim 1. If $n_1 \ne n_2$ are solutions of $x_i^2 - \Delta = 4q^{n_i}$ for $i = 1, 2$ where $|\Delta| = p > 3$, a prime, then $n_1 \not\equiv n_2 \pmod{\phi(p^2)}$.

(3.1.10)This was proved in [340] by C. Skinner who used only elementary methods. He says therein: "Though Ramanujan introduced his problem as essentially one in number theory, it has turned out to have applications in such diverse areas as coding theory (see Shapiro–Slotnik [336]) and differential algebra (see Mead [216]). It is possible that the equation discussed in this paper may have similar applications in these areas." See also the related work on $y^2 = 4q^n + 4q + 1$ by Tzanakis and Wolfskill [369] as well as their sequel [370].

Suppose, to the contrary, that $n_1 \equiv n_2 \pmod{\phi(p^{e+1})}$ where $n_1 > n_2$, and $e \geq 1$ is the *highest* such power of p. Since $2^{\phi(p^{e+1})} \equiv 1 \pmod{p^{e+1}}$, by Euler's Theorem (see footnote (1.6.19)), then $2^{n_1 - n_2} \equiv 1 \pmod{p^{e+1}}$. We leave the reader to verify

Claim 2. $(1+\sqrt{\Delta})^{n_1-n_2} = 2^{n_1-n_2} w_\Delta^{n_1-n_2} \equiv 1 + (n_1 - n_2)\sqrt{\Delta} \pmod{p^{e+1}}$.

We now show that Claim 2 leads to a contradiction. First, we observe that $2^{n_1-n_2} w_\Delta^{n_1-n_2} = 2^{n_1-n_2-1}(V_{n_1-n_2} + U_{n_1-n_2}\sqrt{\Delta})$. Thus, if we show that $V_{n_1-n_2} \equiv 2 \pmod{p^{e+1}}$, and $U_{n_1-n_2} \equiv 0 \pmod{p^{e+1}}$, then we have our contradiction. We now establish these two facts. Since $x_1^2 - x_2^2 = 4(3^{n_1} - 3^{n_2}) = 4 \cdot 3^{n_2}(3^{n_1-n_2} - 1) \equiv 0 \pmod{p^{e+1}}$, and an easy check shows that p does not divide $x_1 + x_2$ (using Exercise 3.1.5(b)), then $x_1 \equiv x_2 \pmod{p^{e+1}}$. Also, by the aforementioned exercise, $2 \cdot 3^{n_2} V_{n_1-n_2} = x_1 x_2 + 11 \equiv x_2^2 + 11 = 4 \cdot 3^{n_2} \pmod{p^{e+1}}$. Hence, $V_{n_1-n_2} \equiv 2 \pmod{p^{e+1}}$. Also, by the aforementioned exercise, $2 \cdot 3^{n_2} U_{n_1-n_2} = x_1 - x_2$. Therefore, $U_{n_1-n_2} \equiv 0 \pmod{p^{e+1}}$. We have verified Claim 1.

We leave the reader with the task of using Exercise 3.1.5 to verify the following:

Claim 3. $2^{n-1} \equiv n \pmod{p}$.

Claim 4. $n \equiv 2 \pmod{q}$, if $n > 1$.

From Claims 1, and 3–4, it is now merely a computer check for $q = 3, 5$, since any solutions n are incongruent modulo $\phi(p^2)$, and are of the form $n = qk + 2$. □

Exercises 3.1

1. (a) Prove that if the Diophantine equation $-7 = \lfloor 2^{n/2} \rfloor^2 - 2^n$ has solutions for $n > 3$, then $U_k + 2V_k = 2^{(n-1)/2}$ for some even $k \in \mathbb{Z}$ where $(1 + \sqrt{2})^k = V_k + U_k\sqrt{2}$, and $|V_k + 4U_k| = \lfloor 2^{n/2} \rfloor$.

   (b) Prove that the Diophantine equation $-3 = x^2 - 13^n$ has no solutions for $n \in \mathbb{N}$ (*Hint*: Look at the equation modulo 7.)

2. Prove that if $\Delta = b^2 - m^t < 0$, $\Delta \equiv 1 \pmod 4$ is a discriminant with $\gcd(m, f_\Delta) = 1$ and $b \neq \lfloor m^t \rfloor$ or $m^{t/2} - 1$, then t divides h_Δ. (*Hint*: Use the same methodology as in the proof of Theorem 3.1.1.)

3. Let $q > 2$ be a prime. Prove that the equation $x^2 + 11 = q^n$ has only the solution $x = 4$. (*Hint*: Use Theorem 3.1.2.)

4. Prove that if $D = 4 - x^t$, $t > 2$ is a radicand, then $t \mid h_\Delta$.

5. (a) Let α and β be the roots of the equation $x^2 - \sqrt{R}x + Q = 0$ for relatively prime integers R and Q. Verify that $\alpha + \beta = \sqrt{R}$, $\alpha\beta = Q$ and $\alpha - \beta = (R - 4Q)^{1/2}$. Set $\delta = \Delta^{1/2} = (R-4Q)^{1/2}$ and show that $2\alpha = R^{1/2} + \delta = R^{1/2} + (R - 4Q)^{1/2}$ and $2\beta = R^{1/2} - \delta = R^{1/2} - (R - 4Q)^{1/2}$.

(b) Let $U_n = (\alpha^n - \beta^n)/(\alpha - \beta)$ and $V_n = \alpha^n + \beta^n$ ($n \in \mathbf{Z}$ non-negative). These are called *Lucas functions*,$^{(3.1.11)}$ (see Lucas [206]).
Verify that

$$U_{n+2} = R^{1/2}U_{n+1} - QU_n,$$

and

$$V_{n+2} = R^{1/2}V_{n+1} - QV_n.$$

Deduce that $U_{2n+1}, V_{2n} \in \mathbf{Z}$, while U_{2n} and V_{2n+1} are integral multiples of $R^{1/2}$. Then prove

$$\begin{aligned} &V_n^2 - \Delta U_n^2 = 4Q^n, \\ &2U_{n+m} = U_n V_m + V_n U_m, \\ &2V_{n+m} = V_m V_n + \Delta U_n U_m, \\ &2Q^m U_{n-m} = U_n V_m - V_n U_m, \\ &2Q^m V_{n-m} = V_n V_m - \Delta U_n U_m, \end{aligned}$$

$$2^{n-1}U_n = \sum_{k=1}^{\lfloor (n+1)/2 \rfloor} \binom{n}{2k-1} V_1^{n-2k+1}\Delta^{k-1}$$

and

$$2^{n-1}V_n = \sum_{k=0}^{\lfloor n/2 \rfloor} \binom{n}{2k} V_1^{n-2k}\Delta^{k}.$$

In what follows, when discussing divisibility properties, to avoid confusion we assume that a factor of $\sqrt{R}$ may be ignored in U_n or V_n. For example, if $R = 5$, $Q = -3$ then $U_3 = 8$ and $U_6 = 112\sqrt{5}$. We say that $\gcd(U_3, U_6) = 8$, since we ignore $\sqrt{5}$. Also, U_6 may be called even, since we ignore $\sqrt{5}$.

(c) Prove that $\gcd(U_n, Q) = 1 = \gcd(V_n, Q)$ and that $\gcd(U_n, V_n)$ divides 2.

(d) Prove that if U_n is even, then one of the following must hold:

$$R \equiv 0 \pmod 4, \quad Q \text{ is odd and } n \text{ is even},$$

or

$$R \equiv 2 \pmod 4, \quad Q \text{ is odd and } n \equiv 0 \pmod 4,$$

or

$$R \text{ is odd}, \quad Q \text{ is odd and } n \equiv 0 \pmod 3.$$

(e) Prove that if V_n is even, then one of the following must hold:

(1) $R \equiv 0 \pmod 4$ and Q is odd.

$^{(3.1.11)}$The area which studies the properties of these functions is now called *Lucas–Lehmer theory* in honour of the contributions of the late Dick Lehmer. The reader who is interested in pursuing deeper aspects of the theory and who is interested in other aspects of his work is advised to get the collected works which appear in three volumes [172]. Derrick Henry Lehmer (known to his friends as Dick) was born February 23, 1905 in Berkeley, California. His first degree was achieved from the University of California at Berkeley in 1927, and he received a Sc.M. degree from Brown University in 1929. As for his work, we quote from the "Forward" by John Selfridge in volume I of the aforementioned collected works: "Yes, Dick has looked at beautiful parts of number theory, combinatorics and computer science, and has shown us this beauty with the sure hand of a master." Dick Lehmer died May 22, 1991 in his eighty-sixth year.

(2) $R \equiv 2 \pmod 4$, Q is odd and n is even.
(3) R and Q are odd and $n \equiv 0 \pmod 3$.

(f) Prove that if $g = \gcd(m, n)$, then $\gcd(U_m, U_n) = U_g$.

(g) Prove that if $m \mid n$, then $U_m \mid U_n$. Furthermore, prove that if n/m is odd, then $V_m \mid V_n$.

(h) Let q be a prime and let w be the first positive integer such that U_w is divisible by q. If w exists (i.e. if q appears as a divisor in the sequence $\{U_n\}$), then it is called the ***rank of apparition*** of q. Prove that $q \mid U_n$ if and only if $n = kw$, for some positive integer k. (Thus, if $q = 2$, then part (d) gives w.)

(i) Use the binomial expansions in part (b) to establish that for an odd prime p, $U_p \equiv \varepsilon \pmod p$ where $\varepsilon = (\Delta/p)$ and $V_p \equiv \gamma R^{1/2} \pmod p$ where $\gamma = (R/p)$ is the Legendre symbol (see Exercise 1.1.4). Prove that:

(1) If p does not divide QR, then the rank of apparition of w is some divisor of $p - \gamma\varepsilon$.
(2) Show that if $p \mid Q$, then p does not divide U_m for any $m \geq 1$.
(3) If $p^2 \mid R$, then $w = 2$.
(4) If p^2 does not divide R but $p \mid R$, then $w = 2p$.
(5) If $p \mid \Delta$, then $w = p$.
(6) If w is odd, then p does not divide V_m for any $m \geq 1$.
(7) If $w = 2k$, then $p \mid V_{(2n+1)k}$ for every $n \geq 0$ but p does not divide V_m for $m \neq (2n+1)k$.

6. (a) Let q be an odd prime not dividing $A \in \mathbb{Z}$ with $A \equiv 7 \pmod 8$ positive, and let d be the least positive integer such that $a^2 + Ab^2 = q^d$ for some odd $a, b \in \mathbb{N}$. Prove that the Diophantine equation $x^2 + A = q^n$ has a solution if and only if $n = d$ and $b = 1$. (*Hint*: As in the proof of Theorem 3.1.2 we invoke Exercise 3.1.5.)

(b) Is the solution in (a) unique when it exists or can there exist others? If so, find the maximum number of solutions possible, and classify (in terms of the form for A) when the solutions are not unique.

7. The *Binomial Theorem* states that $(x+y)^n = \sum_{i=0}^{n} \binom{n}{i} x^{n-i} y^i$ for variables x, y and binomial coefficient $\binom{n}{i} = n!/[(n-i)!i!]$.

(a) Prove that $\sum_{i=0}^{\lfloor n/2 \rfloor} \binom{n}{2i} = 2^{n-1} = \sum_{i=1}^{\lfloor \frac{n+1}{2} \rfloor} \binom{n}{2i-1}$. Conclude that $\sum_{i=0}^{n} \binom{n}{i}$ $= 2^n$. (*Hint*: See Exercise 3.1.5(b).)

(b) If $\binom{r}{k}$ denotes the binomial coefficient, verify that $\binom{r}{k} = \frac{r}{k}\binom{r-1}{k-1}$ for $r > 0$.

(c) Prove that if $\Delta = 1 - 4q$ (q prime), then

$$w_\Delta(w_\Delta + 1)^k - w'_\Delta(w'_\Delta + 1)^k = \sqrt{\Delta} \sum_{j=0}^{k} \binom{k}{j} U_{j+1},$$

where $U_n = (w_\Delta^n - w_\Delta'^n)/\sqrt{\Delta}$ (see Exercise 3.1.5).

(d) Verify that $(q-1)^k/k \equiv 0 \pmod{(q-1)/2}$ if $k \geq 1$.

8. (a) Show that the Diophantine equation $-19 = \lfloor 2 \cdot 5^{n/2} \rfloor^2 - 4 \cdot 5^n$ has only $n = 1, 2$ or 7 as a solution.

   (b) Show that the equation $-11 = \lfloor 2 \cdot 3^{n/2} \rfloor^2 - 4 \cdot 3^n$ has only the solution $n = 5$. (*Hint*: Use Theorem 3.1.3.)

9. Let $\Delta = \Delta_0$ be a fundamental discriminant. Prove that the only solutions to $u_1 u_2 u_3 = u_1 + u_2 + u_3$ for $u_i \in U_\Delta$ $(i = 1, 2, 3)$ occur for $\Delta = -4, 5$ or 8 and that the solutions are given by the following chart (up to order).

Δ	(u_1, u_2, u_3)
-4	$\pm(i, i, -i)$ and $\pm(1, i, -i)$
8	$\pm(1+\sqrt{2}, 1+\sqrt{2}, 1); \pm(1-\sqrt{2}, 1-\sqrt{2}, 1)$ and $\pm(1+\sqrt{2}, 1-\sqrt{2}, -1)$
5	$\pm((1+\sqrt{5})/2, 2+\sqrt{5}, 1); \pm((1-\sqrt{5})/2, 2-\sqrt{5}, 1);$ $\pm((1+\sqrt{5})/2, 2-\sqrt{5}, -1)$ and $\pm((1-\sqrt{5})/2, 2+\sqrt{5}, -1)$

(*Hint*: Since the only non-trivial units for $\Delta < 0$ occur when $\Delta = -4$ or -3, then these are the only negative values which need to be checked, and $\Delta = -3$ is easily eliminated. If $\Delta > 0$, set $\epsilon_\Delta^t = (a_1 + b_1\sqrt{D})^t = a_t + b_t\sqrt{D}$, then the problem becomes solving either

$$\epsilon_\Delta^{t_1+t_2+t_3} = \epsilon_\Delta^{t_1} + \epsilon_\Delta^{t_2} + \epsilon_\Delta^{t_3},$$

or

$$\epsilon_\Delta^{t_1+t_2+t_3} = \epsilon_\Delta^{t_1} - \epsilon_\Delta^{t_2} - \epsilon_\Delta^{t_3},$$

(up to order). Let $\delta = \pm 1$ and verify that

$$[N(u_1) - \delta N(u_2) - \delta N(u_3) + N(u_4)]/2 = N(u_1)[a_{t_2}a_{t_3} + b_{t_2}b_{t_3}D] + a_{t_2}a_{t_3} - b_{t_2}b_{t_3}D$$

where $u_4 = u_1 u_2 u_3$. Deduce that, in order for solutions to occur, either $a_{t_2}a_{t_3} = 1$ or $b_{t_2}b_{t_3} = 0$. Conclude that the above chart must hold.)[(3.1.12)]

10. Prove that if

$$(x-y)^7 = x^5 - y^5 \tag{3.1.7}$$

where $x, y \in \mathbf{Z}$ and $x \neq y$, then there must exist a positive $m \in \mathbf{Z}$ with $F_{6m+3} = (2y+1)^2 + 1$, where F_i is the ith Fibonacci number. (*Hint*: Without loss of generality, assume that $x > y$ and let $d = \gcd(x, y)$ and $x = dX$, $y = dY$. Substituting into (3.1.7), deduce that $d^2 = 5Y^4 + 10Y^3 + 10Y^2 + 5Y + 1$. Set $v = 2d$ and $u = [(2Y+1)^2 + 1]/2$. Deduce that $v^2 - 5u^2 = -1$. Conclude that $u = (F_{6m+3})/2$ for some $m \in \mathbf{Z}$.) [(3.1.13)]

11. In [63], J.H.E. Cohn posed:

(3.1.12) This was proved in [241]. Finding solutions of $u_1 u_2 u_3 = u_1 + u_2 + u_3$ in other number fields of higher degree were left as an open problem at the end of [241]. In [398]–[399], Zhang and Gordon looked at this problem, especially in cubic fields. See also Mohanty [220].

(3.1.13) In [381] Williams showed that the more general equation $F_n = k^2 + 1$ has only the solutions $n = \pm 1, 2, \pm 3, \pm 5$. Thus, the only solutions of (3.1.7) such that $x > y$ are $(1, 0)$ and $(0, -1)$.

Conjecture 3.1.1. *The Diophantine equation $x^2 + 7 = y^n$, $n > 1$ has only the solutions $x = 1, 3, 5, 11$ and 181.*

Prove that if another solution exists, then it must be of the form $-7 = \lfloor y^{n/2} \rfloor^2 - y^n$ when n is odd, y is even and y is not a power of 2.

12. Show that Theorem 3.1.2 fails to hold if a and b are allowed to be even. (*Hint*: Look at $A = 19$ and $q = 7$.)

3.2 Real Quadratics and Diophantine Equations.

When $D > 0$ is a radicand, we will be concerned with solutions to Diophantine equations of the form

$$x^2 - Dy^2 = \pm\sigma^2 t, \tag{3.2.1}$$

where $t \in \mathbb{Z}$ is positive, and σ is defined in Chapter One, section five.

Certain solutions are not very interesting, so we give them a special name.

Definition 3.2.1. If $D > 0$ is a radicand and $t > 0$ is a rational integer, then a rational integer solution $(x, y) = (u, v)$ of $x^2 - Dy^2 = \pm\sigma^2 t$ is called a *trivial solution* if $t = m^2$ where m divides both u and v. Otherwise, (u, v) is called a *nontrivial solution.* If $\gcd(u, v) = 1$, then it is called a *proper* solution.

The following result will prove to be useful in looking at solutions of Diophantine equations.

Lemma 3.2.1. [(3.2.1)] *Let $D > 0$ be a radicand and $t \in \mathbb{Z}$ positive. If $\varepsilon_\Delta = (T + U\sqrt{D})/\sigma$ and there is a non-trivial integer solution (x, y) to the Diophantine equation $x^2 - Dy^2 = \pm\sigma^2 t$, then*

$$t \geq (2T/\sigma - N(\varepsilon_\Delta) - 1)/U^2 = B_\Delta,$$

called the Hasse bound.

Proof. Exercise 3.2.1. □

The applications of Lemma 3.2.1 which we provide are largely for certain types of real quadratic fields which we now define.

Definition 3.2.2. [(3.2.2)] Let $D > 0$ be a radicand which can be written in the form $D = s^2 + r$ where $r \mid 4s$. D is said to be of *Extended-Richaud-Degert-type* (or

[(3.2.1)] This result is due to Davenport, Ankeny and Hasse in the case where the solutions are non-trivial, and $D = D_0$. The proof for the case where the solutions are trivial was communicated in a letter to this author by H. Yokoi and we included that proof in [231]. The proof had been communicated to H. Yokoi by H. Ichimura who was writing on the topic of H. Yokoi's [393, Theorem 3, p. 147] which has a gap in the proof. The proof from Ichimura fills that gap. In fact, the letter from H. Yokoi to this author was in response to a query which we had about that omission. Later, this was generalized to arbitrary orders by Halter-Koch [106].

[(3.2.2)] The authors of [251] extended the notion of RD-types introduced via Richaud [308] and Degert [70]. These authors, however, resisted the temptation to name narrow ERD-types (NERDS). ERD-types may appear rather specialized, but they turn out to have several unexpected applications. We will develop some of them in the balance of this book. However, there are applications, beyond

simply ERD-type). If $|r| \in \{1,4\}$, then D is called *narrow-Richaud-Degert* type (or simply *narrow* RD-type).

We now explicitly determine the continued fraction expansions of the principal surd for ERD-types. As it turns out, $\ell(w_\Delta) \leq 12$ so our task is limited.

Theorem 3.2.1. [3.2.3] *If $\Delta > 0$ is a discriminant with radicand $D = a^2 + r$ ($r \mid 4a$) of ERD-type, then the continued fraction expansion of w_Δ is given by the following chart*

Case A: $\sigma = 1$, $a \geq 3$, $w_\Delta = \sqrt{D}$.

(a) $a = \lfloor\sqrt{D}\rfloor$, $r \mid 2a$.

i	0	1	2
P_i	0	a	a
Q_i	1	r	1
a_i	a	$\frac{2a}{r}$	$2a$

(b) $a = \lfloor\sqrt{D}\rfloor$, *$a$ is odd, r does not divide $2a$.*

i	0	1	2	3	4	5	6	7
P_i	0	a	$\frac{2a-r}{2}$	$1+\frac{r}{4}$	$a-2$	a	a	$a-2$
Q_i	1	r	$a+1-\frac{r}{4}$	$a-1+\frac{r}{4}$	4	$\frac{r}{4}$	4	$a-1+\frac{r}{4}$
a_i	a	$\frac{4a-r}{2r}$	1	1	$\frac{a-1}{2}$	$\frac{8a}{r}$	$\frac{a-1}{2}$	1

8	9	10
$1+\frac{r}{4}$	$\frac{2a-r}{2}$	a
$a+1-\frac{r}{4}$	r	1
1	$\frac{4a-r}{2r}$	$2a$

(c) $a = \lfloor\sqrt{D}\rfloor$, $2 \mid a$, *r does not divide $2a$.*

i	0	1	2	3	4
P_i	0	a	$\frac{2a-r}{2}$	$\frac{r}{4}+1$	$a-2$
Q_i	1	r	$a+1-\frac{r}{4}$	$a-1+\frac{r}{4}$	4
a_i	a	$\frac{4a-r}{2r}$	1	1	$\frac{a}{2}-1$

5	6	7	8
$a-2$	$\frac{r}{4}+1$	$\frac{2a-r}{2}$	a
$a-1+\frac{r}{4}$	$a+1-\frac{r}{4}$	r	1
1	1	$\frac{4a-r}{2r}$	$2a$

(d) $a = \lfloor\sqrt{D}\rfloor + 1$, $r \mid 2a$, $r \neq -2a$.

---

the scope of this book, which have been communicated to this author by Patrick Worfolk. He, Roy Adler, and Charles Tresser are currently writing up a paper relating quadratic forms and matrices from a topological perspective, which involves class numbers of real quadratic fields of ERD-type. In their research field, dynamical systems, the notion of topological conjugacy of automorphisms of the torus (represented by elements of $GL(2, \mathbf{Z})$) is an old problem. The entropy of the map is given by the largest (in magnitude) eigenvalue. Thus, maps which are conjugate in $GL(2, Q)$ have the same entropy, but need not be topologically conjugate (conjugacy in $GL(2, \mathbf{Z})$!) This makes for some nice examples in ideal class theory in order to study the conjugacy classes.

[3.2.3] In [118] M.D. Hendy established this for fundamental radicands and F. Halter-Koch made the calculations for arbitrary discriminants in [105].

i	0	1	2	3	4
P_i	0	$a-1$	$a+r$	$a+r$	$a-1$
Q_i	1	$2a+r-1$	$-r$	$2a+r-1$	1
a_i	$a-1$	1	$\frac{-2a}{r}-2$	1	$2a-2$

(e) $a = \lfloor\sqrt{D}\rfloor + 1$, *$r$ does not divide $2a$, $2 \mid a$ and $a \geq r$.*

i	0	1	2	3	4	5	6	7	8
P_i	0	$a-1$	$a+r$	$a+\frac{r}{2}$	$a-2$	$a-2$	$a+\frac{r}{2}$	$a+r$	$a-1$
Q_i	1	$2a+r-1$	$-r$	$a-1+\frac{r}{4}$	4	$a-1+\frac{r}{4}$	$-r$	$2a+r-1$	1
a_i	$a-1$	1	$-(\frac{4a+3r}{2r})$	2	$\frac{a}{2}-1$	2	$-(\frac{4a+3r}{2r})$	1	$2a-2$

(f) $a = \lfloor\sqrt{D}\rfloor + 1$, *$r$ does not divide $2a$, and a is odd.*

i	0	1	2	3	4	5	6
P_i	0	$a-1$	$a+r$	$a+\frac{r}{2}$	$a-2$	$a-4$	$a+\frac{r}{4}$
Q_i	1	$2a+r-1$	$-r$	$a-1+\frac{r}{4}$	4	$2a-4+\frac{r}{4}$	$\frac{-r}{r}$
a_i	$a-1$	1	$-(\frac{4a+3r}{2r})$	2	$\frac{a-3}{2}$	1	$-\frac{8a}{r}-2$

i	7	8	9	10	11	12
P_i	$a+\frac{r}{4}$	$a-4$	$a-2$	$a+\frac{r}{2}$	$a+r$	$a-1$
Q_i	$2a-4+\frac{r}{4}$	4	$a-1+\frac{r}{4}$	$-r$	$2a+r-1$	1
a_i	1	$\frac{a-3}{2}$	2	$-\left(\frac{4a+3r}{2r}\right)$	1	$2a-2$

(g) $r = -a = \lfloor\sqrt{D}\rfloor + 1$.

i	0	1	2
P_i	0	$a-1$	$a-1$
Q_i	1	$a-1$	1
a_i	$a-1$	2	$2a-2$

(h) $r = -\frac{4a}{3}$, $2 \mid a = \lfloor\sqrt{D}\rfloor + 1$.

i	0	1	2	3	4
P_i	0	$a-1$	$a-2$	$a-2$	$a-1$
Q_i	1	$-\frac{r}{2}-1$	4	$-\frac{r}{2}-1$	1
a_i	$a-1$	3	$\frac{a}{2}-1$	3	$2a-2$

(i) $r = -\frac{4a}{3}$, $a = \lfloor\sqrt{D}\rfloor + 1$ *is odd.*

i	0	1	2	3	4	5	6	7	8
P_i	0	$a-1$	$a-2$	$a-4$	$\frac{2a}{3}$	$\frac{2a}{3}$	$a-4$	$a-2$	$a-1$
Q_i	1	$-\frac{r}{2}-1$	4	$\frac{5a}{3}-4$	$\frac{a}{3}$	$\frac{5a}{3}-4$	4	$-\frac{r}{2}-1$	1
a_i	$a-1$	3	$\frac{a-3}{2}$	1	4	1	$\frac{a-3}{2}$	3	$2a-2$

Case B: $\sigma = 2$, $w_\Delta = (1+\sqrt{D})/2$.

(a) $a = \lfloor\sqrt{D}\rfloor$ *even and* $r \mid a$.

i	0	1	2	3	4	5	6
P_i	1	$a-1$	$\frac{r+1}{2}$	$a-r$	$a-r$	$\frac{r+1}{4}$	$a-1$
Q_i	2	$a+\frac{r-1}{2}$	$a-(\frac{r-1}{2})$	$2r$	$a-(\frac{r-1}{2})$	$a+\frac{r-1}{2}$	2
a_i	$\frac{a}{2}$	1	1	$\frac{a}{r}-1$	1	1	$a-1$

(b) $a = \lfloor\sqrt{D}\rfloor$ *odd and* r *does not divide* a.

i	0	1	2
P_i	1	a	a
Q_i	2	$\frac{r}{2}$	2
a_i	$\frac{a+1}{2}$	$\frac{4a}{r}$	a

(c) $a = \lfloor\sqrt{D}\rfloor + 1$ *even and* $r \mid a$.

i	0	1	2	3	4
P_i	1	$a-1$	$a+r$	$a+r$	$a-1$
Q_i	2	$a+\frac{r-1}{2}$	$-2r$	$a+\frac{r-1}{2}$	2
a_i	$\frac{a}{2}$	2	$-(\frac{a}{r}+1)$	2	$a-1$

(d) $a = \lfloor\sqrt{D}\rfloor + 1$ *odd and* r *does not divide* a.

i	0	1	2	3	4
P_i	1	$a-2$	$a+\frac{r}{2}$	$a+\frac{r}{2}$	$a-2$
Q_i	2	$2a+\frac{r}{2}-2$	$-\frac{r}{2}$	$2a+\frac{r}{2}-2$	2
a_i	$\frac{a-1}{2}$	1	$-\frac{4a}{r}-2$	1	$a-2$

Proof. See Exercise 3.2.2. □

We now illustrate Theorem 3.2.1 with examples given in sequential order to correspond with each instance of Cases A–B therein.

Example 3.2.1. If $D = D_0 = 11 = 3^2 + 2$, $\Delta = 4D$, with $\sigma = \sigma_0 = f_\Delta = g = 1$, then the simple continued fraction expansion of $\sqrt{D}$ is given by

i	0	1	2
P_i	0	3	3
Q_i	1	2	1
a_i	3	3	6

Example 3.2.2. Let $D = 9^2+12 = 3\cdot 31 = D_0$ and $\Delta = 4D$ with $\sigma_0 = 2 = f_\Delta = g$ and $\sigma = 1$. The simple continued fraction expansion of $\sqrt{D}$ is given by

i	0	1	2	3	4	5	6	7	8	9	10
P_i	0	9	3	4	7	9	9	7	4	3	9
Q_i	1	12	7	11	4	3	4	11	7	12	1
a_i	9	1	1	1	4	6	4	1	1	1	18

We see the symmetry properties of the simple continued fraction expansions in the above two examples (see Exercise 2.1.13). This is a result of properties of ambiguous classes which we will study in detail in Chapter Six. For now, we just note that once we reach the "middle of the period", then the balance of each tableaux is known to us via symmetry. Hence, we list only half of the tables in each of the following examples (unless the cycle is very short).

Example 3.2.3. Let $D = 204 = 14^2 + 8 = 2^2 \cdot 317 = 4D_0 = \Delta_0$ and $\Delta = 4D$ with $\sigma_0 = 1 = \sigma = g$ and $f_\Delta = 2$. The continued fraction expansion of $\sqrt{D}$ is given by

i	0	1	2	3	4	5	
P_i	0	14	10	3	12	12	
Q_i	1	8	13	15	4	15	...
a_i	14	3	1	1	6	1	

Example 3.2.4. Let $D = 434 = 2 \cdot 7 \cdot 31 = 21^2 - 7 = D_0$ with $f_\Delta = 1 = \sigma$. For $\sqrt{D}$:

i	0	1	2	3	
P_i	0	20	14	14	
Q_i	1	34	7	34	...
a_i	20	1	4	1	

Example 3.2.5. Let $D = 18^2 - 8 = 316 = 4 \cdot 79 = 4D_0 = \Delta_0$ and $\Delta = 4D$ with $\sigma_0 = 1 = \sigma = g$ and $f_\Delta = 2$. For $\sqrt{D}$:

i	0	1	2	3	4	5	
P_i	0	17	10	14	16	16	
Q_i	1	27	8	15	4	15	...
a_i	17	1	3	2	8	2	

Example 3.2.6. Let $D = 77 = 9^2 - 4 = D_0 = \Delta_0$ and $\Delta = 4D$ with $f_\Delta = 2 = \sigma_0 = g$, $\sigma = 1$. For $\sqrt{D}$:

i	0	1	2	3	4	5	6	7	
P_i	0	8	5	7	7	5	8	8	
Q_i	1	13	4	7	4	13	1	13	...
a_i	8	1	3	2	3	1	16	1	

We note here that $\ell(\sqrt{D}) = 6$ not 12 as predicted by Theorem 3.2.1 Case A(f), since $r = -4$ which means that (since $Q_6 = -r/4$) we complete the cycle in "half the time", i.e. some cases in Theorem 3.2.1 do not say that the period length *must* be that listed, but rather that it is the maximum period length.

Example 3.2.7. Let $D = D_0 = 42 = 7^2 - 7$ with $f_\Delta = 1$. For $\sqrt{D}$:

i	0	1	2
P_i	0	6	6
Q_i	1	6	1
a_i	6	2	12

which is the complete cycle.

Example 3.2.8. Let $D = 28 = 6^2 - 8 = 4D_0 = \Delta_0$ and $\Delta = 4D$ with $f_\Delta = 2$ and $\sigma = \sigma_0 = g = 1$. For $\sqrt{D}$:

i	0	1	2	3	
P_i	0	5	4	4	
Q_i	1	3	4	3	...
a_i	5	3	2	3	

Example 3.2.9. Let $D = 69 = 9^2 - 12 = D_0 = \Delta_0$ and $\Delta = 4D$ with $f_\Delta = 2 = g = \sigma_0$ and $\sigma = 1$. For $\sqrt{D}$:

i	0	1	2	3	4	5	
P_i	0	8	7	5	6	6	
Q_i	1	5	4	11	3	11	...
a_i	8	3	3	1	4	1	

In each of the following examples, $w_\Delta = (1 + \sqrt{D})/2$.

Example 3.2.10. Let $D = 10^2 + 5 = 105 = D_0 = \Delta_0$ with $f_\Delta = 1$. For w_Δ:

i	0	1	2	3	4	
P_i	1	9	3	5	5	
Q_i	2	12	8	10	8	...
a_i	5	1	1	1	1	

Example 3.2.11. Let $D = 5^2 + 4 = 29 = D_0 = \Delta_0$ with $f_\Delta = 1$. For w_Δ:

i	0	1	2
P_i	1	5	5
Q_i	2	2	2
a_i	3	5	5

This is the complete cycle.

Example 3.2.12. Let $D = 6^2 - 3 = 33 = D_0 = \Delta_0$ with $f_\Delta = 1$. For w_Δ:

i	0	1	2	3	
P_i	1	5	3	3	
Q_i	2	4	6	4	...
a_i	3	2	1	2	

Example 3.2.13. Let $D = 7^2 - 4 = 45 = 3^2 \cdot 5 = f_\Delta^2 D_0 = \Delta$ with $f_\Delta = 3$, $g = 1$ and $\sigma = \sigma_0 = 2$.

i	0	1	2	
P_i	1	5	5	
Q_i	2	10	2	...
a_i	3	1	5	

This is the complete cycle with $\ell(w_\Delta) = 2$, instead of 4 as predicted by Case B(d) of Theorem 3.2.1, since $r = -4$ and $Q_2 = -r/2$ thereby completing the cycle in "half the time", as we did for the instance of Case A(f) depicted in Example 3.2.6.

In the hint to Exercise 3.2.6 (where we ask the reader to calculate the fundamental unit for each case of Theorem 3.2.1), we calculated the fundamental unit for each of the above examples together with the unit index. In certain circumstances, there are ways to know the fundamental unit via the form of the discriminant derived from ERD-types (see Exercise 3.2.7).

In order to study solutions of Diophantine equations involving radicands of ERD-type, we need

Theorem 3.2.2. [(3.2.4)] *If $D = a^2 + r > 0$ is a fundamental radicand of ERD-type and $x^2 - Dy^2 = \pm\sigma^2 t$ has a non-trivial solution, then $t \geq A$, where*

$$A = \begin{cases} 2(a-1) & \text{if } r = -1, \\ 2a/\sigma^2 & \text{if } r = 1 \text{ and } D \neq 5, \\ a & \text{if } r = 4, \\ a-2 & \text{if } r = -4, \\ 2(a-2) & \text{if } r = -2a, \\ a/3 & \text{if } r = -4a/3, \\ a-4 & \text{if } r = -4a, \\ |r|/\sigma^2 & \text{if } r \notin \{\pm 1, \pm 4, -2a, -4a/3, -4a\}. \end{cases}$$

Proof. We leave this as an exercise for the reader since it is an easy application of Lemma 3.2.1. □

Theorem 3.2.3. [(3.2.5)] *If $D = a^2 + r > 5$ is a fundamental radicand with*

(a) *$a = st$ where $s > 0$, $t > 1$ and $\gcd(t, r) = 1$ for $s, t \in \mathbb{Z}$,*

(b) *$r \mid 4s$,*

(c) *Either $D \not\equiv 1 \pmod 4$ or $|r| \in \{1, 4\}$,*

(d) *If $|r| = 4$, then $s > 1$ and*

(e) *If $r = 1$ and $2 \mid a$, then $s > 2$,*

then the Diophantine equation $x^2 - Dy^2 = \pm\sigma^2 t$ has a non-trivial solution if and only if $D = 7$ and $t = 3$.

Proof. See Exercise 3.2.3. □

(3.2.4)This was proved in [270].

(3.2.5)A slightly weaker version of this result (for ordinary RD-types) was proved in [231]. The techniques involved lengthy algebraic calisthenics. The greater power and ease of use of the continued fraction approach is implicit in Theorem 2.1.2. The authors of [249] began the development of this particular continued fraction approach, which led to numerous results on class numbers of real quadratic fields which we will see in Chapter Five. In [105], F. Halter-Koch developed a method of proof involving continued fractions which he purported therein to include Theorem 3.2.3. However, in [106] he admitted that certain cases of this result were missed by his method in [105]. Nevertheless, in [106] he developed different techniques which cover the cases of Theorem 3.2.3 which his methods in [105] missed. His results are covered in the next section. There is another version of this result in [224] where we considered discriminants of a special type, namely, those for which the conductor is the *least* positive integer which makes $f_\Delta^2 D_0$ of ERD-type where D_0 is allowed to be any fundamental radicand. The proof uses Exercise 3.2.7.

Exercises 3.2

1. Prove Lemma 3.2.1. (*Hint*: Let $u^2 - Dv^2 = \pm\sigma^2 t$ where (u, v) is a non-trivial solution with $u \geq 0$ and smallest possible $v > 0$. Prove that $((uT - nvU)/\sigma, (uU - vT)/\sigma)$ is a non-trivial solution. Conclude that $|(uU - vT)/\sigma| \geq v$ and deduce that $t \geq ((2T/\sigma) - N(\varepsilon_\Delta) - 1)/U^2$.)

2. Prove Theorem 3.2.1. (*Hint*: This is a routine exercise on expanding w_Δ via the continued fraction algorithm, using equations (2.1.2)-(2.1.5).)

3. Prove Theorem 3.2.3. (*Hint*: Verify that $t < \sqrt{\Delta}/2$, then use Theorem 2.1.2 to conclude that (via Corollary 1.4.3) $t = Q_i/\sigma$ for some $i \in \mathbf{Z}$, with $1 \leq i \leq \ell(w_\Delta) - 1$. The balance of the proof is a simple check using Theorem 3.2.1. For example, when $\sigma = 1$ (the easy case), one can verify that $t = 3$, $s = 1$ and $r = -2$ is forced (i.e. $D = 7$) from $t = 2st + r - 1$ (Case A(b).)

4. Let θ be any irrational number. Put $\overline{\theta}_0 = \theta$ and $\overline{q}_0 = Ne(\overline{\theta}_0)$ where $Ne(\alpha)$ for $\alpha \in \mathbf{R}$ is given in Definition 2.2.1. Define $\overline{\theta}_{k+1} = 1/(\overline{a}_k - \overline{\theta}_k)$, and $\overline{a}_{k+1} = N(\overline{\theta}_{k+1})$, then

$$\theta = \overline{a}_0 - \cfrac{1}{\overline{a}_1 - \cfrac{1}{\overline{a}_2 - \cdots - \cfrac{1}{\overline{\theta}_n}}}.$$

We write $\theta = (\overline{a}_0; \overline{a}_1, \ldots, \overline{\theta}_n)$ called the ***nearest integer continued fraction*** of θ (or *NICF* of θ). $\overline{C}_n = (\overline{a}_0, \overline{a}_1, \ldots, \overline{a}_n)$ is called the nth ***convergent of the NICF*** of θ.

(a) Define two sequences $\{\overline{A}_k\}$ and $\{\overline{B}_k\}$ inductively by

$$\overline{A}_{-2} = 0,\ \overline{A}_{-1} = 1,\ \overline{A}_{k+1} = \overline{a}_{k+1}\overline{A}_k - \overline{A}_{k-1}$$

and,

$$\overline{B}_{-2} = -1,\ \overline{B}_{-1} = 0,\ \overline{B}_{k+1} = \overline{a}_{k+1}\overline{B}_k - \overline{B}_{k-1}$$

for $(k \geq -1)$. Verify that $\overline{C}_k = \overline{A}_k/\overline{B}_k$ and $\overline{A}_{k-1}\overline{B}_k - \overline{B}_{k-1}\overline{B}_k = 1$. (*Hint*: See Exercise 2.1.2.)

(b) Let C_k be a convergent in the ordinary simple continued fraction expansion (OCF) defined in Exercise 2.1.2, and assume $\overline{C}_n = C_k$ $(n, k \geq -1)$ where $\overline{C}_n$ is some convergent in the NICF of θ. Verify that $\overline{C}_{n+1} = C_{k+1}$ or C_{k+2}. Also establish that $\overline{C}_{n+1} = C_{k+2}$ if and only if $a_{k+2} = 1$. (*Hint*: Verify that $\overline{A}_{n+1} = \nu A_{k+1} + tA_k$ and $\overline{B}_{k+1} = \nu B_{k+1} + tB_k$ where $|\nu| = 1$.)[3.2.6]

[3.2.6]This verifies, in particular, that $\overline{C}_n$ is a convergent of the OCF of θ whenever it is a convergent of the NICF of θ. Furthermore, it shows that if C_k and C_{k+1} are two successive convergents in the OCF of θ, then at least one of them must be a convergent in the NICF of θ. These types, NICF and OCF, were compared with yet another type called *singular continued fractions* (SCF) by Williams in [380]. See also Perron [288] and Selenius [323] for more detail on these so-called *semi-regular* expansions.

(c) Let $\theta = \sqrt{D}$ where $D > 0$ is not a perfect square and define the integers $\overline{P}_k$ and $\overline{Q}_k$ by $\overline{\theta}_k = (\overline{P}_k + \sqrt{D})/\overline{Q}_k$. Verify the identities

$$\overline{P}_{k+1} = \overline{a}_k\overline{Q}_k - \overline{P}_k,$$

$$\overline{Q}_{k+1}\overline{Q}_k = \overline{P}^2_{k+1} - D,$$

$$\overline{A}^2_k - D\overline{B}^2_k = \overline{Q}_{k+1}.$$

(d) When is the OCF of θ equal to the NICF of θ? (To the best of this author's knowledge, this is an open question which may turn out to be difficult.)

5. (a) Let $\theta = \sqrt{D}$ where $D > 0$ is not a perfect square and set $\theta_k = (P_k + \sqrt{D})/Q_k$ in the OCF of θ. Show that if $Q_k \mid 2P_k$, then $\ell(\theta) = 2k$. (*Hint*: See Exercise 2.1.13 and 3.2.4.)

   (b) If, in the OCF of $\sqrt{D}$, we have $a_{r+1} = 1$, $D = Q_r^2 + (Q_{r+2}/2)^2$, then $Q_r = Q_{r+1}$ and $\ell(\sqrt{D}) = 2r + 1$. Prove this using equations (2.1.4)–(2.1.5).

6. Calculate the fundamental unit for each ERD-type listed in Theorem 3.2.1. (*Hint*: Use Theorem 2.1.3. For example, we will do Case A(a). $\varepsilon_\Delta =$

$$[(P_1+\sqrt{D})/Q_1][(P_2+\sqrt{D})/Q_2] = [(a+\sqrt{D})/r][a+\sqrt{D}] = (2a^2+r+2a\sqrt{D})/r.$$

Of course, not all of them are this easy, so we provide the following fundamental units of each of the orders in Examples 3.2.1–3.2.13, against which the reader may check the calculations. They are (in order corresponding to the examples) as follows: $\varepsilon_{44} = 10 + 3\sqrt{11} = \varepsilon_{\Delta_0}$, $\varepsilon_{372} = 12151 + 1260\sqrt{93} = \varepsilon^3_{\Delta_0} = (13 + 3w_{\Delta_0})^3$, $\varepsilon_{816} = 4999 + 350\sqrt{204} = \varepsilon^3_{\Delta_0} = (50 + 7\sqrt{51})^3$, $\varepsilon_{1736} = 125 + 6\sqrt{434} = \varepsilon_{\Delta_0}$, $\varepsilon_{1264} = 12799 + 720\sqrt{316} = \varepsilon^2_{\Delta_0} = (80 + 9\sqrt{79})^2$, $\varepsilon_{308} = 351 + 40\sqrt{77} = \varepsilon^3_{\Delta_0} = (4 + w_{\Delta_0})^3$, $\varepsilon_{168} = 3 + 2\sqrt{42} = \varepsilon_{\Delta_0}$, $\varepsilon_{112} = 127+24\sqrt{28} = \varepsilon^2_{\Delta_0} = (8+3\sqrt{7})^2$, $\varepsilon_{276} = 7775+936\sqrt{69} = \varepsilon^3_{\Delta_0} = (11+3w_{\Delta_0})^3$, $\varepsilon_{105} = 37 + 8w_\Delta = \varepsilon_{\Delta_0}$, $\varepsilon_{29} = 2 + w_\Delta = \varepsilon_{\Delta_0}$, $\varepsilon_{33} = 19 + 8w_\Delta = \varepsilon_{\Delta_0}$, $\varepsilon_{45} = 3 + w_\Delta = \varepsilon^4_{\Delta_0} = ((1 + \sqrt{5})/2)^4$.

Note that we have also calculated the unit index for each of the examples as well. A further useful exercise, therefore, would be for the reader to calculate the class number h_Δ of each of these orders (see footnote (1.5.9)).)

7. Let D_0 be a fundamental radicand and let f be the *least* positive integer such that $f^2D_0 = D = a^2 + r$ with $-a < r \le a$ and $r \mid 4a$. Prove that the fundamental unit of $\mathcal{O}_\Delta$ is as follows, where Δ is the discriminant associated with the radicand D:

   (i) $\varepsilon_\Delta = a + f\sqrt{D_0}$ if $|r| = 1$, unless $(D_0, f) = (5, 1)$.

   (ii) $\varepsilon_\Delta = (a + f\sqrt{D_0})/2$ for $|r| = 4$.

   (iii) $\varepsilon_\Delta = (2a^2 + r + 2af\sqrt{D_0})/|r|$, if $|r| \notin \{1, 4\}$.

Furthermore, $N(\varepsilon_\Delta) = -\text{sgn}\,(r) = -r/|r|$ in cases (i)–(ii). (*Hint*: Use Exercise 3.2.6 on these special cases.)[3.2.7]

[3.2.7] This was established by Kutsuna in [153]. Note that not all fundamental units of a given order can be found in this way. For instance, if D_0 is already of ERD-type and $D = f^2D_0$ with $f > 1$, then this method fails to apply and we must resort to the techniques in Exercise 3.2.6. For example, let $D = 3^2 \cdot 3 = f^2D_0$. See also Walsh [374].

8. (a) Let D_0 be a fundamental radicand with $D_0 = a^2 + r$ with $r \mid 4a$. Prove that $x^2 - D_0y^2 = c$ has a proper solution with $|c| < \sqrt{D_0}$, then
   (i) $c \in \{1, |r|, 2a + r - 1\}$ when $\sigma = 1$, or
   (ii) $c \in \{1, |r/4|, (2a \pm r\pm)/4,\ |r|,\ a + r/4 - 1,\ a/2 + (r-1)/4\}$ when $\sigma = 2$. (*Hint*: Since $|c| < \sqrt{D_0}$, then c is the norm of a principal reduced ideal, so we just check the relevant cases of Theorem 3.2.1 for fundamental discriminants.)

   (b) Use (a) to show that the only possible solution to the Diophantine equation in Theorem 3.2.3 can occur when $D_0 = t^2 - 2$, the one case to which (a) does not apply, since $t > \sqrt{D_0}$.

9. (a) Let $D = 2^{2k+1}$ for $k \geq 1$. Classify such radicands which are of ERD-type.
   (b) Prove that if $\Delta = 3^2 \cdot 2^{2k+1}$ is a discriminant, then $\Delta = (17 \cdot 2^{k-2})^2 - 2^k$ for $k \geq 5$, and find the fundamental unit ε_Δ.

10. Let $\Delta = \Delta_0 > 0$ be a fundamental discriminant of ERD-type, with $\ell = \ell(w_\Delta)$ and $h_\Delta = 1$. Verify the following chart by establishing that each form of the radicand D must be associated with the given period length.

ℓ	D	
1	$(2m+1)^2 + 1$	$(m \geq 0)$
1	$(2m+1)^2 + 4$	$(m \geq 0)$
2	$4m^2 - 1$	$(m \geq 1)$
2	$m^2 + 2$	$(m \geq 1)$
2	$(2m+1)^2 - 4$	$(m \geq 2)$
2	$(mn)^2 + 4m$	$(m \geq 2)$
3	$4m^2 + 1$	$(m \geq 2)$
4	$m^2 - 2$	$(m \geq 3)$
4	$(2mn)^2 - m$	$(m \geq 1)$
4	$(mn)^2 - 4m$	$(m \geq 2)$

11. (a) Let $\Delta = \Delta_0 > 0$ be a fundamental discriminant with radicand D and let $\varepsilon_\Delta = (T + U\sqrt{D})/\sigma$ be the fundamental unit. If $B_\Delta = (2T/\sigma - N(\varepsilon_\Delta) - 1)/U^2$ is the *Hasse bound* (see Lemma 3.2.1), then define n_Δ to be its *nearest integer*, i.e.

$$n_\Delta = \begin{cases} \lfloor B_\Delta \rfloor & \text{if } B_\Delta - \lfloor B_\Delta \rfloor < 1/2 \\ \lfloor B_\Delta \rfloor + 1 & \text{if } B_\Delta - \lfloor B_\Delta \rfloor > 1/2, \end{cases}$$

   observing that $B_\Delta - \lfloor B_\Delta \rfloor$ cannot equal 1/2 (see Definition 2.2.1). Prove that if $U > 2$, then the following are equivalent:
   (1) $n_\Delta = 0$.
   (2) $T > 4D/\sigma$.
   (3) $U^2 > 16D/\sigma^2$.

   (*Hint*: This is a straightforward analysis using inequalities involving $T^2 - U^2D = N(\varepsilon_\Delta)\sigma^2$.)

   (b) Conclude from part (a) that if $n_\Delta \neq 0$, then $\varepsilon_\Delta < 8D/\sigma^2$.

   (c) Prove that if p_Δ is the least split prime and if $n_\Delta \neq 0$, then $h_\Delta \geq \log n_\Delta / \log p_\Delta$. (*Hint*: Use Lemma 3.2.1.)

(d) Prove that if $n_\Delta \not\equiv 0$, then $U \not\equiv 0 \pmod{D}$.[(3.2.8)]

(e) Let $D = pq \equiv 5 \pmod{8}$ where $p < q$ are primes with $p \equiv q \equiv 3 \pmod{4}$. Prove that if $T > U^2p + 1$, then $D = p^2U^2 \pm 4p$. (*Hint*: First establish that $x^2 - Dy^2 = \pm 4p$ is solvable, then prove that $y = 1$ and $x = pU$).[(3.2.9)]

12. Let $\Delta = D \equiv 1 \pmod{4}$ be a fundamental radicand with $N(\varepsilon_\Delta) = 1$. Prove that for some proper divisor $d > 1$ of D, we have a solution to $x^2 - Dy^2 = \pm 4d$. (*Hint*: Use Theorem 2.1.3 and Exercise 2.1.13(d).)[(3.2.10)]

3.3 Reduced Ideals and Diophantine Equations.

In this section, we will link the solvability of certain quadratic Diophantine equations with the existence of certain principal reduced ideals when $\Delta > 0$. We already have a sense of where this is headed via Theorem 2.1.2, since the continued fraction algorithm gives us all the norms of principal reduced ideals. We now make this more explicit in terms of the Diophantine equations as follows. First, we need some notation.

Definition 3.3.1. Let $\Delta > 0$ be a discriminant and set

$$\Lambda(\Delta) = \{N(I) : I \text{ as a primitive principal ideal of } \mathcal{O}_\Delta\}.$$

Proposition 3.3.1. *If $\Delta > 0$ is a discriminant and $t \in \mathbb{Z}$ is positive, then the following are equivalent:*

(a) $t \in \Lambda(\Delta)$.

(b) *The Diophantine equation $x^2 - Dy^2 = \pm\sigma^2 t$ has a proper solution.*

Proof. See Exercise 3.3.1. □

Now we specialize to reduced ideals.

Definition 3.3.2. Let $\Delta > 0$ be a discriminant and let

$$\Lambda^*(\Delta) = \{N(I) : I \text{ is a principal reduced ideal of } \mathcal{O}_\Delta\}.$$

Remark 3.3.1. We observe that, via Theorem 2.1.2, $\Lambda^*(\Delta) = \{Q_i/\sigma\}_{i=1}^{\ell}$ where $\ell = \ell(w_\Delta)$, and Q_i is given by (2.1.5), i.e. the Q_i/σ represent the norms of all principal reduced ideals as given by the simple continued fraction expansion of w_Δ. For instance, the following comes from Theorem 3.2.1.

(3.2.8) In particular, the Ankeny–Artin–Chowla conjecture holds if $U_\Delta \not\equiv 0 \pmod{D}$ (see Chapter One, section six). The authors of [243] conjectured that $U_\Delta \not\equiv 0 \pmod{D}$ when $D \equiv 7 \pmod{8}$, so this conjecture also holds when $n_\Delta \not\equiv 0$ (see footnote (1.6.5)).

(3.2.9) All the results of the problem were established in [263].

(3.2.10) Having digested two sections on methods for solving Diophantine equations, the reader may want to see a general overall picture of methods for solving such equations. Such an overview is given for the non-specialist in an article by Stroeker [355].

Example 3.3.1. We can read off all the $\Lambda^*(\Delta)$ from the tables in Theorem 3.2.1 when Δ is of ERD-type. For example, if $\Delta = 4D$ where $D = (\lfloor\sqrt{D}\rfloor + 1)^2 + r$, $\lfloor\sqrt{D}\rfloor$ is even and $r \nmid 2(\lfloor\sqrt{D}\rfloor + 1)$, then for $a = \lfloor\sqrt{D}\rfloor + 1$, we have

$$\Lambda^*(\Delta) = \{1, 2a + r - 1, -r, a - 1 + r/4, 4, 2a - 4 + r/4, -r/4\}.$$

The following result links Definitions 3.3.1–3.3.2 and, therefore (via Proposition 3.3.1), associate solutions of Diophantine equations with norms of reduced ideals (at least in certain moderately restricted cases which will prove sufficient for our purposes).

Theorem 3.3.1. [3.3.1] *If $D = \sigma^2 t^2 + r$ is a radicand with $t > 0$, $r \not\equiv (\sigma - 1) \pmod{\sigma^2 t}$, and $r \geq -2\sigma^2 t + 3\sigma - 1$, then the following are equivalent:*

(a) $t \in \Lambda(\Delta)$.

(b) $t \in \Lambda^*(\Delta)$.

Proof. We will prove the result in the case where $\sigma = 1$, leaving the remaining case as an exercise for the reader since it involves essentially the same argument. Since $\Lambda^*(\Delta) \subseteq \Lambda(\Delta)$, we need only establish that (a) implies (b).

If $r > 0$, then $t < \sqrt{\Delta}/2$, so the result follows from Corollary 1.4.3, and Exercise 1.5.9. If $r < 0$ and $t > \sqrt{\Delta}$, then since $D = t^2 + r > (t-1)^2$, we must have $\sqrt{D} > t - 1 > \sqrt{\Delta} - 1 = 2\sqrt{D} - 1$, a contradiction. In view of Corollary 1.4.2 and Exercise 1.5.9, we may now assume that $\sqrt{\Delta}/2 < t < \sqrt{\Delta}$.

Suppose that $I = [t, b + \sqrt{D}]$ is a primitive ideal, where $0 \leq b < t$ may be assumed by Remark 1.2.1. If I is not reduced, then by Corollary 1.4.4 and Exercise 1.5.9, either $b \geq \sqrt{D}$ or $\sqrt{D} \leq t - b$. If the former occurs, then $(t-1)^2 \geq b^2 \geq D > (t-1)^2$, a contradiction. If the latter occurs, then $(t-b)^2 \geq D > (t-1)^2$. Thus, $t - b > t - 1$, so $b = 0$. However, this contradicts the condition that $r \not\equiv (\sigma - 1) \pmod{\sigma^2 t}$, since $\sigma = 1$ in this case, and $t \mid N(b + \sqrt{D})$. □

An immediate consequence is

Corollary 3.3.1. *If $D = \sigma^2 t^2 + r$ is a radicand with $r \not\equiv (\sigma - 1) \pmod{\sigma^2 t}$, $t > 0$ and $\sqrt{D} > \sigma(t-1)$, then the following are equivalent:*

(a) $t \in \Lambda(\Delta)$.

(b) $t \in \Lambda^*(\Delta)$.

We observe that by Proposition 3.3.1, we have a criterion for the solvability of the Diophantine equation (3.2.1) (for radicands satisfying the hypothesis of Theorem 3.3.1) in terms of the norms of *reduced* ideals (which need not be strictly reduced).

Theorem 3.3.2. *Let $\Delta = t^2 + r$ be a discriminant with $0 \leq t - 2 < \sqrt{\Delta} < t + 2$.*

(i) *If $t \in \Lambda(\Delta)$, then there exists $n \in \Lambda^*(\Delta)$ such that $n < \sqrt{\Delta}/2$ and $\Delta - 4tn$ is a perfect square.*

(ii) *If $n \in \Lambda(\Delta)$ and $\Delta - 4tn$ is a perfect square with all $\mathcal{O}_\Delta$-ideals I having $N(I) = n$ being principal, then $t \in \Lambda(\Delta)$.*

[3.3.1] This generalizes the result of F. Halter-Koch in [106]. He required that $\gcd(t, f_\Delta) = 1$. Similar comments pertain to Theorem 3.3.2 and Proposition 3.3.2 in what follows.

(iii) *If* $\Delta - 4t$ *is a perfect square, then* $t \in \Lambda(\Delta)$.

Proof. If $t \in \Lambda(\Delta)$, then form $I = [t, (P + \sqrt{\Delta})/2]$ for $0 \leq P < 2Q$ where $t = Q$. Form the Lagrange neighbour $I^+ = [Q^+, (P^+ + \sqrt{\Delta})/2]$ via Definition 2.1.7. Thus, $\Delta - 4QQ^+ = (P^+)^2$. If we set $Q^+ = n$, then we need only verify that $Q^+ < \sqrt{\Delta}/2$ in order to complete (i). If $Q^+ = (\Delta - (P^+)^2)/(4Q) > \sqrt{\Delta}/2$, then $\Delta \geq \Delta - (P^+)^2 > 2Q\sqrt{\Delta}$ so $\sqrt{\Delta} > 2Q$. However, $Q + 2 > \sqrt{\Delta}$ so $Q = 1$, a contradiction.

To secure (ii), we let $I = [n, (P + \sqrt{\Delta})/2]$. By Exercise 3.3.2, we have $I \sim [t, (-P + \sqrt{\Delta})/2]$, so $t \in \Lambda(\Delta)$.

Finally, (iii) is just (ii) with $n = 1$. □

Remark 3.3.2. The hypothesis of Theorem 3.3.2, which bounds $\sqrt{\Delta}$ in terms of t, basically forces any primitive principal ideal of norm t into being "within one step" of being reduced. In other words, if $I = [t, (P + \sqrt{\Delta})/2] \sim 1$ and I is not already reduced, then I^+ will be reduced with $t^+ < \sqrt{\Delta}/2$. This is all that the proof of part (i) does, i.e. shows that $n = t^+$, and so naturally $\Delta - 4tt^+$ is a square, namely $(P+)^2$!

Now we specialize to ERD-types.

Proposition 3.3.2. *Let* $\Delta = 4a^2 + r$ *be a discriminant where* $a, r \in \mathbf{Z}$ *with* $1 \leq |r| < a$, $r \mid a$ *and* $r \equiv 1 \pmod 4$.

(i) *If* $a \in \Lambda(\Delta)$, *then* $r = 1$.

(ii) *If* $\gcd(f_\Delta, a) = 1$, *then* $2a \in \Lambda(\Delta)$ *if and only if either* $4a^2 - 8a + r$ *or* $4a^2 - 8a|r| + r$ *is a perfect square.*

(iii) $2a \in \Lambda(4a^2 + 1)$ *if and only if* $a = 2$.

Proof. By Theorem 3.2.1, Case B(a), we get $\Lambda^*(\Delta) = \{1, a \pm (\frac{r-1}{4}), r\}$ if $r > 0$, and if $r < 0$, then $\Lambda^*(\Delta) = \{1, -r, a + (r-1)/4\}$, by Case B(c).

(i) is just an application of Corollary 3.3.1. For (ii), we apply Theorem 3.3.2. If $t = 2a \in \Lambda(\Delta)$, we apply part (i) which yields that $D - 4tn$ is a perfect square for some $n \in \Lambda^*(\Delta)$. If $n = a \pm (\frac{r-1}{4})$, then $D - 4tn < 0$ a contradiction. If $n = |r|$, then $D - 4tn = 4a^2 - 8a|r| + r$ and if $n = 1$, then $D - 4tn = 4a^2 - 8a + r$.

Conversely, if $D - 4tn$ is a perfect square for $n = 1$ or $|r|$, then since there exists exactly one $\mathcal{O}_\Delta$-prime above $n \in \Lambda^*(\Delta)$ (see Exercise 1.5.10), the result follows from Theorem 3.3.2(ii).

Part (iii) is an easy application of (ii). □

We now illustrate Proposition 3.3.2 for norms of ideals *not* prime to the conductor.

If we remove the condition $\gcd(f_\Delta, r) = 1$ from (ii) of Proposition 3.3.2 (the only place where we need such a condition), then the result fails to hold in one direction.

Example 3.3.2. Let $\Delta = D = 1305 = 36^2 + 3^2 = 9 \cdot 145$ with $f_\Delta = 3$, $D_0 = \Delta_0 = 145$, $\sigma_0 = 2 = \sigma$, $r = 9$, $a = 18$, and $g = 1$. Thus, $4a^2 - 8a|r| + r = 9$. However, as we shall see, $36 \notin \Lambda(\Delta)$. The simple continued fraction expansion of w_Δ is given by Theorem 3.2.1 Case B(a):

i	0	1	2	3	4	5	6
P_i	1	35	5	27	27	5	35
Q_i	2	40	32	18	32	40	2
a_i	18	1	1	3	1	1	35

Thus, as predicted by Exercise 2.1.13, $w_\Delta = \langle 18; \overline{1,1,3,1,1,35}\rangle$. In order to verify that $36 \notin \Lambda(\Delta)$, the reader should first establish (via the continued fraction algorithm in conjunction with Theorem 1.2.2 and Exercise 1.5.3) that all of the primitive $\mathcal{O}_\Delta$-ideals of norm 36 are given by: $I_1 = [36, (3+\sqrt{1305})/2]$, $I_2 = [36, (21+\sqrt{1305})/2]$, $I_3 = [36, (27+\sqrt{1305})/2]$, $I_4 = [36, (45+\sqrt{1305})/2]$, $I_5 = [36, (51+\sqrt{1305})/2]$, and $I_6 = [36, (69+\sqrt{1305})/2]$. From their simple continued fraction expansions, the reader will also see that $I_j \not\sim 1$ for $j = 1, 2, \ldots, 6$; whereas, $I_1 \sim I_2 \not\sim I_3 \sim I_4 \not\sim I_5 \sim I_6$. Furthermore, $I_1 \sim [9, (15+\sqrt{1305})/2] = J_1$ and $I_5 \sim [9, (3+\sqrt{1305})/2] = J_1'$. Also, from the simple continued fraction expansion of w_Δ, we have that $J_3 = [9, (9+\sqrt{1305})/2] = J_3' \sim 1$. However, $J_1 \not\sim J_1'$, so we have three (!) distinct $\mathcal{O}_\Delta$-ideals of norm 9, one of which is ambiguous while the other two are not even in an ambiguous class. (Note here that, when referring to a class, we do *not* mean a class of C_Δ, since these ideals are not invertible, but rather to the disjoint equivalence classes established by Definition 1.3.1 and Exercise 1.5.8.) We have illustrated this process here since the reader did not have the power of the continued fraction algorithm to solve Exercise 1.5.10(b) in Chapter One. (Compare this with Example 1.5.4.)

Furthermore, the reader may verify that the class of I_5' is the class of I_1 in the sense that each ideal in the cycle of I_1 is the conjugate of an ideal in the cycle of I_5. Also, for interest sake, the reader may verify that the cycle of I_3 (and so of I_4) is ambiguous, and has an ambiguous ideal in it, namely $[29, (29+\sqrt{1305})/2]$ in the "center" of the periodic part (see Chapter Six).

Getting back to Proposition 3.3.2, we see that the reason for $2a \notin \Lambda(\Delta)$ (despite the fact that $4a^2 - 8a|r| + r$ is a perfect square) is that the hypothesis of Theorem 3.3.2(ii) (which we invoked in the proof of Proposition 3.3.2) is violated. In fact, we demonstrated the existence of *three* distinct ideals of norm r. Thus, uniqueness of norm requires the assumption that $\gcd(r, f_\Delta) = 1$. We note that we could have replaced the gcd assumption with the assumption that r is square-free since the latter implies the former, but it is a stronger assumption than necessary. Also, the gcd assumption is not needed to prove the converse of Proposition 3.3.2(ii) discussed above, since once $2a \in \Lambda(\Delta)$, we employ only the existence of a Lagrange neighbour, which exists whether or not the ideal is invertible.

Proposition 3.3.3. *Let $\Delta = a^2 + 4r$ be a discriminant with $a, r \in \mathbf{Z}$, $a > 1$ odd, $r \mid a$, and $r \neq -a$. If $\gcd(a, f_\Delta) = 1$, then:*

(i) *$a \in \Lambda(\Delta)$ if and only if either $a^2 - 4a + 4r$ or $a^2 - 4a|r| + 4r$ is a perfect square.*

(ii) *$a \in \Lambda(a^2 - 4)$ if and only if $a = 5$.*

Proof. Since $N((a + \sqrt{a^2 + 4a})/2) = -a$, we have $a \in \Lambda(a^2 + 4a)$, so this allows us to assume that $|r| < a$. By Theorem 3.2.1 Case B, we get that $\Lambda^*(\Delta) = \{1, r\}$ if $r > 0$, and $\Lambda^*(\Delta) = \{1, -r, a + r - 1\}$ if $r < 0$.

We apply Theorem 3.3.2. If $a \in \Lambda(\Delta)$, then $\Delta - 4an$ is a perfect square for some $n \in \Lambda^*(\Delta)$. If $n = a + r - 1$, then $\Delta - 4tn < 0$, a contradiction. If $n = |r|$, then $D - 4tn = a^2 - 4a|r| + 4r$, and if $n = 1$, then $D - 4tn = a^2 - 4a + 4r$.

The converse is proved as in Proposition 3.3.2, where we invoked the fact that $\gcd(a, f_\Delta) = 1$ as discussed in Example 3.3.3.

(ii) is an easy exercise. □

An application of Proposition 3.3.3 is

Example 3.3.3. [3.3.2] If $x, y \in \mathbf{Z}$ are positive and (for some choice of sign)

$$c = (x^2 + y^2)/(xy \pm 1) \in \mathbf{Z}, \tag{3.3.1}$$

then either c is a perfect square of $c = 5$.

Proof. Suppose that c is not a perfect square. An easy check also shows that we may assume $c > 2$. Equation (3.3.1) becomes $x^2 - cxy + y^2 = \pm c$. If we set $\Delta = c^2 - 4$, which is a discriminant, then $N((2x - cy + y\sqrt{\Delta})/2) \pm c$. Therefore, $c \in \Lambda(\Delta)$. If 2 divides c, then $c \in \Lambda^*(\Delta)$ by Theorem 3.3.1. A check of Theorem 3.2.1, Case A(d), yields $\Lambda^*(\Delta) = \{1, c-2\}$, a contradiction. Hence, $\gcd(f_\Delta, c) = 1$, and c is odd. By Proposition 3.3.3, $c = 5$. □

Proposition 3.3.4. *Let $\Delta = 4(a^2+r)$ be a discriminant where a and r are integers with $a \geq 3$, $r \mid 2a$ and $r > -a$. If $\gcd(f_\Delta, a) = 1$, then*

(i) *If either a is odd or $a^2 + r$ is not a discriminant, then $a \in \Lambda(\Delta)$ if and only if $a = r$.*

(ii) *If a^2+r is not a discriminant, then $2a \in \Lambda(\Delta)$ if and only if either $a^2 - 2a + r$ or $a^2 - 2a|r| + r$ is a perfect square.*

Proof. Given that this is so similar to the above, we leave this as an easy task for the reader (see Exercise 3.3.3). □

Exercises 3.3

1. Prove Proposition 3.3.1. (*Hint*: $I = ((x + y\sqrt{D})/\sigma)$ has no rational integer factors when it is a primitive ideal of $\mathcal{O}_\Delta$.)

2. Let $\Delta > 0$ be a discriminant and suppose the radicand is $D = P^2 - Q_1Q_2\sigma^2$. Prove that the $\mathcal{O}_\Delta$-ideals $I_1 = [Q_1, (P + \sqrt{D})/\sigma]$ and $[Q_2, (-P + \sqrt{D})/\sigma]$ are equivalent in C_Δ. (*Hint*: $1 \sim [Q_1Q_2, (P+\sqrt{D})/\sigma] = [Q_1, (P+\sqrt{D})/\sigma][Q_2, (P+\sqrt{D})/\sigma]$ via equations (1.2.1)–(1.2.5). Now multiply by I_2'.)

3. (a) Prove Proposition 3.3.4. (*Hint*: Use Theorem 3.2.1 and Theorem 3.3.1 for (i), and Theorem 3.3.2 for (ii).) Conclude that in particular:

   (i) If $r = 1$, then $2a \in \Lambda(\Delta)$.

   (ii) If $r \in \{-1, 2\}$, then $2a \notin \Lambda(\Delta)$.

   (iii) If $r = -2$, then $2a \in \Lambda(\Delta)$ if and only if $a = 3$.

   (b) Derive Theorem 3.2.3 from the results of this section.

[3.3.2] This was posed as Problem number 177 in Bulletin de l'Association des Professeurs de Mathématiques no 374, 1990. It was solved by Halter-Koch in [106, Proposition 5, p. 151].

3.4 Class Numbers and Real Quadratics.

We look at criterion for divisibility of class numbers of real quadratic fields as a complement to the first section where this was done for negative discriminants. We begin with a rather innovative idea of Hendy [121].

Theorem 3.4.1. *Let $\Delta = \Delta_0 > 0$ be a fundamental discriminant with radicand $D = D_0 = \sigma^2 a^m + b^2$, $a > 1$, $m > 1$, then there is a divisor n of m such that $n \mid h_\Delta$ and $n > \log_a(D/\sigma^2)/(\ell + 1)$ where $\ell = \ell(w_\Delta)$.*

Proof. If $I = [a, (b + \sqrt{D})/\sigma]$, then $I^m = [a^m, (b + \sqrt{D})/\sigma] \sim 1$. Thus, if n is the order of I in C_Δ, then $n \mid m$ and $n \mid h_\Delta$.

If $r = \lfloor \log_a(D/\sigma^2)/(2n) \rfloor$, then $\{I^{jn}, I'^{jn}\}_{j=0}^{r}$ are $2r+1$ distinct principal ideals with

$$N(I^j) \le N(I^{rn}) \le a^{rn} < a^{1/2 \log_a(D/\sigma^2)} = \sqrt{D}/\sigma.$$

Thus, $\ell = \ell(w_\Delta) \ge 2r + 1 \ge \log_a(D/\sigma^2)/n - 1$, i.e. $n \ge \log_a(D/\sigma^2)/(\ell + 1)$. □

Corollary 3.4.1. *If $D = \sigma^2 a^p + b^2$ is a fundamental radicand where $\ell < p$ a prime, then $p \mid h_\Delta$.*

Proof. Since $\ell < p$, we have $\ell \le p - 1 = \log_a(a^p) - 1 < \log_a(D/\sigma^2) - 1$, i.e. $\log_a(D/\sigma^2)/(\ell + 1) > 1$. Thus, by Theorem 3.4.1, $p \mid h_\Delta$. □

What we actually proved in Theorem 3.4.1 is formally stronger than what is stated therein, namely C_Δ has a *cyclic subgroup* of order n.(3.4.1) We now provide a criterion for this to occur in the maximal order of a quadratic field. First we need

Definition 3.4.1. Let Δ be a discriminant, and define $\alpha \in \mathcal{O}_\Delta$ to be a *primitive element* of the order if α has no non-trivial rational integer factors, i.e. if $\alpha = n\beta$ where $n \in \mathbf{Z}$ and $\beta \in \mathcal{O}_\Delta$, then $|n| = 1$.

Theorem 3.4.2. (3.4.2) *If $\Delta = \Delta_0$ is a fundamental discriminant and $n > 1$ is a rational integer, then the following are equivalent:*

(1) *C_Δ has a cyclic subgroup of order n.*

(2) *There exists a primitive element $\alpha \in \mathcal{O}_\Delta$ with $|N(\alpha)| = z^n$ for some $z \in \mathbf{Z}$, but there does not exist a primitive element $\beta \in \mathcal{O}_\Delta$ with $|N(\beta)| = z^j$ for any $j \in \mathbf{Z}$ with $1 \le j < n$, and $j \mid n$.*

Proof. First we assume that (1) holds. Thus, there is a class of order n in C_Δ. By Exercise 1.5.14, we may assume that this class contains a prime $\mathcal{O}_\Delta$-ideal $\mathcal{P}$, such that $N(\mathcal{P}) = p$ does not divide $|\Delta|$. By Exercise 3.4.1, there exists a primitive

(3.4.1)Lu [203] also studied the phenomenon, as well as Weinberger [377].

(3.4.2)This was proved for real fields by Washington–Zhang in [376], as was Proposition 3.4.1, following. They were motivated by examining h_Δ for $D = n^2 + 3n + 9$, where 3 often divides h_Δ. Also, Nakahara [278] looked at cyclic 3-power subgroups. In [335], Shanks and Weinberger looked at $D = a^6 + 4b^6$ for C_Δ of large 3-rank. More recently, Nakahara [279] looked at the structure of 3-class groups for $\Delta = \Delta_0 > 0$ with $D < 1200000$, and some values of D between 2000000 and 4033723.

element $\alpha \in \mathcal{O}_\Delta$ such that $\mathcal{P}^n = (\alpha)$, so $|N(\alpha)| = p^n$. If there exists a primitive element $\beta \in \mathcal{O}_\Delta$ with $N(\beta) = p^j$ for some $j \mid n$, then either $(\beta) = \mathcal{P}^j$ or $(\beta) = (\mathcal{P}')^j$, since these are the only two ideals of norm p^j.

If (2) holds, then Exercise 3.4.1(c) tells us that there is a primitive $\mathcal{O}_\Delta$-ideal $I = [z^n, (b+\sqrt{\Delta})/2]$. If $p \mid \gcd(z, |\Delta|)$, then since $b^2 \equiv \Delta \pmod{4z^n}$ and $n > 1$, p^2 divides Δ, so $p = 2$. Hence, $\Delta \equiv 0 \pmod 4$, from which we get that $(b/2)^2 \equiv \Delta/4 = D \pmod 4$, a contradiction since $D \not\equiv 1 \pmod 4$. Now we may invoke Exercise 3.4.3 to conclude that there is a primitive $\mathcal{O}_\Delta$-ideal J with $J^n = I$. If there exists $j \mid n$ such that $J^j \sim 1$, then by Exercise 3.4.1(b), there is a primitive element $\beta \in \mathcal{O}_\Delta$ with $J^j = (\beta)$, a contradiction. □

Now we provide a simple proof of a result established independently by Weinberger [377] and Yamamoto [392].

Theorem 3.4.3. *For all positive $n \in \mathbb{Z}$, there are infinitely many fundamental discriminants $\Delta > 0$ with $n \mid h_\Delta$.*

Proof. First we prove that if $\Delta = a^{2n}+4$ is a fundamental discriminant with $a > 1$ odd, then $n \mid h_\Delta$. Consider the primitive element $\alpha = 2+\sqrt{\Delta}$ in $\mathcal{O}_\Delta$ (see Exercise 3.4.1(a)). By Exercise 3.4.1(c), there exists a primitive ideal $I = [a^{2n}, (b+\sqrt{\Delta})/2]$ with $|b| < a^{2n}$. Clearly $\gcd(a, |\Delta|) = 1$, so we may invoke Exercise 3.4.3 to conclude that there exists a primitive $\mathcal{O}_\Delta$-ideal J with $J^{2n} = I$. If $J^{2j} \sim 1$ for any $j \mid n$, then by Exercise 3.4.1(b), there exists a primitive $\alpha = (u+v\sqrt{\Delta})/2$ with $|N(\alpha)| = a^{2j}$. Since $\varepsilon_\Delta = (a^n+\sqrt{\Delta})/2$, by Theorem 3.2.1 Case B(b), we may invoke Lemma 3.2.1 to get that $a^{2j} \geq a^n$. If n is odd, then $n = j$, so $n \mid h_\Delta$ by Exercise 1.5.13, whereas if n is even, then $(n/2) \mid h_\Delta$, similarly. It remains to show that there exist infinitely many fundamental discriminants of the form $\Delta = x^{2n}+4$ for a fixed $n > 1$. We merely observe that by Thue's Theorem[(3.4.3)] $\mathbb{Q}(\sqrt{\Delta}) = \mathbb{Q}(\sqrt{D})$, where D is square-free, for at most finitely many values of x. The reason for this is that the equality holds only if there is a $y \in \mathbb{Z}$ with $\Delta = x^{2n}+4 = Dy^2$, which Thue's Theorem guarantees to have only finitely many solutions. □

We now illustrate the simple process used in the proof of Theorem 3.4.3.

Example 3.4.1. Let $\Delta = 3^{14}+4 = 4782973$, and consider the primitive element $\alpha = 2+\sqrt{\Delta}$ (which the reader may use Exercise 3.4.1(a) to verify). Also, the reader may verify that the $b \in \mathbb{Z}$, from Exercise 3.4.1(c), is $b = -4782971$. Thus, $I = [3^{14}, \beta] = [3^{14}, (-4782971+\sqrt{\Delta})/2]$ is a primitive $\mathcal{O}_\Delta$-ideal, where $|N(\beta)| = |N(\alpha)N(\alpha_0)| = 3^{14}\cdot 1195743$, and $\alpha_0 = (-1+\sqrt{\Delta})/2$ (from the hint in Exercise 3.4.1(c)). From Theorem 3.2.1, Case B(b), we have that $\varepsilon_\Delta = (3^7+\sqrt{\Delta})/2 = (2187+\sqrt{\Delta})/2$. Also, from Exercise 3.4.3, we have that $J^{14} = [3, w_\Delta]^{14} = [3^{14}, \beta]$, which the reader may verify directly by using the multiplication formulae (1.2.1)–(1.2.5). Since $I \sim 1$ and $J^2 = [3^2, (-4782971+\sqrt{\Delta})/2] \not\sim 1$, then $7 \mid h_\Delta = 84$.

What we have been studying in these last few results are certain ERD-types. We have more from other authors. The following is a sample result from a recent paper of Washington–Zhang [376] on cyclic subgroups of C_Δ for $\Delta > 0$, a

(3.4.3) This is exactly how Weinberger proved the infinitude, although the balance of his proof is more involved than ours. Thue's Theorem says that the equation $f(x,y) = a_0x^n + a_1x^{n-1}y + \cdots + a_ny^n = m \neq 0$ for $n \geq 3$, and $f(x,y)$ irreducible over $\mathbb{Q}$, has only a finite number of solutions $x, y \in \mathbb{Z}$. See Thue [362], published in 1909.

fundamental discriminant of ERD-type.

Proposition 3.4.1. *If $D = (\pm z^n + y - 1)^2 + 4y$ is a fundmental radicand with $x, y \in \mathbb{Z}$ such that $y > 0$ divides $z^n \mp 1$, and x is odd with $|z| \neq 1$, then C_Δ has a cyclic subgroup of order n.*

Proof. Exercise 3.4.4. □

As illustrations, we have

Example 3.4.2. Let $\Delta = 1573 = (z^3 + t - 1)^2 + 4 \cdot t$ with $z = 3$, and $t = 13$. C_Δ has a subgroup of order 3 (and $h_\Delta = 12$).

Example 3.4.3. If $\Delta = 469 = (-z^3 + t + 1) + 4t$ with $z = 3$, $t = 7$, then C_Δ is of order 3.

An illustration of Exercise 3.4.4(b) (which is also taken from Washington-Zhang [376]) is

Example 3.4.4. Let $D = (z^n + t - 4)^2/16 + t$ where $z = 3$, $n = 7$, $t = 37$. Thus, C_Δ has a subgroup of order 7. Here, $h_\Delta = 56$.

Exercises 3.4

1. Let Δ be a fundamental discriminant.

   (a) Prove that the following are equivalent:

   (i) $\beta = (x + y\sqrt{\Delta})/2 \in \mathcal{O}_\Delta$ is a primitive element.

   (ii) either $\gcd(x, y) = 1$, or else $\gcd(x, y) = 2$, $\Delta \equiv 1 \pmod 4$ and $xy \equiv 0 \pmod 8$.

   (*Hint*: If β is primitive, and $g = \gcd(x, y) > 1$, then show that $x/g \not\equiv y/g \pmod 2$, so g is even. Show that $g > 2$ is not possible, so $g = 2$, and $\Delta \equiv 1 \pmod 4$. Demonstrate that $xy \not\equiv 0 \pmod 8$ implies $x/2 \equiv y/2 \pmod 2$. This yields one direction, and the other direction is easy.)

   (b) Prove that J is a primitive principal $\mathcal{O}_\Delta$-ideal if and only if there exists a primitive element $\alpha \in \mathcal{O}_\Delta$ with $J = (\alpha)$.

   (c) Let $\alpha \in \mathcal{O}_\Delta$ be a primitive element with $|N(\alpha)| = c$. Prove that there exists a $b \in \mathbb{Z}$ with $|b| < c$ such that $I = [c, (b + \sqrt{\Delta})/2]$ is a primitive $\mathcal{O}_\Delta$-ideal. (*Hint*: Let $\alpha = (x + y\sqrt{\Delta})/2$ with $g = \gcd(x, y) = 1, 2$. Thus, there exist $x_0, y_0 \in \mathbb{Z}$ with $yy_0/g - xx_0/g = 1$. Set $\alpha_0 = (y_0 - x_0\sqrt{\Delta})/g$ and form $\alpha\alpha_0 = (b + \sqrt{\Delta})/2$.)

2. Let Δ be a discriminant. Prove that if I is a proper, primitive $\mathcal{O}_\Delta$-ideal with $\gcd(N(I), |\Delta|) = 1$, then I^n is primitive for any positive integer $n \in \mathbb{Z}$. (*Hint*: Use Exercise 1.5.2(d) to conclude that $I = \prod_{i=1}^{n} \mathcal{P}_i^{a_i}$ for some uniquely determined $\mathcal{O}_\Delta$-prime ideals $\mathcal{P}_i$. Demonstrate that it suffices to prove that $\mathcal{P}^n$ is primitive whenever $\mathcal{P}$ is primitive (where $N(\mathcal{P}) = p$ is necessarily split).

Assume that $r \in \mathbb{Z}$ is a factor of $\mathcal{P}^n$, and deduce the contradiction that $\mathcal{P} = \mathcal{P}'$.)

3. Let Δ be a fundamental discriminant, and let I be a primitive $\mathcal{O}_\Delta$-ideal with $\gcd(N(I), |\Delta|) = 1$. Prove that if $N(I) = c^n$ for positive $n, c \in \mathbb{Z}$, then there exists a primitive $\mathcal{O}_\Delta$-ideal J, such that $I = J^n$. (*Hint*: Use Exercise 1.5.2 to conclude that $I = \prod_{i=1}^n \mathcal{P}_i^{a_i}$ for prime $\mathcal{O}_\Delta$-ideals $\mathcal{P}_i$. Show that J must be primitive since I is.)

4. (a) Prove Proposition 3.4.1. (*Hint*: Use Theorem 3.2.1 to show that $\pm z^j$ satisfies (2) of Theorem 3.4.2, and observe that $(a \pm 2)^2 - D = \pm 4z^n$ where $a = |\pm z^n + y - 1|$.)

   (b) In a similar fashion, verify that if $D = (z^n + t - 4)/16 + t$ is a fundamental radicand with $z \equiv 4 \pmod t$, $D \neq t$, 37, then C_Δ has a cyclic subgroup of order n.

5. Let $\Delta = 4x^{2n} + 1$ where $n \in \mathbb{Z}$ with $n = 3^e n'$ and $\gcd(3, n') = 1$, for $e \geq 1$. Prove that there are infinitely many fundamental radicands of this shape for which $3 \mid h_\Delta$. (*Hint*: Use the techniques of Theorem 3.4.3.)[3.4.4]

6. (a) Prove the analogue of Theorem 3.4.2 for negative fundamental discriminants.[3.4.5] (*Hint*: Use Theorem 3.1.1 on discriminants of the form $1 - 4m^n$ for $n > 1$.)

   (b) Show that if $\Delta = -14348903$, then C_Δ has a cyclic subgroup of order 15. (*Hint*: Invoke Theorem 3.4.2 with $\alpha = 2 + \sqrt{\Delta}$.)

   (c) For α in part (b), show that there exists a $b \in \mathbb{Z}$ with $|b| < 3^{15}$ such that $I = [3^{15}, (b + \sqrt{\Delta})/2]$. (*Hint*: See the hint for Exercise 3.4.1(c).)

   (d) For b obtained in part (c), show that $J^{15} = [3, w_\Delta]^{15} = I = [3^{15}, (b + \sqrt{\Delta})/2]$. This illustrates Exercise 3.4.3. (*Hint*: Use the multiplication formula (1.2.1)-(1.2.5).)

   (e) In view of Theorem 1.3.2, explain why $J^{15} \sim 1$ in part (d), yet $N((b + \sqrt{\Delta})/2) \neq 3^{15}$.

7. (a) Prove that $7 \mid h_\Delta$ for discriminant $\Delta = 6103515629$. (*Hint*: Consider the ideal $I = [5^2, 2 + \sqrt{\Delta}]$ in the order $\mathcal{O}_{4\Delta}$. Show that $I^7 \sim 1$, and use footnote (1.5.9) to verify that $h_{4\Delta} = 3h_\Delta$.)

   (b) The ideas in the hint for part (a) provide the basis for an alternative proof of Theorem 3.4.2. Develop this proof.

8. (a) Prove that if $D = 4x^{2n} + 1$ is a fundamental radicand, then C_Δ has a cyclic subgroup of order n.

   (b) Let $D = 15629$. Prove that C_Δ has a cyclic subgroup of order 3.

[3.4.4] This was proved by Nakahara in [278], as an alternative to work of Weinberger [377].

[3.4.5] This was originally proved by Nagell [276] in 1922.

3.5 Halfway to a Solution.

In this section, we look at how to use the infrastructure to solve the Diophantine equation $x^2 - Dy^2 = -3$. Although this may seem very specialized, it is in fact a paradigm for solutions to Diophantine equations $x^2 - Dy^2 = -q$ where $q \pm 1$ is a perfect square. It is only for the sake of clarity of presentation that we look at $q = 3$, since the details get rather involved. Also, this will leave a wealth of examples for the reader to work out. *Throughout, $D > 0$ is assumed to be a square-free rational integer.*

Suppose that

$$X^2 - DY^2 = -3 \tag{3.5.1}$$

has a solution $x, y \in \mathbf{Z}$. We will establish the existence of ideals $[Q_h, P_h + \sqrt{D}]$ and $[3Q_h, -P+\sqrt{D}]$ which are "near" each other, and whose product is $H = [3, b+\sqrt{D}]$. Since H cannot be the square of an ideal, then we will establish below that the ideals in the product are "halfway" along the cycle of reduced ideals to $[3, P + \sqrt{D}]$ (in terms of the infastructure's ordering).

Whenever (3.5.1) has a solution, then $(D/p) \neq -1$ for all primes p dividing D. This is equivalent to saying that all primes p dividing D are either $p \equiv 1 \pmod 3$ or $p = 3$. Thus,

$$D = M^2 + MN + N^2 \tag{3.5.2}$$

for positive $M, N \in \mathbf{Z}$ (Exercise 3.5.1). Furthermore, if D has ν distinct prime divisors congruent to 1 modulo 3, then there are $2^{\nu-1}$ distinct representations of D in the form (3.5.2), where distinct means that not only are $M, N > 0$, but also we disregard order of the factors and actions by roots of unity.

We may clearly assume, without loss of generality, that M is odd and N is even. Suppose that $N = 2Q$, then $D = (M+Q)^2+3Q^2$. Conversely, if $D = P^2+3Q^2$, then $D = (P-Q)^2+2(P-Q)Q+(2Q)^2$. Hence, there are $2^{\nu-1}$ distinct representations of D in the form

$$D = P^2 + 3Q^2 \tag{3.5.3}$$

with positive $P, Q \in \mathbf{Z}$.

We now form the primitive ideal $I = [Q, P + \sqrt{D}]$ in the order $\mathcal{O}_\Delta = [1, \sqrt{D}]$ where $\Delta = 4D$. By Exercise 3.5.1(c), either $(P + \sqrt{D})/Q$ is reduced or $(P + Q + \sqrt{D})/Q$ is reduced. Set

$$P_h = P + gQ \tag{3.5.4}$$

where $g = 0$ if $(P + \sqrt{D})/Q$ is reduced, and $g = 1$ otherwise (which is allowed by Theorem 1.2.2). Thus, we may now set $Q = Q_h$ and $I = I_{h+1} = [Q_h, P_h + \sqrt{D}]$ to mark its place in the cycle of $\mathcal{O}_\Delta$ via Theorem 2.1.2. Furthermore, I is reduced. Now consider the ideal $J = [3Q_h, -P + \sqrt{D}]$. In what follows, we describe what it means for I and J to be "near" each other, as mentioned at the outset.

First assume that $(P+\sqrt{D})/Q_h$ is reduced, and look at the Lagrange neighbour of J. By Definition 2.1.7, $J^+ = [Q^+, P^+ + \sqrt{D}]$ where

$$P^+ = P + 3Q_h\lfloor(\sqrt{D} - P)/(3Q_h)\rfloor \text{ and } Q^+ = (D - P^{+2})/(3Q_h).$$

However, if $1 < (\sqrt{D} - P)/(3Q_h)$, then certainly $1 < (\sqrt{D} - P)/Q_h$, contradicting that $(P + \sqrt{D})/Q_h$ is reduced by Definition 2.1.4, so $P^+ = P$ (observing that

$\sqrt{D} > P$ by (3.5.3)). Thus,

$$Q^+ = Q_h \text{ and so } J^+ = I_{h+1}.$$

Next, suppose that $(P + Q_h + \sqrt{D})/Q_h$ is reduced, and look at the Lagrange neighbour of I_{h+1}. Again, using the notation from Definition 2.1.7

$$(I_{h+1})^+ = [P^+, Q^+ + \sqrt{D}]$$

where, in this case,

$$P^+ = -P_h + Q_h\lfloor(P_h + \sqrt{D})/Q_h\rfloor \text{ and } Q^+ = (D - P^{+2})/Q_h.$$

Since $(P_h + \sqrt{D})/Q_h = (P + Q_h + \sqrt{D})/Q_h$ is reduced, then $(P + \sqrt{D})/Q_h$ is not reduced so $Q_h > \sqrt{D} = \sqrt{\Delta}/2$ by Corollary 1.4.3 so $\lfloor(P_h + \sqrt{D})/Q_h\rfloor = 1$. Thus, $P^+ = -P_h + Q_h = -P$ by (3.5.4), and $Q^+ = 3Q_h$ by (3.5.3), so $(I_{h+1})^+ = J$. Hence, we have shown that either $J^+ = I_{h+1}$ or $(I_{h+1})^+ = J$, i.e. they lie "side by side" as Lagrange neighbours in their cycle. Hence, 'near to each other' is now clearly defined. Furthermore, $I_{h+1}J = [3, b + \sqrt{D}]$ where $b \equiv P \pmod 3$ (an easy exercise for the reader using (1.2.1)–(1.2.5)). Hence, the ideals I_{h+1}, if they indeed lie in the principal cycle, i.e. if (3.5.1) has a solution, are roughly halfway along the cycle to an ideal $[3, b + \sqrt{D}]$.

Given the above, we observe something which we now formalize. On the one hand, $N(P_h + \sqrt{D}) = -Q_hQ_{h-1}$ by (2.1.5), while on the other hand

$$N(P_h + \sqrt{D}) = P_h^2 - D = P_h^2 - [(P_h - gQ_h)^2 + 3Q_h^2],$$

by (3.5.4). Hence,

$$3Q_h = Q_{h+1}$$

or (3.5.5)

$$4Q_h = Q_{h-1} + 2P_h.$$

We formalize this as

Definition 3.5.1. We will refer to an event as in (3.5.5) as a *signal*, namely a signal that we may be halfway to a solution of (3.5.1).

We could have played the above game by starting with the conjugate ideal $I' = [Q, -P + \sqrt{D}]$. In this case (by Theorem 1.2.2), we may set

$$P_{h+1} = -P + gQ_h, \tag{3.5.6}$$

so that $I' = [Q_h, P_{h+1} + \sqrt{D}]$ is reduced. An easy exercise shows that if $g = 0$, then $(I')^+ = J'$ and if $g = 1$, then $(J')^+ = I'$.

Since $N(P_{h+1} + \sqrt{D}) = -Q_hQ_{h+1}$ by (2.1.5), and

$$P_{h+1}^2 - D = P_{h+1}^2 - ((P_{h+1} - gQ_h)^2 + 3Q_h^2)$$

by (3.5.3), then $(3+g)^2Q_h = Q_{h+1} + 2gP_{h+1}$, i.e.

$$3Q_h = Q_{h+1} \tag{3.5.7}$$

or

$$4Q_h = Q_{h+1}2P_{h+1}.$$

Thus, altogether from (3.5.6)–(3.5.7) we have

Proposition 3.5.1. [(3.5.1)] *Each representation $D = P^2 + Q_h^2$ with $Q_h > 0$ yields a pair of reduced elements $(P_h + \sqrt{D})/Q_h$, $(P_{h+1} + \sqrt{D})/Q_h$ and either of a conjugate pair of signals*

$$4Q_h = Q_{h-1} + 2P_h \quad or \quad 3Q_h = Q_{h-1},$$

and

$$4Q_h = Q_{h+1} + 2P_{h+1} \quad or \quad 3Q_h = Q_{h+1}.$$

Remark 3.5.1. We hasten to add that Proposition 3.5.1 does *not* rely on a solution to (3.5.1), but rather only upon (3.5.3). Thus, there is an integer b such that $3 \mid D - b^2$, making $[3, b + \sqrt{D}]$ an $\mathcal{O}_\Delta$-ideal.

The conjugate pair of signals in Proposition 3.5.1 arose from the representation as given in (3.5.3). We now turn to the representations given in (3.5.2), i.e. $D = P^2 + PQ_h + Q_h^2$ with $Q_h > 0$ odd (without loss of generality). For $P > 0$, it is an easy check that $(P + \sqrt{D})/Q_h$ is reduced. Thus, we see $P = P_h$ and $I_{h+1} = [Q_h, P_h + \sqrt{D}]$. Hence, from (2.1.5), $D = P_h^2 + Q_hQ_{h-1}$ and from (3.5.2), $D = P_h^2 + P_hQ_h + Q_h^2$, so we have the attractive signal $Q_{h-1} = Q_h + P_h$. If we look at the conjugate case, we get $I'_{h+1} = [Q_h, P_{h+1} + \sqrt{D}]$ and similarly, $Q_{h+1} = Q_h + P_{h+1}$. Altogether we have

Proposition 3.5.2. *Each representation $D = P^2 + PQ_h + Q_h^2$, with $Q_h > 0$ odd, yields a pair of reduced elements $(\sqrt{D} + P_h)/Q_h$ and $(\sqrt{D} + P_{h+1})/Q_h$, and a pair of conjugate signals*

$$Q_{h-1} = Q_h + P_h \quad and \quad Q_{h+1} = Q_h + P_{h+1}.$$

Remark 3.5.2. By Exercise 3.5.2, if $Q_{h-1} = Q_h + P_h$, then $P_{h-1} = Q_h$ and $Q_{h-2} = P_h$. Hence, automatically $Q_{h-1} = P_{h-1} + Q_{h-2}$ which is the conjugate signal slightly displaced. Thus, it actually suffices to emphasize only one of the present pair of signals. Moreover, we emphasize that Proposition 3.5.2 does not rely upon a solution of (3.5.1), but rather upon a representation of the form (3.5.2). Our purpose in alluding to a solution at all was to be able to say that a solution is equivalent to the presence of an ideal $[3, b + \sqrt{D}]$ in the principal cycle. However, we have the representations as soon as $D \equiv b^2 \pmod 3$ and this does not tell us the class of $[3, b + \sqrt{D}]$.

Now assume that the halfway ideal $I = [Q_h, P_h + \sqrt{D}]$ lies in the principal class, and that $D = P_h^2 + P_hQ_h + Q_h^2$. Let $\{A_i\}$ and $\{B_i\}$ be the sequences defined in Exercise 2.1.2 for $\gamma = \sqrt{D}$, i.e. $C_i = A_i/B_i$ is the ith convergent of the simple continued fraction of $\sqrt{D}$. Furthermore, let θ_i be as in the proof of Theorem 2.1.2, i.e. $\theta_i = (A_{i-2} - B_{i-2}\sqrt{D})(-1)^{i-1}$ (and observe that since $I \sim 1$, then Claim 2 of that proof shows $(\theta_{h+1})I_{h+1} = (Q_h)$).

Set

$$\gamma = \theta'^2_{h+1}(2\sqrt{D} - 2P_h - Q_h)/Q_h^2, \tag{3.5.8}$$

observing that $2\sqrt{D} - 2P_h - Q_h$ generates a principal ideal, since $4D - (2P_h + Q_h)^2 = -3Q_h^2$ given the assumed representation for D. Also, by Exercise 2.1.2(g)(iv), we

(3.5.1) This and the entire development in this section were essentially taken from [240]. However, the development in [240] is done in terms of matrix theory and form theory, whereas our development is strictly in terms of continued fractions and reduced ideals.

have that $N(\theta'_{h+1})^2 = Q_h^2$. Hence, $N(\gamma) = -3$. Furthermore, by Exercise 2.1.2(g)(i)

$$\theta'^2_{h+1} = A_{h-1}^2 + DB_{h-1}^2 + 2A_{h-1}B_{h-1}\sqrt{D}$$

so if $\gamma = X + Y\sqrt{D}$, then we have

Theorem 3.5.1. *If $Q_{h-1} = Q_h + P_h$ in the principal cycle, then*

$$X^2 - DY^2 = -3,$$

where

$$Q_h^2 X = 4DA_{h-1}B_{h-1} - (2P_h + Q_h)(A_{h-1}^2 + DB_{h-1}^2),$$

and

$$Q_h^2 Y = 2(A_{h-1}^2 + DB_{h-1}^2) - 2(2P_h + Q_h)A_{h-1}B_{h-1}$$

(where $A_{h-1}^2 - DB_{h-1}^2 = (-1)^h Q_h$ and A_{h-1}/B_{h-1} is the $(h-1)$th convergent of $\sqrt{D}$).

It remains only for the reader to verify that $X, Y \in \mathbf{Z}$.

An illustration is now in order.

Example 3.5.1. The continued fraction expansion of $\sqrt{1729}$ is

i	0	1	2	3	4	5	6	7	8	9	10	11	12	13	14	15	16	
P_i	0	41	7	28	26	13	27	23	25	21	35	37	3	40	41	39	39	
Q_i	1	48	35	27	39	40	25	48	23	56	9	40	43	3	16	13	16	...
a_i	41	1	1	2	1	1	2	1	2	1	8	1	1	27	5	6	5	

with the balance given by symmetry (see Exercise 2.1.13).

Our first signal is $Q_7 = 48 = 23 + 25 = Q_8 + P_8$, so $h = 8$. Via Exercise 2.1.2, we quickly calculate that $A_{h-1} = A_7 = 1788$ and $B_{h-1} = B_7 = 43$. Plugging these values into Theorem 3.5.1, we get $X = 122831$ and $Y = 2954$, i.e.

$$122831^2 - 1729 \cdot 2954^2 = -3.$$

This is a rather pleasant concrete outcome after all the theory we have developed.

We also notice that $43 = Q_{12} = P_{13} + Q_{13} = 40 + 3$ is our second signal, and if we now change h to be 13, we get $A_{h-1} = 122831$ and $B_{h-1} = 2954$, which are our previous X and Y. Furthermore, if we plug these new values into Theorem 3.5.1, i.e. $Q_h = 3$, $P_h = 40$, $A_{h-1} = 122831$ and $B_{h-1} = 2954$, we get a new set of X and Y values, namely $X = 544796401$ and $Y = 13101974$ with

$$544796401^2 - 1729 \cdot 13101974^2 = -3.$$

We quickly calculate, via Exercise 2.1.2, that $A_{16} = 544796401$ and $B_{16} = 13101974$. Of course, none of this should come as a surprise in view of Exercise 2.1.10(b), i.e. the solution (X, Y) of (3.5.1) *must* satisfy that X/Y is a convergent of $\sqrt{D}$. What Theorem 3.5.1 has done is to show us explicitly how to find these solutions, given a signal in the principal class.

We now turn to the other pair of signals given by Proposition 3.5.1, namely

$$(3 + g^2)Q_h = Q_{h-1} + 2gP_h$$

where $g = 0$, or 1. In this case, we have that the element $-P_h + gQ_h + \sqrt{D} = \alpha$ generates a principal ideal since $N(\alpha) = -3Q_h^2$. Thus, we play the same game as in the discussion preceding Theorem 3.5.1, namely set

$$\gamma = \theta_{h+1}'^2(-P_h + gQ_h + \sqrt{D})/Q_h^2,$$

and get $N(\gamma) = -3$ so we have (allowing the reader to verify that $X, Y \in \mathbf{Z}$),

Theorem 3.5.2. *If* $(3+g^2)Q_h = Q_{h-1} + 2gP_h$, *($g = 0$ or 1) in the principal class, then*

$$X^2 - DY^2 = -3,$$

where

$$Q_h^2 X = 2A_{h-1}B_{h-1}D - (P_h - gQ_h)(A_{h-1}^2 + DB_{h-1}^2),$$

and

$$Q_h^2 Y = A_{h-1}^2 + DB_{h-1}^2 - 2(P_h - gQ_h)A_{h-1}B_{h-1},$$

(where $A_{h-1}^2 - DB_{h-1}^2 = (-1)^h Q_h$).

We have worked sufficiently hard that we deserve another example.

Example 3.5.2. If $D = 1891$, then the simple continued fraction expansion of $\sqrt{1891}$ is:

i	0	1	2	3	4	5	6	7	8	9
P_i	0	43	41	39	35	37	21	29	13	28
Q_i	1	42	5	74	9	58	25	42	41	27
a_i	43	2	16	1	8	1	2	1	1	2

10	11	12	13	14	15	16	17	18	19	
26	19	15	34	41	43	41	29	31	31	
45	34	49	15	14	3	70	15	62	15	...
1	1	1	5	6	28	1	4	1	4	

with the balance given by symmetry. Our first signal is $4Q_6 = Q_5 + 2P_6 = 58 + 2 \cdot 21 = 100$, so $h = 6$, $g = 1$, and we calculate, via Exercise 2.1.2, that $A_{h-1} = 15133 = A_5$ and $B_{h-1} = 348 = B_5$. Plugging these values into Theorem 3.5.2, we get $X = 34798636$ and $Y = 800233$ with

$$34798636^2 - 1891 \cdot 800233^2 = -3.$$

We also observe some facts concerning Lagrange neighbours discussed at the outset of this section. Let

$$I_{h+1} = [Q_h, P_h + \sqrt{D}] = I_7 = [25, 21 + \sqrt{D}].$$

Since $P_h = 21$ and $Q_h = 25$ then, by (3.5.4), $P = P_h - Q_h = -4$, so we set $J = [3Q_h, P + \sqrt{D}] = [75, 4 + \sqrt{1891}]$ and look at

$$J^+ = [25, -4 + \sqrt{1891}] = [25, 21 + \sqrt{1891}] = I_{h+1},$$

as we predicted.

The next signal in the principal cycle is $Q_{12} = 49 = 15 + 34 = Q_{13} + P_{13}$ which is a signal from Theorem 3.5.1. Thus, in this case, we have $h = 13$, $A_{h-1} = 1000126$ and $B_{h-1} = 22999$. Hence,

$$\begin{aligned} Q_h^2 X &= 4DA_{h-1}B_{h-1} - (2P_h + Q_h)(A_{h-1}^2 + DB_{h-1}^2) \\ &= 4 \cdot 1891 \cdot 1000126 \cdot 22999 - (2 \cdot 34 + 15)(1000126^2 + 1891 \cdot 22999^2) \\ &= 15^2 \cdot 35308981699, \\ Q_h^2 Y &= 2(A_{h-1}^2 + DB_{h-1}^2) - 2(2P_h + Q_h)A_{h-1}B_{h-1} \\ &= 2(1000126^2 + 1891 \cdot 22999^2) - 2(2 \cdot 34 + 15)1000126 \cdot 22999 \\ &= 15^2 \cdot 811968962, \end{aligned}$$

and

$$35308981699^2 - 1891 \cdot 811968962^2 = -3.$$

We also observe that $4Q_6 = 2P_7 + Q_7$ which is the conjugate of the signal $4Q_6 = Q_5 + 2P_6$ for a pair as predicted by Proposition 3.5.1. We deal with solutions generated by these conjugates in

Theorem 3.5.3. *If* $(3 + g^2)Q_h = Q_{h+1} + 2gP_{h+1}$ *in the principal cycle with* $g = 0$ *or* 1, *then*

$$X^2 - DY^2 = -3,$$

where

$$Q_h^2 X = 2A_{h-1}B_{h-1}D + (P_{h+1} - gQ_h)(A_{h-1}^2 + DB_{h-1}^2),$$

and

$$Q_h^2 Y = A_{h-1}^2 + DB_{h-1}^2 + 2(P_{h+1} - gQ_h)A_{h-1}B_{h-1},$$

(with $A_{h-1}^2 - DB_{h-1}^2 = (-1)^h Q_h$ *and* A_{h-1}/B_{h-1} *is the* $(h-1)$*th convergent of* $\sqrt{D}$*).*

Proof. We play the same game as in the previous two theorems. Let

$$\gamma = \theta_{h+1}'^2 (P_{h+1} - gQ_h + \sqrt{D})/Q_h^2,$$

where θ'_{h+1} is as above. As before, $N(\gamma) = -3$ since $P_{h+1} - gQ_h + \sqrt{D}$ generates a principal ideal, given that $D = (P_{h+1} - gQ_h)^2 + 3Q_h^2$. The balance is an easy exercise for the reader. □

Example 3.5.3. If $D = 5719$, then the reader may verify, by looking at the simple continued fraction expansion of $\sqrt{5719}$, that $Q_{h+1} + 2gP_{h+1} = Q_7 = 27 = 3 \cdot 9 = 3Q_6 = (3 + g^2)Q_h$ so $g = 0$, $h = 6$. Via Exercise 2.1.2, the reader may also verify that $P_{h+1} = P_7 = 74$, $A_{h-1} = A_5 = 605$ and $B_{h-1} = B_5 = 8$. Thus, the reader may now plug these values into Theorem 3.5.3 to get that $X = 1352234$ and $Y = 17881$ for

$$1352234^2 - 5719 \cdot 17881^2 = -3.$$

We also have $Q_9 = 75 = 3 \cdot 25 = 3 \cdot Q_{10}$ so, in this case, $h = 10$, and we are in Theorem 3.5.2. Here $A_{h-1} = 111848 = A_9$, $B_{h-1} = 1479 = B_9$, and $P_h = P_{10} = 62$. Plugging these values into Theorem 3.5.2, we get $X = 545397694$ and $Y = 7211959$ for

$$545397694^2 - 5719 \cdot 7211959^2 = -3.$$

The above illustrates the case where $g = 0$ in Theorem 3.5.3. We now illustrate the case where $g = 1$.

Example 3.5.4. If $D = 9139$, then the reader may verify that $(3 + g^2)Q_h = 4Q_{16} = 4 \cdot 49 = 10 + 2 \cdot 93 = Q_{17} + 2P_{17}$ so $h = 16$. We easily compute that $A_{h-1} = A_{15} = 21578407$ and $B_{h-1} = B_{15} = 225720$. Plugging these values into Theorem 3.5.3, we get $X = 54144725563676$ and $Y = 566378577169$ for

$$54144725563676^2 - 9139 \cdot 566378577169^2 = -3.$$

We observe that $A_{24} = 54144725563676$ and $B_{24} = 566378577169$, so we were roughly half-way to a solution via our signal. We also may verify that $Q_{13} = 107 = 30+77 = Q_{14}+P_{14}$, which yields $h = 14$ and puts us back into Theorem 3.5.1. Here $A_{h-1} = 3259609$ and $B_{h-1} = 34097$. Plugging these values into Theorem 3.5.1, we get

$$X^2 - DY^2 = -3 = 169911899891^2 - 9139 \cdot 1777356134^2.$$

We observe that in fact $A_{22} = 169911899891$ and $B_{22} = 946807765$, so the signal put us roughly half-way to the above solution.

Definition 3.5.2. We have seen signals arising in two cases; one from the representations in (3.5.3) which we will call *signals of the first kind*; and those arising from (3.5.2) which we will call *signals of the second kind* .

We have been working strictly in the order $\mathcal{O}_\Delta = [1, \sqrt{D}]$. However, when $D \equiv 1 \pmod 4$, then this is not the maximal order, so a few comments are warranted here.

Proposition 3.5.3. *If $D \equiv 1 \pmod 4$, then any ideal $I = [2S, R + \sqrt{D}]$ which is ambiguous in $\mathcal{O}_{4D}$ is an ideal in $\mathcal{O}_D = [1, (1+\sqrt{D})/2]$, and any ideal $[Q, P+\sqrt{D}] \sim I$ in $\mathcal{O}_D$ must have Q even. Conversely, if an ideal $[Q, P + \sqrt{D}]$ has Q even but is* not *an ideal of the order $\mathcal{O}_D$, then $4 \mid Q$ and 4 properly divides $D - P^2$.*

Proof. Since $I = [2S, R + \sqrt{D}]$ is an ideal in $\mathcal{O}_{4\Delta}$ where $\Delta = D$, then $R^2 \equiv D \pmod{2S}$. Therefore, R must be odd. If $I = I'$ in $\mathcal{O}_{4\Delta}$, then $2Sn - R = R$ for some $n \in \mathbf{Z}$ by Exercise 1.5.3(c), so S is odd as well. Hence, $I = (2)\ [S, (R-1)/2 + w_\Delta]$ which is an ideal in $\mathcal{O}_\Delta$. Similarly, if $[Q, P + \sqrt{D}] \sim I$ in $\mathcal{O}_\Delta$, then Q is forced to be even.

Conversely, $[Q, P+\sqrt{D}]$ is not an ideal of $\mathcal{O}_\Delta$, so $\frac{1}{2}Q$ does not divide $(P^2-D)/4$, although $Q \mid (P^2 - D)$. Hence, $4 \mid Q$, but 4 properly divides $P^2 - D$. □

Remark 3.5.3. In the case of the signal $Q_{h-1} = Q_h + P_h$ or $Q_{k+1} = Q_h + P_{h+1}$, we have that Q_h is odd. For the remaining signals, we have $D = P^2 + 3Q_h^2$, and an easy exercise verifies that Q_h is even precisely when $D \equiv 1 \pmod 4$.

Assume that either $3Q_h = Q_{h-1}$ or $3Q_h = Q_{h-1}$. Since $(P_h^2 - D) = -Q_{h-1}Q_h$, respectively $(P_{h+1}^2 - D) = -Q_hQ_{h+1}$, then clearly $4 \mid Q_h$ cannot be accompanied by 4 properly dividing $(P_h^2 - D)$ nor, respectively, 4 properly dividing $(P_{h+1}^2 - D)$. Similarly, as the reader may verify, $4Q_h = Q_{h-1} + 2P_h$ does not allow both $4 \mid Q_h$ and 4 properly dividing $P_h^2 - D$, and that $4Q_h = Q_{h+1} + 2P_{h+1}$ does not allow 4 properly dividing $P_{h+1}^2 - D$. Thus, if Q_h is even, our signals must occur in cycles of the order $[1,(1+\sqrt{D})/2]$. Hence, if $D\equiv 1 \pmod 4$, at most, the signal $Q_{h-1}=Q_h+P_h$ and its conjugate signal $Q_{h+1}=Q_h+P_{h+1}$, can occur in the principal cycle of the order $[1,\sqrt{D}]$. Example 3.5.1 bears witness to this elucidation.

We have been dealing strictly with the principal cycle. However, we could work within any ambiguous cycle. We leave this discussion for Chapter Six, where we will acquire the relevant machinery, since the central topic of that chapter is ambiguous cycles of ideals (see Exercise 6.1.8).

We have verified

Theorem 3.5.4. *If* (3.5.1) *has* $s>0$ *primitive solutions (i.e. we do* not count associates *as distinct solutions), then*

(a) *If* $3 \mid D$, *then* $s=1$ *and the solution belongs to the ambiguous ideal halfway along the principal cycle.*

(b) *If* 3 *does not divide* D, *then*

  (i) *If* $D\equiv 3 \pmod 4$, *then* $s=2$. *The two solutions are conjugate, so there is just one belonging to an ideal in the first half of the principal cycle.*

  (ii) *If* $D\equiv 1 \pmod 4$, *then* $s=1$ *or* 2 *according as* $N(\varepsilon_\Delta)=-1$ *or* $+1$. *Also, there is at least one signal of the second kind in the principal cycle of* $[1,\sqrt{D}]$, *and there is at least one signal of the first kind in the principal cycle of* $[1,(1+\sqrt{D})/2]$.

Proof. It remains to verify the assertions about s. See Exercise 3.5.3. □

Remark 3.5.4. Example 3.5.1 shows the signal of the second kind in the principal cycle of $[1,\sqrt{D}]$. The reader may verify that in the principal cycle of $[1,(1+\sqrt{D})/2]$, we have $3Q_9=60=Q_{10}$, a signal of the first kind, as well as its conjugate signal $Q_{16}=60=3Q_{17}$.

We conclude by noting that there are signals in the order $[1,(1+\sqrt{D})/2]$ signalling solutions to $(2X-1)^2-DY^2=-12$, rather than (3.5.1). Generalizations for $q=u^2\pm 1$, rather than just $q=-3$, are explored in the Exercises 3.5.4.

Exercises 3.5

1. (a) Prove that if a solution to (3.5.1) exists, then $(D/p)\neq -1$ for all primes $p\mid D$ and that D has a representation as $D=M^2+MN+N^2$ for $M,N\in\mathbf{Z}$.

   (b) Prove that if D has ν distinct prime factors congruent to 1 modulo 3, then there are $2^{\nu-1}$ distinct representations of D as in (a) where $M,N>0$, order of the factors is not counted, and actions by roots of unity are also discounted.

   (c) Prove that if $D=P^2+3Q^2$, then either $(P+\sqrt{D})/Q$ or $(P+Q+\sqrt{D})/Q$ is reduced.

2. Prove the allegations made in Remark 3.5.2. (*Hint*: Look at the fact that if $D=P_h^2+P_hQ_h+Q_h^2$, then $(\sqrt{D}+Q_h)/(P_h+Q_h)$ is reduced.)

3. Prove Theorem 3.5.4. (*Hint*: Since we are only considering non-associated solutions, then $s > 0$ means that in the principal cycle we can get at most two solutions.)

4. Let $1 - q = u^2 \equiv 0 \pmod 4$, and develop a theory for the solution of $x^2 - Dy^2 = q$ as we have done in this section. (*Hint*: $x^2 - Dy^2 = q$ has a solution $x, y \in \mathbf{Z}$ if and only if $N(\alpha) = q$ for some $\alpha \in \mathbf{Z}[\sqrt{D}]$. Assuming that D is square-free, then $(q/p) \neq -1$ for all primes $p \mid D$. Let $\alpha = (\sqrt{q} - 1)/2$ and observe that $\alpha^2 - \mathrm{Tr}(\alpha)\alpha + N(\alpha) = 0$, $D = N(P+Q\alpha) = P^2 + PQ + N(\alpha)Q^2$, i.e. D has the form $x^2 - \mathrm{Tr}(\alpha)xy + N(\alpha)y^2 = (x + \alpha y)(x - \alpha' y)$, and $N(\alpha) = (u/2)^2$. Assume Q odd and look at $D = (P + (u^2/2)Q)^2 + qQ(P + N(\alpha)Q)$.)

5. Using the development in Exercise 3.5.4,

   (a) find a solution to $x^2 - 31y^2 = -99$, and

   (b) find a solution to $x^2 - 65y^2 = -35$.

Chapter 4

Prime-Producing Polynomials.

For centuries, there has been a fascination with prime-producing quadratic polynomials. In this chapter, we will provide necessary and sufficient conditions for quadratic polynomials of both positive and negative discriminant, $F_\Delta(x)$, to generate consecutive, distinct prime values in an initial range of values, i.e. we will classify those $F_\Delta(x)$ which are prime for $x = 0, 1, 2, \ldots, B$, where B is either a Minkowski or a Rabinowitsch bound. We shall also look at some spin-offs of this investigation in terms of representing the elementary abelian 2-subgroup of the class group of a maximal complex quadratic order via differences of squares.

4.1 Complex Prime-Producers.

The most celebrated of the quadratic prime-producing polynomials is $f(x) = x^2 - x + 41$, discovered by Euler [85] in 1772.[(4.1.1)] This polynomial is prime for $x = 1, 2, 3, \ldots, 40$. Similarly, Legendre [167] observed in 1798 that the polynomial $g(x) = x^2 + x + 41$ is prime for all integers $x = 0, 1, \ldots, 39$. In any case, the prime-producing capacity of these polynomials has less to do with their specific form than it does with their discriminant $\Delta = -163$. We can find numerous polynomials of discriminant -163 which generate consecutive prime values for at least 40 values of x. To see this, set

$$F_n(x) = x^2 - (2n + 1)x + n^2 + n + 41$$

which has discriminant $\Delta = -163$, and is prime for $x = 0, 1, 2, \ldots, 40 + n$ $(0 \leq n \leq 39)$ (Exercise 4.1.1). For example, if $n = 39$, then $F_{39}(x) = x^2 - 79x + 1601$ is prime for the 80 values $x = 0, 1, 2, \ldots, 79$ with each prime repeated twice.[(4.1.2)] However, this is for us, in a sense, "cheating" because we want to investigate quadratic polynomials which generate consecutive *distinct* prime values.[(4.1.3)] We will be studying quadratics $f_\Delta(x) = ax^2 + bx + c$ with discriminant $\Delta < 0$ which produce distinct primes for a string of values of x beginning with $x = 0$, i.e. for what we call

(4.1.1)See Ribenboim [304] for an elementary survey article.

(4.1.2)This was discovered by Escott [84] in 1899. There are numerous polynomials in the literature which purport to have long strings of prime values but are merely other quadratic polynomials of discriminant $-163f^2$, e.g. see Higgins [123], and see Exercise 4.1.5.

(4.1.3)The reasons for this will become obvious when we tie the existence of such polynomials to the class group structure.

an *initial string of prime values.*[(4.1.4)] Therefore, we assume that $c > 0$. We begin our investigation with monic polynomials ($a = 1$). If $\Delta \equiv 0 \pmod 4$ and c is even, then $f_\Delta(2)$ is even and composite. If c is odd, then $f_\Delta(1)$ is even and composite (unless $f_\Delta(x) = x^2 + 1$, in which case $f_\Delta(3)$ is even and composite). Thus, when $\Delta \equiv 0 \pmod 4$, we cannot get a consecutive prime-producing quadratic which goes beyond a couple of values. Henceforth, we therefore assume that $\Delta \equiv 1 \pmod 4$ for the monic case.

The transformation $x \rightarrow x + \frac{(1-b)}{2}$ sends $x^2 + bx + c$ to

$$x^2 + x + A \text{ where } A = \frac{(1-b)^2}{4} + \frac{(1-b)b}{2} + c \in \mathbf{Z},$$

which is a polynomial of the same discriminant as the original.[(4.1.5)] Thus, we have lost no generality in the sense that we will be classifying quadratic prime-producers in terms of their discriminant, rather than any of the numerous forms which the polynomial can take, as illustrated by $F_n(x)$ above for $\Delta = -163$.

Before proceeding any further, we wish to instill in the reader an appreciation of the polynomial $f_\Delta(x) = x^2 + x + A$ and our quest to classify such prime-producers. We may ask: What is the *largest* number of consecutive primes values which $f_\Delta(x)$ can assume? The following will give evidence that the answer is: *Any* number of consecutive prime values may be assumed. To do this, we need to understand something called the "prime k-tuples conjecture". This is essentially a generalization of the *twin primes conjecture* which says that p and $p+2$ are both prime infinitely often.[(4.1.6)] However, can we have $p, p+2$ and $p+4$ simultaneously prime infinitely often? The answer is no, since at least one of them must be a multiple of 3. A similar argument proves that in the sequence $p, p+2, p+6, p+8, p+24$, one of the values is always divisible by 5 (Exercise 4.1.2). Thus, we must look further in attempting a generalization of the twin prime conjecture, since it is not so straightforward. We do this as follows.

Let $R = \{r_1, \ldots, r_k\}$ with $r_i \in \mathbf{Z}$ for $i = 1, 2, \ldots, k$. Clearly, if q is a prime such that, for each $n \in \mathbf{Z}$ with $1 \leq n \leq q$, we have $\prod_{i=1}^{k}(n + r_i) \equiv 0 \pmod q$, then there cannot exist infinitely many values p such that $\{p + r_i\}_{i=1}^{k}$ are all simultaneously prime. If such a prime q exists, then call R *inadmissible*, and otherwise call R *admissible*. Another way of looking at this is that R is admissible if and only if, for all primes q, there exists an integer a_q with $1 \leq a_q \leq q$ such that $\prod_{i=1}^{k}(a_q + r_i) \not\equiv 0 \pmod q$.

Hardy and Littlewood[(4.1.7)] felt that if there is no good reason why $p + r_1, p + r_2, \ldots, p + r_k$ cannot all be prime infinitely often, then they should be, or more precisely

Conjecture 4.1.1. (The Prime k-Tuples Conjecture.)[(4.1.8)] *If R is an admissible set, then there are infinitely many integers n such that $n + r$ is prime for each $r \in R$. (The twin prime conjecture is the case $R = \{0, 2\}$).*

(4.1.4)Recall that $\Delta = b^2 - 4ac$, Lagrange's notion of a discriminant [160] which differs from Gauss (see footnote (4.1.15) and Appendix E).

(4.1.5)This Eulerian form is one of the most studied in the literature (see Lehmer [169] and Szekeres [358] for instance).

(4.1.6)There is strong evidence to support the validity of this conjecture (see Guy [101]).

(4.1.7)They are responsible for the prime k-tuples conjecture (see [109]).

(4.1.8)There is a conflict between the prime k-tuples conjecture and the conjecture A : $\pi(x + y) \leq \pi(x) + \pi(y)$ where $x, y \in \mathbf{Z}$; $x, y \geq 2$ and $\pi(x)$ denotes the number of primes not exceeding x. At least one of these must be false (almost certainly conjecture A). See Hensley and Richards [122].

We are now in a position to provide the evidence that the number of consecutive prime values in an initial range which can be assumed by $x^2 + x + A$ is unbounded.

Theorem 4.1.1. (4.1.9) *If the prime k-tuples conjecture is true, then for any integer $B > 0$, there exists a quadratic polynomial of the form $f_\Delta(x) = x^2 + x + A$ such that $f_\Delta(x)$ is prime for all integers x with $1 \leq x \leq B$.*

Proof. Let $r_j = f(j) = j^2 + j$ for $j = 1, 2, 3, \ldots, B$.

Claim. The set $\{r_i\}_{i=1}^{B}$ is admissible.

If $q = 2$, then let $a_q = 1$. Since each r_j is even, then $\prod_{j=1}^{B}(r_j + 1)$ is odd. For each odd prime q, let b_q be any quadratic non-residue modulo q, and set $a_q \equiv (1 - b_q)/4 \pmod q$. If $\prod_{j=1}^{B}(r_j + a_q) \equiv 0 \pmod q$, then $r_j + a_q \equiv 0 \pmod q$ for some j with $1 \leq j \leq B$, i.e. $r_j \equiv -a_q \pmod q$. Therefore, $(2j+1)^2 \equiv 4r_j + 1 \equiv 1 - 4a_q \equiv b_q \pmod q$, a contradiction to the fact that b_q is a quadratic nonresidue. This establishes the claim.

By the prime k-tuples conjecture, there exist arbitrarily large values of A for which $\{f_j + A\}_{j=1}^{B}$ are primes. For such an A, we have that $f_\Delta(x) = x^2 + x + A$ is prime for $x = 1, 2, \ldots, B$. □

Remark 4.1.1. No unconditional proof of Theorem 4.1.1 has yet been found. In fact, a search for an $f_\Delta(x) = x^2 + x + A$ with 41 consecutive, distinct, initial prime values has failed to come up with a value of A.(4.1.10) Furthermore, as we shall see very shortly, if we require that $B = A - 2$ in Theorem 4.1.1, then we must have $\Delta > 0$. Note as well that $f_\Delta(A-1) = A^2$, so although the number of prime values which can be taken by such polynomials $f_\Delta(x)$ can be unbounded, the number of consecutive distinct prime values, in an initial string, taken on by a *fixed* $f_\Delta(x)$ is bounded by $A - 1$.

We now focus our attention upon finding those $f_\Delta(x) = x^2 + x + A$ which are prime for $x = 0, 1, \ldots, A-2$. Questions concerning prime-producing quadratic polynomials become interesting only if we look at those polynomials which are *irreducible*. Even more than that, we must assume that there does not exist a prime p which divides $f_\Delta(x)$ for all $x \in \mathbb{Z}$ (since, for example, $x^2 + x + 4$ is irreducible but *never* prime).(4.1.11)

Now we look at the criterion which links our search for prime-producing quadratics to Gauss' class number one problem for complex quadratic fields. First, we need some notation and a technical lemma.

Definition 4.1.1. For a discriminant Δ, set

$$F_\Delta(x) = x^2 + (\sigma - 1)x + (\sigma - 1 - \Delta)/4.$$

(4.1.9)We are indebted to A. Granville for providing this proof which appeared in Louboutin *et al.* [200].

(4.1.10)In Lukes *et al.* [207], it is shown that if the value of A exists, then $A > 10^{17}$.

(4.1.11)In fact, we note that W. Bouniakowsky [32] conjectured that, if a polynomial $p(x) \in \mathbb{Z}[x]$ is irreducible and has N as the gcd of $p(x)$ where x runs over all integers, then $p(x)/N$ takes on prime values for infinitely many $x \in \mathbb{Z}$. The best attack in the problem is due to Richert [309] who showed that if $p(x)$ is irreducible, then there exist infinitely many integers m such that $p(m)/N$ is the product of at most $g + 1$ primes, where g is the degree of $p(x)$.

We call $F_\Delta(x)$ the *Euler-Rabinowitsch polynomial.* [(4.1.12)] Also set

$$S_\Delta = \{x \in \mathbf{Z} : 0 \le x \le \lfloor |\Delta|/4 - 1 \rfloor\}.$$

We call $\lfloor |\Delta|/4 - 1 \rfloor$ the *Rabinowitsch bound.*

Lemma 4.1.1. *If $\Delta < 0$ is a discriminant ($\Delta \neq -3, -4$) and $x \ge 0$, then $F_\Delta(x) < N(w_\Delta)^2$ if and only if $x \in S_\Delta$.*

Proof. Since $F_\Delta(x) = ((\sigma x + \sigma - 1)^2 - D)/\sigma^2$, then for $\sigma = 1$, $\lfloor |\Delta|/4 - 1 \rfloor = -D - 1$, so

$$F_\Delta(x) \le F_\Delta(-D-1) = (D+1)^2 - D < D^2 = N(w_\Delta)^2,$$

unless $\Delta = -4$, which is excluded by hypothesis. If $\sigma = 2$, then $\lfloor |\Delta|/4 - 1 \rfloor = -(D+7)/4$, so

$$F_\Delta(x) \le F_\Delta(-(D+7)/4) = (D^2 + 6D + 25)/16 < (D^2 - 2D + 1)/16 < N(w_\Delta)^2,$$

unless $\Delta = -3$ which is excluded by hypothesis.

Conversely, assume that $F_\Delta(x) < N(w_\Delta)^2$. If $\sigma = 1$, then

$$F_\Delta(|\Delta|/4) = F_\Delta(-D) = D^2 - D < D^2 = N(w_\Delta)^2.$$

Thus, $x \le |\Delta|/4 - 1$. If $\sigma = 2$, then $F_\Delta(-(D+3)/4) = N(w_\Delta)^2$. Since $-(D+3)/4 = \lceil |\Delta|/4 - 1 \rceil$, then $x \le -(D+7)/4 = \lfloor |\Delta|/4 - 1 \rfloor$. □

Remark 4.1.2. It is worth observing that the proof of Lemma 4.1.1 actually shows that $x = \lfloor |\Delta|/4 - 1 \rfloor$ is the *largest* integer such that $F_\Delta(x)$ is less than $N(w_\Delta)^2$.

Now we may state the criterion.

Theorem 4.1.2. [(4.1.13)] (Rabinowitsch's Criterion.) *If $\Delta < 0$ is a discriminant with $\Delta \equiv 1 \pmod 4$ and conductor f_Δ satisfying $\gcd(F_\Delta(x), f_\Delta) = 1$ for all $x \in S_\Delta$, then $F_\Delta(x) = x^2 + x + (1-\Delta)/4$ is prime for all $x \in S_\Delta$ if and only if $h_\Delta = 1$.*

Proof. If $h_\Delta = 1$ and $F_\Delta(x) \equiv 0 \pmod p$ for some prime p and some $x \in S_\Delta$, then by Exercise 1.5.7 and Lemma 4.1.1, $F_\Delta(x) = p$. Conversely, if $h_\Delta > 1$, then by Exercise 2.2.2(d) there is a non-principal ideal $I = [N(I), b + w_\Delta]$ with

---

(4.1.12) Often this polynomial is called the *Frobenius-Rabinowitsch polynomial* since Frobenius' work [88] is considered to be the seminal work on the topic. However, for reasons delineated above, we hold that our terminology is more suitable in terms of our development and goals concerning prime-producing quadratics.

(4.1.13) Rabinowitsch proved this for the maximal order in [298]-[299] as did Szekeres in [358], but both of these proofs were lengthy. Ayoub and Chowla [10] gave a simple proof for the maximal order based only on elementary properties of quadratic fields. Here we give a simple proof for *arbitrary* orders based only on properties of principal ideals in complex quadratic orders.
An interesting anecdote about Rabinowitsch comes from the entertaining and amusing article by Mordell [274]. "In 1923, I attended a meeting of the American Mathematical Society held at Vasser College in New York State. Some one called Rainich from the University of Michigan at Ann Arbor, gave a talk upon the class number of quadratic fields, a subject in which I was very much interested. I noticed that he made no reference to a rather pretty paper written by Rabinowitz from Odessa and published in Crelle's journal. I commented upon this. He blushed and stammered and said, "I am Rabinowitz". He had moved to the U.S.A. and changed his name... ". Thanks go to Alf van der Poorten for bringing the paper to this author's attention. The spelling of Rabinowitsch used in this book coincides with that which appears in Crelle [299].

$0 \le b < N(I) < M_\Delta \le |\Delta|/4 - 1$ (unless $\Delta \ge -11$ for which the result trivially holds). Since $N(b + w_\Delta) = F_\Delta(b)$ and $b \in \mathcal{S}_\Delta$, then $F_\Delta(b)$ cannot be prime by Exercise 1.5.7. □

We see that Euler's polynomial $F_{-163}(x) = x^2 + x + 41$ fits quite nicely into the criterion. In point of fact, it is the last one to do so, as the following solution of the class number one problem for complex quadratics shows. This was solved independently by Baker [14] and Stark [346], anticipated by Heegner [114].(4.1.14)

Theorem 4.1.3. (4.1.15) (Class Number One for Complex Quadratic Orders.) *If $\Delta < 0$ is a discriminant, then $h_\Delta = 1$ if and only if $-\Delta \in \{3, 4, 7, 8, 11, 12, 16, 19, 27, 28, 43, 67, 163\}$.*

Proof. See Cox [67, Theorem 7.30, p. 149]. □

Observe that, for instance, $F_{27}(x) = x^2 + x + 7$ is not prime at $x = 1 \in \mathcal{S}_\Delta$, but $\gcd(f_\Delta, F_\Delta(x)) = 3$.

Now we see from Theorems 4.1.2–4.1.3 that $f_\Delta(x) = x^2 + x + A$ cannot be consecutively prime for $x = 0, 1, \ldots, A-2$ when $A > 41$. This essentially completes our search for the monic prime-producing quadratics. The Euler polynomial tops the list with 40 consecutive, distinct, initial prime values. We now tabulate the rest of them.

Table 4.1.1. *Monic Prime-Producing Quadratics in Complex Orders.*

$-\Delta$	$F_\Delta(x)$	$F_\Delta(x)$ for $x \in \mathcal{S}_\Delta$.
3	x^2+x+1	–
7	x^2+x+2	2.
11	x^2+x+3	3, 5.
19	x^2+x+5	5, 7, 11, 17.
43	x^2+x+11	11, 13, 17, 23, 31, 41, 53, 67, 83, 101.
67	x^2+x+17	17, 19, 23, 29, 37, 47, 59, 73, 89, 107, 127, 149, 173, 199, 227, 257.
163	x^2+x+41	41, 43, 47, 53, 61, 71, 83, 97, 113, 131, 151, 173, 197, 223, 251, 281, 313, 347, 383, 421, 461, 503, 547, 593, 641, 691, 743, 797, 853, 911, 971, 1033, 1097, 1163, 1231, 1301, 1373, 1447, 1523, 1601.

There is more to this than meets the eye at an initial glance. However, in order

(4.1.14)For a brief overview of the history up to the solution of this problem, see Chapter Five, section four. See also Goldfeld [95] for a history of the solution of Gauss' general class number problem: Find an algorithm which determines all negative discriminants of a given class number. Gauss conjectured in [91, Article 303, pp. 361–363] that the number of negative discriminants with a given class number is finite. Furthermore in [91, Article 302, p. 360–361] Gauss gave empirical evidence for the way h_Δ gets large. This was made explicit by Siegel [338] (ineffectively) and later by Goldfeld, Gross, and Zagier and Oesterlé (effectively), see [95]. Also see Chapter Eight.

(4.1.15)It is not very well-known that Gauss' class number one problem was solved by Landau [162] in 1902. Although Theorem 4.1.3 is known as the solution of Gauss' class number one problem, that is not entirely accurate. The problem is essentially one of interpretation. Gauss considered forms of the type $ax^2+2bxy+cy^2$ and defined the determinant as $\Delta = b^2-ac$. This is quite different from Lagrange's definition (see footnote (4.1.4)). Thus, Landau proved $h_\Delta > 1$ for $\Delta = b^2 - ac < -7$. Since Gauss' definition makes Δ even, then the problem solved by Landau is much simpler than the one solved in Theorem 4.1.3. See Appendix E.

to dig out the answers, we need to generalize the Rabinowitsch criterion. It turns out to be only the tip of the iceberg. First, we need some more notation.

Definition 4.1.2. If $n = \prod_{i=1}^{t} p_i^{e_i}$ is the canonical prime factorization of a positive $n \in \mathbb{Z}$, then set

$$\Omega(n) = \sum_{i=1}^{t} e_i,$$

and let

$$\tau(n) = \prod_{i=1}^{t} (e_i + 1).$$

Also set

$$F(\Delta) = \max\{\Omega(|F_\Delta(x)|) : x \in \mathcal{S}_\Delta\},$$

where Δ is a discriminant. ($F(\Delta)$ is sometimes called an *Ono invariant* (see [314]).)[(4.1.16)]

Now we are in a position to generalize the Rabinowitsch criterion. Recall that the *exponent* e_Δ of C_Δ is the least positive integer such that $I^{e_\Delta} \sim 1$ for all $\{I\} \in C_\Delta$.

Theorem 4.1.4. [(4.1.17)] *Let $\Delta = \Delta_0 < 0$ ($\Delta \neq -3, -4$) be a fundamental discriminant. If Δ is divisible by exactly $N + 1$ ($N \geq 0$) distinct primes, then the following are equivalent:*

(1) *C_Δ has exponent $e_\Delta \leq 2$.*

(2) *$h_\Delta = 2^{F(\Delta)-1}$ and $F(\Delta) = N + 1$.*

Proof. By Theorem 1.3.3, we immediately have that (2) implies (1). Now assume that (1) holds. First we establish that $2^{F(\Delta)-1} \geq h_\Delta$. If $N = 0$, then the hypothesis, in conjunction with Theorem 1.3.3, forces $h_\Delta = 1$. Since $F(\Delta) = 0$ if and only if $\Delta = -3$ or -4, then $F(\Delta) \geq 1$. Thus, $2^{F(\Delta)-1} \geq h_\Delta$, so we may assume that $N \geq 1$. Let $C_\Delta = C_{\Delta,2} = \langle Q_1 \rangle \times \langle Q_2 \rangle \times \cdots \times \langle Q_N \rangle$ with $h_\Delta = 2^N$ where Q_i is the unique $\mathcal{O}_\Delta$-prime above the prime divisor q_i of Δ for $i = 1, 2, \ldots, N$. Since Theorem 1.3.3 tells us that Δ is divisible by exactly one more distinct prime, we may assume that prime $q_{N+1} \geq q_i$ for $i = 1, 2, \ldots, N$. Set $Q = \prod_{i=1}^{N} Q_i$, which we may write as $Q = [q, b + w_\Delta]$, where $0 \leq b < q = \prod_{i=1}^{N} q_i$ by Theorems 1.2.1–1.2.2 and Remark 1.2.1. Moreover, $q < |\Delta|/4 - 1$ (since $\Delta \neq -3, -4$ by hypothesis and $\Delta \neq -7$ since $N > 0$). There cannot be a relationship among the generators of a group, so Q is not principal. Since q divides $N(b + w_\Delta) = F_\Delta(b) > q$, we have $F(\Delta) \geq N + 1$, i.e. $2^{F(\Delta)-1} \geq h_\Delta$. Now we show that $h_\Delta \geq 2^{F(\Delta)} - 1$.

Let $F(\Delta) = n$ and let $x \in \mathcal{S}_\Delta$ such that $F_\Delta(x) = \prod_{i=1}^{m} p_i^{e_i} = r$, say, with $\Omega(F_\Delta(x)) = n$ for distinct primes p_i ($1 \leq i \leq m$). It follows that $\mathcal{P} = \prod_{i=1}^{m} \mathcal{P}_i^{e_i} = [r, x + w_\Delta]$, where $\mathcal{P}_i$ is a fixed choice of $\mathcal{O}_\Delta$-ideal above p_i ($1 \leq i \leq m$). Since $N(x + w_\Delta) = F_\Delta(x) = r$, then $\mathcal{P} \sim 1$.

Let $S' = \{i : 1 \leq i \leq m$ and e_i is odd $\}$. Since $e_\Delta \leq 2$, then $1 \sim \mathcal{P} \sim \prod_{i \in S'} \mathcal{P}_i$. Moreover, by the multiplication formulae (1.2.1)–(1.2.5), we have that $[r, x + w_\Delta] =$

(4.1.16) $\tau(n)$ is called the *divisor function*, and measures the number of positive divisors of n, whereas $\Omega(n)$ measures the number of (not necessarily distinct) prime divisors of n.

(4.1.17) This result, as well as the balance of the results of this section, and some of the previous results first appeared in [232] or [234].

$[r', x + w_\Delta][r/r', x + w_\Delta]$, where $r' = \prod_{i \in S'} p_i$, so $[r, x + w_\Delta] \sim [r', x + w_\Delta]$. By Lemma 4.1.1 and Theorem 1.3.2, either $r' = 1$ (in which case e_i is even for all $i \in \{1, 2, \ldots, m\}$), or $r = r'$ (in which case $e_i = 1$ for all $i \in \{1, 2, \ldots, m\}$ and $n = m$). If the former occurs, then $1 \sim \mathcal{P} \sim \mathcal{P}_1^2 = [p_1^2, x + w_\Delta]$. By Theorem 1.3.2, $p_1^2 = N(x + w_\Delta) = r$, i.e. $m = 1$ and $e_1 = 2 = n$. Thus, $p_1 < N(w_\Delta)$ and $h_\Delta \geq 2 = 2^{F(\Delta)-1}$, by Theorem 1.3.3. Thus, for the remainder of the proof we may assume the latter above, i.e. that $n = F(\Delta) = m$ and $e_i = 1$ with $1 \leq i \leq m$.

Suppose that a_1 and a_2 are two positive relatively prime divisors of r. If $I_1 = [a_1, x + w_\Delta] \sim I_2 = [a_2, x + w_\Delta]$, then $J = I_1 I_2 = [a_1 a_2, x + w_\Delta] \sim I_2^2 \sim 1$ since $e_\Delta \leq 2$. Moreover, J is a principal, primitive ideal since $\gcd(a_1, a_2) = 1$. Therefore, $N(x + w_\Delta) = a_1 a_2$ by Theorem 1.3.2. It follows that the exact number of equivalences, among ideals whose norms are relatively prime positive divisors of r, is equal to the number of distinct factorizations of r into two positive factors, where the *order* of the factors is *not* taken into account. By Exercise 4.1.4, this number is 2^{n-1}. Hence, we have shown that there are $\tau(r) - 2^{n-1} = 2^n - 2^{n-1} = 2^{n-1}$ pairwise inequivalent ideals, i.e. $h_\Delta \geq 2^{F(\Delta)-1}$. □

The proof of Theorem 4.1.4 actually contains some hidden information which we isolate below.

Definition 4.1.3. For a positive $n \in \mathbb{Z}$ with $n = \prod_{i=1}^m p_i^{e_i}$ being its canonical prime factorization set

$$d(n) = m,$$

and for a discriminant Δ, let

$$G(\Delta) = \max\{d(|F_\Delta(x)|) : x \in \mathcal{S}_\Delta\}.$$

(Thus, $d(n)$ is the number of distinct prime divisors of n. See Definition 4.1.2.)

Corollary 4.1.1. *If $\Delta = \Delta_0 < 0$ ($\Delta \neq -3, -4$) is a fundamental discriminant and $e_\Delta \leq 2$, then $F(\Delta) = G(\Delta) = N + 1 = d(\Delta)$.*

Proof. The proof of Theorem 4.1.4 shows that the result holds except possibly when $F(\Delta) = 2$, so we show that if $F(\Delta) = 2$, then $G(\Delta) = 2$. By Theorem 4.1.4, $h_\Delta = 2$. If $\Delta \equiv 0 \pmod 8$, then $F_\Delta(0) = -\Delta/4$ is even composite. If $\Delta \equiv 4 \pmod 8$, then $F_\Delta(1) = 1 - \Delta/4$ is even composite (since $\Delta < -4$ when $h_\Delta = 2$). If $\Delta \equiv 1 \pmod 8$, then $F_\Delta(0) = (1 - \Delta)/4$ is even composite. If $\Delta \equiv 5 \pmod 8$, then $h_\Delta = 2$ implies $\Delta = -q_1 q_2$, where $q_1 \not\equiv q_2 \pmod 4$ by Theorem 1.3.3. Thus, $F_\Delta((q-1)/2) = q_1(q_1 + q_2)/4$ is composite. □

Corollary 4.1.2. *For any fundamental discriminant $\Delta = \Delta_0 < 0$ ($\Delta \neq -3, -4$) $F(\Delta) \geq N + 1 = d(\Delta)$.*

Proof. By Theorem 1.3.3, C_Δ has an elementary abelian 2 subgroup of order 2^N. □

Now we show how the Rabinowitsch criterion follows.

Corollary 4.1.3. *If $\Delta = \Delta_0 < 0$ ($\Delta \neq -3, -4$) is a discriminant, then $h_\Delta = 1$ if and only if $F(\Delta) = 1$.*

Proof. If $h_\Delta = 1$, then by Theorem 4.1.4 $F(\Delta) = 1$. If $F(\Delta) = 1$ and $p < M_\Delta = \sqrt{-\Delta/3}$ is any non-inert prime, we may form the ideal $\mathcal{P} = [p, b + w_\Delta]$ where $0 \le b < p < M_\Delta < |\Delta|/4$ (since $\Delta \ne -3, -4$). Since $p \mid N(b + w_\Delta)$, then $F_\Delta(b) = p = N(b + w_\Delta)$ so $\mathcal{P} \sim 1$. This secures the result by Theorem 1.3.1. □

There is also a more recent class number 2 phenomenon which we get from Theorem 4.1.4.

Corollary 4.1.4. (Sasaki [314]). *If $\Delta = \Delta_0 < 0$ ($\Delta \ne -3, -4$) is a discriminant, then $h_\Delta = 2$ if and only if $F(\Delta) = 2$.*

Proof. If $h_\Delta = 2$, then $F(\Delta) = 2$ by Theorem 4.1.4. If $F(\Delta) = 2$, then let $\mathcal{P}$ and $\mathcal{Q}$ be any non-principal prime ideals. We may assume that $\Delta < -12$, since an easy check shows that the result holds if $\Delta > -12$. Thus, $I = [pq, b + w_\Delta]$ with $0 \le |b| < pq/2 < -\Delta/6 < -\Delta/4 - 1$. Therefore, $N(b + w_\Delta) = F_\Delta(b) = pq$ forcing $\mathcal{PQ} \sim 1$. In particular, if $\mathcal{P} = \mathcal{Q}$, then $\mathcal{P}^2 \sim 1$, i.e. $e_\Delta = 2$. Furthermore, if $h_\Delta \ge 4$, then let $\mathcal{P}$ and $\mathcal{Q}$ be distinct generators of C_Δ. Since $\mathcal{PQ} \sim 1$, then $\mathcal{P} \sim \mathcal{Q}' \sim \mathcal{Q}$ (since the conjugate must be in the same class when $e_\Delta = 2$), a contradiction. □

Remark 4.1.3. In [314], Sasaki observed that there are fields for which $h_\Delta > F(\Delta)$ and cites $\Delta = -84$ as an example where $h_\Delta = 4$ and $F(\Delta) = 3$. However, Theorem 4.1.4 asserts that $e_\Delta \le 2$ if and only if $h_\Delta = 2^{F(\Delta)-1} = 2^{N-1}$. Thus, $h_\Delta = F(\Delta)$ if and only if $h_\Delta = 1$ or 2 (when $e_\Delta \le 2$). Thus, these fields are uniquely characterized by Corollaries 4.1.3–4.1.4.

Now that we have demonstrated that the Rabinowitsch criterion underlies the deeper phenomenon, Theorem 4.1.4, which is a criterion for the class group of a complex quadratic field to have exponent 2, we can look at those polynomials which satisfy it.

It follows from work of Weinberger [378] that under the assumption of a suitable Riemann hypothesis(4.1.18) for $\Delta_0 < 0$ and $e_\Delta = 2$, $|\Delta| < 2 \cdot 10^7$ (see Louboutin [196] as well). Thus, a computer search can provide us with a list of them.(4.1.19) Here they are together with h_Δ, $F(\Delta)$ and $F_\Delta(x)$.

(4.1.18)See Chapter Five, section four, for a description of the GRH.

(4.1.19)In [41, p. 81], Buell states that Gauss' problem of determining all maximal orders with one class per genus is solved. Such discriminants are called "numeri idoneal", or suitable numbers of Euler (see Borevich–Shafarevich [30, p. 426] for the 65 known values). However, his statement is incorrect. The validity of the statement would imply that our Table 4.1.2 is complete without the GRH assumption. To the best of this author's knowledge, this remains an open problem. See Appendix E. This is mentioned only so that the reader will not be led astray when comparing with [41].

Table 4.1.2.

$-D$	h_Δ	$F(\Delta)$	$F_\Delta(x)$
5	2	2	x^2+5
6	2	2	x^2+6
10	2	2	x^2+10
13	2	2	x^2+13
15	2	2	x^2+x+4
21	4	3	x^2+21
22	2	2	x^2+22
30	4	3	x^2+30
33	4	3	x^2+33
35	2	2	x^2+x+9
37	2	2	x^2+37
42	4	3	x^2+42
51	2	2	x^2+x+13
57	4	3	x^2+57
58	2	2	x^2+58
70	4	3	x^2+70
78	4	3	x^2+78
85	4	3	x^2+85
91	2	2	x^2+x+23
93	4	3	x^2+93
102	4	3	x^2+102
105	8	4	x^2+105
115	2	2	x^2+x+29
123	2	2	x^2+x+31
130	4	3	x^2+130
133	4	3	x^2+133
165	8	4	x^2+165
177	4	3	x^2+177
187	2	2	x^2+x+47
190	4	3	x^2+190
195	4	3	x^2+x+49
210	8	4	x^2+210
235	2	2	x^2+x+59
253	4	3	x^2+253
267	2	2	x^2+x+67
273	8	4	x^2+273
330	8	4	x^2+330
345	8	4	x^2+345
357	8	4	x^2+357
385	8	4	x^2+385
403	2	2	$x^2+x+101$
427	2	2	$x^2+x+107$
435	4	3	$x^2+x+109$
462	8	4	x^2+462
483	4	3	$x^2+x+121$
555	4	3	$x^2+x+139$
595	4	3	$x^2+x+149$
627	4	3	$x^2+x+157$
715	4	3	$x^2+x+179$
795	4	3	$x^2+x+199$
1155	8	4	$x^2+x+289$
1365	16	5	x^2+1365
1435	4	3	$x^2+x+359$
1995	8	4	$x^2+x+499$
3003	8	4	$x^2+x+751$
3315	8	4	$x^2+x+829$

Although Theorem 4.1.4 is the canonical generalization of the Rabinowitsch criterion in that it is essentially a criterion for when C_Δ has exponent $e_\Delta \le 2$, in terms of the factorization of the Euler–Rabinowitsch polynomial, our goal stated at the outset was to look at *consecutive* (initial) prime-producing quadratic polynomials. We now show how Table 4.1.1 can be transformed into such polynomials by now turning to the *non-monic* quadratic polynomials. For example, in [372] Van der Pol and Speziali (via C. Coxe) in 1951, observed that $6x^2+6x+31$ is prime for $x=0,1,\ldots,28$. Here $\Delta=-4\cdot 177$ and $D=-177$ appears in Table 4.1.1. This is no accident, as the following demonstrates.

In order to switch our attention to the non-monic polynomials, we need a more general setting which we now describe.

Definition 4.1.4. Let Δ be a discriminant and let $q \ge 1$ be a square-free divisor of Δ. Let $\alpha_\Delta = 1$ if $4q$ divides Δ and $\alpha_\Delta = 2$ otherwise. We call

$$F_{\Delta,q}(x) = qx^2 + (\alpha_\Delta - 1)qx + ((\alpha_\Delta - 1)q^2 - \Delta)/(4q)$$

the qth *Euler–Rabinowitsch polynomial.* (Here the first Euler–Rabinowitsch polynomial is the content of Definition 4.1.1.)

The following also generalizes the Ono invariant.

Definition 4.1.5. Let Δ and q be as in Definition 4.1.4, and let

$$F(\Delta, q) = \max\{\Omega(|F_{\Delta,q}(x)|) : x \in \mathcal{S}_{\Delta,q}\},$$

where

$$\mathcal{S}_{\Delta,q} = \{x : 0 \le x \le \lfloor |\Delta|/(4q) - 1 \rfloor\}.$$

($\lfloor |\Delta|/(4q) - 1 \rfloor$ is called the qth *Rabinowitsch bound.*)

We now provide a generalization of Theorem 4.1.4, which will be our device for examining the prime-producing non-monic quadratic polynomials.

Theorem 4.1.5. *Let $\Delta = \Delta_0 < 0$ ($\Delta \ne -3, -4$) be a fundamental discriminant divisible by exactly $N + 1$ ($N \ge 0$) distinct primes q_i ($1 \le i \le N + 1$) with $q_1 < q_2 < \cdots < q_{N+1}$. If $q \ge 1$ is a square-free divisor of Δ, divisible by exactly $M \ge 0$ of the primes q_i for $i = 1, 2, \ldots, N$, then the following are equivalent:*

(1) $e_\Delta \le 2$.

(2) $F(\Delta, q) = N + 1 - M$ *and* $h_\Delta = 2^{F(\Delta,q)+M-1}$.

Proof. If (2) holds, then $h_\Delta = 2^N$, so Theorem 1.3.3 tells us that (1) holds. If (1) holds and $M = 0$, then this is just Theorem 4.1.4, so we now assume that $M \ge 1$. Let $Q = \prod_{i=1}^{N} q_i$ and form the ideal $I = [Q, b + w_\Delta]$ where

$$b = \begin{cases} 0 & \text{if } D \equiv 0 \pmod 2, \\ D/q_{N+1} & \text{if } D \equiv 3 \pmod 4, \\ (D/q_{N+1} - 1)/2 & \text{if } D \equiv 1 \pmod 4, \end{cases}$$

then an easy exercise verifies that $Q \mid N(b + w_\Delta) \ne Q$. The latter forces, in view of Theorem 4.1.4, $N(b + w_\Delta) = F_\Delta(b)$ to be divisible by exactly $N + 1$ primes.

Claim. $2b+\sigma-1 = q(2x_0+\alpha_\Delta-1)$ for some non-negative integer $x_0 \le |\Delta|/(4q)-1$.

First we observe that, if the equality holds, then $0 \le x_0 \le |\Delta|/4q - 1$ follows immediately from the fact that $0 \le b < |\Delta|/4$. Now, if $\sigma = \alpha_\Delta$, then clearly q divides $2b + \alpha_\Delta - 1$, so $2b + \sigma - 1 = q(2x_0 + \alpha_\Delta - 1)$. If $\alpha_\Delta \ne \sigma$, then we must have $\alpha_\Delta = 2$, $\sigma = 1$ and q even. Therefore, q divides $2b = 2b + \sigma - 1$ where b is odd, i.e. $2b + \sigma - 1 = q(2x_0 + \alpha_\Delta - 1)$, securing the Claim.

By the Claim, $F_\Delta(b) = (q^2(2x_0 + \alpha_\Delta - 1)^2 - \Delta)/4$. Thus, $F_\Delta(b)/q = F_{\Delta,q}(x_0)$ is divisible by exactly $N + 1 - M$ primes. Finally, we show that there cannot exist an integer x with $0 \le x \le |\Delta|/(4q) - 1$ such that $F_{\Delta,q}(x)$ is divisible by more than $N + 1 - M$ primes. Suppose, to the contrary, that one such value does exist. Let

$$y = \begin{cases} qx & \text{if } \alpha_\Delta = 1, \\ qx + (q-1)/2 & \text{if } \alpha_\Delta = 2,\ q \text{ odd}, \\ qx + q/2 & \text{if } \alpha_\Delta = 2,\ q \text{ even}, \end{cases}$$

then $qF_{\Delta,q}(x) = F_\Delta(y)$ with $0 \le y \le |\Delta|/4 - 1$ is divisible by more than $N + 1$ primes, contradicting Theorem 4.1.4. □

Although Theorem 1.3.3 tells us that $e_\Delta \le 2$ if and only if $h_\Delta = 2^N$ when $\Delta < 0$, what Theorem 4.1.5 reveals is that the factorization of the qth Euler–Rabinowitsch polynomial is intimately linked to M and N via condition (2). We now illustrate that this is indeed a "tight fit" (see also Theorem 4.1.8).

Example 4.1.1. If $\Delta = -9867$ and $q = 429$, then $M = N = 3$ and $F(\Delta, q) = N + 1 - M = 1$. However, $h_\Delta = 2^4$ and $4 \neq F(\Delta, q) + M - 1 = 3$. In fact, $C_\Delta = C_2 \times C_2 \times C_4$ where C_i denotes a cyclic group of order i. Yet $F_{\Delta,q}(x) = 429x^2 + 429x + 113$ is prime whenever $0 \leq x \leq \lfloor |\Delta|/4q - 1 \rfloor = 5$.

Example 4.1.2. If $\Delta = -184$, then $N = 1$, $M = 0$ and $F(\Delta, 1) = 3 \neq N + 1 - M$. Here $h_\Delta = 2^{F(\Delta,1)+M-1} = 4$ and $C_\Delta = C_4$.

If we look through Table 4.1.2, we see that $\Delta = -15$ is the only discriminant congruent to 1 modulo 8. Furthermore, we know from Theorem 4.1.3 that the only $\Delta \equiv 1 \pmod 8$, with $h_\Delta = 1$, is $\Delta = -7$.

We now show that these are the only possibilities for *any* fundamental discriminant.

Theorem 4.1.6. *Let* $\Delta = \Delta_0 < 0$ *be a fundamental discriminant with* $\Delta \equiv 1 \pmod 8$, *then the following are equivalent:*

(1) $e_\Delta \leq 2$.

(2) $\Delta = -7$ *or* -15.

Proof. Since $(2) = \mathcal{P}\mathcal{P}'$ with $\mathcal{P} \not\sim \mathcal{P}'$ and $\mathcal{P}^2 \sim 1$, then there exists an $\alpha \in \mathcal{O}_\Delta$, $\alpha \notin \mathbf{Z}$ with $N(\alpha) = 4$, i.e. $16 = x^2 - \Delta y^2$ for some $x, y \in \mathbf{Z}$, $y \neq 0$, so $|\Delta| \leq 15$ follows. □

In view of Theorem 4.1.6, discriminants $\Delta \equiv 1 \pmod 8$ for $\Delta < 0$ are uninteresting from the perspective of both $e_\Delta \leq 2$ and prime-producing quadratics.

Remark 4.1.4. Although we could list numerous examples to illustrate Theorem 4.1.5, given its generality, the most interesting applications for our purposes, in determining non-monic polynomials which generate initial, consecutive distinct prime values, occur when $M = N$. Before illustrating this case with numerous motivating examples, which will lead us to the prime-producing non-monic quadratic polynomials, we make an assumption which we state now so that we may invoke it by name in order to avoid unwanted repetition. This is merely a notational device.

Assumption 4.1.1. Let $\Delta = \Delta_0$ be a fundamental discriminant ($\Delta \neq -3, -4$) divisible by exactly $N + 1$ ($N \geq 0$) distinct primes with q_{N+1} the largest and set $q = \prod_{i=1}^{N} q_i$, the product of the remaining prime divisors of Δ (with $q = 1$ if $N = 0$).

Application 4.1.1. If $\Delta \equiv 4 \pmod 8$ with $\Delta < 0$ satisfies Assumption 4.1.1, then

$$F_{\Delta,q}(x) = qx^2 + qx^2 + (q^2 - \Delta)/(4q)$$

is prime for all non-negative integers $x < q_{N+1}/2 - 1$ whenever $e_\Delta \leq 2$. Under GRH, the largest string of primes occurs when $D = -177$ and $q = 6$ where

$$F_{\Delta,q}(x) = 6x^2 + 6x + 31$$

is prime for $x = 0, 1, ..., 28$ (see Table 4.1.3). This is the example from [372] cited above and which motivated this result.

Application 4.1.2. If $\Delta \equiv 0 \pmod 8$ with $\Delta < 0$ satisfies Assumption 4.1.1, then if $e_\Delta \leq 2$,

$$F_{\Delta,q}(x) = qx^2 + q_{N+1}$$

is prime whenever $0 \leq x \leq q_{N+1} - 1$. Under GRH, the largest such string is given by $D = -58$ and $q = 2$ where

$$2x^2 + 29$$

is prime for $0 \leq x \leq 28$ (see Table 4.1.4). This example was cited by Sierpiński in [339] (probably known to Euler) and motivated this result.

Application 4.1.3. If $\Delta \equiv 1 \pmod 4$ with $\Delta < 0$ satisfies Assumption 4.1.1, then whenever $e_\Delta \leq 2$, we have that

$$F_{\Delta,q}(x) = qx^2 + qx + (q^2 - \Delta)/(4q)$$

is prime for all non-negative integers $x < \lfloor q_{N+1}/4 - 1 \rfloor$. Under GRH, the largest string occurs when $D = -267$ and $q = 3$ where

$$3x^2 + 3x + 23 = F_{\Delta,3}(x)$$

is prime whenever $0 \leq x \leq 21$ (see Table 4.1.5). A version of this example was noticed in 1914 by Levy in [185], and motivated our result.

The following three tables (which together comprise all values in Table 4.1.2) give all $D < 0$, by congruence modulo 4, together with their non-monic, consecutive, prime-producing quadratics for an initial string of values of x. Furthermore, we list the largest prime q_{N+1} and the number of initial, consecutive, distinct prime values (the column labelled B) generated by the associated quadratic as predicted by Theorem 4.1.6. However, it is possible to have a given quadratic, listed in one of the tables, which goes on to be prime beyond the value of B. For example, in Table 4.1.5, $\Delta = D = -1995$ has $B = 4$ listed. Yet, $F_{\Delta,q}(x)$ for $q = 105$ goes on to be prime for 8 values $0 \leq x \leq 7$, rather than just the 4 predicted by Theorem 2.5. Nevertheless, the B value given in Theorem 4.1.6 is a 'sharp' bound in the generality given. For example, in Table 4.1.3, if $D = -177$, then $F_{\Delta,6}(x)$ is prime for $0 \leq x \leq 28$, i.e. $B = 29$, and $F_{\Delta,6}(29)$ is composite.

It is also interesting to observe that in Theorem 4.1.6, we have $F_{\Delta,q}(x) = F_{\Delta,q}(-\alpha_\Delta + 1 - x)$. Hence, we may duplicate its prime-producing capacity (albeit with each prime repeated) by letting x take on the values $x = -1, -2, ..., -\lceil |\Delta|/4q - 1 \rceil$. For example, in Table 4.1.4, when $D = -58$, we have that $F_{\Delta,2}(x) = 2x^2 + 29$ is prime for $-28 \leq x \leq 28$ (as noted in Sierpiński [339]).

Table 4.1.3. $D \equiv 3 \pmod 4$

D	q_{N+1}	$F_{\Delta,q}(x)$	B
5	5	$2x^2+2x+3$	2
13	13	$2x^2+2x+7$	6
21	7	$6x^2+6x+5$	3
33	11	$6x^2+6x+7$	5
37	37	$2x^2+2x+19$	18
57	19	$6x^2+6x+11$	9
85	17	$10x^2+10x+11$	8
93	31	$6x^2+6x+17$	15
105	7	$30x^2+30x+11$	3
133	19	$14x^2+14x+13$	9
165	11	$30x^2+30x+13$	5
177	59	$6x^2+6x+31$	29
253	23	$22x^2+22x+17$	11
273	13	$42x^2+42x+17$	6
345	23	$30x^2+30x+19$	11
357	17	$42x^2+42x+19$	8
385	11	$70x^2+70x+23$	5
1365	13	$210x^2+210x+59$	6

Table 4.1.4. $D \equiv 2 \pmod 4$

D	$q_{N+1}=B$	$F_{\Delta,q}(x)$
6	3	$2x^2+3$
10	5	$2x^2+5$
22	11	$2x^2+11$
30	5	$6x^2+5$
42	7	$6x^2+7$
58	29	$2x^2+29$
70	7	$10x^2+7$
78	13	$6x^2+13$
102	17	$6x^2+17$
130	13	$10x^2+13$
190	19	$10x^2+19$
210	7	$30x^2+7$
330	11	$30x^2+11$
462	11	$42x^2+11$

Table 4.1.5. $D \equiv 1 \pmod 4$

D	q_{N+1}	$F_{\Delta,q}(x)$	B
15	5	$3x^2+3x+2$	1
35	7	$5x^2+5x+3$	2
51	17	$3x^2+3x+5$	4
91	13	$7x^2+7x+5$	3
115	23	$5x^2+5x+7$	5
123	41	$3x^2+3x+11$	10
187	17	$11x^2+11x+7$	4
195	13	$15x^2+15x+7$	3
235	47	$5x^2+5x+13$	12
267	89	$3x^2+3x+23$	22
403	31	$13x^2+13x+11$	7
427	61	$7x^2+7x+17$	16
435	29	$15x^2+15x+11$	7
483	23	$21x^2+21x+11$	5
555	37	$15x^2+15x+13$	9
595	17	$35x^2+35x+13$	4
627	19	$33x^2+33x+13$	4
715	13	$55x^2+55x+17$	3
795	53	$15x^2+15x+17$	13
1155	11	$105x^2+105x+29$	2
1435	41	$35x^2+35x+19$	10
1995	19	$105x^2+105x+31$	4
3003	13	$231x^2+231x+61$	3
3315	17	$195x^2+195x+53$	4

What we have accomplished thus far is a complete list (under the assumption of the GRH[(4.1.20)]) of all polynomials of Euler–Rabinowitsch type which generate consecutive, distinct prime values for an initial range of input values up to the Rabinowitsch bound. This task is thus completed.

[(4.1.20)] See Chapter Five, section four.

We now turn to a search for necessary and sufficient conditions for prime-producing quadratics of negative discriminant, in an initial range, up to a Minkowski bound (Definition 1.3.2), another canonical bound.[(4.1.21)] First, we need a few useful technical lemmata.

Lemma 4.1.2. *Let Δ be a discriminant with associated conductor f_Δ and with fundamental discriminant Δ_0 divisible by a square-free positive $q \in \mathbf{Z}$, where* $\gcd(f_\Delta, q) = 1$. *If p is a prime, then the following are equivalent:*

(1) $F_{\Delta,q}(x) \equiv 0 \pmod p$ *for some* $x \geq 0$.

(2) $(\Delta/p) \neq -1$ *and p does not divide q.*

Proof. See Exercise 4.1.6. □

The above tells us that only non-inert primes divide $F_{\Delta,q}(x)$, and the following tells us certain bounded non-inert primes *must* divide it.

Lemma 4.1.3. *Let Δ be a discriminant and let $q \geq 1$ be a square-free divisor of it. If $p < B$ is any non-inert prime which does not divide q and B is any positive real number, then there exists a non-negative integer $x < (B - \alpha_\Delta + 1)/2$ such that p divides $F_{\Delta,q}(x)$.*

Proof. If $p = 2$, then q is odd. If $\alpha_\Delta = 1$ and $D \equiv 3 \pmod 4$, then $F_{\Delta,q}(1)$ is even. If $D \equiv 0 \pmod 2$, then $F_{\Delta,q}(0)$ is even. If $\alpha_\Delta = 2$, then $\sigma = 2$, so $\Delta \equiv 1 \pmod 8$, so $F_{\Delta,q}(0)$ is even. We may now assume that $p > 2$. By Lemma 4.1.2, there is an integer $x \geq 0$ such that p divides $F_{\Delta,q}(x)$, so $q^2(2x + \alpha_\Delta - 1)^2 \equiv \Delta \pmod p$. Therefore, we may assume without loss of generality that $0 \leq 2x + \alpha_\Delta - 1 < p$ (since we may take the least non-negative residue modulo p, and when $\alpha_\Delta = 2$, we may assume that the residue is odd since p is odd). Hence, $0 \leq x < (B - \alpha_\Delta + 1)/2$. □

The following shows how values of $F_{\Delta,q}(x)$ lead to equivalence between that value and the $\mathcal{O}_\Delta$-ideal above q.

Lemma 4.1.4. *Let Δ be a discriminant with $q \geq 1$, a square-free divisor of $|\Delta|$ with* $\gcd(f_\Delta, q) = 1$. *If $a > 0$ is an integer with $F_{\Delta,q}(x) = a$ where x is any non-negative integer, then $\mathcal{Q} \sim \mathcal{A}$, an $\mathcal{O}_\Delta$-ideal with norm a, and $\mathcal{Q}$ is the unique $\mathcal{O}_\Delta$-prime above q.*

Proof. Form the ideal $\mathcal{AQ} = [aq, (b + \sqrt{\Delta})/2]$ where $b = (2x + \alpha_\Delta - 1)q$, then $N\left((b + \sqrt{\Delta})/2\right) = qF_{\Delta,q}(x) = aq$. Therefore, $\mathcal{AQ} = \left((b + \sqrt{\Delta})/2\right)$, i.e. $\mathcal{Q} \sim \mathcal{A}$. □

Now we can give the precursor result on the Minkowski bound.

Theorem 4.1.7. *If $\Delta < 0$ satisfies Assumption 4.1.1, then (1) $\Rightarrow$ (2) in what follows.*

(1) $F_{\Delta,q}(x)$ *is prime for all integers x with* $0 \leq x < (M_\Delta - \alpha_\Delta + 1)/\alpha_\Delta$.

(4.1.21) This search is motivated by results in the literature such as that of Hendy [119], which the first result in this direction, Theorem 4.1.7, generalizes.

(2) $e_\Delta \le 2$ *and* $M_\Delta < q_{N+1}$.

Proof. By Lemma 4.1.3, there is an integer x with $0 \le x < (M_\Delta - \alpha_\Delta + 1)/2$ such that $F_{\Delta,q}(x) \equiv 0 \pmod p$ for any $p < M_\Delta$ with $(\Delta/p) \ne -1$. Thus, by hypothesis (1), we have that $F_{\Delta,q}(x) = p$ (observing that $(M_\Delta - \alpha_\Delta + 1)/2 \le (M_\Delta - \alpha_\Delta + 1)/\alpha_\Delta$). By Lemma 4.1.4, this implies that $\mathcal{Q} \sim \mathcal{P}$ where $\mathcal{Q}$ (respectively $\mathcal{P}$) has norm q (respectively p) in $\mathcal{O}_\Delta$. Therefore, $\mathcal{P}^2 \sim 1$ and so $e_\Delta \le 2$, by Theorem 1.3.1 (see Chapter One, section five).

Now we show that $M_\Delta < q_{N+1}$. If $x = (q_{N+1} - \alpha_\Delta + 1)/\alpha_\Delta$ which is an integer, then $F_{\Delta,q}(x) = q_{N+1}(qq_{N+1} + \alpha_\Delta/\sigma)/\alpha_\Delta^2$ which is composite (since Assumption 4.1.1 rules out $\Delta = -3$). Hence, $x > \lfloor (M_\Delta - \alpha_\Delta + 1)/\alpha_\Delta \rfloor$. If $M_\Delta > q_{N+1}$, then $(q_{N+1} - \alpha_\Delta + 1)/\alpha_\Delta > \lfloor (M_\Delta - \alpha_\Delta + 1)/\alpha_\Delta \rfloor > (M_\Delta - \alpha_\Delta + 1)/\alpha_\Delta - 1 > (q_{N+1} - \alpha_\Delta + 1)/\alpha_\Delta - 1$. However, there cannot exist an integer strictly between x and $x - 1$. □

The converse of Theorem 4.1.7 fails for $\Delta \equiv 1 \pmod 4$. For instance, if $\Delta = -195$ (from Table 4.1.5), then $\lfloor M_\Delta \rfloor = 8$ and $q_{N+1} = 13$. However, $F_{\Delta,q}(3) = 11 \cdot 17$ where $q = 21$ and $3 < (\lfloor M_\Delta \rfloor - \alpha_\Delta + 1)/\alpha_\Delta = 3.5$. Nevertheless, the converse of Theorem 4.1.7 does hold for $\Delta \equiv 0 \pmod 4$.

Theorem 4.1.8. *If* $\Delta \equiv 0 \pmod 4$ *with* $\Delta < 0$ *satisfies Assumption* 4.1.1, *then the following are equivalent:*

(1) $F_{\Delta,q}(x)$ *is prime for all integers* x *with* $0 \le x < (M_\Delta - \alpha_\Delta + 1)/\alpha_\Delta$.

(2) $e_\Delta \le 2$ *and* $M_\Delta < q_{N+1}$.

Proof. In view of Theorem 4.1.7, we need only prove (2) $\Rightarrow$ (1). If $e_\Delta \le 2$, then by Theorem 4.1.5, we have $F(\Delta, q) = 1$, i.e. $F_{\Delta,q}(x)$ is prime whenever $0 \le x \le |\Delta|/(4q) - 1 = q_{N+1}/\alpha_\Delta - 1$. If $\lfloor (M_\Delta - \alpha_\Delta + 1)/\alpha_\Delta \rfloor > (q_{N+1} - \alpha_\Delta + 1)/\alpha_\Delta - 1$, then by hypothesis, $(q_{N+1} - \alpha_\Delta + 1)/\alpha_\Delta > (M_\Delta - \alpha_\Delta + 1)/\alpha_\Delta > \lfloor (M_\Delta - \alpha_\Delta + 1)/\alpha_\Delta \rfloor > (q_{N+1} - \alpha_\Delta + 1)/\alpha_\Delta - 1$, a contradiction. Hence,

$$\lfloor (M_\Delta - \alpha_\Delta + 1)/\alpha_\Delta \rfloor \le (q_{N+1} - \alpha_\Delta + 1)/\alpha_\Delta - 1 \le q_{N+1}/\alpha_\Delta - 1$$

and (1) holds. □

Remark 4.1.5. We actually know, unconditionally, all values of Δ for which $h_\Delta \le 4$ and $e_\Delta \le 2$. By the $h_\Delta \le 2$ solution of Baker [14]–[15], Stark [346]–[347], and the solution of the $h_\Delta = 4$ problem by Arno [8], we know that all values of Δ in Table 4.1.2, with $h_\Delta \le 4$, comprise a complete list thereof.(4.1.22) Hence, the results of Hendy [119] and Louboutin [198] are reduced to a numerical verification on the initial values of the relevant polynomials. Moreover, attempts to replace M_Δ with a different bound fails for $N > 2$, as noted by the example $\Delta = -2737 \equiv 3 \pmod 4$ found by Louboutin in [198], who pointed out that for $q_{N+1} = 23$, we have that $F_{\Delta,q}(x) = 238x^2 + 238x + 71$ is prime for $0 \le x \le 10 = (q_{N+1} - 3)/2$. However, $C_\Delta = C_2 \times C_2 \times C_4$. Here $\lfloor M_\Delta \rfloor = 60$.

(4.1.22) In fact, in recent correspondence with this author, Steven Arno has indicated that he, Wheeler and Robinson, jointly solved the *odd* class number $h_\Delta = m$ problem when $\Delta < 0$ for $5 \le m \le 23$, and that he now has some new ideas for working on the even cases $h_\Delta \equiv 2 \pmod 4$ (see Arno *et al.* [9].)

Remark 4.1.6. From the unconditional classification for $N \leq 2$ mentioned in Remark 4.1.5, we may list those values for which Theorem 4.1.8 holds when $N \leq 2$. They are

$$-D \in \{6, 10, 22, 58, 78, 102, 190\} \text{ when } \Delta \equiv 0 \pmod 8,$$

and

$$-D \in \{5, 13, 21, 33, 37, 57, 85, 93, 133, 177\} \text{ for } \Delta \equiv 4 \pmod 8.$$

(The only other value in Table 4.1.3 for which Theorem 4.1.8 holds is $\Delta = -345$ where $h_\Delta = 8$.)

We observe just how 'tight' our criterion in Theorem 4.1.8 happens to be by noting that for the values $-D = 42$ or 130, we have that $\lfloor M_\Delta \rfloor = q_{N+1}$, and Theorem 4.1.8 (1) fails for these values. Thus, condition (2) of Theorem 4.1.8 has the most precise bound possible, in that it is within a fraction of an integer from failing for certain values such as the two given above. In fact, $F_{\Delta,q}(q_{N+1})$ is composite if $\Delta \equiv 0 \pmod 8$ (see the proof of Theorem 4.1.7). Furthermore, since attempts to use q_{N+1} as a bound (instead of M_Δ) fail at $N = 3$ as remarked above, then the criterion in Theorem 4.1.8 is the most accurate general criterion for prime quadratics when $\Delta \equiv 0 \pmod 4$ up to a Minkowski bound. Now we turn to a criterion for $\Delta \equiv 1 \pmod 4$.

Conjecture 4.1.2. *If* $\Delta \equiv 1 \pmod 4$ *with* $\Delta < 0$ *satisfies Assumption* 4.1.1, *then the following are equivalent:*

(1) $e_\Delta \leq 2$ *and* $M_\Delta < (q_{N+1} + 3)/2$.

(2) $F_{\Delta,q}(x)$ *is prime whenever* $0 \leq x < (M_\Delta - 1)/2$.

Conjecture 4.1.2 is valid if we assume the GRH, since we just check the list in Table 4.1.5 (and those $h_\Delta = 1$) to verify it. Furthermore, the values for which it holds in Table 4.1.5 are certain values for which $N \leq 2$.

The bound in (1) of Conjecture 4.1.2 is the best possible. For example, if $\Delta = -483$, then (2) of Conjecture 4.1.2 holds and $\lfloor M_\Delta \rfloor = 12 < M_\Delta < (q_{N+1} + 3)/2 = (23+3)/2 = 13$. Also, $F_{\Delta,q}(6) = F_{\Delta,q}((q_{N+1} + 1)/4) = 19 \cdot 47$ where $q = 21$ and $h_\Delta = 4$. $F_{\Delta,q}(x)$ is prime whenever $0 \leq x \leq 5 < (M_\Delta - 1)/2 < 6$. Also, if $\Delta = -195$, then if $q = 15$, we have $M_\Delta > (q_{N+1} + 3)/2 = (13 + 3)/2 = 8 = \lfloor M_\Delta \rfloor$ and $F_{\Delta,q}(3) = F_{\Delta,q}((q_{N+1} - 1)/4) = 11 \cdot 17$, so (2) of Conjecture 4.1.2 fails since $3 < (M_\Delta - 1)/2$. Hence, the condition $M_\Delta < (q_{N+1} + 3)/2$ in (1) of Conjecture 4.1.2 is as sharp as possible, as shown by the above two examples which show that it fails if we lower the bound to $(q_{N+1} + 2)/2$ (a counterexample is $\Delta = -483$), and it fails if we raise the bound to $(q_{N+1} + 4)/2$ (a counterexample is $\Delta = -195$).

Remark 4.1.7. We just miss being able to directly prove that (2) $\Rightarrow$ (1) of Conjecture 4.1.2 holds. We know that $F_{\Delta,q}(p-1) = p(pq - q + 1)$ where $p = (q + q_{N+1})/4$. Hence, if (2) of Conjecture 4.1.2 holds, then $(q_{N+1} + q)/4 - 1 \geq (M_\Delta - 1)/2$, i.e. $(q_{N+1} + q)/2 - 1 \geq M_\Delta$. However, we cannot get any closer than this in general. Observe that, if $h_\Delta = 2$, then $x = p - 1$ is the first value for which $F_{\Delta,q}(x)$ is composite in Table 4.1.5. We note that, in fact, if $M_\Delta < (q_{N+1} + 3)/2$, then $h_\Delta \leq 4$. We also observe that (under Assumption 4.1.1), if $\Delta \equiv 1 \pmod 4$, then all $h_\Delta \leq 4$ in Table 4.1.5 which satisfy $M_\Delta < (q_{N+1} + 3)/2$ have the property that $q = m^2 + r$ where $|r| \in \{1, 2, 4\}$. If we could, unconditionally, verify that (2) of Conjecture

4.1.2 must imply this condition on q, then we would have a constructive proof that (2) $\Rightarrow$ (1). For example, if $q+1$ is a square, then

$$F_{\Delta,q}((q_{N+1}-1)/4) = [(q_{N+1}+1)/4]^2(q+1) - [(q_{N+1}-1)/4]^2$$

is a difference of squares, so $M_\Delta \leq (q_{N+1}+1)/2$. The values in Table 4.1.2 for which $q+1$ is a square, $h_\Delta \leq 4$ and $M_\Delta \leq (q_{N+1}+1)/2$, are $-\Delta \in \{15, 51, 123, 267, 435, 555, 795\}$. In fact, one observes that

$$\begin{aligned} F_{\Delta,q}(p-1) &= F_{\Delta,q}((q_{N+1}+q)/4-1) \\ &= [((q_{N+1}+q-4)(q+1)+8)/8]^2 - [(q-1)(q_{N+1}+q-4)/8]^2 . \end{aligned}$$

Furthermore, each of the values in Table 4.1.2 first becomes composite for an x value, such that $F_{\Delta,q}(x)$ is a difference of squares in a unique way depending on the aforementioned special form for q. That this phenomenon even occurs is of interest in its own right, so we tabulate the values as follows. The column labelled x_0 represents the smallest non-negative value of x with $F_{\Delta,q}(x_0)$ composite. The column labelled $F_{\Delta,q}(x_0)$ lists the value as a difference of squares, and the column labelled q lists it in its special form m^2+r as above. A column for q_{N+1} is also given. To prove that q is one of these special forms, say $q = m^2 - 4$, we must verify that $q = x_0 x_1$ where $x_0 - x_1 = 4$, since $((x_0+x_1)/2)^2 - ((x_0-x_1)/2)^2 = q$ (see Exercise 4.1.7).

Table 4.1.6.

$-\Delta$	x_0	$F_{\Delta,q}(x_0)$	q	q_{N+1}
15	$1 = (q_{N+1}-1)/4$	$3^2 - 1^2$	$3 = 2^2 - 1$	5
35	$2 = (q_{N+1}+1)/4$	$7^2 - 4^2$	$5 = 3^2 - 4$	7
51	$4 = (q_{N+1}-1)/4$	$9^2 - 4^2$	$3 = 2^2 - 1$	17
91	$4 = (q_{N+1}+3)/4$	$17^2 - 12^2$	$7 = 3^2 - 2$	13
115	$6 = (q_{N+1}+1)/4$	$19^2 - 12^2$	$5 = 2^2 + 1$	23
123	$10 = (q_{N+1}-1)/4$	$21^2 - 10^2$	$3 = 2^2 - 1$	41
187	$6 = (q_{N+1}+7)/4$	$37^2 - 30^2$	$11 = 3^2 + 2$	17
235	$12 = (q_{N+1}+1)/4$	$37^2 - 24^2$	$5 = 2^2 + 1$	47
267	$22 = (q_{N+1}-1)/4$	$45^2 - 22^2$	$3 = 2^2 - 1$	89
403	$10 = (q_{N+1}+9)/4$	$71^2 - 60^2$	$13 = 3^2 + 4$	31
427	$16 = (q_{N+1}+3)/4$	$65^2 - 48^2$	$7 = 3^2 - 2$	61
435	$7 = (q_{N+1}-1)/4$	$30^2 - 7^2$	$15 = 4^2 - 1$	29
483	$6 = (q_{N+1}+1)/4$	$33^2 - 14^2$	$21 = 5^2 - 4$	23
555	$9 = (q_{N+1}-1)/4$	$38^2 - 9^2$	$15 = 4^2 - 1$	37
795	$13 = (q_{N+1}-1)/4$	$54^2 - 13^2$	$15 = 4^2 - 1$	53

Remark 4.1.8. It is worth pointing out that Lemma 4.1.4 tells us that when $q = 1$, then $\mathcal{A} \sim 1$. What this means is that the factorization of the Euler–Rabinowitsch polynomial $F_{\Delta,1}(x) = F_\Delta(x)$ up to the Rabinowitsch bound $(|\Delta|/4 - 1)$, as given by Theorem 4.1.4, tells us the equivalence classes in C_Δ. For instance, we have

Example 4.1.3. If $\Delta = -3315$, then $F_\Delta(x) = x^2 + x + 829$ and $h_\Delta = 8$ from Table 4.1.2. Also, the Rabinowitsch bound is $\lfloor |\Delta|/4 - 1 \rfloor = 827$. The only split primes $p < M_\Delta$ are 29 and 31 (see Table 4.1.7 below), and since $F_\Delta(19) = 3 \cdot 13 \cdot 31$, then $\mathcal{Q}_3\mathcal{Q}_{13} \sim \mathcal{Q}_{31}$ where $\mathcal{Q}_q$ lies over q in $\mathcal{O}_\Delta$. Also, since $F_\Delta(25) = 3 \cdot 17 \cdot 29$, then $Q_3Q_{17} \sim Q_{29}$. In fact, Theorem 4.1.10 below says a great deal about $e_\Delta \leq 2$ and the number of split primes $p < M_\Delta$. First, we have the following introductory result on a Rabinowitsch bound.

Theorem 4.1.9. *If $\Delta < 0$ satisfies Assumption 4.1.1, $F(\Delta, q) = 1$ and $M_\Delta \leq |\Delta|/(2q) - 1$, then we have $e_\Delta \leq 2$.*

Proof. This is an easy consequence of Theorem 4.1.5. □

Remark 4.1.9. The condition $M_\Delta \leq |\Delta|/2q - 1$ is 'stronger' than $h_\Delta = 2^N$ which, of course, gives us a necessary and sufficient condition for $e_\Delta \leq 2$ in Theorem 4.1.5. The above condition is clearly not necessary. In fact, $M_\Delta \leq |\Delta|/2q - 1$, for example, implies that $4q < 3q_{N+1}$, if $\sigma = 1$ or $\alpha_\Delta = 1$.

We now present a necessary and sufficient condition for $e_\Delta \leq 2$ for a fundamental discriminant Δ.

Theorem 4.1.10. *If $\Delta = \Delta_0 < 0$ is a fundamental discriminant, then the following are equivalent:*

(1) $e_\Delta \leq 2$.

(2) *For every split prime $p < M_\Delta$, there exists a square-free divisor, $q > p$, of $|\Delta|$ such that $\Delta = q^2 - 4pq$.*

Proof. If (2) holds, then (1) follows from Theorem 1.3.1 and Lemma 4.1.4, since $F_{\Delta,q}(0) = p$ when $\Delta \not\equiv 0 \pmod 8$, which we may assume, given that such a split prime does not otherwise exist.

Conversely, if (1) holds and $p < M_\Delta$ is any split prime, then by Exercise 1.2.8, there exists an $\mathcal{O}_\Delta$-ideal $I = [p, (b + \sqrt{\Delta})/2]$ for some $b \in \mathbf{Z}$ with $|b| < p$. The result now follows from Corollary 1.4.5. □

Remark 4.1.10. Notice that in the above proof, the question of the "reduction" of I did not arise. However, it turns out that there is a rather remarkable fact, i.e. that for any split prime $p < M_\Delta$, I *must* be reduced when $e_\Delta \leq 2$. This is the content of (1) in the following consequence of Theorem 4.1.10.

Corollary 4.1.5. *If $\Delta < 0$ and $e_\Delta \leq 2$ for a fundamental discriminant Δ, then*

(1) *If $p < M_\Delta$ is a split prime, then the $\mathcal{O}_\Delta$-primes above p are reduced, i.e. there are no* non-reduced *$\mathcal{O}_\Delta$-ideals of norm $p < M_\Delta$.*

(2) *If $\Delta \equiv 4, 5 \pmod 8$, then for any split prime $p < M_\Delta$, the $\mathcal{O}_\Delta$-ideals above p are $\mathcal{P} = [p, (q - 2p + \sqrt{\Delta})/2]$, and $\mathcal{P}' = [p, (q' - 2p + \sqrt{\Delta})/2]$, where $q' = |\Delta|/q$, and $\mathcal{Q} \sim \mathcal{P}$ is the unique ideal above $q > p$.*

(3) *If $\Delta \equiv 0 \pmod 8$, then there are no split primes $p < M_\Delta$.*

(4) *If $\Delta \equiv 1 \pmod 8$, then $\Delta = -7$ or -15.*

(5) *If $h_\Delta = 1$, then there are no split primes $p < M_\Delta$.*

Proof. Let $p < M_\Delta$ be a split prime such that $I = [p, (b + \sqrt{\Delta})/2]$, with $|b| < p$, is not reduced. By Theorem 1.4.2(b), $p > (b^2 - \Delta)/(4p) = c$. Form the Lagrange neighbour, $I^+ = [c, (-b + \sqrt{\Delta})/2]$ of I, which is reduced by Algorithm 2.2.1. The reader may verify that since $p < M_\Delta$ implies $|b| < c$, then c does not divide $|\Delta|$. Therefore, by Theorem 1.4.2(e), $b^2 - \Delta = 4c^2$ forcing $c = p$, a contradiction. This secures (1). (2) follows from (1) and Theorem 1.4.2(e)–(f). (3) is contained in Corollary 1.4.5. (4) is clear, and (5) is straightforward. □

Conjecture 4.1.3. *If $\Delta = \Delta_0 < 0$ is a fundamental discriminant with $\Delta \equiv 5 \pmod 8$ and $e_\Delta \le 2$, then there are at most 2 split primes $p < M_\Delta$, and if $\Delta \equiv 4 \pmod 8$, then there is at most 1 such prime.*

We can, however, give the following unconditional proof for a certain case.

Theorem 4.1.11. *If $\Delta < 0$ satisfies Assumption 4.1.1 with $\Delta \not\equiv 0 \pmod 8$ and $F_{\Delta,q}(x)$ is prime whenever $0 \le x < (M_\Delta - 1)/2$, then there is at most one split prime $p < M_\Delta$.*

Proof. If $F_{\Delta,q}(1) < M_\Delta$, then $2q + (q^2 - \Delta)/(4q) = (9q + q_{N+1})/4 < M_\Delta$ if $\Delta \equiv 1 \pmod 4$. Thus, $243q^2 + 38qq_{N+1} + 3q_{N+1}^2 < 0$, a contradiction. Therefore, by Lemma 4.1.3, the only possible split prime $p < M_\Delta$ is $p = (q + \alpha_\Delta q_{N+1}/\sigma)/4$. The other case is similar. □

To illustrate Theorems 4.1.10–4.1.11 and Conjecture 4.1.3, we provide the following list of fundamental discriminants for which a split prime $p < M_\Delta$ exists together with a value of q such that $\Delta = q^2 - 4pq$ (and its complement q' where $qq' = \prod_{i=1}^{N+1} q_i$, with the q_i's being all the distinct primes dividing Δ_0). The table is split into 2 parts, one for $\Delta_0 \equiv 4 \pmod 8$, and one for $\Delta_0 \equiv 5 \pmod 8$.

Table 4.1.7.

(I) $\Delta_0 \equiv 4 \pmod 8$

$-D_0$	p	q	q'
21	5	14	6
105	11	30	14
133	13	38	14
165	13	30	22
273	17	42	26
345	19	46	30
357	19	42	34
1365	37	78	70

(II) $\Delta_0 \equiv 5 \pmod 8$

$-\Delta_0$	p	q	q'
35	3	7	5
91	5	13	7
187	7	17	11
195	7	15	13
403	11	31	13
435	11	29	15
483	11	23	21
555	13	37	15
595	13	35	17
627	13	33	19
1155	17	35	33
1155	19	55	21
1995	23	57	35
3003	29	77	39
3003	31	91	33
3315	29	65	51
3315	31	85	39

We now show how the structure of $C_{\Delta,2}$ for $\Delta = \Delta_0 < 0$ can be completely determined by the representation of Δ as a difference of two squares. This has some interesting consequences when $e_\Delta = 2$, which we illustrate after the result. (See Exercise 4.1.7.)

Theorem 4.1.12. *If $\Delta \equiv 5 \pmod 8$ with $\Delta < 0$ satisfies Assumption 4.1.1, then Δ is a difference of squares in exactly 2^N distinct ways, namely*

$$\Delta = a_i^2 - 4b_i^2$$

with $b_i \leq N(\omega_\Delta)$, $a_i = (q_{N+1}q^{(i)} \mp q_1^{(i)})/2$ and $b_i = (q_{N+1}q^{(i)} \pm q_1^{(i)})/4$, where

$$q = q^{(i)}q_1^{(i)}$$

runs through all 2^{N-1} distinct factorizations for $q^{(i)} \geq 1$. Furthermore, the primitive ideals $[b_i, (a_i + \sqrt{\Delta})/2]$ comprise the elementary abelian 2-subgroup $C_{\Delta,2}$ of C_Δ. Also, for each $i \leq 2^{N-1}$,

$$\Delta = (q_{N+1}q^{(i)})^2 \mp 4b_i q_{N+1}q^{(i)}$$

(where "$\mp$" corresponds to "$\pm$" in the definitions of b_i above).

Proof. q_{N+1} being prime has exactly one representation as a difference of two squares, namely

$$q_{N+1} = [(q_{N+1}+1)/2]^2 - [(q_{N+1}-1)/2]^2 .$$

Moreover, q has exactly 2^{N-1} distinct such representations, namely

$$q = \left[(q^{(i)} + q_1^{(i)})/2\right]^2 - \left[(q^{(i)} - q_1^{(i)})/2\right]^2$$

for each of the 2^{N-1} distinct factorizations $q = q^{(i)}q_1^{(i)}$ for $i = 1, 2, ..., 2^{N-1}$ with $q^{(i)} \geq 1$. Furthermore, each such representation for q yields 2 distinct such representations for Δ as follows (where $x = (q_{N+1}-1)/2$, $y = (q_{N+1}+1)/2$, $u = (q^{(i)}+q_1^{(i)})/2$ and $v = (q^{(i)} - q_1^{(i)})/2$),

$$\Delta = (x^2 - y^2)(u^2 - v^2) = (xu \pm yv)^2 - (yu \pm xv)^2.$$

Moreover, since $\Delta \equiv 5 \pmod 8$, then $yu \pm xv$ is divisible by 4 and $yu + xv = (q_{N+1}q^{(i)} + q_1^{(i)})/2$ has its largest possible value at $q^{(i)} = q$ and $q_1^{(i)} = 1$, i.e. $(yu + xv)/2 = (1 - \Delta)/4 = N(\omega_\Delta)$. Thus, we have that for

$$b_i = \left((q_{N+1}q^{(i)})^2 - \Delta\right)/(4q_{N+1}q^{(i)}),$$

and

$$b_i^{(1)} = \left((q_{N+1}q^{(i)})^2 - \Delta\right)/(4q_{N+1}q_1^{(i)}),$$

$\mathcal{B}_i \sim \mathcal{Q}_{N+1}\mathcal{Q}^{(i)}$ and $\mathcal{B}_i^{(1)} \sim \mathcal{Q}_{N+1}\mathcal{Q}_1^{(i)}$ where $\mathcal{B}_i$, $\mathcal{B}_i^{(1)}$, $\mathcal{Q}_{N+1}$, $\mathcal{Q}_1^{(i)}$ and $\mathcal{Q}^{(i)}$ are $\mathcal{O}_\Delta$-primes above b_i, $b_i^{(1)}$, q_{N+1}, $q_1^{(i)}$ and $q^{(i)}$ respectively, by Lemma 4.1.4. If $\mathcal{B}_i \sim \mathcal{B}_j$ or $\mathcal{B}_i \sim \mathcal{B}_j^{(1)}$ for any $i \neq j$ (say $\mathcal{B}_i \sim \mathcal{B}_j$ for convenience), then $\mathcal{Q}^{(i)}\mathcal{Q}^{(j)} \sim 1$. However, after possibly removing square factors of ideals from $\mathcal{Q}^{(i)}\mathcal{Q}^{(j)}$ (since squares are necessarily principal), we are left with a nontrivial principal $\mathcal{O}_\Delta$-prime whose norm divides q. This is a contradiction, since the value q is the product of those ramified primes whose $\mathcal{O}_\Delta$-primes generate $C_{\Delta,2}$. Hence, the $[b_i, (a + \sqrt{\Delta})/2]$ comprise $C_{\Delta,2}$. The last statement of the theorem is an easy check. □

We illustrate Theorem 4.1.12 with the following example from Table 4.1.5.

Example 4.1.4. If $\Delta = -3315$, then $h_\Delta = 8$, $(N = 3)$, $q_{N+1} = 17 = q^2 - 8^2$ and $q = 195 = 14^2 - 1^2 = 22^2 - 17^2 = 34^2 - 31^2 = 98^2 - 97^2$. Hence,

$$\begin{aligned}
\Delta &= (8^2 - 9^2)(14^2 - 1^2) \\
&= (8 \cdot 14 + 9 \cdot 1)^2 - (9 \cdot 14 + 8 \cdot 1)^2 \\
&= 121^2 - 4 \cdot 67^2 && (4.1.1) \\
&= (8 \cdot 14 - 9 \cdot 1)^2 - (9 \cdot 14 - 8 \cdot 1)^2 \\
&= 103^2 - 4 \cdot 59^2 && (4.1.2) \\
&= (8^2 - 9^2)(22^2 - 17^2) \\
&= (8 \cdot 22 + 9 \cdot 17)^2 - (9 \cdot 22 + 8 \cdot 17)^2 \\
&= 329^2 - 4 \cdot 167^2 && (4.1.3) \\
&= (8 \cdot 22 - 9 \cdot 17)^2 - (9 \cdot 22 - 8 \cdot 17)^2 \\
&= 23^2 - 4 \cdot 31^2 && (4.1.4) \\
&= (8^2 - 9^2)(34^2 - 31^2) \\
&= (8 \cdot 34 + 9 \cdot 31)^2 - (9 \cdot 34 + 8 \cdot 31)^2 \\
&= 551^2 - 4 \cdot 277^2 && (4.1.5) \\
&= (8 \cdot 34 - 9 \cdot 31)^2 - (9 \cdot 34 - 8 \cdot 31)^2 \\
&= 7^2 - 4 \cdot 29^2 && (4.1.6) \\
&= (8^2 - 9^2)(98^2 - 97^2) \\
&= (8 \cdot 98 + 9 \cdot 97)^2 - (9 \cdot 98 + 8 \cdot 97)^2 \\
&= 1657^2 - 4 \cdot 829^2 && (4.1.7) \\
&= (8 \cdot 98 - 9 \cdot 97)^2 - (9 \cdot 98 - 8 \cdot 97)^2 \\
&= 89^2 - 4 \cdot 53^2. && (4.1.8)
\end{aligned}$$

We note that each of the $8 = 2^N$ distinct representatives (4.1.1)–(4.1.8) of Δ as a difference of squares yields the class group C_Δ, via the primitive ideals $I_1 = [67, (11^2 + \sqrt{\Delta})/2]$, $I_2 = [59, (103 + \sqrt{\Delta})/2]$, $I_3 = [167, (329 + \sqrt{\Delta})/2]$, $I_4 = [31, (23 + \sqrt{\Delta})/2]$, $I_5 = [277, (551 + \sqrt{\Delta})/2]$, $I_6 = [29, (7 + \sqrt{\Delta})/2]$, $I_7 = [829, (1657 + \sqrt{\Delta})/2]$ and $I_8 = [53, (89 + \sqrt{\Delta})/2]$. These ideals are not necessarily reduced, but the two representing the split prime $p < M_\Delta$, i.e. I_4 and I_6, are reduced. Furthermore, via the last statement of Theorem 4.1.12, we have

$$\Delta = (q_{N+1}q^{(i)})^2 - 4b_i q_{N+1} q^{(i)} = (17 \cdot 5)^2 - 4 \cdot 31 \cdot 17 \cdot 5 = (17 \cdot 3)^2 - 4 \cdot 29 \cdot 17 \cdot 3$$

as predicted by Theorem 4.1.10. Moreover, it can be shown, via the factorization of the Euler-Rabinowitsch polynomial as given in Theorem 4.1.4, that $I_1 \sim \mathcal{Q}_{13}$, $I_2 \sim \mathcal{Q}_3\mathcal{Q}_5$, $I_3 \sim \mathcal{Q}_5$, $I_4 \sim \mathcal{Q}_3\mathcal{Q}_{13}$, $I_5 \sim \mathcal{Q}_3$, $I_6 \sim \mathcal{Q}_3\mathcal{Q}_{17}$, $I_7 \sim 1$, and $I_8 \sim \mathcal{Q}_{17}$. Thus, $C_\Delta = \langle\{I_1\}\rangle \times \langle\{I_3\}\rangle \times \langle\{I_5\}\rangle$.

Thus, Theorem 4.1.12 provides a description of the class group C_Δ via differences of squares when $e_\Delta = 2$. Note that by Theorem 4.1.10, if $e_\Delta = 2$, then necessarily all $p < M_\Delta$ which are split primes must appear as some b_i in Theorem 4.1.12.

Remark 4.1.11. If we let $q^{(i)} = q/q_i$ and $q_1^{(i)} = q_i$ in Theorem 4.1.12 for $i = 1, 2, \ldots, N$, then we get that N differences of squares $(xu + yv)^2 - (yu + xv)^2$ (for

x, y, u and v as given in the proof of Theorem 4.1.12) give rise to N generators of $C_{\Delta,2}$. Thus, we have

Corollary 4.1.6. *If* $\Delta \equiv 5 \pmod 8$ *with* $\Delta < 0$ *satisfies Assumption* 4.1.1, *then* $C_{\Delta,2}$ *is generated by the classes containing*

$$[b_i, (a_i + \sqrt{\Delta})/2],$$

for $i = 1, 2, \ldots, N$, *where*

$$b_i = (q_i^2 - \Delta)/(4q_i) \text{ and } a_i = -(q_i^2 + \Delta)/(2q_i).$$

We also have similar results for $\Delta \equiv 0 \pmod 4$.

Theorem 4.1.13. *If* $\Delta \equiv 4 \pmod 8$, $\Delta < 0$ *is a fundamental discriminant divisible by exactly* $N + 1$ *distinct primes with* $q_1 = 2 < q_2 < q_3 < \ldots < q_{N+1}$, *then* $C_{\Delta,2}$ *is generated by the classes containing the ideals*

$$[b_i, a_i + \sqrt{D}],$$

where

$$b_i = (q_i^2 - D)/(2q_i) \text{ and } a_i = -(q_i^2 + D)/(2q_i),$$

for $i = 2, 3, \ldots, N$, *and*

$$b_1 = (1 - D)/2, \text{ with } a_1 = -(1 + D)/2.$$

Proof. Exercise 4.1.8. □

Note that the ideals in Theorem 4.1.13 arise from D as a difference of 2 squares in much the same way as Theorem 4.1.12. For instance, we have

Example 4.1.5. Let $\Delta = -4 \cdot 1365 = -4 \cdot 3 \cdot 5 \cdot 7 \cdot 13$ (from Table 4.1.3) where $105 = q = 3 \cdot 5 \cdot 7 = q_2 q_3 q_4$ and $q_5 = q_{N+1} = 13$. Also, $q = 19^2 - 16^2 = 11^2 - 4^2 = 53^2 - 52^2 = 13^2 - 8^2$ and $q_{N+1} = 7^2 - 6^2$. Hence,

$$\begin{aligned} D &= (6^2 - 7^2)(19^2 - 16^2) = 226^2 - 229^2 \\ &= (6^2 - 7^2)(11^2 - 4^2) = 94^2 - 101^2 \\ &= (6^2 - 7^2)(53^2 - 52^2) = 682^2 - 683^2 \\ &= (6^2 - 7^2)(13^2 - 8^2) = 134^2 - 139^2 \end{aligned}$$

and these four representations give rise to the generators of C_Δ, namely $I_1 = [229, 226 + \sqrt{D}]$, $I_2 = [101, 94 + \sqrt{D}]$, $I_3 = [683, 682 + \sqrt{D}]$, $I_4 = [139, 134 + \sqrt{D}]$. Also, we observe (from the factorization of the Euler–Rabinowitsch polynomial as given in Theorem 4.1.4) that $I_1 \sim Q_2 Q_3$, $I_2 \sim Q_2 Q_7$, $I_3 \sim Q_2$, and $I_4 \sim Q_2 Q_5$, where Q_q is the unique $\mathcal{O}_\Delta$-ideal above q.

We now present a result for $\Delta \equiv 0 \pmod 8$.

Theorem 4.1.14. *If* $\Delta \equiv 0 \pmod 8$ *with* $\Delta < 0$, *is a fundamental discriminant divisible by exactly* $N + 1$ *distinct primes* q_i, *then* $C_{\Delta,2}$ *is generated by the classes of the ideals*

$$I_i = [b_i, a_i + \sqrt{D}],$$

for $i = 2, 3, ..., N+1$, *where*

$$b_i = 2\bar{q}/q_i + q_i \text{ and } a_i = -q_i,$$

with $\bar{q}$ *being the product of the* N *distinct* *odd* *primes* $q_2, q_3, ..., q_{N+1}$.

Proof. Exercise 4.1.8. □

Theorem 4.1.14 is illustrated by the following, which shows how the ideals arise from Δ as a difference of squares.

Example 4.1.6. Let $\Delta = 1848 = 8 \cdot 3 \cdot 7 \cdot 11$ (from Table 4.1.4). Here $8 = 3^2 - 1$ and $231 = 3 \cdot 7 \cdot 11 = 40^2 - 37^2 = 20^2 - 13^2 = 16^2 - 5^2$. Thus

$$\begin{aligned}\Delta &= (1^2 - 3^2)(40^2 - 37^2) = 151^2 - 157^2 \\ &= (1^2 - 3^2)(20^2 - 13^2) = 59^2 - 73^2 \\ &= (1^2 - 3^2)(16^2 - 5^2) = 31^2 - 53^2.\end{aligned}$$

If we set $q_2 = 3$, $q_3 = 7$, $q_4 = 11$, and $\bar{q} = 231$, then $b_1 = 157$, $b_2 = 73$, and $b_3 = 53$. Thus, the generators are $I_1 = [157, -3 + \sqrt{D}] = [157, 154 + \sqrt{D}]$, $I_2 = [73, -7 + \sqrt{D}] = [73, 66 + \sqrt{D}]$, and $I_3 = [53, -11 + \sqrt{D}] = [53, 42 + \sqrt{D}]$. We observe that, unlike the cases in Examples 4.1.4–4.1.5, D cannot be represented as a difference of two squares (Exercise 4.1.7(b)). Also, we cannot form, for example, $J = [157, 151 + \sqrt{\Delta}]$, since this is *not* an ideal in $\mathcal{O}_\Delta$ (see Theorem 1.2.1), although it *is* an ideal in $\mathcal{O}_{4\Delta}$. Thus, the best that we can hope to accomplish is I_i for $i = 1, 2, 3$ above. In each case, $D = q_i^2 - b_i q_i$ as predicted by Theorem 4.1.14. Furthermore, by using Theorem 1.2.2, we achieve in each case that $I_i = [b_i, b_i - q_i + \sqrt{D}]$, as above.

This concludes our study of prime-producing quadratics of negative discriminant up to both a Rabinowitsch and a Minkowski bound, together with the spin-offs we discussed involving class group structure via differences of squares. In the next section, we will investigate prime-producing quadratics of positive discriminant.

Exercises 4.1

1. Prove that $F_n(x) = x^2 - (2n+1)x + n^2 + n + 41$ is prime for $x = 0, 1, 2, \ldots, 40+n$ whenever $0 \leq n \leq 39$. How many of the primes are repeated? Find the values of x in the above range for which $F_n(x_1) = F_n(x_2)$. Show that there cannot exist a longer string of consecutive primes for $n \geq 40$.

2. Prove that $p, p+2, p+6, p+8, p+24$ cannot be simultaneously prime infinitely often.

3. Does Theorem 4.1.1 hold if we allow $0 \leq x \leq B$?

4. Prove that if $r \in \mathbf{Z}$ is positive where $r = \prod_{i=1}^{M} p_i^{e_i}$ is its canonical prime factorization, then the number of distinct factorizations of r into two positive factors is 2^{M-1}.

5. (a) Show that the polynomials $f(x) = 9x^2 - 231x + 1523$ and $g(x) = 9x^2 - 471x + 6203$ generate 40 consecutive primes and that both have the discriminant $\Delta = -3^2 \cdot 163$. Show that $k(x) = x^2 - 2999x + 2248541$ is prime for 80 consecutive values of x.

   (b) Let $\Delta = -1467$ and prove that $h_\Delta = 4$. (*Hint*: See Chapter One, footnote (1.5.9).)

   (c) Prove that C_Δ is cyclic and find generators by using Algorithm 2.2.1. (*Hint*: See Exercise 1.5.12.)

6. Prove Lemma 4.1.2. (*Hint*: For $(\Delta/p) \neq -1$, establish the existence of an $x \in \mathbf{Z}$ such that $\Delta \equiv q^2(2x + \alpha_\Delta - 1)^2 \pmod{p}$. The converse is easy.)

7. (a) Fermat proved in 1643 that odd $n > 2$ is prime if and only if n is the difference of two squares in exactly one way, i.e. $n = x^2 - y^2$ for $x, y > 0$. Prove this. (*Hint*: $p = (\frac{p+1}{2})^2 - (\frac{p-1}{2})^2$ and $ab = (\frac{a+b}{2})^2 - (\frac{a-b}{2})^2$.) Show that this fails if n is allowed to be even.

   (b) Prove that $n \in \mathbf{Z}$ is a difference of two squares if and only if $n \not\equiv 2 \pmod{4}$.

   (c) Prove that $n \in \mathbf{Z}$ is a difference of relatively prime squares if and only if $n \not\equiv 2, 4, 6 \pmod{8}$.

   (d) Prove that odd $n \in \mathbf{Z}$ is a difference of relatively prime squares in exactly one way (i.e. $n = x^2 - y^2$ $(x, y > 0)$) if and only if $n = p^e$ where p is prime and $e \geq 1$.

   (e) Prove that if $a = x^2 - y^2$ and $b = u^2 - v^2$, then $ab = (xu + yv)^2 - (yu + xv)^2 = (xu - yv)^2 - (yu - xv)^2$.

   (f) Assume that n is representable as a difference of squares. Prove that the number of ways that $n \in \mathbf{Z}$ can be represented as a difference of relatively prime squares is $2^{d(n)-1}$ where $d(n)$ is given in Definition 4.1.3. Here we are counting only those differences $x^2 - y^2$ where $x, y > 0$.

   (g) Prove that n is square-free if and only if $2^{d(n)} = \tau(n)$. (See Definition 4.1.2.)

8. Prove Theorems 4.1.13–4.1.14. (*Hint*: Use the same techniques as in the proof of Theorem 4.1.12.)

9. Let Δ be a discriminant with conductor f_Δ and let q be a positive square-free divisor of Δ with $\gcd(f_\Delta, q) = 1$. Prove that if $p < M_\Delta$ is prime, then $F_\Delta(x) \equiv 0 \pmod{p}$ for some x with $0 \leq x < (M_\Delta - \sigma + 1)/2$ if and only if $(\Delta/p) \neq -1$ and p does not divide q. (*Hint*: Use Lemmas 4.1.2–4.1.3.)

10. Does Theorem 4.1.6 hold for arbitrary orders? If it does, then prove it. If not, then find a discriminant $\Delta < 0$, $\Delta \equiv 1 \pmod{8}$ with $e_\Delta = 2$ and $\Delta \neq -7$ or -15.

11. Provide a version of Theorem 4.1.10 for arbitrary complex quadratic orders. (*Hint*: see Exercise 2.2.2(d).)

4.2 Real Prime-Producers.

In this section, we have the same goals as in section one of this chapter, but we now look at quadratic polynomials of positive discriminant. This allows us to use the full power of the continued fraction algorithm from Chapter Two. We begin with some technical results on continued fractions, which we will have occasion to use later. First, we set some notation. When referring to the continued fraction expansion of $(P+\sqrt{D})/Q$, for $D > 0$, a radicand with discriminant Δ, then for $I = [Q/\sigma, (P+\sqrt{D})/\sigma]$, we use the notation $P_{I(j)}$, $Q_{I(j)}$ and $a_{I(j)}$, from (2.1.3)–(2.1.5). This will avoid confusion, since we will be referring to several ideals within a given argument.

Lemma 4.2.1. *Let $\Delta = \Delta_0 > 0$ be a fundamental discriminant and assume that I and J are reduced ambiguous ideals in $\mathcal{O}_\Delta$. If $I \sim J$ and $N(I) \neq N(J)$, then $Q_{\ell(I)/2}/\sigma = N(J)$.*

Proof. See Exercise 4.2.1. □

Lemma 4.2.2. *Let $\Delta > 0$ be a discriminant. If I is a reduced ideal in O_Δ, then*

$$P_{I(i)} = [a^2_{I(i)}Q_{I(i)} - Q_{I(i-1)} + Q_{I(i+1)}]/(2a_{I(i)}).$$

Proof. See Exercise 4.2.2. □

With Definitions 4.1.3–4.1.4 in mind, we have

Lemma 4.2.3. *If $\Delta = \Delta_0 > 0$ $(\Delta \neq 5)$ is a fundamental discriminant with $e_\Delta \leq 2$, then $2^{G(\Delta)-1} \geq h_\Delta$.*

Proof. If $t_\Delta = 0$ (see Definition 1.3.3), then $h_\Delta = 1$ by Theorem 1.3.3 (observing that $G(\Delta) = 0$ if and only if $\Delta = 5$). Thus, $2^{G(\Delta)-1} \geq h_\Delta$. Now assume $t_\Delta \geq 1$, i.e. $h_\Delta = 2^{t_\Delta} \geq 2$ (and $e_\Delta = 2$), so by Theorem 1.3.3, $C_\Delta = \langle \mathcal{Q}_1 \rangle \times \langle \mathcal{Q}_2 \rangle \times \cdots \times \langle \mathcal{Q}_{t_\Delta} \rangle$. The $\mathcal{Q}_i$ lie above the q_i which divide Δ. Form $\mathcal{Q} = \prod_{i=1}^{t_\Delta} \mathcal{Q}_i = [q, b + w_\Delta]$, where $0 \leq b < q = \prod_{i=1}^{t_\Delta} q_i \leq \Delta/4 - 1$ (since $\Delta \neq 5$). $\mathcal{Q}$ cannot be principal and q divides $|N(b + w_\Delta)| = |F_\Delta(b)| > q$. Hence, $G(\Delta) \geq t_\Delta + 1$, i.e. $2^{G(\Delta)-1} \geq h_\Delta$. □

What we would like to see is an analogue of Theorem 4.1.4 for $\Delta > 0$ (with Corollary 4.1.1 in mind). However, $G(\Delta)$ is too general in the real case. We need a restriction as follows.

Definition 4.2.1. Let $\Delta > 0$ be a discriminant, with associated conductor f_Δ, and set

$$G_1(\Delta) = \max\{d(|F_\Delta(x)|) : x \in S_\Delta, \gcd(f_\Delta, |F_\Delta(x)|) = 1, \text{ and } m \notin \Lambda(\Delta) \text{ whenever } m \mid |F_\Delta(x)|, \text{ for } 1 < m < |F_\Delta(x)|\}.$$

Now we may prove

Theorem 4.2.1. [4.2.1] *If $\Delta > 0$ is a discriminant, then $h_\Delta \geq 2^{G_1(\Delta)-1}$.*

[4.2.1] This result is an extended version of that which appeared in [199].

Proof. Let $G_1(\Delta) = d(|F_\Delta(x)|) = M \geq 1$, and set $|F_\Delta(x)| = \prod_{i=1}^M p_i^{e_i} = a$, say (for distinct primes p_i). Therefore, $I = [a, x + w_\Delta]$ is a primitive, principal ideal of $\mathcal{O}_\Delta$. Since $\gcd(f_\Delta, a) = 1$, then by Exercise 1.5.2(d), there are $\mathcal{O}_\Delta$-ideals $\mathcal{P}_i$ with $N(\mathcal{P}_i) = p_i$ such that $\prod_{i=1}^M \mathcal{P}_i^{e_i} \sim 1$.

Claim 1. If p divides Δ, then p^2 does not divide $|F_{\Delta,1}(x)|$ for any non-negative integer x. (Exercise 4.2.3.)

Set $S = \{1, 2, \ldots, M\}$ and divide it into two subsets S_1 and S_2 as follows. We let $i \in S_1$ if and only if $\gcd(p_i, \Delta) = 1$ and $i \in S_2$ otherwise. Thus, by Claim 1, $a = \prod_{i \in S_1} p_i^{e_i} \prod_{j \in S_2} p_j$. Any divisor of a may be written in the form $\prod_{i \in S_1} p_i^{f_i} \prod_{j \in S_3} p_j$ where $0 \leq f_i \leq e_i$ and $\emptyset \subseteq S_3 \subseteq S_2$. Thus, any combination $c = ((f_i)_{i \in S_1}, S_3)$ of an $|S_1|$-tuple (f_i) of integers and a subset S_3 of S_2 represents a divisor of a. Thus, the set of all these combinations has cardinality $\tau(a)$ (see Definition 4.1.2). With $\mathcal{P}_i$'s fixed as above, we let $\mathcal{F}(c)$ denote the ideal class of $\prod_{i \in S_1} \mathcal{P}_i^{f_i} \prod_{j \in S_3} \mathcal{P}_j$ in C_Δ. Hence, $\mathcal{F}$ is a map of S into the ideal class group of $\mathcal{O}_\Delta$.

Claim 2. $\mathcal{F}(c_1) = \mathcal{F}(c_2)$ for exactly 2^{M-1} distinct pairs (c_1, c_2) with $c_1 \neq c_2$.

Let $\mathcal{F}(c_1) = \mathcal{F}(c_2)$ where

$$\mathcal{F}(c_1) = \prod_{i \in S_1} \mathcal{P}_i^{f_i} \prod_{j \in S_3} \mathcal{P}_j \sim \prod_{i \in S_1} \mathcal{P}_i^{g_i} \prod_{j \in S_4} \mathcal{P}_j = \mathcal{F}(c_2) \tag{4.2.1}$$

with $\emptyset \subseteq S_3, S_4 \subseteq S_2$ and $0 \leq f_i, g_i \leq e_i$. We now subdivide S_1 into two disjoint sets T_1 and T_2 with $i \in T_1$ if and only if $f_i \geq g_i$ and $i \in T_2$ otherwise. Moreover, we form the subset T_3 of $S_3 \cup S_4$ with $j \in T_3$ if and only if $j \notin S_3 \cap S_4$. Thus, (4.2.1) may be rewritten as

$$J = \prod_{i \in T_1} \mathcal{P}_i^{f_i - g_i} \prod_{j \in T_2} \mathcal{P}_j'^{g_i - f_i} \prod_{k \in T_3} \mathcal{P}_k \sim 1. \tag{4.2.2}$$

Also, note that $N(J) \mid a$, so by hypothesis $N(J) = a$ (since $c_1 \neq c_2$ implies $N(J) \neq 1$). Hence, $T_1 \cup T_2 = S_1$, $T_3 = S_2$ and $f_i = e_i$ with $g_i = 0$ for all $i \in T_1$; whereas, $g_i = e_i$ and $f_i = 0$ for all $i \in T_2$. Thus, (4.2.2) becomes

$$\prod_{i \in T_1} \mathcal{P}_i^{e_i} \sim \prod_{i \in T_2} \mathcal{P}_i^{e_i} \prod_{j \in S_2} \mathcal{P}_j \tag{4.2.3}$$

where $T_1 \cup T_2 = S_1$ and $T_1 \cap T_2 = \emptyset$. There are 2^{M-1} distinct such relationships (Exercise 4.2.4), so $h_\Delta \geq \tau(a) - 2^{M-1} = \prod_{i=1}^M (e_i + 1) - 2^{M-1} \geq 2^M - 2^{M-1} = 2^{M-1}$. $\square$

Now we are in a position to give an analogue of Theorem 4.1.4 (see Definition 1.3.3 for the definition of t_Δ).

Theorem 4.2.2. *If $\Delta = \Delta_0 > 0$ is a fundamental discriminant, then the following are equivalent:*

(1) $e_\Delta = 2$.

(2) $h_\Delta = 2^{G_1(\Delta)-1}$ *and* $G_1(\Delta) = t_\Delta + 1$.

Proof. See Exercise 4.2.5. □

An illustration is

Example 4.2.1. Let $\Delta = 451605$ for which $h_\Delta = 16$, $e_\Delta = 2$, $N = 5$, $t_\Delta = 4$ and $G_1(\Delta) = 5$ with $h_\Delta = 2^{G_1(\Delta)-1}$.

Actually, Theorem 4.2.2 is better illustrated by a smaller value, so that we can better depict what happens under the influence of Definition 4.2.1 as follows.

Example 4.2.2. If $\Delta = 165$, then $N = 2$, $t_\Delta = 1$ and $G_1(\Delta) = 2$, since the only non-negative values of $x \leq 40 = \lfloor \Delta/4 - 1 \rfloor$, for which $F_\Delta(x)$ is divisible by *more* than two distinct primes, are $x = 16, 22, 25, 27, 32, 37, 40$, each of which have $F_\Delta(x)$ divisible by a value m with $1 < m < F_\Delta(x)$ where $m \in \Lambda(\Delta)$, namely

$$(x, m) \in \{(16, 21), (22, 15), (25, 21), (27, 11), (32, 35), (37, 15), (40, 39)\}.$$

Here, $h_\Delta = 2 = 2^{G_1(\Delta)-1}$.

Remark 4.2.1. We cannot hope for a class number one criterion such as that of Theorem 4.1.2, since it is too much to ask for $|F_\Delta(x)|$ to be 1 or prime for all $x \in \mathcal{S}_\Delta$. In fact, the proof of Theorem 4.2.3 below shows that the latter can only hold if $\Delta = 5, 8$ or 12. To verify that Theorem 4.2.2 holds when $h_\Delta = 1$, it suffices to show that $G_1(\Delta) \neq 0$ when $\Delta \neq 5$ (see section three of this chapter).

A remark on notation is in order before we proceed. The unique ideals over the primes q_i of Assumption 4.1.1 will be denoted by $\mathcal{Q}_i$ and that above q by $\mathcal{Q}$. If we wish to refer to an $\mathcal{O}_\Delta$-ideal above another prime p, we will denote it by $\mathcal{P}$. If we wish to refer to a specific $\mathcal{O}_\Delta$-prime above an explicitly given prime, such as 3 say, we will denote it by $\mathcal{Q}_3$.

The following result tells us what must necessarily follow from the qth Euler-Rabinowitsch polynomial producing consecutive, distinct initial values up to a Minkowski bound.

Theorem 4.2.3. *If $\Delta > 0$ satisfies Assumption 4.1.1, then (1) $\Rightarrow$ (2) in what follows.*

(1) *$|F_{\Delta,q}(x)|$ is 1 or prime for any integer x with $0 \leq x < (M_\Delta - \alpha_\Delta + 1)/\alpha_\Delta$.*

(2) *$e_\Delta \leq 2$ and $M_\Delta < q_{N+1}$.*

Proof. By Lemma 4.1.3, every split prime $p < M_\Delta$ must divide $|F_{\Delta,q}(x)|$ for some non-negative integer $x < (M_\Delta - \alpha_\Delta + 1)/2$. Hence, by (1), $|F_{\Delta,q}(x)| = p$ which implies that $\mathcal{P} \sim \mathcal{Q}$ by Lemma 4.1.4. Therefore, $\mathcal{P}^2 \sim 1$, so by Theorem 1.3.1 $e_\Delta \leq 2$. We now show that $M_\Delta < q_{N+1}$.

If $x_0 = (q_{N+1} - \alpha_\Delta + 1)/\alpha_\Delta$, then $F_{\Delta,q}(x_0) = q_{N+1}(qq_{N+1}/\alpha_\Delta^2 - \Delta/(4qq_{N+1}))$ which will be composite except in the cases cited below, which we will handle separately. First, we note that when $F_{\Delta,q}(x_0)$ is composite, then $x_0 > \lfloor (M_\Delta - \alpha_\Delta + 1)/\alpha_\Delta \rfloor$. If $M_\Delta > q_{N+1}$, then $(q_{N+1} - \alpha_\Delta + 1)/\alpha_\Delta > \lfloor (M_\Delta - \alpha_\Delta + 1)/\alpha_\Delta \rfloor > (M_\Delta - \alpha_\Delta + 1)/\alpha_\Delta - 1 > (q_{N+1} - \alpha_\Delta + 1)/\alpha_\Delta - 1$. However, there cannot exist an integer strictly between x_0 and $x_0 - 1$. Thus, to complete the proof, we need only verify the special cases where $F_{\Delta,q}(x_0)$ is not composite as follows.

Case 1. $\alpha_\Delta = 1$ and $qq_{N+1} - \Delta/(4qq_{N+1}) = 1$. It is an easy check to show that the only values satisfying this case is $\Delta = 8$, for which our result trivially holds.

Case 2. $\alpha_\Delta = 2$ and $qq_{N+1} - \Delta/(qq_{N+1}) = 4$. Again, an easy check shows that the only values in this case are $\Delta = 5$, 12 for which our result trivially holds. □

The following result shows that when $\Delta \equiv 1 \pmod 8$ and (1) of Theorem 4.2.3 holds, then we are reduced to a single value. This is the analogue of Theorem 4.1.6.

Corollary 4.2.1. *If $\Delta \equiv 1 \pmod 8$ and (1) of Theorem 4.2.3 holds, then $\Delta = 33$.*

Proof. If $\Delta \equiv 1 \pmod 8$, then $F_{\Delta,q}(0) = (q^2 - \Delta)/(4q) = (q - q_{N+1})/4$ is even and it is composite unless either $q = 8 + q_{N+1}$ or $q_{N+1} = 8 + q$, both of which are violated if $F_{\Delta,q}(1) = 2 + (q^2 - \Delta)/(4q)$ is even and composite. Hence, $(M_\Delta - 1)/2 < 1$, i.e. $\Delta = 17$ or 33. However, $\Delta = 17$ has $|F_{\Delta,q}(0)| = 4$, so only $\Delta = 33$ satisfies the hypotheses. □

We now turn to the case where $\Delta \equiv 0 \pmod 8$.

Theorem 4.2.4. *Let $\Delta \equiv 0 \pmod 8$ $\Delta > 0$ satisfy Assumption 4.1.1. If $|F_{\Delta,q}(x)|$ is 1 or prime for all non-negative integers $x < \sqrt{D}$, then one of the following holds:*

(1) *$D = p^2 + 1 = 2q_2$ with either $D = 2$ or else $q_2 \equiv 1 \pmod 4$ is prime, $h_\Delta = 2$, and p is the only split prime less than $\sqrt{D}$.*

(2) *$D = \lfloor\sqrt{D}\rfloor^2 + 2 = 2q_2$ where $q_2 \equiv 3 \pmod 4$ is prime, $h_\Delta = 1$ and there are no split primes less than $\sqrt{D}$.*

(3) *$D = q^2\lfloor\sqrt{D}/q\rfloor^2 + q$, $h_\Delta = 2^{N-1} = 2^{t_\Delta}$, and there are no split primes less than $\sqrt{D}$.*

(4) *$D = [(p+3)/2]^2 - 2 = 2q_2$ with $q_2 = 2[(p+3)/4]^2 - 1$ prime, $h_\Delta = 1$, $p > \sqrt{D}$ is prime and there are no split primes less than $\sqrt{D}$.*

Proof. If $q = 1$, then $D = 2$, so we assume henceforth that $D > 2$, i.e. $q \geq 2$ and $N \geq 1$.

Let $p < \sqrt{D}$ be a split prime, then by Lemma 4.1.3, there exists a non-negative integer $x < \sqrt{D}/2$ such that p divides $F_{\Delta,q}(x)$. Thus, by hypothesis, $F_{\Delta,q}(x) = \epsilon p$ where $\epsilon = \pm 1$. Also, $|F_{\Delta,p}(|p - x|)| = |q(p-x)^2 - q_{N+1}| = |p(qp - 2xq + \epsilon)| = p$, since $0 < |p - x| < \sqrt{D}$. Therefore, $qp - 2xq + \epsilon = \pm 1$. If $qp - 2xq = 0$, then $p = 2x$, a contradiction. If $qp - 2xq = 1 - \epsilon = 2$, then $q = 2$ and $p = 2x + 1$. However, if $\epsilon = -1$, then $qx^2 - q_{N+1} = -p$ so $q((p-1)/2)^2 + p = q_{N+1}$, i.e. $(p^2 + 1)/2 = q_{N+1} = q_2$. Hence, $D = p^2 + 1$ and $h_\Delta = 2$ (since $t_\Delta = 1$ is forced by the fact that the form of D dictates $q_2 \equiv 1 \pmod 4$). A similar argument holds for $\epsilon = 1$. No other split prime can exist, since the above argument forces $D = p^2 + 1$ for any such p. This secures (1).

We now assume that there are no split primes $p < \sqrt{D}$. In this case, there *cannot* exist an ambiguous class of reduced ideals *without* an ambiguous ideal in it. Consider the ideal $I = [q, \sqrt{D}]$, then in the continued fraction expansion of $\sqrt{D}/q$ we have that $D = P_{I(1)}^2 + qQ_{I(1)}$. Therefore, $-Q_{I(1)} = (P_{I(1)}^2 - D)/q = q(P_{I(1)}/q)^2 - \Delta/(4q) = F_{\Delta,q}(x)$, where $x = P_{I(1)}/q = \lfloor\sqrt{D}/q\rfloor$. We note that $0 < x < M_\Delta = \sqrt{D}$ so, by hypothesis, either $Q_{I(1)} = 1$ or $Q_{I(1)} = p$, a prime.

If $Q_{I(1)} = 1$, then $D = P_{I(1)}^2 + q = \lfloor\sqrt{D}\rfloor^2 + q$ since $P_{I(1)}$ is forced to be $\lfloor\sqrt{D}\rfloor$ by Theorem 3.2.1. If $t_\Delta = N$, then Q represents the product of the generators of C_Δ and since Q is in the principal class, then $h_\Delta = 1 = 2^N$, i.e. $N = 0$ and $D = 2$, a contradiction to our initial assumption. We therefore assume that $t_\Delta = N - 1$. Let $Q' = \Pi_{i=1}^{N-1} Q_i$ represent the product of the generators of C_Δ (after possibly renumbering the Q_i if necessary). Consider the ideal $J = [q', \sqrt{D}]$, then in the continued fraction expansion of $\sqrt{D}/q'$, $Q_{J(1)}$ can only be divisible by an unramified prime if it is prime and $Q_{J(1)} > \sqrt{D}$. However, by Lemma 4.1.3, (taking $B = 2\sqrt{D}$) there must be a non-negative integer $x < \sqrt{D}$ such that $Q_{J(1)}$ divides $F_{\Delta,q}(x)$. Thus, by hypothesis, $|F_{\Delta,q}(x)| = Q_{J(1)}$ which implies that $Q' \sim Q$ by Lemma 4.1.4. Hence, $q' = 1$, $q = 2$ and $N = 1$, i.e. $D = \lfloor\sqrt{D}\rfloor^2 + 2 = 2q_2$ with $q_2 \equiv 3 \pmod 4$ prime and $h_\Delta = 1$. This secures (2).

If $Q_{J(1)}$ is not divisible by any unramified primes, then as above, $Q_{J(1)} = q_N$ is forced by Lemma 4.2.1, i.e. $D = \lfloor\sqrt{D}\rfloor^2 + q$ with $h_\Delta = 2^{N-1}$. Observing that we have also shown $\lfloor D\rfloor = q\lfloor\sqrt{D}/q\rfloor$, then we have (3).

Now we assume that $Q_{I(1)} = p$, a prime. Moreover, $p > \sqrt{D}$, so $a_{I(1)} = 1$ by Exercise 4.2.6. Therefore, $Q_{I(2)} < \sqrt{D}$ and since there are no split primes less than $\sqrt{D}$, then it follows that $Q_{I(2)}$ divides D. However, $q_{N+1} > \sqrt{D}$ by Theorem 4.2.3, so it does not divide $Q_{I(2)}$, which must therefore divide q. Since $D = P_{I(2)}^2 + pQ_{I(2)}$, then $Q_{I(2)}$ divides $P_{I(2)}$, thereby forcing $Q_{I(2)} = 1$. Hence, from Lemma 4.2.2, we must have that $P_{I(1)} = (p - q + 1)/2$. Therefore, $D = ((p - q + 1)/2)^2 + pq$, i.e.

$$D = ((p + q + 1)/2)^2 - q. \tag{4.2.4}$$

Since $Q_{I(2)} = 1$, then $t_\Delta = N - 1$, so we again consider $J = [q', \sqrt{D}]$ as above. We now show that $N = 1$. If $Q_{J(1)} > \sqrt{D}$ is prime, then as above, $N = 1$ is forced, so we may assume that $Q_{J(1)} < \sqrt{D}$ which means that $Q_{J(1)}$ can only be divisible by ramified primes and is square-free, since $D = P_{J(1)}^2 + q'Q_{J(1)}$. By Lemma 4.2.1, then $Q_{J(1)} = q_N$, and hence, $D = P_{J(1)}^2 + q$. However, from (4.2.4), we have that $D = A^2 - q$ where $A = (p + q + 1)/2$. Therefore, $2D = A^2 + P_{J(1)}^2 = q^2[(A/q)^2 + (P_{J(1)}/q)^2]$, forcing $q = 2$, $N = 1$ and $h_\Delta = 1$. Also, from (4.2.4), $D = [(p + 3)/2]^2 - 2 = 2q_2$, where $q_2 \equiv 3 \pmod 4$, is prime. This is (4), which secures the entire result. □

The following table illustrates Theorem 4.2.4. In fact, we will establish in Chapter Six that this list is complete with one possible exceptional value remaining.

Table 4.2.1.[(4.2.2)] $D \equiv 0 \pmod 2$ with $|F_{\Delta,q}(x)| = 1$ or prime for all $x \in T = \{0, 1, 2, ..., \lfloor\sqrt{D}\rfloor\}$.

(a) $D = p^2 + 1 > 2$, $p < \sqrt{D}$, $h_\Delta = 2$.

D	p	$F_{\Delta,q}(x)$	$\|F_{\Delta,q}(x)\|$ for all $x \in T$
10	3	$2x^2 - 5$	5,3,3,13
26	5	$2x^2 - 13$	13,11,5,5,19,37
122	11	$2x^2 - 61$	61,59,53,43,29,11,11,37,67,101,139,181
362	19	$2x^2 - 181$	181,179,173,163,149,131,109,83,53,19
			19,61,107,157,211,269,331,397,467,541

(b) $D = \lfloor\sqrt{D}\rfloor^2 + 2$, $h_\Delta = 1$.

D	$F_{\Delta,q}(x)$	$\|F_{\Delta,q}(x)\|$ for all $x \in T$
6	$2x^2 - 3$	3,1,5
38	$2x^2 - 19$	19,17,11,1,13,31,53

(c) $D = [(p + 3)/2]^2 - 2$, $p > \sqrt{D}$, $h_\Delta = 1$.

D	p	$F_{\Delta,q}(x)$	$\|F_{\Delta,q}(x)\|$ for all $x \in T$
14	5	$2x^2 - 7$	7,5,1,11
62	13	$2x^2 - 31$	31,29,23,13,1,19,41,67
398	37	$2x^2 - 199$	199,197,191,181,167,149,127,101,71,37,
			1,43,89,139,193,251,313,379,449,523

Remark 4.2.2. It is also worth observing that for $D = 362$ in Table 4.2.1(a), we actually get $|F_{\Delta,q}(x)| =$ prime up to $x = 27$. This then matches the Karst polynomial (see Karst [139]) $2x^2 - 199$ from Table 4.2.1(c). Hence, both $2x^2 - 199$ and $2x^2 - 181$ are actually 1 or prime for all integers x with $-27 \le x \le 27$. Furthermore, $D = 122$ from Table 4.2.1(a) has $|F_{\Delta,q}(x)|$ prime for all integers x with $-15 \le x \le 15$.

We now turn to $\Delta \equiv 4 \pmod 8$.

Theorem 4.2.5. *Let* $\Delta \equiv 4 \pmod 8$, $\Delta > 0$ *satisfy Assumption* 4.1.1. *If* $|F_{\Delta,q}(x)|$ *is* 1 *or prime for all non-negative integers* $x < (\sqrt{D} - 1)/2$, *then* $e_\Delta \le 2$, *and one of the following holds:*

(1) $D = \lfloor\sqrt{D}\rfloor^2 + q$ *with* $h_\Delta = 2^{N-1}$, *and there are no split primes* $p < \sqrt{D}$.

(2) $D = (\lfloor\sqrt{D}\rfloor + 1)^2 - 2$, *with* $h_\Delta = 1$ *and* $p = 2\lfloor\sqrt{D}\rfloor - 1$, *is the only split prime less than* $\sqrt{\Delta}$.

(3) $D = (\lfloor\sqrt{D}\rfloor + 1)^2 - q_2 = q_2 q_3$ *with* $q_2 < q_3$ *primes,* $h_\Delta = 2$ *and* $p = \lfloor\sqrt{D}\rfloor + 1 - (q_2 + 1)/2$, *is the only split prime less than* $\sqrt{D}$. *Moreover,* $C_\Delta = \langle Q_2 \rangle$.

[(4.2.2)] In [197, Theorem 5, p. 324] Louboutin *assumed* $q = 2$ and came up with the same values as in this table. Thus, Theorem 4.2.4 generalizes Louboutin's result by proving that for $\Delta \equiv 0 \pmod 8$, the assumption that $|F_\Delta, q(x)| = 1$ or prime actually *forces* $q = 2$ and the values in this table. For $\Delta \not\equiv 0 \pmod 8$, the story changes. We also see that there are no known values for which Theorem 4.2.4(3) occurs. From Tables A4 in Appendix A, and the techniques of Chapter Five, we will see that this is known to be true with one GRH-ruled-out exception. Similar comments apply to Tables 4.2.2–4.2.5 below.

(4) $D = (\lfloor\sqrt{D}\rfloor + 1)^2 - q_N$ *with* $h_\Delta = 2^{N-1}$.

(5) $D = \lfloor\sqrt{D}\rfloor^2 + q_N$ *with* $h_\Delta = 2^{N-1}$.

(6) $D = (\lfloor\sqrt{D}\rfloor + 1)^2 - q$, *with* $h_\Delta = 2^{N-1}$, *and* $p = 2\lfloor\sqrt{D}\rfloor - q + 1$, *is the only split prime less than* $\sqrt{\Delta}$.

Proof. If $q = 1$, then $D = 1$, a contradiction, so we assume that $q \geq 2$. Consider the ideal $I = [q, q/2 + \sqrt{D}]$, then in the continued fraction expansion of $(q/2 + \sqrt{D})/q$, we have that $D = P_{I(1)}^2 + qQ_{I(1)}$, where $P_{I(1)} = \lfloor(q/2 + \sqrt{D})/q\rfloor q - q/2$. Thus, $-Q_{I(1)} = (P_{I(1)}^2 - D)/q = [q^2(2x_0 + 1)^2 - \Delta]/(4q) = F_{\Delta,q}(x_0)$, where $x_0 = \lfloor(q/2 + \sqrt{D})/q\rfloor - 1$. We note that it is not possible for $x_0 = -1$ since, in that case, $\sqrt{D} < q/2$, i.e. $q_{N+1} < q/2$. However, by Theorem 4.2.3, $q_{N+1} > M_\Delta = \sqrt{D}$ which implies that $q_{N+1} > q/2$. Hence, $x_0 \geq 0$ and since $q \geq 2$, then $x_0 < (M_\Delta - 1)/2$. Therefore, by hypothesis, either $Q_{I(1)} = 1$ or a prime.

Claim 1. If $p < \sqrt{D}$ is any split prime, then $\mathcal{P} \sim \mathcal{Q}$ where $\mathcal{P}$ and $\mathcal{Q}$ are the $\mathcal{O}_\Delta$-primes above p and q respectively.

By Lemma 4.1.3, there is a non-negative integer $x < (\sqrt{D} - 1)/2$ such that p divides $|F_{\Delta,q}(x)|$. Therefore, $|F_{\Delta,q}(x)| = p$ by hypothesis. By Lemma 4.1.4, we have that $\mathcal{P} \sim \mathcal{Q}$, which secures Claim 1.

If $Q_{I(1)} = 1$, then by Claim 1 and Theorem 1.3.1, there are no split primes less than $\sqrt{D}$. It follows from Theorem 3.2.1 that $D = \lfloor\sqrt{D}\rfloor^2 + q$. This secures (1), so we now assume that $Q_{I(1)} = p$.

We observe that p must be a split prime since it would otherwise divide D which is impossible, since $D = P_{I(1)}^2 + pq$.

Case 1. $p < \sqrt{D}/2$.

Recall that $F_{\Delta,q}(x_0) = -p$. If $p < x_0 + 1$, then $|F_{\Delta,q}(x_0 - p)| =$

$$|q(x_0 - p)^2 + q(x_0 - p) + (q^2 - \Delta)/(4q)| = |qp^2 - 2pqx_0 - qp - p| = p|qp - 2qx_0 - q - 1|.$$

Since $0 \leq x_0 - p < (\sqrt{D} - 1)/2$, then by hypothesis $qp - 2qx_0 - q - 1 = \pm 1$. If $qp - 2qx_0 - q - 1 = -1$, then $p = 2x_0 + 1 > x_0 + 1$, a contradiction. If $qp - 2qx_0 - q - 1 = 1$, then $q = 2$ and $p = 2 + 2x_0$, a contradiction. Thus, $0 \leq p - x_0 - 1 < (\sqrt{D} - 1)/2$ and

$$\begin{aligned}|F_{\Delta,q}(p - x_0 - 1)| &= |q(p - x_0 - 1)^2 + q(p - x_0 - 1) + (q^2 - \Delta)/(4q)| \\ &= |-p + qp^2 - 2pqx_0 - pq| = p|-1 + qp - 2qx_0 - q|.\end{aligned}$$

By hypothesis, $-1 + qp - 2qx_0 - q = \pm 1$. If $-1 + qp - 2qx_0 - q = 1$, then $q = 2$ and $p = 2x_0 + 2$, a contradiction. If $-1 + qp - 2qx_0 - q = -1$, then $p = 2x_0 + 1$. Therefore, $D = (q/2)^2p^2 + qp$ forcing p to divide D, a contradiction. We have shown that Case 1 cannot occur, so we assume henceforth that $p > \sqrt{D}/2$.

We now show that there is at most one other split prime $p_1 < \sqrt{D}$.

Claim 2. There is at most one split prime $p_1 \neq p$ with $p_1 < \sqrt{D}$, and when it exists, then $p_1 = F_{\Delta,q}(x_0 + 1)$, and $p_1 > \sqrt{D}/2$.

By Lemma 4.1.3, if such a prime $p_1 < \sqrt{D}$ exists, then p_1 divides $|F_{\Delta,q}(x_1)|$ for some non-negative integer $x_1 < (\sqrt{D}-1)/2$. Thus, $|F_{\Delta,q}(x_1)| = p_1$ by hypothesis. We now show that $x_1 = x_0 + 1$ and so no other such prime can exist.

If $F_{\Delta,q}(x_0+1) < 0$, then $2q(x_0+1) < p < \sqrt{D}$, since

$$F_{\Delta,q}(x_0+1) = F_{\Delta,q}(x_0) + 2q(x_0+1) = -p + 2q(x_0+1).$$

If $q > \sqrt{D}$, then $2(x_0+1) < 1$, a contradiction. If $q < \sqrt{D}$, then

$$\sqrt{D} > p > 2q(x_0+1) > 2q((q/2+\sqrt{D})/q - 1) = 2\sqrt{D} - q$$

yields a contradiction. Hence, $F_{\Delta,q}(x_0+1) > 0$. If $x_1 > x_0+1$, then $F_{\Delta,q}(x_0+2) < \sqrt{D}$. However,

$$F_{\Delta,q}(x_0+2) = F_{\Delta,q}(x_0+1) + 2q(x_0+3) > 2q((q/2+\sqrt{D})/q+1) = 3q + 2\sqrt{D},$$

a contradiction. Hence, if $x_1 > x_0$, then $x_1 = x_0 + 1$. On the other hand, if $0 \le x_1 \le x_0$, then $F_{\Delta,q}(x_0-1) = -p > -\sqrt{D}$. However, $F_{\Delta,q}(x_0-1) = F_{\Delta,q}(x_0) - 2qx_0 = -p - 2qx_0$. Thus, $2qx_0 < \sqrt{D} - p < \sqrt{D}/2$, since $p > \sqrt{D}/2$.

Therefore, $\sqrt{D}/4 > qx_0 > \sqrt{D} - 3q/4$, i.e. $q > \sqrt{D}$. Hence, $\lfloor (q/2+\sqrt{D})/q \rfloor < 2$, forcing $x_0 = 0$, which contradicts that $0 \le x_1 < x_0$. To see that $p_1 > \sqrt{D}/2$, we reproduce the proof of Case 1, with the beginning statement being that $F_{\Delta,q}(x_0+1) = p_1$ rather than $F_{\Delta,q}(x_0) = p$. The result follows mutatis mutandis. This secures Claim 2.

Claim 3. If a split prime $p_1 < \sqrt{D}$ exists with $p_1 \ne p$, then $p + p_1 = 2P_{I(1)} + q = 2q(x_0+1)$.

By Claim 2, $F_{\Delta,q}(x_0+1) = p_1$ so $p_1 + p = F_{\Delta,q}(x_0+1) - F_{\Delta,q}(x_0) = F_{\Delta,q}(x_0) + 2q(x_0+1) - F_{\Delta,q}(x_0) = 2q(x_0+1) = 2q\lfloor (q/2+\sqrt{D})/q \rfloor = 2P_{I(1)} + q$, completing Claim 3.

Case 2. $p > \sqrt{D}/2$ and $h_\Delta = 1$.

We have that $N = 1$, $q = q_1 = 2$ and $D = q_2 \equiv 3 \pmod 4$, by Theorem 1.3.3.

Claim 4. $Q_{I(j)}$ is not even for any integer j with $0 < j < \ell(I)$, unless $Q_{I(j-1)} = 1$ and $Q_{I(j)}/2$ is prime. (Exercise 4.2.7(a).)

By Exercise 4.2.7(b), we may assume that p does not divide $Q_{I(2)}$ and $\sqrt{D} > 6$. Also, by Exercise 4.2.7(b), we may assume that p_1 does not divide $Q_{I(2)}$. Hence, $Q_{I(2)}$ is not divisible by either p or p_1. If $Q_{I(1)}$ is divisible by any other unramified prime, then $Q_{I(2)} = p_2 > \sqrt{D}$ is prime in which case $a_{I(2)} = 1$. Hence, $P_{I(3)} = p_2 - P_{I(2)}$ is odd, and so $Q_{I(3)}$ is even, which contradicts Claim 4. Therefore, $Q_{I(2)} < \sqrt{D}$, and is divisible only by ramified primes. Since $D = q_2 > \sqrt{D}$, then only 2 can divide $Q_{I(2)}$, which contradicts Claim 4, so $Q_{I(2)} = 1$. Thus, $D = P_{I(2)}^2 + p = P_{I(1)}^2 + 2p$, and since $P_{I(2)} = a_{I(1)}p - P_{I(1)}$, then $P_{I(1)}^2 + 2p = (a_{I(1)}p - P_{I(1)})^2 + p$, i.e. $p = a_{I(1)}^2 p^2 - 2a_{I(1)} - 2a_{I(1)}pP_{I(1)}$ forcing $a_{I(1)} = 1$ and $p = 2P_{I(1)} + 1$. Thus, $D = ((p+3)/2)^2 - 2 = (\lfloor \sqrt{D} \rfloor + 1)^2 - 2$, by Theorem 3.2.1. Moreover, $p > \sqrt{D}$ and is the only split prime less than $\sqrt{\Delta}$ (observing that $p = 2\lfloor \sqrt{D} \rfloor - 1$, and there are no split primes $< \sqrt{D}$). This secures (2).

Case 3. $p > \sqrt{D}/2$ and $h_\Delta > 1$.

Let $Q' = \Pi_{i=1}^{N-1} Q_i$ where (after possible renumbering the Q_i) Q' represents the product of the generators of C_Δ (observing that $t_\Delta = N-1$ since $D \equiv 3 \pmod 4$) and $t_\Delta > 0$ since $h_\Delta > 1$, forcing $N \geq 2$).

Case 3(a). $Q \sim Q'$ where Q and Q' are the $\mathcal{O}_\Delta$-primes above q and q' respectively. Since $Q \sim Q'$, then $Q_N \sim 1$. Consider $J = [q_N, \alpha + \sqrt{D}]$, where $\alpha = 0$ if q_N is odd and $\alpha = q_N/2 = 1$ if $q_N = 2$. By Exercise 4.2.7(c), we have (3), (4) and (5).

Case 3(b). $Q \not\sim Q'$.

Consider the ideal $G = [q, \beta + \sqrt{D}]$ where $\beta = 0$ if q is odd and $\beta' = q'/2$ if q' is even. By Exercise 4.2.7(d), $Q_{G(2)} = q_N$. Hence, $Q' \sim Q_N$, i.e. $Q \sim 1$. By Exercise 4.2.7(e), neither p nor p_1 divides $Q_{I(2)}$. Furthermore, if $Q_{I(2)}$ is divisible by any other unramified primes, then arguments similar to the above yield a contradiction, so $Q_{I(2)} = 1$. By Lemma 4.2.2,

$$P_{I(1)} = (a_{I(1)}^2 p - q + 1)/(2a_{I(1)}).$$

However, by Exercise 4.2.7(f), $a_{I(1)} = 1$. Hence,

$$D = ((p+q+1)/2)^2 - q = \lfloor\sqrt{D}\rfloor^2 - q,$$

with $h_\Delta = 2^{N-1}$ and $p = 2\lfloor\sqrt{D}\rfloor - q + 1$ is the only split prime less than $\sqrt{\Delta}$. This is (6), and hence the full result. $\square$

Table 4.2.2. $D \equiv 3 \pmod 4$ with $|F_{\Delta,q}(x)| = 1$ or prime for all integers x with $0 \leq x < (\sqrt{D}-1)/2$.

(a) $D = \lfloor\sqrt{D}\rfloor^2 + q$ with $h_\Delta = 2^{N-1}$ where $N \leq 2$ and there are no split primes less than $\sqrt{D}$.

D	h_Δ	$F_{\Delta,q}(x)$	$\lvert F_{\Delta,q}(x)\rvert$ for $0 \leq x < (\sqrt{D}-1)/2$
3	1	$2x^2+2x-1$	1
11	1	$2x^2+2x-5$	5, 1
15	2	$6x^2+6x-1$	1, 11
35	2	$10x^2+10x-1$	1, 19, 59
83	1	$2x^2+2x-41$	41, 37, 29, 17
87	2	$6x^2+6x-13$	13, 1, 23, 59, 107
143	2	$22x^2+22x-1$	1, 41, 131, 263, 439, 659
227	1	$2x^2+2x-113$	113, 109, 101, 89, 73, 53, 29, 1
447	2	$6x^2+6x-73$	73, 61, 37, 1, 47, 107, 179, 263, 359, 467, 587
635	2	$10x^2+10x-61$	61, 41, 1, 59, 139, 239, 359, 499, 659, 839, 1039, 1259, 1499

(b) $D = (\lfloor\sqrt{D}\rfloor+1)^2 - 2 = ((p+3)/2)^2 - 2$ with $h_\Delta = 1$ and $p > \sqrt{D}$ is the only split prime less than $2\sqrt{D} = \sqrt{\Delta}$.

D	$F_{\Delta,q}(x)$	p	$\lvert F_{\Delta,q}(x)\rvert$ for $0 \leq x < (\sqrt{D}-1)/2$
7	$2x^2+2x-3$	3	3
23	$2x^2+2x-11$	7	11, 7
47	$2x^2+2x-23$	11	23, 19, 11
167	$2x^2+2x-8$	23	83, 79, 59, 43, 23, 1

(c) $D = (\lfloor\sqrt{D}\rfloor + 1)^2 - q_2 = q_2q_3$ with $q_2 < q_3$, $h_\Delta = 2$, $C_\Delta = \langle Q_2\rangle$ and $p = \lfloor\sqrt{D}\rfloor + 1 - (q_2+1)/2$ is the only split prime less than $\sqrt{D}$.

D	p	$F_{\Delta,q}(x)$	$\lvert F_{\Delta,q}(x)\rvert$ for $0 \le x < (\sqrt{D}-1)/2$
95	7	$10x^2 + 10x - 7$	7, 53, 113, 193, 293
395	17	$10x^2 + 10x - 37$	37, 17, 23, 83, 163, 263, 383, 523, 683, 863

(d) $D = \lfloor\sqrt{D}\rfloor^2 + q_N$ where $h_\Delta = 2 = 2^{N-1}$.

D	h_Δ	$F_{\Delta,q}(x)$	$\lvert F_{\Delta,q}(x)\rvert$ for $0 \le x < (\sqrt{D}-1)/2$
39	2	$6x^2 + 6x - 5$	5, 7, 31
203	2	$14x^2 + 14x - 11$	11, 17, 73, 157, 269, 409, 577
327	2	$6x^2 + 6x - 53$	53, 41, 17, 19, 61, 127, 199, 283, 379

The case $h_\Delta = 1$ occurs in (a)–(b), so we do not repeat them here. A similar comment holds for (e) below.

(e) $D = (\lfloor\sqrt{D}\rfloor + 1)^2 - q$ with $h_\Delta = 2^{N-1}$ and $p = 2\lfloor\sqrt{D}\rfloor - q + 1$ is the only split prime less than $\sqrt{\Delta}$.

D	p	$F_{\Delta,q}(x)$	$\lvert F_{\Delta,q}(x)\rvert$ for $0 \le x < (\sqrt{D}-1)/2$
215	19	$10x^2 + 10x - 19$	19, 1, 41, 101, 181, 281, 401

The remaining case where $\Delta \equiv 1 \pmod 4$, reduces to the case $D = \Delta \equiv 5 \pmod 8$ by virtue of Corollary 4.2.1. The case where $h_\Delta = 1$ is special. In fact, in a search for a real quadratic field analogue of Theorem 4.1.2, we have the following Rabinowitsch-type criterion for real quadratic fields.

Theorem 4.2.6. *If $\Delta > 0$ is a fundamental discriminant, then the following are equivalent:*

(1) *$|F_{\Delta,1}(x)|$ is 1 or prime for all integers x with $3 - \sigma \le x < \sqrt{\Delta - \sigma + 1}/2$.*

(2) *Either $D \in \{2, 3, 6, 7, 11, 17\}$ or $h_\Delta = 1$, $\Delta \equiv 5 \pmod 8$, and $\Delta = n^2 + r$ where $r \in \{1, \pm 4\}$.*

Proof. See Exercise 4.2.8. □

Table 4.2.3. *Fundamental Radicands* $\Delta = \Delta_0 > 0$ *satisfying Theorem* 4.2.6.

Δ	$F_{\Delta,1}(x)$	$\lvert F_{\Delta,1}(x)\rvert$ for $3-\sigma \le x < (\sqrt{\Delta}-\sigma+1)/2$
2	x^2-2	–
3	x^2-3	–
5	x^2+x-1	–
6	x^2-6	2
7	x^2-7	3
11	x^2-11	7, 2
13	x^2+x-3	1
17	x^2+x-4	2
21	x^2+x-5	3, 1
29	x^2+x-7	5, 1
37	x^2+x-9	7, 3
53	x^2+x-13	11, 7, 1
77	x^2+x-19	17, 13, 7, 1
101	x^2+x-25	23, 19, 13, 5
173	x^2+x-43	41, 37, 31, 23, 13, 1
197	x^2+x-49	47, 43, 37, 29, 19, 7
293	x^2+x-73	71, 67, 61, 53, 43, 31, 17, 1
437	$x^2+x-109$	107, 103, 97, 83, 79, 67, 53, 37, 19, 1
677	$x^2+x-169$	167, 163, 157, 149, 139, 127, 113, 97, 79, 59, 37, 13

In view of Theorem 4.2.6, we wish to expand the range of values which x can take to include $x = 0$, when $\Delta \equiv 5 \pmod 8$.

Proposition 4.2.1. [(4.2.3)] *If* $\Delta \equiv 5 \pmod 8$ *is a positive fundamental discriminant, then the following are equivalent:*

(1) $|F_{\Delta,1}(x)|$ *is* 1 *or prime for all non-negative integers* $x < (\sqrt{D}-2)/4$.

(2) $D = p^2+4$ *or* $D = (p+2)^2-4$ *where* p *is prime, and there are no non-inert primes less than* $\sqrt{D}/2$, *(so* $h_\Delta = 1$*).*

Proof. If (1) holds and $p < \sqrt{D}/2$ is non-inert, then by Lemma 4.1.3 there is a non-negative integer $x < (\sqrt{D}-2)/4$ such that p divides $F_{\Delta,1}(x)$. Hence, $|F_{\Delta,1}(x)| = p$, i.e. $D = (2x+1)^2+4p$. By Exercise 4.2.9, if $I_0 = [Q_{I(0)}/2, (P_{I(0)}+\sqrt{D})/2]$ is a reduced ideal in $\mathcal{O}_\Delta$, then for any j with $0 \le j < l(I)$ and $I_j = [Q_{I(j)}/2, (P_{I(j)}+\sqrt{D})/2]$, we have $I_j' = [Q_{I(j)}/2, (P_{I(j+1)}+\sqrt{D})/2]$. Thus, we may assume that if $I = [p, (2x+1+\sqrt{D})/2]$, then $P_{I(1)} = x$ (since (1) implies $e_\Delta \le 2$ by Theorem 4.2.3). However, $D = P_{I(1)}^2+4p$, and $D = P_{I(1)}^2 + Q_{I(1)}Q_{I(0)}$, so $Q_{I(1)} = 2$. However, the form for period $l(I) = 2$ is known from Theorem 3.2.1, so $P_{I(1)} = \lfloor\sqrt{D}\rfloor$, a contradiction, since $2x+1 < \sqrt{D}/2$. Hence, there are no non-inert primes less than $\sqrt{D}/2$. By Exercise 4.2.10, we know that $D = m^2 \pm 4r$, with $1 < r < m/2$, has a unique representation with $m = \lfloor\sqrt{D}\rfloor$ for the + sign and $m = \lfloor\sqrt{D}\rfloor+1$ for the − sign. Since any odd prime dividing r would be a non-inert prime less than $\sqrt{D}/2$, then $r = \pm 4$. Furthermore, m must be a prime p since the principal class I has period 1 or 2 when $D = m^2 \pm 4$. In the case where the period is 2, $Q_{I(1)}/2$ must be a prime bigger than $\sqrt{D}/2$, so $p = \lfloor\sqrt{D}\rfloor = Q_{I(1)}/2 = P_{I(1)}$ and $D = p^2+4p = (p+2)^2-4$. If the period length is 1, then $D = \lfloor\sqrt{D}\rfloor^2+4$, and so any prime dividing $\lfloor\sqrt{D}\rfloor$ would split, i.e. $\lfloor\sqrt{D}\rfloor = p$, so $h_\Delta = 1$.

[(4.2.3)] This was proved by different techniques in Louboutin [197] and Mollin–Williams [248]. The proof given here provides a simple version using only continued fractions.

Conversely, if (2) holds and $|F_{\Delta,1}(x_0)|$ is composite for some non-negative $x_0 < (\sqrt{D}-2)/4$, then one of the primes dividing $|F_{\Delta,1}(x_0)|$ is less than $\sqrt{D}/2$. However, $(2x_0+1)^2 \equiv D \pmod{p}$, a contradiction. □

Remark 4.2.3. Proposition 4.2.1 tells the tale when $N = t_\Delta = 0$. However, when $t_\Delta = N-1$, the situation is not so easy to determine. In fact, we posed the following in [246].

Conjecture 4.2.1. *If $\Delta = q_1 q_2 \equiv 5 \pmod{8}$ with $q_1 \equiv q_2 \equiv 3 \pmod{4}$ primes with $q_1 < q_2$, then the following are equivalent:*

(1) *$|F_{\Delta,q_1}(x)|$ is 1 or prime for all non-negative integers $x < (\sqrt{D}-2)/4$.*

(2) *$D = q_1^2 s^2 \pm 4q_1$ or $4q_1^2 s^2 - q_1$, and $h_\Delta = 1$.*

All of the known values for which Conjecture 4.2.1 holds are tabulated below.

Table 4.2.4. *Values of $\Delta = \Delta_0 > 0$ for which Conjecture 4.2.1 holds.*

Δ	q_1	q_2	$\lvert q_1x^2 + q_1x + (q_1^2 - \Delta)/(4q_1)\rvert$ for $0 \le x < (\sqrt{D}-2)/4$
33	3	11	2
69	3	23	5, 1
93	3	31	7, 1
141	3	47	11, 5, 7
213	3	71	17, 11, 1, 19
237	3	79	19, 13, 1, 17
413	7	59	13, 1, 29, 71, 127
453	3	151	37, 31, 19, 1, 23
573	3	191	47, 41, 29, 11, 13, 43
717	3	239	59, 53, 41, 23, 1, 31, 67
1077	3	359	89, 83, 71, 53, 29, 1, 37, 79
1133	11	103	23, 1, 43, 109, 197, 307, 439, 593
1253	7	179	43, 29, 1, 41, 97, 167, 251, 349, 461
1293	3	431	107, 101, 89, 71, 47, 17, 19, 61, 109
1757	7	251	61, 47, 19, 23, 79, 149, 233, 331, 443, 569

In [197], Louboutin proved all conjectures posed in [246] *except* Conjecture 4.2.1, and commented (see [197, p. 334]) that if Conjecture 4.2.1 "holds numerically, it is thanks to nothing but the work of chance, and there is no hope to ever settling it by algebraic means." (sic)

Theorem 4.2.7. *Let $\Delta \equiv 5 \pmod{8}$ be a positive discriminant with $h_\Delta > 1$. If Δ satisfies Assumption 4.1.1 and $|F_{\Delta,q}(x)|$ is 1 or prime for all non-negative integers x with $x < (\sqrt{D}-2)/4$, then $e_\Delta \le 2$, and one of the following must hold:*

(1) *$D = \lfloor\sqrt{D}\rfloor^2 + 4q$, $h_\Delta = 2^{N-1}$, and there are no split primes less than $\sqrt{D}/2$.*

(2) *$D = (3p + q_N + 1)^2 - 4q$ with $h_\Delta = 2 = 2^{N-1}$, $q_N \equiv 1 \pmod{3}$, and p is prime.*

(3) *$D = l^2 - 4q_N$, with $h_\Delta = 2^{N-1}$.*

(4) *$D = l^2 + 4q_N$, with $h_\Delta = 2^{N-1}$.*

(5) $D = l^2 \pm 4$, *with* $h_\Delta = 2$.

(6) $D = l^2 - 4q$, *with* $h_\Delta = 2^{N-1}$.

(7) $D = l^2 + 4q$, *with* $h_\Delta = 2^{N-1}$.

Proof. If $I = [q, (q+\sqrt{D})/2]$, then $D = P_{I(1)}^2 + 2qQ_{I(1)}$, i.e. $-Q_{I(1)}/2 = (P_{I(1)}^2 - D)/(4q) = [q^2(2x_0+1)^2 - \Delta]/(4q)$, where $x_0 = \lfloor (q+\sqrt{D})/(2q) \rfloor - 1$. Since $0 \leq x_0 < (\sqrt{D}-2)/4$, then $Q_{I(1)}/2 = 1$ or p a prime. The following three claims are Exercise 4.2.11(a).

Claim 1. If $Q_{I(1)} = 2p$, then $p > \sqrt{D}/4$ and there is at most one other split prime $p_1 < \sqrt{D}/2$ (and $p_1 > \sqrt{D}/4$).

Claim 2. $p + p_1 = P_{I(1)} + q = 2q(x_0+1)$ if p and p_1 exist.

Claim 3. If $Q_{I(1)} = 2$, then $h_\Delta = 2^{N-1}$ and $D = \lfloor \sqrt{D} \rfloor^2 + 4q$ with no split primes less than $\sqrt{D}/2$.

Claim 3 is (1), so we may now assume that $Q_{I(1)} = 2p > \sqrt{D}/2$.

Case 1. $\mathcal{Q} \not\sim \mathcal{Q}'$, where $\mathcal{Q}' = \Pi_{i=1}^{t_\Delta} \mathcal{Q}_i$.

First, we assume that $t_\Delta = N-1$, so $\mathcal{Q}' = \Pi_{i=1}^{N-1}\mathcal{Q}_i$ after possibly renumbering the $\mathcal{Q}_i$. Thus $\mathcal{Q}_N \sim 1$. Consider $J = [q_N, (q_N + \sqrt{D})/2]$. If p divides $Q_{J(1)}$, then $Q_{J(1)} \leq 6p$. If $Q_{J(1)} = 6p$, then either $p = 3$, $p_1 = 3$ or $3 \mid D$. If p or p_1 is 3, then $\sqrt{D} < 12$. A check of these values yields only $D = 85$ which is a special case of (7). Therefore, we assume henceforth that $\sqrt{D} > 12$. Thus, $3 \mid D$ and $1 \sim \mathcal{Q}_N \sim \mathcal{Q}_3\mathcal{P} \sim \mathcal{Q}_3\mathcal{Q} \sim \mathcal{Q}_3\mathcal{Q}'$, so $q' = 3$. If $p \mid Q_{J(2)}$, then $Q_{J(2)} = 2p$ since $Q_{J(2)} < \sqrt{D}$. Therefore, $1 \sim \mathcal{Q}_N \sim \mathcal{P} \sim \mathcal{Q} \sim \mathcal{Q}'$, a contradiction. Similarly, p_1 does not divide $Q_{J(2)}$. It follows from Lemma 4.2.1 that $Q_{J(2)} = 2$. Hence, by Lemma 4.2.2, $P_{J(1)} = 3p - q_N + 1$, so $D = (3p-q_N+1)^2 + 12pq_N = (3p+q_N+1)^2 - 4q_N$ with $q_N \equiv 1 \pmod 3$. This secures (2).

We now assume that $Q_{J(1)} \neq 6p$. Clearly, $Q_{J(1)} \neq 4p$, so assume $Q_{J(1)} = 2p$. Thus, $1 \sim \mathcal{Q}_N \sim \mathcal{P} \sim \mathcal{Q} \sim \mathcal{Q}'$, a contradiction. A similar analysis to the above holds if p_1 divides $Q_{J(1)}$, so we may assume that $Q_{J(1)}$ is not divisible by either p or p_1. If $Q_{J(1)}$ is divisible by any other unramified prime, then $Q_{J(1)} > \sqrt{D}$ with $Q_{J(1)}/2$ a prime. If p divides $Q_{J(2)}$, then $Q_{J(2)} = 2p$, so $1 \sim \mathcal{Q}_N \sim \mathcal{P} \sim \mathcal{Q} \sim \mathcal{Q}'$, a contradiction. Similarly, p_1 does not divide $Q_{J(2)}$, so $Q_{J(2)} = 2$ by Lemma 4.2.1. From Lemma 4.2.2, it follows that $P_{J(1)} = Q_{J(1)}/2 - q_N + 1$. Thus, $D = (Q_{I(1)}/2 - q_N + 1)^2 + 2q_N Q_{J(1)} = (Q_{(1)}/2 + q_N + 1)^2 - 4q_N$. This secures (3), so we now assume that $Q_{J(1)} = 2$. In this case, $D = \lfloor \sqrt{D} \rfloor^2 + 4q_N$ which secures (4).

Now we assume that $t_\Delta = N$, so $q = q'$. Let q_i be any prime dividing q and set $H = [q/q_i, (q/q_i + \sqrt{D})/2]$. By Exercise 4.2.11(b), $Q_{H(1)}/2$ is prime and $Q_{H(1)} > \sqrt{D}$. Also, $Q_{H(2)}$ cannot be divisible by either p or p_1, so it follows from Lemma 4.2.1 that $Q_{H(2)} = 2q/q_i$. Thus, $P_{H(1)} = Q_{H(1)}/2$ from Lemma 4.2.2, so $D = (Q_{H(2)}/2)^2 + Q_{H(1)}Q_{H(0)} = (Q_{H(1)}/2 + 2q/q_i)^2 - 4(q/q_i)^2$. This forces $q = q_i = q_1$, so $N = 1$ and $h_\Delta = 2$ with $D = (\lfloor \sqrt{D} \rfloor + 1)^2 - 4$. This is (5). If $Q_{H(1)}$ is not divisible by any unramified primes, then $Q_{H(1)} = 2q/q_i$, again forcing $q = q_i$ and $D = \lfloor \sqrt{D} \rfloor^2 + 4$, which is (5).

Case 2. $\mathcal{Q} \sim \mathcal{Q}'$.

In this case, $t_\Delta = N - 1$ is manifest. So $G = [q', (q' + \sqrt{D})/2]$. By Exercise 4.2.11(c), neither p nor p_1 divides $Q_{G(1)}$. If $Q_{G(1)}$ is divisible by any other unramified prime, then $Q_{G(1)}/2 > \sqrt{D}/2$ is prime. Thus, $Q_{G(2)} < \sqrt{D}$. If p or p_1 divides $Q_{G(2)}$, then $Q_{G(2)} = 2p$ or $2p_1$ is forced. Thus, $\mathcal{Q}' \sim \mathcal{P} \sim \mathcal{Q}$, a contradiction. Hence, $Q_{G(2)} = 2q_N$ from Lemma 4.2.1. Therefore, $\mathcal{Q}' \sim \mathcal{Q}_N$, i.e. $\mathcal{Q} \sim 1$. From Lemma 4.2.2, $P_{G(1)} = Q_{G(1)}/2 + q_N - q'$, so $D = (Q_{G(1)}/2 + q_N - q')^2 + Q_{G(1)}2q' = (Q_{G(1)}/2 + q_N + q')^2 - 4q$, which is (6).

Finally, we assume that $Q_{G(1)}$ is divisible only by ramified primes. From Lemma 4.2.1, we get that $Q_{G(1)} = 2q_N$, so $D = P_{G(1)}^2 + 4q$ which is (7). □

Table 4.2.5. $D = \Delta \equiv 5 \pmod 8$ and $|F_{\Delta,q}(x)|$ is 1 or prime for all non-negative $x < b(D) = (\sqrt{D} - 2)/4$, and $h_\Delta > 1$.

D	q	h_Δ	$\lfloor b(D) \rfloor$	$F_{\Delta,q}(x)$	$\lvert F_{\Delta,q}(x) \rvert$	Type
85	5	$2 = 2^N$	1	$5x^2 + 5x - 3$	3, 7	$\ell^2 + 4$
165	5	$2 = 2^{N-1}$	2	$5x^2 + 5x - 7$	7, 2, 23	$\ell^2 - 4$
285	15	$2 = 2^{N-1}$	3	$15x^2 + 15x - 1$	1, 29, 89, 179	$\ell^2 + 4q$
357	3	$2 = 2^{N-1}$	4	$3x^2 + 3x - 29$	29, 23, 11, 7, 31	$\ell^2 - 4$
365	15	$2 = 2^N$	4	$5x^2 + 5x - 17$	17, 7, 13, 43, 83,	$\ell^2 - 4$
645	$15 = 3q_N$	$2 = 2^{N-1}$	5	$15x^2 + 15x - 7$	7, 23, 83, 173, 293, 443	$\ell^2 + 4q_N$
957	3	$2 = 2^{N-1}$	7	$3x^2 + 3x - 79$	79, 73, 61, 43, 19, 11, 47, 89	$\ell^2 - 4$
965	5	$2 = 2^N$	7	$5x^2 + 5x - 47$	47, 37, 17, 13, 53, 103, 163, 233	$\ell^2 - 4$
1085	5	$2 = 2^{N-1}$	7	$5x^2 + 5x - 53$	53, 43, 23, 7, 47, 97, 157, 227	$\ell^2 - 4$
1245	$15 = 3q_N$	$2 = 2^{N-1}$	8	$15x^2 + 15x - 17$	17, 13, 73, 163, 283, 433, 613, 823, 1063	$\ell^2 + 4q_N$
1685	5	$2 = 2^N$	9	$5x^2 + 5x - 83$	83, 73, 53, 23, 17, 67, 127, 197, 277, 367	$\ell^2 - 4$
1965	15	$2 = 2^{N-1}$	10	$15x^2 + 15x - 29$	61, 151, 271, 421, 601, 811, 1051, 1321, 1621	$\ell^2 - 4q$
2085	15	$2 = 2^{N-1}$	10	$15x^2 + 15x - 31$	31, 1, 59, 149, 269, 419, 599, 809, 1049, 1319, 1619	$\ell^2 + 4q$
2373	$21 = 3q_N$	$2 = 2^{N-1}$	11	$21x^2 + 21x - 23$	23, 19, 103, 229, 397, 607, 859, 1153, 1489, 1867, 2287, 2749	$\ell^2 - 4q_N$
2397	3	$2 = 2^{N-1}$	11	$3x^2 + 3x - 199$	199, 193, 181, 163, 139, 109, 73, 31, 17, 71, 131, 197	$\ell^2 - 4$
4245	$15 = 3q_N$	$2 = 2^{N-1}$	15	$15x^2 + 15x - 67$	67, 37, 23, 113, 233, 383, 563, 773, 1013, 1283, 1583, 1913, 2273, 2663, 3083, 3533	$\ell^2 + 4q_N$

This completes the project of finding all of the qth Euler–Rabinowitsch polynomials which produce initial, consecutive, distinct prime values, insofar as we have

Conjecture 4.2.2. *Tables* 4.2.1–4.2.5 *represent all positive fundamental discriminants for which Theorems* 4.2.4–4.2.7 *and Conjecture* 4.2.1 *hold.*

Exercises 4.2

1. Prove Lemma 4.2.1. (*Hint*: Use the continued fraction algorithm to conclude that there exists an integer i with $0 < i < \ell = \ell(I)$ such that $N(J) = Q_{I(i)}/\sigma$. Then use induction to establish symmetry with $P_{\ell/2} = P_{\ell/2-1}$. Then use (2.1.3)–(2.1.4) with Exercise 1.2.5.)

2. Prove Lemma 4.2.2. (*Hint*: This is routine, using formulas (2.1.2)–(2.1.4).)

3. Prove Claim 1 in the proof of Theorem 4.2.1. (*Hint*: Consider $F_{\Delta,1}(x) \equiv 0 \pmod{p^2}$ for which $p = 2$ is forced by hypothesis. Show that this leads to a contradiction.)

4. Verify that (4.2.3) in the proof of Theorem 4.2.1 has 2^{M-1} distinct possibilities. (*Hint*: Use the binomial coefficients (see Exercise 3.1.7).)

5. (a) Prove Theorem 4.2.2. (*Hint*: Use Theorem 1.3.3 to prove (2) $\Rightarrow$ (1) and use Lemma 4.2.3 and Theorem 4.2.1 to prove the converse.)
   (b) Show that if $G_1(\Delta) \neq 0$, then (1) of Theorem 4.2.2 may be replaced by $e_\Delta \leq 2$.

6. Let $\Delta > 0$ be a discriminant with associated radicand D and assume that $I = [N(I), b + w_\Delta]$ is a reduced ideal of $\mathcal{O}_\Delta$. Prove that if $Q_i \geq \sqrt{D}$ in the continued fraction expansion of $(b + w_\Delta)/N(I)$, then $a_i = 1$. (*Hint*: Establish the inequality $2\sqrt{D} > P_i + \lfloor\sqrt{D}\rfloor \geq a_i Q_i \geq a_i\sqrt{D}$.)

7. In the proof of Theorem 4.2.5:
   (a) Establish Claim 4. (*Hint*: Use 2.1.15(e)–(f).)
   (b) Prove that if $p \mid Q_{I(2)}$ in Case 2, then $D = 3, 7, 11$, or 23. (*Hint*: Use Claim 2 and Claim 4.) Also verify that if $p_1 \mid Q_{I(2)}$, then $Q_{I(2)} = p_1$ and $a_{I(1)} = 1$. Then prove that this is not possible so that p_1 does not divide $Q_{I(2)}$.
   (c) Prove, in Case 3(a), that if $Q_{J(1)} = 2p$ we get (3), and if $Q_{J(1)} > \sqrt{D}$ is an unramified prime, then we get (4). Finally, show that if $Q_{J(1)}$ is not divisible by any unramified primes, then we get (5). (*Hint*: Use techniques already established.)
   (d) In Case 3(b), prove that $Q_{G(2)} = q_N$. (*Hint*: Use arguments similar to those given to prove that $P_{G(2)} = P_{G(3)}$, so $P_{G(1)} = (Q_{G(1)} - q' + Q_{G(2)})/2$ by Lemma 4.2.2.)

(e) In Case 3(b), prove that neither p nor p_1 divides $Q_{I(2)}$. (*Hint*: Using arguments similar to those given to deduce that if $p_1 \mid Q_{I(2)}$ then $Q_{I(3)} = 1$, $a_{I(1)} = 1$, $a_{I(2)} = 2$ and

$$D = P_{I(1)}^2 + pq = ((9p + 4q - 1)/12)^2 + q/9. \qquad (4.2.5)$$

Then show that this leads to a contradiction. The latter is done by considering the ideal $H = [q/2, \sqrt{D}]$, and deducing that

$$D = (Q_{H(1)} + 2 + q/2)^2/4 - q. \qquad (4.2.6)$$

Set $A = (9p + 4q - 1)/4$ and $B = (Q_{H(1)} + 2 + q/2)$ in (4.2.5)-(4.2.6) to get that

$$10D = A^2 + B^2 = (q/2)^2[(2A/q)^2 + (2B/q)^2].$$

Show that this forces $q = 2$ and $h_\Delta = 1$. A similar argument holds for p.)

(f) Prove the final fact that $a_{I(1)} = 1$. (*Hint*: $a_{I(1)} \leq 3$. If $a_{I(1)} = 3$, then $D = (9p + q + 1)^2/36 - q/9$. Then consider $H = [q/2, \sqrt{D}]$ as above, and deduce $8D = C^2 - B^2$ where $C = (9p + q + 1)/2$ which, together with the above representation for D, yields $10D = C^2 + P_{H(1)}^2$. Show that this yields a contradiction.)

8. Prove Theorem 4.2.6. (*Hint*: That (2) implies (1) is an easy check when $D \in \{2, 3, 6, 7, 11, 17\}$. If $D \equiv 5 \pmod 8$, with $D = n^2 + r$, $r \in \{1, \pm 4\}$, then $\ell(w_\Delta) \leq 3$, by Theorem 3.2.1. Suppose that $F_\Delta(x) = pn$ where $n > 1$, p is prime, $p \leq n$ and $1 < x < (\sqrt{D} - 1)/2$, then $p = Q_i/2$ for some positive $i < \ell(w_\Delta)$. Show that this leads to a contradiction, using the tables in Theorem 3.2.1. Conversely assume (1). The cases are all easy checks except $D \equiv 5 \pmod 8$. In this case, $D = P_i^2 + Q_iQ_{i-1}$ where $0 < i \leq \ell(w_\Delta) = \ell$. If $P_i = 2x_i + 1$, then $0 \leq x_i \leq (\sqrt{D} - 1)/2$ and $F_\Delta(x_i) = Q_iQ_{i-1}/4$. Assume $\ell > 3$ and get a contradiction to the primality of $F_\Delta(x)$. For $\ell \leq 3$, use the tables in Theorem 3.2.1.)

9. Let $\Delta > 0$ be a discriminant and $I = [Q/\sigma, (P + \sqrt{D})/\sigma]$ a reduced ideal in $\mathcal{O}_\Delta$. Prove that in the continued fraction expansion of $(P + \sqrt{D})/Q$, we have that

$$(I_i)' = [Q_i/\sigma, (P_i + \sqrt{D})/\sigma]' = [Q_i/\sigma, (P_{i+1} + \sqrt{D})/\sigma].$$

(*Hint*: Use (2.1.4).)

10. Let $\Delta > 0$ be a discriminant. Prove that Δ can be written in exactly one way as $\Delta = m^2 \pm 4r$ with $1 < r < m/2$, namely $m = \lfloor\sqrt{\Delta}\rfloor$ in the case of the $+$ sign and $m = \lfloor\sqrt{\Delta}\rfloor + 1$ for the $-$ sign.

11. In the proof of Theorem 4.2.7:

(a) Verify Claims 1-3. (*Hint*: Use the same methodology as in the proof of Theorem 4.2.5.)

(b) In Case 1, verify that $Q_{H(1)}/2$ is prime and $Q_{H(1)} > \sqrt{D}$. (*Hint*: First show that p does not divide $Q_{H(1)}$ in a manner similar to the proof of Theorem 4.2.5. Similarly, p_1 does not divide $Q_{H(1)}$, so if any other unramified prime divides $Q_{H(1)}$, then $Q_{H(1)}/2$ is prime and $Q_{H(1)} > \sqrt{D}$.)

(c) In Case 2, show that neither p nor p_1 divides $Q_{G(1)}$. (*Hint*: The hint in (b) applies, mutatis mutandis.)

12. Does Corollary 4.2.1 hold for arbitrary orders? If it does, then prove it. If not, then find a discriminant $\Delta > 0$, $\Delta \equiv 1 \pmod 8$, with $e_\Delta = 2$ and $\Delta \neq 33$.

.3 Density of Primes.

In the previous two sections, we were concerned with consecutive prime-producing quadratics in an initial range of values, and we essentially solved the problem for all qth Euler–Rabinowitsch polynomials. It is natural to ask how many primes are produced by a given quadratic polynomial up to a given bound. For instance, motivated by Euler's polynomial and subsequent investigations of related polynomials, we would like to know the number $P_A(n)$ of primes assumed by $F_\Delta(x) = x^2 + x + A$ for $x = 0, 1, 2, \ldots, n$ and $\Delta = 1 - 4A < 0$. From section one of this chapter, we know that $P_A(A-2) = A - 1$ if and only if $A \in \{2, 3, 5, 11, 17, 41\}$, i.e. $\Delta \in \{-7, -11, -19, -43, -67, -163\}$. Before it was known that $A = 41$ is the largest value for which this phenomenon occurs, Lehmer [169] in 1936 attempted to find a larger such value of A. First, he observed that in order for $F_\Delta(x)$ to be prime, then A must be odd, so that $-\Delta \equiv 3 \pmod 8$. What he did then was to let N_p, for an odd prime p, be the least positive integer such that $N_p \equiv 3 \pmod 8$ and $(N_p/q) = -1$, for all odd primes $q \leq p$, where $(/)$ is the Legendre symbol. If we think back to section one of this chapter, we can see why he did this. By Lemma 4.1.2, a prime q with $(\Delta/q) = -1$ cannot divide $F_\Delta(x)$. Hence, if $(\Delta/q) = -1$ for "enough" small primes q, then $F_\Delta(x)$ is less likely to be composite, i.e. $F_\Delta(x)$ would be prime a larger percentage of the time. Lehmer determined that if $P_A(A-2) = A - 1$, then $N_{109} > -\Delta$, and $A > 1.25 \cdot 10^9$, which told him that the existence of $A > 41$ with $P_A(A-2) = A - 1$ was highly unlikely. Thus, Lehmer thought that it might be more fruitful to search for an A with $P_A(40) = 41$. As noted in section one of this chapter, we know that if this occurs, as Theorem 4.1.1 suggests it must, then $A > 10^{17}$.

In 1939, Beegner [20] found $F_\Delta(x) = x^2 + x + A$ with $A = 19421, 27941$, and 72491. He did this by computing all positive $\Delta \in \mathbf{Z}$ with $\Delta < 10^6$, $\Delta \equiv 3 \pmod 8$ such that $(-\Delta/q) = -1$, for all odd primes $q \leq 43$. The only numbers satisfying this criterion are $\Delta = 77683, 111763$, and 289963, i.e. corresponding to $(\Delta + 1)/4 = A = 19421$, 27941, and 72491 respectively. In 1939, Poletti [293] computed values of $P_A(n)$ for these values of A and some $N < 11 \cdot 10^3$ such as $P_{41}(11000) = 4605$, $P_{27941}(11000) = 4819$ and $P_{72491}(11000) = 4923$. Thus, Beegner's polynomial seemed to him to be better at prime-production than Euler's polynomial. What is behind all of this is a result of Hardy and Littlewood [109] from 1923, of which the above authors ostensibly were surprisingly not aware. It is (what Hardy and Littlewood called "Conjecture F"):

Conjecture 4.3.1. [4.3.1] *If $F_\Delta(x) = x^2 + x + A$ (with $\Delta = 1 - 4A$ not a perfect*

[4.3.1] Actually, Conjecture F is more general than this, but we have boiled it down to suit our particular situation. Note that $f(n) \sim g(n)$ means that $\lim_{n\to\infty} \frac{f(n)}{g(n)} = 1$. See Exercise 4.3.1.

square), then

$$P_A(n) \sim C(\Delta)L_A(n),$$

where

$$L_A(n) = 2\int_0^n dx/\log F_\Delta(x),$$

and

$$C(\Delta) = \prod_{p\geq 3}(1-(\Delta/p)/(p-1)),$$

where p ranges over all odd primes and $(/)$ is the Legendre symbol.

For instance, we know from Fung and Williams [89] that $C(-163) = 3.3197732$ for $A = 41$, $C(-111763) = 3.631998$ for $A = 27941$, and $C(-289963) = 3.6947081$ for $A = 72491$, confirming Beegner's observations. Thus, Conjecture 4.3.1 allows us to find examples of $F_\Delta(x)$ with a ***high asymptotic density of prime values.*** Also, as calculated by Fung and Williams [89],

$$P_{41}(10^6)/L_{41}(10^6) = 3.3203421,$$

and

$$P_{27941}(10^6)/L_{27941}(10^6) = 3.6396821,$$

demonstrating Conjecture 4.3.1's prediction of the proximity of $P_A(10^6)/L_A(10^6)$ and $C(\Delta)$. How $C(\Delta)$ is computed, and extensions of Beegner's ideas to find new $F_\Delta(x)$ with high asymptotic density of primes, can be found in the recent paper [207] of Lukes *et al.* They use a new number sieve called the MSSU developed at the University of Manitoba.(4.3.2) By using this machine, they not only found new values of A which we present below, but also extended the list of known pseudosquares which are linked to the problem of whether primality testing can be accomplished in deterministic polynomial time.(4.3.3)

Before listing the new "record" value of A computed by Lukes *et al.* in [207], we note that the values of N_p defined above were calculated by Lehmer *et al.* [174], Shanks [334] and Fung and Williams [89]. In the latter, they use a method of calculation which is dependent upon the GRH. Nevertheless, in [89], Fung and Williams achieved the largest value of $C(\Delta)$ theretofore known, namely

$$C(\Delta) = 5.0870883 \text{ for } \Delta = -531497118115723,$$

with

$$F_\Delta(x) = x^2 + x + 132874279528931.$$

Up until that paper in 1990, the largest known value was that of Shanks [333] with $C(\Delta) = C(-991027) = 4.1237067$. We note that in the example of Fung and Williams, $(\Delta/q) = -1$ for all primes q with $2 < q \leq 179$. Moreover, they required not only Conjecture 4.3.1, but also the GRH.

(4.3.2)This acronym means *The Manitoba Scalable Sieve Unit* which employs *Very Large Scale Integration* (VLSI) circuits designed by Patterson [285]-[287], at the University of Calgary as part of his dissertation (upon whose Ph.D. committee this author sat). The raw aggregate sieving rate of the MSSU is in excess of 6.4 billion integers per second.

(4.3.3)See Chapter Eight for a description of such time measures. Also, a *pseudosquare* L_p is defined as follows: (1) $L_p \equiv 1 \pmod 8$, (2) The Legendre symbol $(L_p/q) = 1$ for all odd primes $q \leq p$, and (3) L_p is the least positive non-square integer satisfying (1) and (2).

In Lukes *et al.* [207], they used the MSSU to search for all $\Delta \equiv 5 \pmod 8$ with $|\Delta| < 10^{19}$ and $(\Delta/q) = -1$, for all odd primes less than 200. The largest value of $C(\Delta)$ which they found is

$$C(\Delta) = 5.3384021,$$

for

$$\Delta = -6849319464662435083.$$

Hence, under the assumption of the validity of Conjecture 4.3.1, $F_\Delta(x) = x^2+x+A$, with

$$A = 1712329866165608771$$

has the largest asymptotic density of primes for any polynomial of this type known to date.

Other authors have looked at less demanding problems involving prime-producing quadratics. For example, Boston *et al.* [31] wanted to find a quadratic polynomial $f(x)$ which represents the most *distinct* primes $0 \le x \le 99$. Of course, we have observed earlier that if we allow repetitions, then we can get longer strings than if we impose the restriction of distinctness. Boston observes that $x^2 - 69x + 1231$ represents 95 primes for $x = 0, \ldots, 99$, albeit not distinct. His largest value found, which generates distinct values, is given by $41x^2 + 33x - 43321$ with $\Delta = 7105733$.

In looking at class number one problems for real quadratic fields, questions arose regarding the polynomials

$$g_1(x) = 47x^2 - 2247x + 21647, \quad g_2(x) = 103x^2 - 3945x + 34381$$

(called the Fung polynomials) and

$$g_3(x) = 36x^2 - 810x + 2753$$

(called the Ruby polynomial), where g_1 and g_2 generate primes for $x = 0, 1, \ldots, 42$ and, g_3 generates primes for $x = 0, 1, \ldots, 44$ (see Mollin–Williams [253]). To date, these are the largest known consecutive distinct prime-producers.(4.3.4)

Exercises 4.3

1. With respect to the notation $\sim$ introduced in footnote (4.3.1), prove the following:

   (a) $f \sim f$.

   (b) If $f \sim g$ then $g \sim f$.

   (c) If $f \sim g$ and $g \sim h$ then $f \sim h$.

   (d) If $f_1 \sim g_1$ and $f_2 \sim g_2$ then $f_1 f_2 \sim g_1 g_2$.

   (e) $\displaystyle\sum_{i=1}^{x} \frac{1}{i} \sim \log x$.

(4.3.4)The challenge left at the end of Boston's paper [31], where he cites these polynomials from [253], is to find an example that gives 49 or even 50 distinct primes. Theorem 4.1.1 says one should be able to do so. Also, note that $C(\Delta) = 3.7006266$, where $\Delta = 979373$, for $g_1(x)$, $C(\Delta) = 3.9727065$ where $\Delta = 1398053$ for g_2, and $C(\Delta) = 3.9727065$, where $\Delta = 259668$, for $g_3(x)$ (see Lukes *et al.* [207]).

Chapter 5

Class Numbers: Criteria and Bounds.

In the previous chapter, we saw the relationship between orders of class number one and prime-producing quadratics. In this chapter, we bring in the theory of continued fractions in a stronger way to show the interrelationships. Since the class number one problem for complex quadratic orders has been completely solved (Theorem 4.1.3), and the class number one problem for real quadratic orders remains open, we concentrate upon the latter. Other class number criteria will also be explored, as well as bounds on class numbers for both the real and complex case.

5.1 Factoring Rabinowitsch.

First, we describe the motivation behind the study which gave rise to the results in this section. For a given discriminant Δ, there are ideals of norms a and b, whose classes are inverses of one another in C_Δ, if and only if there is a principal ideal of norm ab, i.e. an $\alpha \in \mathcal{O}_\Delta$ with $\alpha = x + yw_\Delta$ whose norm, $ab = N(\alpha) = F_\Delta(x, y) = x^2+(\sigma-1)xy+(\sigma-1-\Delta)y^2/4$ (compare with the principal form in Appendix E). The ideals of norms a, b are either both principal, or both non-principal, so ultimately the structure of the class group is reflected in the factoring of $F_\Delta(x, y)$, and only the composite values of $F_\Delta(x, y)$ lead to a non-trivial class group. If $I = [a, x + w_\Delta]$ is a reduced $\mathcal{O}_\Delta$-ideal, then $y = 1$ and $F_\Delta(x, 1) = F_\Delta(x) = N(x + w_\Delta)$. When $\Delta < 0$, the Rabinowitsch criterion (Theorem 4.1.2) says that $F_\Delta(x)$ is prime whenever $x \in \mathcal{S}_\Delta$, i.e. no factorization of $F_\Delta(x)$ occurs up to a Rabinowitsch bound, so $h_\Delta = 1$. On the other hand, if $F_\Delta(x)$ factors, then we can count the number of such factors, and if we do this properly, then we can count all of the classes of C_Δ and all of the reduced ideals of $\mathcal{O}_\Delta$. This is the reason for the title of this section, and this is what we now show how to do concretely for $\Delta > 0$.

We therefore begin with a result which counts the number of divisors of the Euler–Rabinowitsch polynomial.

Theorem 5.1.1. [(5.1.1)] *Let $I = [a, b + w_\Delta]$ be a reduced ideal in $\mathcal{O}_\Delta$, where $\Delta = \Delta_0 > 0$ is a fundamental discriminant with $-a < b + w'_\Delta < 0$, and set*

$$d(a, F) = |\{x \in \mathbb{Z} : 2 - \sigma \leq x \leq \lfloor w_\Delta \rfloor - \sigma + 1 \text{ and } a \mid F_\Delta(x)\}|.$$

[(5.1.1)]This was established in [225], where we used a slightly different form of $F_\Delta(x)$.

If $q = \lfloor (b + w_\Delta)/a \rfloor$ and a is prime, then

$$d(a, F) = \begin{cases} q+1 & \text{if } a \nmid \Delta \\ \lfloor (q+1)/2 \rfloor & \text{if } a \mid \Delta. \end{cases}$$

If a is composite, then $d(a, F)$ is at least this number.

Proof. First, we observe that $F_\Delta(x) = N(x + w_\Delta)$. Also, we remark that the case $\sigma = 1$ and $b = 0$ cannot occur, by Corollary 1.4.4.

Claim 1. If $x \equiv b \pmod a$ or $x \equiv -b - \sigma + 1 \pmod a$, then $F_\Delta(x) \equiv 0 \pmod a$. Conversely, if a is prime, then $F_\Delta(x) \equiv 0 \pmod a$ implies that either $x \equiv b \pmod a$, or $x \equiv -b - \sigma + 1 \pmod a$ (see Exercise 5.1.1(a)).

Case 1. $x \equiv -b - \sigma + 1 \pmod a$, i.e. $x = -b - \sigma + 1 + ga$, where $g \in \mathbf{Z}$. Since

$$2 - \sigma \le x \le \lfloor w_\Delta \rfloor + 1 - \sigma,$$

then

$$(b+1)/a \le g \le (b + \lfloor w_\Delta \rfloor)/a,$$

i.e. $\lceil (b+1)/a \rceil \le g \le q$.

Claim 2.

$$\lceil (b+1)/a \rceil = \begin{cases} q/2 \text{ or } q/2+1 & \text{if } q \text{ is even,} \\ (q+1)/2 & \text{if } q \text{ is odd.} \end{cases}$$

(See Exercise 5.1.1(b).)

Now, let $\#g$ denote the number of distinct values taken on by g.

Claim 3.

$$\#g = \begin{cases} q/2+1 & \text{if } \lceil (b+1)/a \rceil = q/2, \\ (q+1)/2 & \text{if } \lceil (b+1)/a \rceil = (q+1)/2, \\ q/2 & \text{if } \lceil (b+1)/a \rceil = q/2+1. \end{cases}$$

Since $g = \lceil (b+1)/a \rceil, \lceil (b+1)/a \rceil + 1, \ldots, q$, then Claim 3 follows from Claim 2.

Case 2. $x \equiv b \pmod a$, so $x = b + ha$ where $h \in \mathbf{Z}$. Since $2 - \sigma \le x \le \lfloor w_\Delta \rfloor - \sigma + 1$, then $(2 - \sigma - b)/a \le h \le (\lfloor w_\Delta \rfloor - b - \sigma + 1)/a < 1$, so $\lceil (2 - \sigma - b)/a \rceil \le h \le 0$, unless $\lceil (2 - \sigma - b)/a \rceil > 0$, i.e. $\sigma = 1$ and $b = 0$, which is ruled out by our initial remark.

Claim 4. $\lceil (2 - \sigma - b)/a \rceil = -\lceil (b+1)/a \rceil + 1$. (See Exercise 5.1.1(c).)

Now, let $\#h$ be the number of distinct values taken on by h.

Claim 5. $\#h + \#g = q + 1$.

Since $h = 0, -1, -2, \ldots, \lceil (2 - \sigma - b)/a \rceil$, then $\#h = -\lceil (2 - \sigma - b)/a \rceil + 1 = \lceil (b+1)/a \rceil$ by Claim 4. Claim 5 now follows from Claim 3.

At this juncture, we do not know whether or not these $q + 1$ values are indeed distinct, since there may be duplications between Cases 1 and 2. We now examine this possibility.

Claim 6. If $a \mid \Delta$, then we have exactly $\lfloor (q+1)/2 \rfloor$ values of x being distinct (when $2-\sigma \le x \le \lfloor w_\Delta \rfloor - \sigma + 1$) and, if $a \nmid \Delta$, then there are exactly $q+1$ such values of x which are distinct (see Exercise 5.1.1(d)).

We have shown that if a is a prime, then (by Claims 1 and 6) there are exactly $q+1$ values of x with $2-\sigma \le x \le \lfloor w_\Delta \rfloor - \sigma + 1$ such that $F_\Delta(x) \equiv 0 \pmod a$ provided that $a \nmid \Delta$. If $a \mid \Delta$ and is prime, then there are exactly $\lfloor (q+1)/2 \rfloor$ such values. If a is composite, then (by Claims 1 and 6) there are at least that many values. □

An illustration of Theorem 5.1.1 is

Example 5.1.1. Let $\Delta = \Delta_0 = 433$ and consider the reduced ideal $I = [18, (19+\sqrt{433})/2]$. The simple continued fraction expansion of $(19+\sqrt{433})/36$ is given by the following:

i	0	1	2	3	4	5	6	7	8	9	10	11
P_i	19	17	19	17	15	11	13	9	7	17	19	17
Q_i	36	4	18	8	26	12	22	16	24	6	12	12
a_i	1	9	2	4	1	2	1	1	1	6	3	3

with the balance given by symmetry.

Thus, $\ell(I) = 23$ and $q = 1$. Also, for $0 = 2-\sigma \le x \le \lfloor w_\Delta \rfloor - \sigma + 1 = 9$, we have

x	1	2	3	4	5	6	7	8	9
$\lvert F_\Delta(x) \rvert$	$2 \cdot 53$	$2 \cdot 3 \cdot 17$	$2^5 \cdot 3$	$2^3 \cdot 11$	$2 \cdot 3 \cdot 13$	$2 \cdot 3 \cdot 11$	$2^2 \cdot 13$	$2^2 \cdot 3^2$	$2 \cdot 3^2$

Thus, $d(18, F) = q + 1 = 2$ as predicted by Theorem 5.1.1.

A further demonstration of the applicability of Theorem 5.1.1 to solving class number problems is given in Exercise 5.1.2 for $\Delta \equiv 1 \pmod 8$.

We have provided a table of all ERD-types with $\Delta = \Delta_0 \equiv 1 \pmod 8$ and $h_\Delta \le 23$ in Appendix A, Table A8.(5.1.2)

Remark 5.1.1. In Theorem 5.1.1, $d(a, F)$ counts the number of times the given norm of a reduced ideal divides the Euler–Rabinowitsch polynomial over an initial string of non-negative values bounded by the floor of the principal surd. Thus, what this theorem reveals is that the first partial quotient of the simple continued fraction expansion of the given reduced ideal is intimately related to $d(a, F)$. This allows us to determine class numbers of certain fields such as those of ERD-type, since the number of reduced ideals in a given class, especially the principal class, is small. This enables us to use $d(a, F)$, for a equal to the norm of these ideals, in order to bound the discriminant, and hence, list all such discriminants of a given class number. When $\Delta = \Delta_0 \equiv 1 \pmod 8$ is of ERD-type, this becomes an especially rewarding tool because then, the ideal above 2 can be analyzed since $2 \in \Lambda^*(\Delta)$, when $h_\Delta = 1$. To determine such fields with $h_\Delta = 1$, we rely upon the fact that

(5.1.2)In fact, this author has a list of all such $\Delta \le 2 \cdot 20^9$.

the Euler–Rabinowitsch polynomial always has even values when $\Delta \equiv 1 \pmod 8$, so $d(a, F)$, for $a = 2$, is the floor of the principal surd. This restricts the number of possibilities for D quite nicely. Similarly, for $h_\Delta = 2, 3, 4, 5$ (etc.) (see Exercise 5.1.2), we can look at the square, cube, etc. of the ideal above 2, so that power must be the norm of a reduced principal ideal. Thus, we use $d(a, F)$ to bound the possibilities for D, so that we may list these as well. This describes how to exploit $d(a, F)$ in the principal class. It can, however, be used in any class given its generality, and the reader may solve Exercise 5.1.2, for example, by this alternative method. See also Exercise 5.1.3.

We are now in a position to show how continued fractions and the factorization of the Euler–Rabinowitsch polynomial can be used to determine $h_\Delta = 1$ for a fixed period length of w_Δ. We begin with $\ell(w_\Delta) = 3$ (see Exercise 5.1.4).

Theorem 5.1.2. [(5.1.3)] *If $\Delta = \Delta_0 > 0$ is a fundamental discriminant with $\ell(w_\Delta) = 3$ and $\Delta \not\equiv 1 \pmod 4$, then $h_\Delta > 1$. If $\Delta \equiv 1 \pmod 8$, then $h_\Delta = 1$ if and only if $\Delta = 17$. If $\Delta \equiv 5 \pmod 8$, then $w_\Delta = \langle a, \overline{b, b, 2a-1}\rangle$ $(a, b \in \mathbb{Z})$,*

$$\Delta = (b + c(b^2+1))^2 + 4(bc+1) \text{ with } a = (b + c(b^2+1)+1)/2 \quad (c \in \mathbb{Z}),$$

and

$$h_\Delta = 1 \text{ if and only if } bc+1 \text{ is prime and } |F_\Delta(x)| \text{ is } 1 \text{ or prime,}$$

whenever $\gcd(|F_\Delta(x)|, bc+1) = 1$ (i.e. whenever $x \equiv 2^{-1}(\pm c - 1) \pmod{bc+1}$ and $0 \le x \le a-1$).

Proof. If $\Delta \not\equiv 1 \pmod 4$, then Δ is a sum of two squares (Exercise 5.1.5). Also, by Corollary 1.3.2, if $h_\Delta = 1$, then $D = p$ or $2p$ where p is a prime with $p \equiv 3 \pmod 4$. However, by Exercise 1.1.4(c), Δ cannot be a sum of two squares, if such a prime $p \equiv 3 \pmod 4$ divides Δ, a contradiction. Hence, $h_\Delta > 1$ when $\Delta \not\equiv 1 \pmod 4$.

If $\Delta \equiv 1 \pmod 4$, then $w_\Delta = \langle a, \overline{b, b, 2a-1}\rangle$ by Exercise 2.1.13(b) where $a, b \in \mathbb{Z}$. Moreover, by Exercise 5.1.6, the following chart holds for the simple continued fraction expansion of w_Δ.

i	0	1	2	3
P_i	1	$b+c(b^2+1)$	$b+c(b^2-1)$	$b+c(b^2+1)$
Q_i	2	$2(bc+1)$	$2(bc+1)$	2
a_i	$\frac{b+c(b^2+1)+1}{2}$	b	b	$b+c(b^2+1)$

Hence, $\Delta = D = (b + c(b^2+1))^2 + 4(bc+1)$ (by equation 2.1.5). Thus, if $\Delta \equiv 1 \pmod 8$, and $h_\Delta = 1$, then $bc + 1 = 2 \in \Lambda^*(\Delta)$ (see Definition 3.3.2 and Theorem 2.1.2), i.e. $b = c = 1$, so $\Delta = 17$. Now assume that $\Delta \equiv 5 \pmod 8$ and $h_\Delta = 1$. If $bc + 1 > \sqrt{\Delta}/2$, then by Exercise 4.2.6, $b = 1$ and $\Delta = 4(c+1)^2 + 1$. By Theorem 4.2.6, $|F_\Delta(x)|$ is 1 or prime for all $x \in \mathbb{Z}$ with $1 \le x \le c+1$, which is the desired result since $|F_\Delta(0)| = (c+1)^2$. We may now assume that $bc + 1 < \sqrt{\Delta}/2$. Since Theorem 2.1.2 tells us that any split prime $p < \sqrt{\Delta}/2$ must be in $\Lambda^*(\Delta)$, then $bc + 1$ is forced to be prime (and, in fact, there can be no *other* split primes less than $\sqrt{\Delta}/2$). If $\gcd(|F_\Delta(x)|, bc+1) = 1$, then $F_\Delta(x)$ must be prime. To see this, we observe that $|F_\Delta(x)| < D/4$, so if $|F_\Delta(x)|$ were composite, it would have to be divisible by a prime $p < \sqrt{\Delta}/2$ which is impossible, unless $p = bc+1$. Furthermore,

---

(5.1.3) This was established in [249] with far more algebraic calisthenics performed than necessary in the proof since we did not have the advantage of Theorems 4.2.6 and 5.1.1.

by Theorem 5.1.1, $bc+1$ divides $|F_\Delta(x)|$ precisely when $2x \equiv \pm c - 1 \pmod{bc+1}$ and $0 \le x \le a-1$. □

An illustration of Theorem 5.1.4 is

Example 5.1.2. If $\Delta = D = 317$, then $w_\Delta = \langle 9, \overline{2, 2, 17}\rangle$. Here, $a = 9$, $b = 2$ and $c = 3$. Also, $bc + 1 = 7$ and $7 \mid F_\Delta(x)$ for exactly $x \equiv 2^{-1}(\pm 3 - 1) \pmod 7$ with $0 \le x \le 8$, i.e. $x = 1, 5$. Hence, since $h_\Delta = 1$, we see that $|F_\Delta(x)|$ is prime whenever it is not divisible by 7, for $0 \le x \le 8$.

x	0	1	2	3	4	5	6	7	8
$\lvert F_\Delta(x)\rvert$	79	$7 \cdot 11$	73	67	59	7^2	37	23	7

We can also look at period 4 from the same perspective (Exercise 5.1.7). We now turn to period 5. We first note that if $\ell(w_\Delta)$ is odd, then $h_\Delta = 1$ implies that $\Delta \equiv 1 \pmod 4$ and Δ is prime (see Exercise 5.1.5(b)).

Theorem 5.1.3. [(5.1.4)] *If $\Delta = \Delta_0 \equiv 1 \pmod 4$ is a fundamental discriminant with $\ell(w_\Delta) = 5$, then $w_\Delta = \langle a, \overline{b, c, c, b, 2a-1}\rangle$ where $2a - 1 = b(c^2+1)^2 + c(c^2+1) - f((bc+1)^2 + b^2)$ and $\Delta = (2a-1)^2 + 4r$ with $r = (c^2+1)^2 - (bc^2 + c + b)f$ for some positive $a, b, c \in \mathbb{Z}$ and $f \in \mathbb{Z}$, $f \ne 0$. If $s = c(c^2+1) - f(bc+1)$, then:*

(a) *If $\Delta \equiv 1 \pmod 8$, then $h_\Delta = 1$ if and only if $\Delta = 41$.*

(b) *If $\Delta \equiv 5 \pmod 8$, then $h_\Delta = 1$ if and only if each of the following holds:*

  (i) *r is prime or $r = (bs+1)^2$.*

  (ii) *$bs + 1$ is prime.*

  (iii) *If $r = (bs+1)^2$, then $|F_\Delta(x)|/(bs+1)$ is prime whenever $0 \le x \le a-1$,*

$$x \equiv 2^{-1}(\pm s - 1) \pmod{bs+1}$$

  and

$$x \not\equiv 2^{-1}(\pm s - 1) \pmod{(bs+1)^2}.$$

  Also, $|F_\Delta(x)|/(bs+1)^2$ is 1 or prime whenever $0 \le x \le a-1$, and

$$x \equiv 2^{-1}(\pm s - 1) \pmod{(bs+1)^2}.$$

  (iv) *If $r \ne (bs+1)^2$, then $|F_\Delta(x)|/(bs+1)$ is prime, r^2 or $(bs+1)^2$ whenever $0 \le x \le a-1$ and $x \equiv 2^{-1}(\pm(br - s) - 1) \pmod{bs+1}$.*

  (v) *If $r \ne (bs+1)^2$, then $|F_\Delta(x)|/r$ is 1, prime, r^2 or $r(bs+1)$ whenever $0 \le x \le a-1$, and $x \equiv 2^{-1}(\pm s - 1) \pmod r$.*

  (vi) *$|F_\Delta(x)|$ is prime whenever $0 \le x \le a-1$, and x does not satisfy any of the congruences in* (iii)–(v).

Proof. The first statement is Exercise 5.1.8. To prove (a), we set $2a - 1 = br + s$ where $0 < s < r$. If $\Delta \equiv 1 \pmod 8$, then r is even and since $h_\Delta = 1$ implies $2 \in \Lambda^*(\Delta)$, then either $r = 2$ or $Q_2/2 = 4ab - 2b^2r - 2b = 2$. However, if $r = 2$,

(5.1.4) This was established in [252], and a similar comment to that given in footnote (5.1.3) applies here.

then $\Delta = (2a-1)^2 + 8$, and Exercise 5.1.9 tells us that this discriminant has $\ell(w_\Delta) \neq 5$. Thus, $4ab - 2b^2r - 2b = 2$, i.e. $2ab - b^2r - b = 1$, so $b = 1$. Using the first statement of the theorem, together with the fact that $c + 1 = a$ by Theorem 5.1.1, we deduce that $c = 2$, $f = 3$, and $r = 4$, i.e. $D = 41$.

Now assume that $D \equiv 5 \pmod 8$ and $h_\Delta = 1$. In the simple continued fraction expansion of w_Δ, we calculate that $P_2 = br - s$ and $Q_2 = 2(bs+1)$. If $bs+1$ is not prime, then it must have a prime divisor $p < \sqrt{\Delta}/2$, so $p \in \Lambda^*(\Delta)$ forcing $p = r$, but $r > bs + 1$, a contradiction which secures (ii).

If r is not prime, then by the same argument as above, it must have a prime divisor $p = bs+1$. Also, $r/(bs+1) < \sqrt{\Delta}/2$, since $Q_1 < 2\sqrt{\Delta}$, so $r/(bs+1) = bs+1$ as above, i.e. $r = (bs+1)^2$. By Theorem 5.1.1, $F_\Delta(x) \equiv 0 \pmod{bs+1}$ if and only if $2x + 1 \equiv \pm(br - s) \pmod{bs+1}$; and $F_\Delta(x) \equiv 0 \pmod r$ if and only if $2x + 1 \equiv \pm(br + s) \equiv \pm s \pmod r$. Also, $F_\Delta(x) \equiv 0 \pmod{(bs+1)^2}$ if and only if $2x + 1 \equiv \pm s \pmod{(bs+1)^2}$. An analysis of these facts yields (iii)–(vi). □

Table 5.1.1 below illustrates Theorem 5.1.4.

Table 5.1.1: $\Delta_0 = D \equiv 5 \pmod 8$, $h_\Delta = 1$ *and* $\ell(w_\Delta) = 5$

D	a	b	c	f	r	$bs+1$	$\lvert F_\Delta(x)\rvert$ for $0 \leq x \leq a-1$
149	6	1	1	-1	7	5	37, 35, 31, 25, 17, 7
157	6	1	3	7	9	3	39, 37, 33, 27, 19, 9
181	7	4	2	1	3	5	45, 43, 39, 33, 25, 15, 3
269	8	1	2	2	11	5	67, 65, 61, 55, 47, 37, 25, 11
397	10	2	6	17	9	3	99, 97, 93, 87, 79, 69, 57, 43, 27, 9
941	15	1	5	21	25	5	235, 233, 229, 223, 215, 205, 193, 179, 163, 145, 125, 103, 79, 53, 25
1013	16	2	2	1	13	11	253, 251, 247, 241, 233, 223, 211, 197, 181, 163, 143, 121, 97, 71, 43, 13
2477	25	2	1	-3	19	23	619, 617, 613, 607, 599, 589, 577, 563, 547, 529, 509, 487, 463, 437, 409, 379, 347, 313, 277, 239, 199, 157, 113, 67, 19
2693	26	2	4	7	23	11	673, 671, 667, 661, 653, 643, 631, 617, 601, 583, 563, 541, 517, 491, 463, 433, 401, 367, 331, 293, 253, 211, 167, 121, 73, 23
3533	30	4	1	-1	13	29	883, 881, 877, 871, 863, 853, 841, 827, 811, 793, 773, 751, 727, 701, 673, 643, 611, 577, 541, 503, 463, 421, 377, 331, 283, 233, 181, 127, 71, 13
4253	33	9	3	1	7	19	1063, 1061, 1057, 1051, 1043, 1033, 1021, 1007, 991, 973, 953, 931, 907, 881, 853, 823, 791, 757, 721, 683, 643, 601, 557, 511, 463, 413, 361, 307, 251, 193, 133, 71, 7

Conjecture 5.1.1. *The values in Table* 5.1.1 *represent all* Δ *with* $h_{\Delta_0} = 1$ *and* $\ell(w_{\Delta_0}) = 5$.

We will prove this conjecture later (with a GRH-ruled out exceptional value).

In fact, we will do more than that. We shall verify a table of such values for $\ell(w_{\Delta_0}) \leq 24$, with $h_{\Delta_0} = 1$ or 2 (see Appendix A, Tables A1 and A3).

Remark 5.1.2. Theorems 5.1.2–5.1.3 show the work involved in characterizing $h_{\Delta_0} = 1$ in terms of the factorization of the Euler–Rabinowitsch polynomial. It remains an open problem to find an algorithm for such a parameterized solution in the general case.

Exercises 5.1

1. With reference to the proof of Theorem 5.1.1:

   (a) Prove Claim 1. (*Hint*: Use the initial observation in the proof, and note that square roots may be taken across a congruence when the modulus is prime.)

   (b) Prove Claim 2. (*Hint*: Use the fact that $-a < b + w'_\Delta < 0 \leq b < a$.)

   (c) Prove Claim 4. (*Hint*: Use inequalities established in (b).)

   (d) Prove Claim 6. (*Hint*: Set $-b - \sigma + 1 + ga = b + ha$ and, keeping the initial remark of the proof in mind, verify that $a|\Delta$.)

2. Let $\Delta = \Delta_0 \equiv 1 \pmod 8$ be a fundamental discriminant of ERD-type.[(5.1.5)]

   (a) Prove that $h_\Delta = 1$ if and only if $\Delta = 17$ or 33. (*Hint*: Use Theorem 3.2.1 and Theorem 2.1.2 to show that the only principal reduced ideals have norm $a/2 \pm (r-1)/4$, and use the fact that $2 \in \Lambda^*(\Delta)$ if $h_\Delta = 1$. For the balance of this exercise, such techniques, together with Theorem 5.1.1, can be used.)

   (b) Prove that $h_\Delta = 2$ if and only if $\Delta = 65$ or 105.

   (c) Prove that $h_\Delta = 3$ if and only if $\Delta = 257$, 321 or 473.

   (d) Prove that $h_\Delta = 4$ if and only if $\Delta \in \{145, 689, 777, 897, 905, 1025, 1305\}$.

   (e) Prove that $h_\Delta = 5$ if and only if $D \in \{401, 4353\}$.

3. Prove that if Δ is a fundamental discriminant, then the following are equivalent:

   (i) $h_\Delta = 1$.

   (ii) $\sum_{p<\sqrt{\Delta}/2} d(p, F) = \sum_{i=1}^{\lfloor \ell/2 \rfloor} \beta_i$,

   where the left hand sum ranges over primes $p < \sqrt{\Delta}/2$, $\ell = \ell(w_\Delta)$, and

$$\beta_i = \begin{cases} a_i + 1 & \text{if } i \neq \ell/2 \text{ and } Q_i/2 \text{ is prime,} \\ \lfloor (a_i+1)/2 \rfloor & \text{if } i = \ell/2 \text{ and } Q_i/2 \text{ is prime,} \\ 0 & \text{otherwise,} \end{cases}$$

   with $w_\Delta = \langle a; \overline{a_1, a_2, \ldots, a_\ell} \rangle$. (*Hint*: Apply Theorem 5.1.1 to w_Δ with Theorem 2.1.2 in mind.)[(5.1.6)]

[(5.1.5)] These facts were established in [225].

[(5.1.6)] This was established in [238].

4. Establish class number one criteria for fundamental discriminants $\Delta = \Delta_0 > 0$ when $\ell(w_\Delta) \leq 2$. (*Hint*: Follow the example of the proof of Theorem 5.1.2.)

5. (a) Prove that if $\Delta = \Delta_0 > 0$ is a fundamental discriminant and $\ell(w_\Delta)$ is odd, then Δ is a sum of two squares. (*Hint*: See Exercise 2.1.13(d).)

   (b) Furthermore, prove that if $h_\Delta = 1$, then $\Delta \equiv 1 \pmod 4$ and Δ is prime. (*Hint*: See Corollary 1.3.2.)

6. Verify the chart in the proof of Theorem 5.1.2. (*Hint*: Use Theorem 2.1.2 and symmetry via Exercise 2.1.13(d).)

7. (a) Let $\Delta = \Delta_0 \equiv 1 \pmod 4$ and assume $w_\Delta = \langle a, \overline{b, c, b, 2a-1} \rangle$. Prove that

$$\Delta = (2a-1)^2 + 4(c(fb-c)+f) \text{ and } 2a-1 = b^2cf - bc^2 - c + 2bf$$

   for some positive $a, b, c, f \in \mathbf{Z}$. Prove that $h_\Delta = 1$ if and only if each of the following holds:

   (i) $b(fb-c)+1$ is prime.
   (ii) $c(fb-c)+f$ is prime.
   (iii) $|F_\Delta(x)|/(b(fb-c)+1)$ is prime whenever $0 \leq x \leq a-1$ and $x \equiv -2^{-1} \pmod{b(fb-c)+1}$.
   (iv) $|F_\Delta(x)|/(c(fb-c)+f)$ is prime whenever $0 \leq x \leq a-1$ and $x \equiv -2^{-1}(fb-c+1) \pmod{c(fb-c)+f}$.
   (v) $|F_\Delta(x)|/(c(fb-c)+f)$ is 1 or prime whenever $0 \leq x \leq a-1$ and

$$x \equiv 2^{-1}(fb-c-1) \pmod{c(fb-c)+f}.$$

   (vi) $|F_\Delta(x)|$ is prime whenever $0 \leq x \leq a-1$ and x does *not* satisfy any of the congruences in (iii)–(v). (*Hint*: Use the techniques of proof in Theorems 5.1.2–5.1.3.)

   (b) Conclude from (a) that if $\Delta \equiv 1 \pmod 8$ and $\ell(w_\Delta) = 4$, then $h_\Delta = 1$ if and only if $\Delta = 33$.

   (c) Let $\Delta = \Delta_0 \not\equiv 1 \pmod 4$ with $\ell(w_\Delta) = 4$. Prove that $w_\Delta = \langle a; \overline{b, c, b, 2a} \rangle$, $D = a^2 - c^2 + f(bc+1)$ and $2a = b^2cf + 2fb - bc^2 - c$ for positive $a, b, c, f \in \mathbf{Z}$. Verify that $h_\Delta = 1$ if and only if $D = (c+2)^2 - 2$. (*Hint*: A comment similar to that in (a) applies.)[5.1.7]

8. Verify the first statement of Theorem 5.1.4. (See the hint to Exercise 5.1.6.)

9. Let $\Delta = \Delta_0 = (2a-1)^2 + 8$. Show that $\ell(w_\Delta) \neq 5$.

10. (a) Let $D \not\equiv 1 \pmod 4$ be a fundamental radicand of ERD-type such that $D = \ell^2 + r > 3$, and assume that D cannot be written in the form $D = a^2 \pm 2$. Prove that $h_\Delta = 2$ if and only if for $1 \leq x \leq \ell$, $|F_\Delta(x)|$ factors in the following ways:

   (i) (for $|r|$ odd) $|F_\Delta(x)|$ is a prime, twice a prime, $|r|$ times a prime, $2|r|$ times a prime, or the product of two primes $(\ell + (r-1)/2)(\ell - (r-1)/2)$, the latter occurring at $x = |r+1|/2$, and $|r|$ is prime or 1.

[5.1.7] The facts in this problem were established in [245].

(ii) (for $|r|$ even) $|F_\Delta(x)|$ is a prime, twice a prime, $|r|/2$ times a prime or $|r|$ times a prime with $|r|/2$ prime. Moreover, if $r < 0$, then additionally $(r + 4\ell - 4)/2$ times a prime where $(r + 4\ell - 4)/2$ is prime. (*Hint*: Assume $h_\Delta = 2$ and use Theorem 3.2.1 to analyze each case with Theorem 2.1.2 in mind. For the converse, it suffices to show for a given reduced ideal $I = [N(I), b + w_\Delta]$ that either $I \sim 1$ or $I \sim [2, \alpha + \sqrt{D}]$, where $\alpha = 1$ if $D \equiv 3 \pmod 4$ and $\alpha = 0$ otherwise. Do this by looking at $Q_i Q_{i+1}$ in the simple continued fraction expansion of $(b + w_\Delta)/N(I)$.) [(5.1.8)]

(b) Apply part (a) to $D = 477 = 21^2 + 6$ and $D = 215 = 15^2 - 10$.

11. (a) Let $D = a^2 + r$ with $|r| \in \{1, 4\}$ be a fundamental radicand. Prove that p is inert for all primes p with $p^{h_\Delta} < w_\Delta$, unless h_Δ is even, in which case p may be ramified. (*Hint*: Prove that if p splits, then $p^{h_\Delta} \geq w_\Delta$ by using Lemma 3.2.1.)

(b) Let $D = 4m^2 + 1 \equiv 5 \pmod 8$, and let p be the least prime quadratic residue modulo D. Prove that $D \leq 4p^{2h_\Delta} + 1$.

(c) Let $D = a^2 + r$ with $|r| \in \{1, 4\}$ and $D \not\equiv 5 \pmod 8$. Prove that if h_Δ is odd, then $D \leq 4^{h_\Delta}$.

12. Let $\Delta = \Delta_0 > 0$ be a fundamental discriminant with $\Delta \neq 5, 8, 12, 24$. Prove that there is at least one representative n of a residue class modulo Δ such that $0 < n < \Delta/4$ and $(\Delta/n) = -1$, where (/) denotes the Kronecker symbol (see Chapter One).[(5.1.9)] (*Hint*: For $\Delta \equiv 0 \pmod 4$ set

$$n = \begin{cases} D - 8 & \text{if } \Delta = 4D \text{ with } D \equiv 3 \pmod 8,\ D > 3, \\ D - 2 & \text{if } \Delta = 4D \text{ with } D \equiv 7 \pmod 8, \\ d + 8 & \text{if } \Delta = 8d \text{ with } d \equiv 3 \pmod 4,\ d > 7, \\ d + 2 & \text{if } \Delta = 8d \text{ with } d \equiv 1 \pmod 4,\ d > 1. \end{cases}$$

Use multiplicative properties of $\chi(n) = (\Delta/n)$ to verify that $\chi(n) = -1$. For the case $\Delta \equiv 1 \pmod 8$, assume first that there is a prime $p \equiv 3 \pmod 4$ dividing Δ and set $\Delta = pq$. Set $n = (p + q)/2$ and deduce that $\chi(n) = -1$ with $n \leq \frac{1}{2}\left(\frac{\Delta}{3} + \frac{\Delta}{11}\right) < \frac{\Delta}{4}$. In the case where $\Delta \equiv 1 \pmod 8$ and there is no such prime p dividing Δ, it is impossible to give a desired integer n explicitly for every Δ. However, if $\Delta \equiv 2 \pmod 3$, show that $\chi(3) = -1$ with $3 < \Delta/4$ must hold. If $\Delta \equiv 1 \pmod 3$ and $\chi(m) = -1$ for $0 < m < \Delta/2$ with $\gcd(m, 6) = 1$, then set

$$n = \begin{cases} (\Delta - m)/6 & \text{if } m \equiv 1 \pmod 6, \\ (\Delta + m)/6 & \text{if } m \equiv 5 \pmod 6 \end{cases}.$$

Deduce that $\chi(n) = -1$ and $n < (\Delta + \Delta/2)/6$. If $\Delta \equiv 5 \pmod 8$, then $\chi(2) = -1$.)

(5.1.8) This was established in [222], as was part (b) and Exercise 5.1.11.

(5.1.9) This is taken from Mitsuhiro *et al.* [219].

5.2 Class Number One Criteria.

Given that Gauss [91, Article 304, p. 363] conjectured that there are infinitely many fundamental discriminants $\Delta_0 > 0$ with $h_{\Delta_0} = 1$, and that this remains an open and deeply difficult problem to this day, it is of interest and value to have class number one criteria. We discuss many of them in this section, especially those which pertain to continued fractions. The most notable and well-known(5.2.1) of these is implicit in the continued fraction algorithm, Theorem 2.1.2. It is

Theorem 5.2.1. *If $\Delta = \Delta_0 > 0$ is a fundamental discriminant, then the following are equivalent:*

(1) $h_\Delta = 1$.

(2) *For each prime $p < \sqrt{\Delta}/2$ with p dividing $|F_\Delta(x)|$ where $0 \leq x < \sqrt{\Delta}/2$, there exists an $i \in \mathbb{Z}$ with $1 \leq i \leq \ell(w_\Delta) = \ell$ such that $p = Q_i/\sigma$ in the simple continued fraction expansion of w_Δ.*

Proof. If $h_\Delta = 1$ and $p < \sqrt{\Delta}/2$ divides $|F_\Delta(x)|$, then the result follows from Theorem 2.1.2 and Lemma 4.1.3 (with $B = M_\Delta$). Conversely, if (2) holds, then the result follows from Theorem 2.1.2 and Theorem 1.3.1. □

In Theorem 4.2.6, we already have seen a class number one criterion for certain ERD-types (see Exercise 5.2.2 as well). There is a more general class number one criterion based upon the following result.

Lemma 5.2.1. *Let $\Delta > 0$ be a discriminant and let $B_\Delta = (2T/\sigma - N(\varepsilon_\Delta) - 1)/U^2$ be the Hasse bound (see Lemma 3.2.1). If $h_\Delta = 1$, then $(\Delta/p) = -1$ for all $p < B_\Delta$.*

Proof. If $h_\Delta = 1$, then there are $x, y \in \mathbb{Z}$ with $x^2 - \Delta y^2 = \pm\sigma^2 p$. Thus, $p \geq B_\Delta$ by Lemma 3.2.1. □

Theorem 5.2.2. *Let B_Δ be the Hasse bound. If $\Delta = \Delta_0 \equiv 1 \pmod 4$ is a fundamental discriminant such that $\sqrt{\Delta - 1}/2 \leq B_\Delta$, then the following are equivalent:*

(1) $h_\Delta = 1$.

(2) $(\Delta/p) = -1$ *for all primes $p < B_\Delta$.*

(3) $F_\Delta(x) \not\equiv 0 \pmod p$ *for all integers x and primes p with $0 \leq x < p < \sqrt{\Delta - 1}/2$.*

(4) $|F_\Delta(x)|$ *is prime for all $x \in \mathbb{Z}$ such that $1 \leq x < \sqrt{\Delta - 1}/2$.*

Proof. See Exercise 5.2.1. □

We now use this to give a class number one criterion for *wide ERD-types*, i.e. those of the form $\Delta = a^2 + r$ with $r \mid 4a$ and $|r| \notin \{1, 4\}$.

(5.2.1)However, this seems to have been rediscovered by Louboutin in [195].

Theorem 5.2.3. [5.2.2] *Let $\Delta = \Delta_0 = a^2 + r > 7$ with $r \mid 4a$, $|r| \notin \{1,4\}$ and $h_\Delta = 1$.*

(a) *If $\Delta \equiv 1 \pmod 8$, then $\Delta = 33$.*

(b) *If $\Delta \equiv 5 \pmod 8$, then $-r$ is a prime and $(\Delta/p) = -1$ for all primes $p < |r|/4$.*

(c) *If $\Delta \not\equiv 1 \pmod 4$, then all of the following hold:*

(i) $|r| = 2$.

(ii) *$(\Delta/p) = -1$ for all primes $p \mid a$.*

(iii) *If $r = 2$ then $3 \mid a$.*

(iv) *If $r = -2$ then $3 \nmid a$.*

Proof. (a) follows from Exercise 5.1.2. If $|r|$ is not a prime, then there is a prime $p \mid |r|$ with $2 < p < |r|$, when $\Delta \equiv 5 \pmod 8$. Thus, there are $x, y \in \mathbf{Z}$ with $x^2 - \Delta y^2 = \pm 4p$. By Theorem 3.2.2, $4p \geq |r|$. Hence, $|r| = 2p$, $3p$ or $4p$. However, $\Delta \equiv 5 \pmod 8$, so $|r|$ is odd, i.e. $|r| = 3p$, which contradicts $h_\Delta = 1$ via Corollary 1.3.2(2). Hence, $|r|$ is prime. By Corollary 1.3.2(2) again, $(a^2+r)/|r| \equiv 3 \pmod 4$ is prime with $|r| \equiv 3 \pmod 4$. If $r > 0$, then $a^2 \equiv 2r \pmod 4$, a contradiction. Thus, $r < 0$. If $(\Delta/p) \neq -1$ and $p < |r|/4$, then there are $u, v \in \mathbf{Z}$ with $u^2 - \Delta v^2 = \pm 4p$. By Theorem 3.2.2, $p \geq |r|/4$, a contradiction which secures (b).

If $\Delta \not\equiv 1 \pmod 4$, then since $(\Delta/2) \neq -1$, there are $x, y \in \mathbf{Z}$ such that $x^2 - Dy^2 = \pm 2$. By Theorem 3.2.2, $2 \geq |r|$, so $|r| = 2$, which secures (i).

If $(\Delta/p) \neq -1$ and $p \mid a$, then there are $u, v \in \mathbf{Z}$ with $u^2 - Dv^2 = \pm p$. By Theorem 3.2.3, $D = 7$ and $p = 3$, contradicting the hypothesis. This secures (ii).

Since $3 < \sqrt{\Delta}/2$, then $3 \in \Lambda^*(\Delta)$ whenever $(\Delta/3) \neq -1$, so by Theorem 3.3.2, $|r| = 2$. If $r = 2$, then $3 \mid D$, contradicting Corollary 1.3.2(2). If $r = -2$, then $D = 7$ by Theorem 3.2.3, a contradiction. Therefore, $(\Delta/3) = -1$ must hold. Thus, $a \equiv 0 \pmod 3$ if $r = 2$, and $r = -2$ otherwise. □

As observed earlier, we will soon be able to list all ERD-types of class number one.

We now provide another class number one criterion which employs continued fractions. First, we need a key result involving reduced ideals.

Theorem 5.2.4. [5.2.3] *Let $\Delta = \Delta_0 > 0$ be a fundamental discriminant. If $I = [a, b + w_\Delta]$ is a primitive ideal of $\mathcal{O}_\Delta$ with $0 \leq b < a$, then the following are equivalent:*

(a) *I is reduced.*

(b) $a - w_\Delta < b < -w'_\Delta$.

(c) $\lfloor -(b + w'_\Delta)/a \rfloor a > a - b - w_\Delta$.

Proof. The equivalence of (a) and (c) is Theorem 1.4.1. Furthermore, the equivalence of (a) and (b) follows from Corollary 1.4.4, when $\sqrt{\Delta}/2 \leq a < \sqrt{\Delta}$. Thus, it remains to prove the equivalence of (a) and (b), when a is not in this

[5.2.2] Theorems 5.2.2–5.2.3 were established in [248] and [223].

[5.2.3] This was verified in [268]–[269] together with Theorem 5.2.5, which follows.

range. If $a > \sqrt{\Delta}$, then by Corollary 1.4.2, I is not reduced. Furthermore, if $a - w_\Delta < b < -w'_\Delta$, then $a < w_\Delta - w'_\Delta \leq \sqrt{\Delta}$, a contradiction. Therefore, we may assume that $a < \sqrt{\Delta}/2$, so that I is necessarily reduced. It remains to show that (b) holds under the assumption that $a < \sqrt{\Delta}/2$. Suppose $b \geq -w'_\Delta$, then $a > -w'_\Delta$, so $\sqrt{\Delta}/2 > a > b > (-(\sigma - 1) + \sqrt{\Delta})/2$, which is impossible. Clearly $a - w_\Delta < b$, and the result is secured. □

Corollary 5.2.1. *The set*

$$S = \{[a, b + w_\Delta] : a \mid N(b + w_\Delta) \text{ and } 0 \leq b < a < b + w_\Delta < \sqrt{\Delta}\}$$

represents all of the reduced ideals in $\mathcal{O}_\Delta$ and $|S|$ is the number of reduced ideals therein.

Before we present our class number one criterion, we need a definition.

Definition 5.2.1. Let $t(b) = |\{a \in \mathbf{Z} : a \mid N(b + w_\Delta) \text{ and } b < a < b + w_\Delta\}|$.

Remark 5.2.1. We will see that, if $h_\Delta = 1$, then $\sum_{0 \leq b < -w'_\Delta} t(b)$ is just the number of distinct reduced ideals which are equivalent to I (including I itself), where $I = [1, w_\Delta]$.

The class number one criterion will be an immediate consequence of the following criterion for determining C_{Δ_0}.

Theorem 5.2.5. *If $\Delta = \Delta_0 > 0$ is a fundamental discriminant and $I_j = [a_j, b_j + w_\Delta]$ are pairwise primitive inequivalent ideals in $\mathcal{O}_\Delta$ for $1 \leq j \leq n = h_\Delta$, then*

$$C_\Delta = \bigcup_{j=1}^{n} \{I_j\} \text{ if and only if } \sum_{0 \leq b < -w'_\Delta} t(b) = \sum_{j=1}^{n} \ell_j \text{ where } \ell_j = \ell(I_j).$$

Proof. By Corollary 5.2.1, $\sum_{0 \leq b \leq -w'_\Delta} t(b) = |S|$ is the total number of reduced ideals in $\mathcal{O}_\Delta$. By Theorems 1.3.1 and Theorem 1.4.1, the total number of reduced ideals is represented by the I_j's precisely when $C_\Delta = \bigcup_{j=1}^{n} \{I_j\}$. □

A sufficiently complicated example to illustrate Theorem 5.2.5 is

Example 5.2.1. Let $\Delta = \Delta_0 = 35112 = 2^3 \cdot 3 \cdot 7 \cdot 11 \cdot 19$ and consider the ideals: $I_1 = [2, \sqrt{8778}]$, $I_2 = [3, \sqrt{8778}]$, $I_3 = [11, \sqrt{8778}]$, $I_4 = [6, \sqrt{8778}]$, $I_5 = [22, \sqrt{8778}]$, $I_6 = [33, \sqrt{8778}]$, $I_7 = [66, \sqrt{8778}]$ and $I_8 = [1, w_\Delta]$. For these ideals, we have $\ell(I_i) = 8, 8, 10, 12, 10, 14, 16, 8$ for $i = 1, 2, \ldots, 8$ respectively, so $\sum_{j=1}^{8} \ell_j = 86$. Also, since $\lfloor w_\Delta \rfloor = 93$, then an examination of the $t(b)$'s for $0 \leq b \leq 93$ shows that $\sum_{0 \leq b \leq -w'_\Delta} t(b) = 86$. Hence, by Theorem 5.2.5, $C_\Delta = \bigcup_{i=1}^{8} \{I_i\}$. Moreover, an analysis of the equivalences (an easy exercise, at this point, for the reader) among elements of the set shows that, in fact, $C_\Delta = \langle\{I_1\}\rangle \times \langle\{I_2\}\rangle \times \langle\{I_3\}\rangle$ with $h_\Delta = 8$.

We are finally in a position to state the promised class number one criterion.

Corollary 5.2.2. *If $\Delta = \Delta_0 > 0$ is a fundamental discriminant, then the following are equivalent:*

(1) $h_\Delta = 1$.

(2) $\sum_{0 \leq b \leq -w'_\Delta} t(b) = \ell(w_\Delta)$.

This is a particularly useful and easily implemented criterion, as the following examples depict.

Example 5.2.2. Let $D = \Delta/4 = 491 \equiv 3 \pmod 4$, then we have $w_\Delta = \sqrt{491} = \langle 22; \overline{6, 3, 4, 8, 1, 1, 1, 2, 1, 1, 21^*, \ldots, 44} \rangle$ where * indicates that 21 represents the middle of the period and $\ell(w_\Delta) = 22$ with the balance given by symmetry. On the other hand, $\sum_{0 \leq b \leq -w'_\Delta} t(b) = 22$. By Corollary 5.2.2, $h_\Delta = 1$.

Example 5.2.3. Let $\Delta = \Delta_0 = 393 \equiv 1 \pmod 4$. In this case, $w_\Delta = (1 + \sqrt{393})/2 = \langle 10; \overline{2, 2, 2, 1, 9, 4, 1, 5^*, \ldots, 19} \rangle$ with $\ell(w_\Delta) = 16$, and $\sum_{0 \leq b \leq -w'_\Delta} t(b) = 16$, so $h_\Delta = 1$.

Another class number one criterion involving continued fractions was produced by Lu [202], as follows.

Theorem 5.2.6. (5.2.4) *If $\Delta = \Delta_0 > 0$ is a fundamental discriminant, then $h_\Delta = 1$ if and only if*

$$c + \sum_{i=1}^{\ell} a_i = \lambda_1(\Delta) + \lambda_2(\Delta),$$

where $\ell = \ell(w_\Delta)$ with $w_\Delta = \langle a; \overline{a_1, a_2, \ldots, a_\ell} \rangle$,

$\lambda_1(\Delta)$ is the number of solutions of $x^2 + 4yz = \Delta$ for non-negative $x, y, z \in \mathbf{Z}$,

$\lambda_2(\Delta)$ is the number of solutions of $x^2 + 4y^2 = \Delta$ with non-negative $x, y \in \mathbf{Z}$,

and

$$c = \begin{cases} 0 & \text{if } \ell \text{ is even, } a_{\ell/2} \text{ odd when } \Delta \equiv 1 \pmod 4, \\ 1 & \text{otherwise if } \Delta \equiv 1 \pmod 4, \\ 1 & \text{if } \ell \text{ is even, } a_{\ell/2} \text{ odd when } \Delta \equiv 0 \pmod 4, \\ 2 & \text{otherwise, if } \Delta \equiv 0 \pmod 4. \end{cases}$$

We reserve the proof of this fact for Chapter Six, where we will show that the "palindromic index", which we introduce therein, provides a means of describing this phenomenon. In fact, a consequence of what we prove in Chapter Six is the following.

Theorem 5.2.7. (5.2.5) *(Lu reinterpreted). If $\Delta = \Delta_0 > 0$ is a fundamental discriminant, then the following are equivalent:*

(1) $h_\Delta = 1$.

(5.2.4)This was motivated by a similar result for class numbers of complex quadratic fields proved by Hirzebruch [124] and Zagier [396]. The result is of interest in itself: If p is a prime and $p \equiv 3 \pmod 4$, $p > 3$ then $\sum_{i=1}^{\ell} (-1)^i a_i = 3h_\Delta$ where $\Delta = -p$ and $\ell = \ell(\sqrt{p})$, where $\sqrt{p} = \langle a_0; \overline{a_1, \ldots, a_\ell} \rangle$. It is worth mentioning the result of Dirichlet here, which says that, for such a Δ, $h_\Delta = \frac{1}{p}(N - R)$, where N is the sum of the quadratic nonresidues modulo p, and R is the sum of the quadratic residues modulo p. For example, if $p = 11$, then $N = 2 + 6 + 7 + 8 + 10 = 33$, and $R = 1 + 3 + 4 + 5 + 9 = 22$, so $h_\Delta = \frac{1}{11}(33 - 22) = 1$.

(5.2.5)This is established in [238].

(2) $\sum_{P_0 \le x \le \lfloor w_\Delta \rfloor} \tau(F_\Delta(x)) = \sum_{i=1}^{\ell} a_i + c$.
where c is as in Theorem 5.2.6, τ is the divisor function and $P_0 = \sigma - 1$.

For instance, we have

Example 5.2.4. If $\Delta = 4 \cdot 94$, then $w_\Delta = \langle 9; \overline{1,2,3,1,1,5,1,8,1,5,1,1,3,2,1,18} \rangle$ so $\sum_{i=1}^{\ell} a_i + c = \sum_{i=1}^{16} a_i + 2 = 56$ whereas, since

x	0	1	2	3	4
$\lvert F_\Delta(x) \rvert$	$2 \cdot 47$	$3 \cdot 31$	$2 \cdot 3^2 \cdot 5$	$5 \cdot 17$	$2 \cdot 3 \cdot 13$

5	6	7	8	9
$3 \cdot 23$	$2 \cdot 29$	$3^2 \cdot 5$	$2 \cdot 3 \cdot 5$	13

then $\sum_{P_0 \le x \le \lfloor w_\Delta \rfloor} \tau(F_\Delta(x)) = 56$, so $h_\Delta = 1$.

Remark 5.2.2. What is really involved in Theorem 5.2.6 is the divisor function. All we have to do is recognize that $\lambda_1(\Delta) + \lambda_2(\Delta)$ is just $\sum_{P_0 \le x \le \lfloor w_\Delta \rfloor} \tau(F_\Delta(x))$. To see this, we observe that $x_1^2 + 4yz = \Delta$ implies that $yz = (\Delta - x_1^2)/4 = |F_\Delta(x)|$, where $x_1 = \sigma x + \sigma - 1$ and $x_1^2 + 4y^2 = \Delta$ is just $x_1^2 + 4yz = \Delta$ with $y = z$. Finally, observe that since $y, z \in \mathbf{Z}$ are non-negative, then $P_0 \le x \le \lfloor w_\Delta \rfloor$.

In the next section, we will look more closely at the divisor function in terms of bounding the class number from below. For now, we conclude this section with a class number one criterion which is unfortunately not as useful as are the above criteria based on continued fractions.

Theorem 5.2.8. [(5.2.6)] *If $\Delta = \Delta_0 > 0$ is a fundamental discriminant, then $h_\Delta = 1$ if and only if, for all non-zero $\alpha, \beta \in \mathcal{O}_\Delta$ with β not dividing α and $|N(\alpha)| \ge |N(\beta)|$, there exist $\sigma, \delta \in \mathcal{O}_\Delta$ with $0 < |N(\alpha\sigma - \beta\delta)| < |N(\beta)|$.*

Exercises 5.2

1. Prove Theorem 5.2.2. (*Hint*: Use Lemma 5.2.1 and Theorem 4.2.6.)
2. (a) Let $\Delta = \Delta_0 = 4m^2 + 1$ be a fundamental discriminant. Prove that the following are equivalent:[(5.2.7)]
   (i) $h_\Delta = 1$.
   (ii) $(\Delta/p) = -1$ for all primes $p < m$.
   (iii) $F_\Delta(x) \not\equiv 0 \pmod{p}$ for all $x \in \mathbf{Z}$ and primes p with $0 \le x < p < m$.
   (iv) $|F_\Delta(x)|$ is prime for all $x \in \mathbf{Z}$ with $1 \le x < m$.
   (*Hint*: Use Theorem 5.2.2.)
   (b) Deduce from (a) that if m is composite, then $h_\Delta > 1$.
   (c) Deduce from (a) that if Δ is composite, then $h_\Delta > 1$. (Observe as well that the equivalence of (i) and (iv) is a special case of Theorem 4.2.6.)

[(5.2.6)] This is attributed to Dedekind and Hasse (see Pollard and Diamond [294, Theorem 9.5, p. 124]). However, it seems to have been rediscovered by Kutsuna [154] in 1980, and again by Queen [297] in 1993.
[(5.2.7)] This was proved by strictly elementary means in [229].

3. Let $\Delta = \Delta_0 = a^2 \pm 4 > 5$ be a fundamental discriminant. Prove that if $(\Delta - 1)/4$ is not a prime, then $h_\Delta > 1$. Also, show that the following are equivalent:

   (i) $h_\Delta = 1$

   (ii) $(\Delta/q) = -1$ for all primes $p < m$ (if $\Delta = m^2 + 4$) and $p < m - 2$ (if $n = m^2 - 4$) respectively.

   (iii) $F_\Delta(x) \not\equiv 0 \pmod q$ for all $x \in \mathbb{Z}$ with $0 \leq x < q < \sqrt{p}$ with $p = \sqrt{\Delta - 1}/4$.

   (iv) $|F_\Delta(x)|$ is prime for all integers x satisfying $1 \leq x < \sqrt{\Delta - 1}/2$. (*Hint*: Use techniques as in Exercise 5.2.2.)

4. (a) Let $\Delta = \Delta_0 > 0$ be a fundamental discriminant. Prove that $h_\Delta = 1$ if and only if $m \in \Lambda^*(\Delta)$ for exactly $\nu(m)$ values Q_i/σ for $1 \leq i \leq \ell(w_\Delta)$, and all m with $1 \leq m < \sqrt{\Delta/5}$, where $\nu(m)$ is the number of primitive ideals of norm m.[(5.2.8)]

   (b) Illustrate (a) by calculating $\nu(m)$ for each m with $1 \leq m < \sqrt{\Delta/5}$, where Δ is a value in Theorem 5.1.3.

5.3 Class Number Bounds Via The Divisor Function.

In the preceding section, we saw how the divisor function played a crucial role in some class number one criteria. We now show how it also plays a role in determining when class numbers are bigger than one. First, we have the following generalization of Definitions 3.3.1–3.3.2.

Definition 5.3.1. Let $\Delta > 0$ be a discriminant and set $I = I_1 = [N(I), b + w_\Delta]$, a primitive ideal of $\mathcal{O}_\Delta$. The ideals $I_i = [Q_{I(i-1)}/\sigma, (P_{I(i-1)} + \sqrt{D})/\sigma]$, for $1 \leq i \leq \ell(I)$, are determined by (2.1.1)–(2.1.5) in the continued fraction expansion of $(b + w_\Delta)/N(I)$. Set

$$\Lambda_I^*(\Delta) = \{Q_{I(i)} : 1 \leq i \leq \ell(I)\},$$

and

$$\Lambda_I(\Delta) = \{N(J) : \text{ is a primitive ideal of } \mathcal{O}_\Delta \text{ and } I \sim J \text{ in } C_\Delta\}.$$

Observe that when I is principal, then we are reduced to Definitions 3.3.1–3.3.2, and we write $\Lambda_1(\Delta)$ for $\Lambda(\Delta)$. We also need

Definition 5.3.2. Let $\Delta > 0$ be a discriminant with radicand D and $P = \{p_i\}_{i=1}^n$ be a set of $n \geq 1$ distinct primes and let $A \in \mathbb{Z}$ be positive. Set

$$\mathcal{P}_\Delta(A) = \{s = \prod_{i=1}^{n} p_i^{b_i} : b_i \geq 0,\ s \leq A \quad \text{and if } p_i \mid D, \text{ then } b_i \leq 1\}.$$

(5.2.8) See, for example, Hendy [118].

Theorem 5.3.1. [5.3.1] *Let $\Delta = \Delta_0 > 0$ be a fundamental discriminant and P be a finite set of primes with $(\Delta/p) \neq -1$, for each $p \in P$, $A \in \mathbb{Z}$ positive and I a primitive product of ramified ideals (possibly $I = 1$), i.e. no ramified prime appears to an exponent bigger than 1. If $\mathcal{P}_\Delta(A) \cap \Lambda_I(\Delta) = \{A, N(I)\}$, then*

$$h_\Delta \geq \begin{cases} \tau(A) - 2^m & \text{if } N(I) \mid A, \\ \tau(A) & \text{otherwise,} \end{cases}$$

where $m = d(A/N(I))$ (see Definition 4.1.3).

Proof. Let $\{p_1, p_2, \ldots\}$ be the set of distinct primes which divide A. The set of indices $\{1, 2, \ldots\}$ of these primes will be divided into two disjoint subsets X and Y as follows: $i \in X$ if p_i is unramified and $i \in Y$ otherwise. Letting $A = \prod_{i \in X} p_i^{\nu_i} \prod_{j \in Y} p_j$, then any divisor of A is of the form $\prod_{i \in X} p_i^{\mu_i} \prod_{i \in Y} p_j$, where $0 \leq \mu_i \leq \nu_i$ and $\emptyset \subseteq Y_0 \subseteq Y$. Thus, as in the proof of Theorem 4.2.1, a combination $c = ((\mu_i)_{i \in X}, Y_0)$ of an $|X|$-tuple (μ_i) of integers and a subset Y_0 of Y represents a divisor of A and $\mathcal{F}(c)$, the ideal class of $\prod_{i \in X} \mathcal{P}_i^{\mu_i} \prod_{j \in Y_0} \mathcal{P}_j$ in $\mathcal{O}_\Delta$, yields a map from the set S of all such combinations into C_Δ.

Claim 1. If $N(I) \nmid A$, then $\mathcal{F}$ is one-to-one (Exercise 5.3.1).

Claim 2. If $N(I) \mid A$, then $\mathcal{F}(c_1) = \mathcal{F}(c_2)$ for exactly 2^m distinct pairs (c_1, c_2) with $c_1 \neq c_2$ (Exercise 5.3.2). □

Corollary 5.3.1. *If $\Delta = \Delta_0 > 0$ is a fundamental discriminant of ERD-type with $D \not\equiv 1 \pmod 4$ and $D = a^2 + r$ where $|r| < 2a$, r odd, then $h_\Delta \geq \tau((2a - |r-1|)/2)$.*

Proof. See Exercise 5.3.3. □

Corollary 5.3.2. *If $\Delta = \Delta_0 = a^2 + r \equiv 1 \pmod 4$ is a fundamental discriminant of ERD-type with $|r| < 2a$ and r odd, then $h_\Delta \geq \tau((2a - |r-1|)/4) - 2^m$ where $m = d(\gcd((2a - |r-1|)/4, \Delta))$.*

Proof. See Exercise 5.3.4. □

The following example illustrates the above.

Example 5.3.1. If $\Delta = 4b^2 + r$ where $r \mid b$, $r > 0$ odd, then $h_\Delta \geq \tau(b - (r - 1)/4) - 2^m$ where $m = d(\gcd(b - (r-1)/4, \Delta))$.

Corollary 5.3.3. *If $A < \sqrt{\Delta}/2$, and I and P are as in Theorem 5.3.1 with $\mathcal{P}_\Delta(A) \cap \Lambda_I^*(\Delta) = \{N(I), A\}$, then $h_\Delta \geq \tau(A) - 2^m$ where m is the number of ramified prime divisors of A.*

Proof. Since $A < \sqrt{\Delta}/2$, then I is reduced so $\mathcal{P}_\Delta(A) \cap \Lambda_I(\Delta) \subseteq \mathcal{P}_\Delta(A) \cap \Lambda_I^*(\Delta)$. □

[5.3.1] This was established in [230] which was inspired by work of Halter-Koch [105] and [108]. A similar comment applies to Theorem 5.3.2 and Exercise 5.3.5, which were established in [268] and [230] respectively. This also generalizes results of Azuhata [11]–[12], Hasse [111] and Yokoi [393]–[394].

The following result for fundamental discriminants has the flavour of Theorem 4.2.1.

Theorem 5.3.2. *Let* $\Delta = \Delta_0 > 0$ *be a fundamental discriminant and let* $A > 0$ *be in* $\Lambda(\Delta)$ *with* $A < \sqrt{\Delta}$. *Furthermore, assume that no divisor* m *of* A *with* $1 < m < A$ *is in* $\Lambda^*(\Delta)$. *Then,* $h_\Delta \geq \tau(A) - 2^{d-1}$ *where* $d = d(A)$.

Proof. See Exercise 5.3.7. □

Corollary 5.3.4. *With the hypothesis of Theorem* 5.3.2, *if* $d > 1$, *then* $h_\Delta > 1$.

Proof. $\tau(A) = \prod_{i=1}^{d}(e_i+1) \geq 2^d$ implies $h_\Delta \geq \tau(A) - 2^{d-1} \geq 2^d - 2^{d-1} = 2^{d-1} > 1$. □

For instance, we have

Example 5.3.2. Let $\Delta = \Delta_0 = 385 = 20^2 - 15 = 5 \cdot 7 \cdot 11$. Consider the continued fraction expansion of w_Δ:

i	0	1	2	3	4	5	6	7
P_i	1	19	17	15	5	13	11	11
Q_i	2	12	8	20	18	12	22	...
a_i	10	3	4	1	1	2	1	...

with the balance given by symmetry. The hypothesis of Theorem 5.3.2 is satisfied with $A = 6$. Here, $h_\Delta = 2 = \tau(A) - 2^{d-1}$.

Corollary 5.3.5. *Let* $\Delta = \Delta_0 = a^2 + r$ *with* $|r| < 2a$ *and set*

$$A = \begin{cases} 2a/\sigma - |r/\sigma^2 - 1| & \text{if } r \text{ is even,} \\ (2a - |r-1|)/\sigma^2 & \text{if } r \text{ is odd.} \end{cases}$$

If $m \notin \Lambda^*(\Delta)$ *for any divisor* m *of* A *with* $1 < m < A$, *then* $h_\Delta \geq \tau(A) - 2^{d-1}$ *where* $d = d(A)$.

Proof. See Exercise 5.3.8. □

For ERD-types, we have

Corollary 5.3.6. *If* d, A *and* Δ *are as in Corollary* 5.3.5 *and* Δ *is of ERD-type, then*

$$h_\Delta \geq \tau(A) - 2^{d-1}.$$

Proof. This merely involves a check of the simple continued fraction expansion of w_Δ via Theorem 3.2.1 to verify that $m \notin \Lambda^*(\Delta)$ for any divisor m of A with $1 < m < A$, an easy exercise. □

Now we provide some illustrations of the above corollaries.

Example 5.3.3. Let $\Delta = 777 = 3 \cdot 7 \cdot 37 = 28^2 - 7$. Here, $A = 12$ and $h_\Delta = 4 = \tau(A) - 2$.

Example 5.3.4. If $D = 5482 = 74^2 + 6$, then $A = 143 = 11 \cdot 13$, $\ell(\sqrt{D}) = 53$ and $11, 13 \notin \Lambda^*(\Delta)$. Therefore, we see that $h_\Delta = 2 = \tau(A) - 2$.

Example 5.3.5. If $\Delta = 21037 = 145^2 + 12$, then $A = 143$ and $\ell(w_\Delta) = 31$ with $11, 13 \notin \Lambda^*(\Delta)$. Here, $h_\Delta = 2 = \tau(A) - 2$.

Now we turn to the Lagrange neighbour to get better bounds for fundamental discriminants of ERD-type.

Lemma 5.3.1. [5.3.2] *Let $\Delta = \Delta_0 > 0$ be a fundamental discriminant, and let $I = [a, b + w_\Delta]$ be a primitive ideal in $\mathcal{O}_\Delta$ with $1 < a < -N(b + w_\Delta)$, $|b| < (\sqrt{\Delta} - \sigma + 1)/2$. If $N(b + w_\Delta)$ has no proper divisor $c > 1$ with $c \in \Lambda^*(\Delta)$, then I is not principal. Furthermore, $I \sim I^+$ and I^+ is reduced and not principal.*

Proof. If I is reduced, then I cannot be principal since $a \mid N(b + w_\Delta)$. Henceforth, assume that I is not reduced. Therefore, $a > \sqrt{\Delta}/2$ by Corollary 1.4.3.

Claim. $\lfloor (b + w_\Delta)/(\sigma a) \rfloor = 0$.

If $b < 0$, then $\lfloor (b + w_\Delta)/(\sigma a) \rfloor \leq (b + w_\Delta)/(\sigma a) \leq (w_\Delta - 1)/(\sigma a) < 1$. If $b \geq 0$ and $a < \sqrt{\Delta}$, then we may invoke Corollary 1.4.4 to get that $a - w_\Delta > b$, i.e. $\sigma a - \sigma + 1 - \sqrt{D} > \sigma b$. Therefore, $1 > (\sigma b + \sigma - 1 + \sqrt{D})/(\sigma a) = (b + w_\Delta)/a \geq (b + w_\Delta)/(\sigma a)$. If $b \geq 0$ and $a > \sqrt{\Delta}$, then in order that $b + w_\Delta \geq a > \sqrt{\Delta}$, we must have $b > (\sqrt{\Delta} - \sigma + 1)/2$, a contradiction which secures the claim.

We may now explicitly display the Lagrange neighbour I^+ of I as

$$I^+ = [-N(b + w_\Delta)/(\sigma^2 a), -\sigma b - \sigma + 1 + w_\Delta]$$

which is reduced, since $1 < -N(b + w_\Delta)/(\sigma^2 a) < \sqrt{\Delta}/2$. Thus, I^+ is not principal since $-N(b + w_\Delta)/(\sigma^2 a)$ is a proper divisor of $N(b + w_\Delta)$. Since $I \sim I^+$, this secures the result. □

We need two more technical lemmata before we can state the result for ERD-types.

Lemma 5.3.2. *Let $\Delta = \Delta_0$ be a fundamental discriminant. If*

$$I_1 = [ca_1, b + w_\Delta] \sim [ca_2, b + w_\Delta] = I_2$$

are primitive $\mathcal{O}_\Delta$-ideals, then

$$J_1 = [a_1, b + w_\Delta] \sim [a_2, b + w_\Delta] = J_2.$$

Proof. By Exercise 1.2.10, $I_i = [a_i, (b + \sqrt{\Delta})/2][c, (b + \sqrt{\Delta})/2]$, using (1.2.1)-(1.2.5). Hence,

$$J_1 = [a_1, (b + \sqrt{\Delta})/2] \sim [a_2, (b + \sqrt{\Delta})/2]. \square$$

Lemma 5.3.3. *Let $\Delta = \Delta_0 > 0$ be a fundamental discriminant with radicand $D = D_0 = a^2 + r$, $r \mid 4a$, and let $I_i = [a_i, b + w_\Delta]$ for $i = 1, 2$ be two primitive ideals in $\mathcal{O}_\Delta$ with $a_1 \neq a_2$. If $a_i = ga_i'$ for $i = 1, 2$ where $g = \gcd(a_1, a_2)$ and $1 < a_1' a_2' < A$ (see Theorem 3.2.2), then $I_1 \not\sim I_2$.*

[5.3.2] This was established in [270], as were Lemmas 5.3.2–5.3.3.

Proof. Let $J_i = [a_i', b + w_\Delta]$ for $i = 1, 2$. If $I_1 \sim I_2$, then $J_1 \sim J_2$ by Lemma 5.3.2, i.e. $J_1 J_2' \sim 1$. From (1.2.1)–(1.2.5), $N(J_1 J_2') = a_1' a_2'$. Therefore, by Exercise 3.4.1(b), there exist $x, y \in \mathbb{Z}$ with gcd (x, y) dividing 2, and $x^2 - \Delta y^2 = \pm 4a_1' a_2'$. By Theorem 3.2.2, this can happen only if $a_1' a_2' = t^2$ for some $t \in \mathbb{Z}$ dividing both x and y, a contradiction. □

Now we can state our result for ERD-types. In what follows, $\Lambda^*(\Delta)$ is as in Definition 3.3.2; $\Omega(n)$ is as in Definition 4.1.2; and A is as in Theorem 3.2.2.

Theorem 5.3.3. *Let $\Delta = \Delta_0 > 0$ be a fundamental discriminant with radicand $D = D_0 = a^2 + r$, $r \mid 4a$ and set $-N(b + w_\Delta) = mn$, where $|b| < (\sqrt{\Delta} - \sigma + 1)/2$ and $m < A$. If mn has no divisor ℓ with $1 < \ell < mn$ which is in $\Lambda^*(\Delta)$, then*

$$h_\Delta \geq \max\{\tau(m), \tau(m) + \Omega(n) - 1\}.$$

Proof. Let $1 < a_1 < a_2 < \cdots < a_t = m$ be all of the divisors of m and set $t = \tau(m)$. By Lemma 5.3.3, $[a_i, b + w_\Delta] \not\sim [a_j, b + w_\Delta]$ for any $i \neq j$. Thus, $h_\Delta \geq \tau(m)$, keeping in mind that no proper divisor $s > 1$ of m is in $\Lambda^*(\Delta)$, since $m < A$.

Let $n = p_1 \cdots p_r$ be a product of not necessarily distinct primes. Since no proper divisor $\ell > 1$ of $-N(b + w_\Delta) = mn$ is in $\Lambda^*(\Delta)$, then by Exercise 5.3.11 and Lemma 5.3.2, for $1 \leq i, j \leq r - 1 = \Omega(n) - 1$, we find that $[mp_1 \cdots p_i, b + w_\Delta] \not\sim [mp_1 \cdots p_j, b + w_\Delta]$, and for $1 \leq i \leq t$, $1 \leq j \leq r - 1$, we find that $[mp_1 \cdots p_j, b + w_\Delta] \not\sim [a_i, b + w_\Delta]$. Therefore, $h_\Delta \geq t + r - 1 = \tau(m) + \Omega(n) - 1$. □

Remark 5.3.1. We observe that in Theorem 5.3.3, the assumption that $-N(b + w_\Delta)$ has no divisor $\ell > 1$ in $\Lambda^*(\Delta)$ is always satisfied when $D = a^2 + \sigma^2$ if a is odd, because $\Lambda^*(\Delta) = \{1\}$ in that case. Furthermore, if $D = a^2 + r$ with $|r| \in \{1, 4\}$, then $h_\Delta = 1$ necessarily implies that $(\Delta/p) = -1$ for all primes $p < A$, by Theorem 5.3.3 (compare, for instance, with Exercises 5.2.2–5.2.3). Also, if $D \not\equiv 1 \pmod 4$ is of ERD-type, and $|r| > 2$, then $h_\Delta > 1$ by Theorem 5.3.3 (compare with Theorem 5.2.3). Hence, what the above shows in particular is that if $D = a^2 + 1$ with a odd, then $h_\Delta = 1$ if and only if $D = 2$. Theorem 5.3.3 also shows that if $D \equiv 1 \pmod 8$ is of ERD-type and $D > 33$, then $h_\Delta > 1$ (compare with Exercise 5.1.2).

To further illustrate the power of Theorem 5.3.3, we have

Example 5.3.6. Let $D = 170 = 13^2 + 1$. By Theorem 5.3.3, choosing $b = 4$, $D - b^2 = 154 = 2 \cdot 7 \cdot 11$ and $m = 14$, we find that $h_{170} \geq 4$. In fact, $h_{170} = 4$.

Example 5.3.7. Let $D = 442 = 21^2 + 1 = 2 \cdot 13 \cdot 17$. By Theorem 5.3.3, choosing $b = 8$, $D - b^2 = 378 = 2 \cdot 3^3 \cdot 7$, $n = 3 \cdot 7$, $m = 2 \cdot 3^2$, we have $h_\Delta \geq \tau(m) + \Omega(n) - 1 = 6 + 2 - 1 = 7$. Actually, $h_\Delta = 8$.

Example 5.3.8. Let $D = 226 = 15^2 + 1 = 2 \cdot 113$. By Theorem 5.3.3, choosing $b = 8$, $D - b^2 = 162 = 2 \cdot 3^4$ and $m = 18$, we have that $h_\Delta \geq 7$, whereas $h_\Delta = 8$.

Example 5.3.9. Let $D = 1373 = 37^2 + 4$. Taking $-N(b + w_\Delta) = 343 = 7^3$ where

$b = 0$, $m = 7$ and $n = 7^2$, by Theorem 5.3.3, we find that $h_{1373} \geq \tau(m) + \Omega(n) - 1 = 2 + 2 - 1 = 3$. Actually, $h_{1373} = 3$.

Example 5.3.10. Let $\Delta = 3485 = 59^2 + 4$. Taking $b = 12$ and $m = 55$, we get $-N(b + w_\Delta) = 715 = 5 \cdot 11 \cdot 13$, so by Theorem 5.3.3, we achieve $h_\Delta \geq \tau(m) = 4$. Actually, $h_\Delta = 4$.

Example 5.3.11. Let $\Delta = 2405 = 49^2 + 4$ and take $b = 2$ and $m = 35$ in $-N(b + w_\Delta) = 595 = 5 \cdot 7 \cdot 17$. By Theorem 5.3.3, $h_\Delta \geq \tau(m) = 4$. Actually, $h_\Delta = 4$.

Example 5.3.12. Let $D = 25935 = 161^2 + 14$. Taking $N(w_\Delta) = 25935 = 3 \cdot 5 \cdot 7 \cdot 13 \cdot 19$ and $m = 3 \cdot 5 \cdot 7$ with $n = 13 \cdot 19$, by Theorem 5.3.3, we get $h_\Delta \geq \tau(m) + \Omega(n) - 1 = 8 + 2 - 1 = 9$, when in fact $h_\Delta = 16$.

Example 5.3.13. Let $D = 14^2 + 7 = 203$ and take $m = 2$, $n = 61$ and $-N(9 + w_\Delta) = 2 \cdot 61$. By Theorem 5.3.3, we get $h_\Delta \geq \tau(m) + \Omega(n) - 1 = 2$. In fact, $h_\Delta = 2$.

We conclude this section with a result for negative discriminants which has implications for results in the preceding chapter.

Theorem 5.3.4. (5.3.3) *Let $\Delta < 0$ be a fundamental discriminant. If $b \in \mathbb{Z}$, and M is any divisor of $N(b + w_\Delta)$ with $M < N(w_\Delta)$, then $h_\Delta \geq \tau(M)$.*

Proof. It suffices to show that if $a_1 \neq a_2$ are both divisors of M, then $I_1 = [a_1, b + w_\Delta] \not\sim I_2 = [a_2, b + w_\Delta]$. Suppose, to the contrary, that $I_1 \sim I_2$. Therefore, $J_1 \sim J_2$, where $J_i = [a_i', b + w_\Delta]$, with $a_i = a_i' g'$ and $g' = \gcd(a_1, a_2)$, by Lemma 5.3.2. Therefore, $J_1 J_2' = [a_1' a_2', b + w_\Delta] \sim 1$, by (1.2.1)–(1.2.5). Hence, $a_1' a_2' < N(w_\Delta)$ since $a_1' a_2' \mid M$. Thus, $a_1' a_2' = 1$, by Exercise 1.2.11. This forces $a_1' = a_2' = 1$, and so $a_1 = a_2$, a contradiction. □

Corollary 5.3.7. *If $b \in \mathbb{Z}$ with $|\sigma b + \sigma - 1| < \sqrt{-D}$ and M is any proper divisor of $N(b + w_\Delta)$, then $h_\Delta \geq \tau(M)$.*

Proof. If $M \geq N(w_\Delta)$ and $N(b + w_\Delta)/2 \geq N(w_\Delta)$, i.e.

$$((\sigma b + \sigma - 1)^2 - D)/(2\sigma^2) \geq ((\sigma - 1)^2 - D)/\sigma^2,$$

then $(\sigma b + \sigma - 1)^2 \geq 2(\sigma - 1)^2 - D$, i.e. $|\sigma b + \sigma - 1| \geq \sqrt{-D}$, a contradiction. The result now follows from Theorem 5.3.4. □

We also show how to get Theorem 4.1.2 for maximal orders via Theorem 5.3.4.

Corollary 5.3.8. *(Rabinowitsch) If $\Delta = \Delta_0 < 0$ is a fundamental discriminant, then $h_\Delta = 1$ if and only if $F_\Delta(x)$ is prime for all $x \in S_\Delta$.*

Proof. By Lemma 4.1.1, if $F_\Delta(b) = N(b + w_\Delta)$ is not prime for some $b \in S_\Delta$, then there exists a divisor $M > 1$ of $F_\Delta(b)$ with $M < N(w_\Delta)$. Hence, by Theorem 5.3.4, $h_\Delta \geq \tau(M) \geq 2$. Conversely, if $h_\Delta > 1$, then there exists a primitive reduced ideal $I = [a, b + w_\Delta]$ with $0 \leq b < a < M_\Delta = \sqrt{-\Delta/3} \leq |\Delta|/4 - 1$, so

(5.3.3) This was established in [267], along with Table 5.3.1.

$N(b + w_\Delta) \leq N(w_\Delta)^2$. Set $F_\Delta(b) = N(b + w_\Delta)$. Since $a \mid F_\Delta(b)$ and I is not principal, then $F_\Delta(b)$ cannot be prime. $\square$

Finally, we illustrate the sharpness of our bounds and also verify some of our findings in Chapter Four.

Table 5.3.1: *Lower bounds for h_Δ when $\Delta_0 = \Delta < 0$, and structure for C_Δ.*

$-D_0$	σ	b	$N(b+w_\Delta)$	M	$N(w_\Delta)$	$\tau(M)$	h_Δ	C_Δ
14	1	2	18	6	14	4	4	C_4
23	2	1	8	4	6	3	3	C_3
26	1	8	90	18	26	6	6	$C_2 \times C_3$
41	1	7	90	30	41	8	8	C_8
110	1	40	1710	90	110	12	12	$C_2 \times C_2 \times C_3$
111	2	4	48	24	28	8	8	C_8
230	1	20	630	210	230	16	20	$C_2 \times C_2 \times C_5$
303	2	4	96	48	76	10	10	$C_2 \times C_5$
337	1	53	3146	286	337	8	8	C_8
357	1	4	112	56	357	8	8	$C_2 \times C_2 \times C_2$
379	2	5	125	25	95	3	3	C_3
411	2	16	375	75	103	6	6	$C_2 \times C_3$
443	2	11	243	81	111	5	5	C_5
466	1	22	950	190	466	8	8	C_8
467	2	26	819	63	117	6	7	C_7
473	1	11	594	198	473	12	12	$C_2 \times C_2 \times C_3$
485	1	55	3510	270	485	16	20	$C_2 \times C_2 \times C_5$
499	2	24	725	25	125	3	3	C_3
555	2	7	195	15	139	4	4	$C_2 \times C_2$
1155	2	52	3045	105	289	8	8	$C_2 \times C_2 \times C_2$
1365	1	105	12390	210	1365	16	16	$C_2 \times C_2 \times C_2 \times C_2$
3315	2	97	10335	195	829	8	8	$C_2 \times C_2 \times C_2$

Remark 5.3.2. We see that the last four entries in Table 5.3.1 are from Table 4.1.2 as well as -357.

Exercises 5.3

1. Prove Claim 1 of Theorem 5.3.1. (*Hint*: Let $\mathcal{F}(c_1) = \mathcal{F}(c_2)$ with $c_1 \neq c_2$, and show $N(I) \mid A$.)

2. Prove Claim 2 of Theorem 5.3.1. (*Hint*: If $N(I) \mid A$ and $\mathcal{F}(c_1) = \mathcal{F}(c_2)$, then $\prod_{i \in X} \mathcal{P}_i^{\gamma_i} \prod_{j \in Y} \mathcal{P}_j \sim 1$.)

3. Prove Corollary 5.3.1. (*Hint*: Let $P = \{\text{primes } p \text{ dividing } A = (2a - |r-1|)/2\}$ and let I be the $\mathcal{O}_\Delta$-ideal with $N(I) = 2$. Prove that $\mathcal{P}_\Delta(A) \cap \Lambda_I(\Delta) \subseteq \mathcal{P}_\Delta(A) \cap \Lambda_I^*(\Delta)$. Then, use the tables in Theorem 3.2.1 to establish that $\mathcal{P}_\Delta(A) \cap \Lambda_I^*(\Delta) = \{2, A\}$.)

4. Prove Corollary 5.3.2. (*Hint*: Let $A = (2a - |r-1|)/4$ and $P = \{ \text{ primes } p \mid A$ and $p \nmid r\}$.)

5. (a) Let $A > 0$ be a real number and $P = \{p_i\}_{i=1}^n$ be a set of primes such that $(\Delta/p_i) = 1$ for $i = 1, 2, \ldots, n$ where $\Delta > 0$ is a fundamental discriminant. Prove that if $\mathcal{P}_\Delta(A) \cap \Lambda_1(\Delta) = \{1\}$, then $h_\Delta \geq \tau(q)$ for all $q \in \mathcal{P}_\Delta(A)$. (*Hint*: Use the method of proof of Theorem 5.3.1.).
   (b) Prove that if $P = \{p\}$ and j is maximal with respect to $p^j \leq A$, and $\mathcal{P}_\Delta(A) \cap \Lambda_1(\Delta) = \{1\}$, then $h_\Delta > \log A/\log p$. (*Hint*: $j = \lfloor \log A/\log p \rfloor$.)
   (c) Prove that if $\Delta = \Delta_0 = a^2 + r$ is of ERD-type, $A = \sqrt{D}/2$ and
   $$P = \{\text{primes } p : p \mid a,\ (r/p) = 1,\ r \not\equiv 1 \pmod{p}\},$$
   then $h_\Delta \geq \tau(q)$ for all $q \in P(A)$. (*Hint*: All that needs to be shown is that $\mathcal{P}(A) \cap \Lambda_1(\Delta) = \{1\}$.)

6. (a) Prove that if $\Delta = \Delta_0 = p^2 + 2$, where $p \equiv \pm 1 \pmod 8$ is a prime fundamental discriminant, then $h_\Delta \geq 2$. (*Hint*: Let $P = \{p\}$ and use Exercise 5.3.5.)
   (b) Prove that if $\Delta = \Delta_0 = 9b^2 - 2$ is a fundamental discriminant, then $h_\Delta \geq 2$. (*Hint*: Let $P = \{3\}$ and use Exercise 5.3.5.)

7. Prove Theorem 5.3.2. (*Hint*: The method of proof of Theorem 4.2.1 applies mutatis mutandis.)

8. Prove Corollary 5.3.5. (*Hint*: Verify that $A < \sqrt{\Delta}$ and $A \in \Lambda(\Delta)$.)

9. Show that the d in Theorem 5.3.2 can be replaced by $1 + d(\gcd(\Delta, A))$ if "$\Lambda^*(\Delta)$" is replaced by "$\Lambda_I^*(\Delta)$ for any ambiguous class $\{I\}$".

10. Let $\Delta = \Delta_0 > 0$ be a fundamental discriminant, with radicand $D = D_0$ and let
   $$\varepsilon_\Delta = (T + U\sqrt{D})/\sigma$$
   be the fundamental unit with $(\Delta/p) = 1$ for some prime p.
   (a) Prove that
   $$h_\Delta \geq \frac{\log[2\sqrt{U^2 D + \sigma^2 N(\varepsilon_\Delta)}/\sigma - N(\varepsilon_\Delta) - 1] - 2\log U}{\log p}.$$
   (b) Using (a), prove the following. If $D = D_0 = a^2 + r$ and $(\Delta/p) = 1$ for some prime p, then
      (i) if $r = 1$, a is even and $a > 2p$, then $h_\Delta \geq \frac{\log(a/2)}{\log p} > 1$.
      (ii) if $r = 1$, a is odd and $a > p/2$, then $h_\Delta \geq \frac{\log(2a)}{\log p} > 1$.
      (iii) if $r = 4$, $a > p$, then $h_\Delta \geq \frac{\log a}{\log p} > 1$.
      (iv) if $r = -4$, $a > p + 2$, then $h_\Delta \geq \frac{\log(a-2)}{\log p} > 1$.
      (v) if $|r| \neq 1$ or 4, $r \mid 4a$, $|r| > \sigma^2 p$ and $r > -\sigma a$ if $r < 0$, then
      $$h_\Delta \geq \frac{\log[(2a^2|r| + r|r| - r^2)/(2\sigma^2 a^2)]}{\log p} > 1.$$
   (*Hint*: For each of the above, use Lemma 3.2.1.)[5.3.4]

[5.3.4] The results of this question were established in [228]. See also Mordell [271].

11. (a) Let $D = D_0 = a^2 + r$ be a fundamental radicand with fundamental discriminant $\Delta = \Delta_0$. Assume $r \mid 4a$ and $|r| < 2a$. Moreover, let $I = [m, b + w_\Delta]$ be a primitive ideal of $\mathcal{O}_\Delta$ with $1 < m < -N(b + w_\Delta)$ and $b < (\sqrt{\Delta} - \sigma + 1)/2$. Prove that if no proper divisor $n > 1$ of $N(b + w_\Delta)$ is in $\Lambda^*(\Delta)$, then I is not principal. (*Hint*: Just invoke Theorem 3.2.1.)

    (b) Apply (a) to the example $D = 226$ with $I = [63, 10 + \sqrt{226}]$. Also, look at $J = [63, 17 + \sqrt{226}]$ to show that (a) does not apply.

12. Let $\Delta = \Delta_0 > 0$ be a fundamental discriminant with radicand $D = D_0 \not\equiv 1 \pmod 4$. Prove that $h_\Delta > 1$ in each of the following cases:[5.3.5]

    (a) $D = (r^k - a)^2 + r$ where one of the following holds:
        (i) $a \equiv 2 \pmod 3$ and k is even.
        (ii) $a \equiv 1 \pmod 3$ with $a \geq 4$.
        (iii) r is not prime.

    (b) $D = (r^k + a + 1)^2 - r$ and one of the following holds:
        (i) $a \not\equiv 2 \pmod 3$ and $a > 2$.
        (ii) r is not prime.

    (c) $D = (r^k - a - 1)^2 - r$, $a \geq 2$, $k \geq 2$ and one of the following holds:
        (i) $a \equiv 1 \pmod 3$, $a \geq 4$, $k \geq 2$.
        (ii) r is not prime.

    (d) $D = (2ar^k + a)^2 - 2ar^k$, $r = 4a - 1$, and $a > 1$ odd.

    (e) $D = (ar^k + a)^2 - ar^k$, $r = 4a - 1$, $a > 2$ square-free.

    (f) $D = (ar^k - a)^2 + ar^k$, $r = 4a - 1$ $a > 2$ square-free.

    (g) $D = (ar^k + a)^2 - r^k$, $r = 4a^2 - 1$ k even, $D \equiv 3 \pmod 4$.

    (h) $D = (ar^k - a)^2 + r^k$, k odd $D \equiv 3 \pmod 4$ and $D > 7$.

    (i) $D = (ar^k - a)^2 + 2ar^k$, a odd and square-free.

    (j) $D = (ar^k + a)^2 - 2ar^k$, $r = 2a - 1$ $a > 1$ odd and square-free.

    (k) $D = (ar^k + a)^2 + 2ar^k$, $a > 1$ odd and square-free.

    (l) $D = (ar^k - a)^2 - 2ar^k$, $a > 1$ odd and square-free.

    (m) $D = (2ar^k + a)^2 - ar^k$, $r = 8a - 1$ $a > 1$ square-free.

    (n) $D = (2ar^k - a)^2 + ar^k$, $r = 8a - 1$, $a > 1$ square-free.

13. Using the results of section two of this chapter, establish class number one criteria for each of the parameterized forms in Exercise 5.3.12.

[5.3.5] This was established by Dubois–Levesque in [76] using techniques involving Theorem 5.2.6. This exercise is designed to allow the reader to apply the results of this section.

5.4 The GRH: Relevance of the Riemann Hypothesis.

In this section, we look at how the generalized Riemann hypothesis (GRH) can be used to solve some outstanding and difficult problems. Also, we show how to use a result of Tatuzawa [359] to show that these results hold unconditionally with only the proviso: there may be one more value remaining, whose existence can be shown to be a counterexample to the GRH. We begin with Gauss.

Gauss conjectured that if $\Delta = \Delta_0 < 0$ is a fundamental discriminant, then $h_\Delta \to \infty$ as $|\Delta| \to \infty$. As we shall see later, the combined efforts of Hecke, Deuring and Heilbronn proved in 1934 that $h_\Delta \to \infty$ as $\Delta \to -\infty$, which solved Gauss' problem for negative fundamental discriminants. In 1935, Siegel [338] proved that

$$\log(h_\Delta) \sim \log(\sqrt{|\Delta|})$$

for a negative fundamental discriminant $\Delta = \Delta_0 < 0$, i.e. $\lim_{\Delta \to -\infty} \frac{\log(h_\Delta)}{\log\sqrt{|\Delta|}} = 1$, so we know how big h_Δ can be as $\Delta \to -\infty$.

The case where $\Delta = \Delta_0 > 0$ is still an open and deep problem. Siegel's result for the real side is

$$h_\Delta R \sim \sqrt{\Delta}$$

where $R = \log(\varepsilon_\Delta)$ is called the *regulator* (5.4.1) of the real quadratic field $K = \mathbb{Q}(\sqrt{\Delta})$ for $\Delta = \Delta_0 > 0$, i.e. $\lim_{\Delta\to\infty} \frac{\log(h_\Delta R)}{\log(\sqrt{\Delta})} = 1$. However, Siegel's result does not help us to solve Gauss' problem, because we don't know enough about the behaviour of the regulator. Thus, the regulator is the crux of the problem.

In order to understand more about the relationship between the class number, the regulator, and ultimately the GRH, we need to understand the analytic class number formula. For this, we need

Definition 5.4.1. Let s be a complex variable and $\Delta = \Delta_0 > 0$ a fundamental discriminant with $\chi(n) = (\Delta/n)$ denoting the Kronecker symbol.(5.4.2) The *Dirichlet L-function* is

$$L(s,\chi) = \sum_{n=1}^{\infty} \chi(n) n^{-s}.$$

Definition 5.4.2. The *Riemann zeta function* is

$$\zeta(s) = \sum_{n=1}^{\infty} n^{-s}$$

for a complex variable s.

If $K = \mathbb{Q}(\sqrt{\Delta})$ where $\Delta = \Delta_0 > 0$ is a fundamental discriminant, then the *Riemann zeta function over K* is

$$\zeta_K(s) = \sum N(I)^{-s}$$

(5.4.1)Note that, by Schur [322], we have $R < \sqrt{\Delta}\log\Delta$. Also, by Hua [126], we have that $L(1,\chi) < 1 + \log\sqrt{\Delta}$, from which it follows that $h_\Delta < \sqrt{\Delta}$, first proved by Slavutsky [342]. Furthermore, a special case of the Ankeny-Brauer-Chowla Theorem (e.g. see Narkiewicz [280, Theorem 8.5, p. 389]) is that for every $\epsilon > 0$, there are infinitely many $\Delta = \Delta_0 > 0$ with $h_\Delta > \Delta^{1/2-\epsilon}$.

(5.4.2)See Exercise 1.1.4. Also, see Appendix F.

where the sum ranges over all non-zero ideals of $\mathcal{O}_\Delta$. Also, $\zeta_K(s)/\zeta(s) = L(s,\chi)$.

The *analytic class number formula*[(5.4.3)] for $\Delta = \Delta_0 > 0$ (or simply for K) is

$$2h_\Delta R = \sqrt{\Delta}\lim_{s\to 1}(s-1)\zeta_K(s) = \sqrt{\Delta}L(1,\chi). \tag{5.4.1}$$

Remark 5.4.1. Formula (5.4.1) tells us that we can compute h_Δ if we can determine $L(1,\chi)$ to sufficient accuracy and find R. Both of these problems have difficulties (see Chapter Eight for algorithms). Some of the problems are lessened by the assumption of GRH, which allows for improved estimates for the value of $L(1,\chi)$. We now give a brief account of the GRH.

Since 1859, it has been an open conjecture that if s is any non-trivial zero of $\zeta(s)$ (i.e. $s \neq -2,-4,-6\ldots$), then the real part of s, $\mathrm{Re}\,(s)$ must be 1/2. Because of a functional equation satisfied by $\zeta(s)$ (see Appendix F, equation (F.2)), it is known that all such zeros must lie in the *critical strip*, $0 < \mathrm{Re}\,(s) < 1$, and that the zeros are symmetrically located about the *critical line* $\mathrm{Re}\,(s) = 1/2$. Thus, the *Riemann hypothesis* (RH) asserts that $\zeta(s) \neq 0$ for any value of s such that $\mathrm{Re}\,(s) > 1/2$. Although the problem remains open, we do know that an infinitude of zeros of $\zeta(s)$ must lie on the critical line,[(5.4.4)] and the numerical evidence to date overwhelmingly supports the Riemann hypothesis (e.g. see Edwards' book [80] devoted to the Riemann hypothesis, and see more recent developments by Deninger [71]).

The hypothesis analogous to Riemann's concerning $L(s,\chi)$ is that for any character χ (see Appendix F), $L(s,\chi) > 0$ for any value of s such that $\mathrm{Re}\,(s) > 1/2$. This is the *GRH*.

To illustrate the historical significance of the role of the GRH in class number problems, let's return to the solution of Gauss' problem for $\Delta = \Delta_0 < 0$, discussed at the outset of this section.

In a result attributed by Landau [163] to Hecke in 1918, it is shown that, under the assumption of the truth of GRH, we get Gauss' conjecture (i.e. that $h_\Delta \to \infty$ as $\Delta \to -\infty$ for $\Delta = \Delta_0 < 0$) from the fact that, if the only non-trivial zeros of $L(s,\chi)$ lie on the critical line, then h_Δ must grow with $|\Delta|$, since Hecke's result bounded h_Δ from below in terms of $|\Delta|$. In 1933, Deuring [72] proved that if RH is false, then $h_\Delta \geq 2$ for $-\Delta$ sufficiently large. Mordell [272] did more in 1934 when he established that, if the RH is *false*, then $h_\Delta \to \infty$ as $\Delta \to -\infty$. This was stepped up by Heilbronn [115] in 1934, who proved that if the GRH is false, then $h_\Delta \to \infty$ as $\Delta \to -\infty$. Hence, under the assumption of either the truth or the falsity of GRH, Gauss' result follows so we have an *unconditional* proof that $h_\Delta \to \infty$ as $\Delta \to -\infty$ by the combined work of the above. Note that assuming the falsity of GRH assumes the existence of a zero *off* the line $\mathrm{Re}\,(s) = 1/2$. This zero (which is presumed not to exist) has come to be known as a *Siegel zero*.

In 1934, Heilbronn and Linfoot [116] showed that Gauss' class number one problem for $\Delta = \Delta_0 < 0$ is solved if the GRH is true, or more precisely, that only one more (a tenth) discriminant $\Delta = \Delta_0 < 0$ could exist with $h_\Delta = 1$, and such existence would be a counterexample to GRH.

[(5.4.3)] See Appendix F for analytic details.

[(5.4.4)] Conrey [64] proved that at least 2/5 of all zeros of $\zeta(s)$ must lie on the critical line. Also, van de Lune *et al.* [371] showed that the first 1.5×10^9 zeros of $\zeta(s)$ lie on the critical line. See Appendix F.

In 1935, Siegel [338][5.4.5] proved that for every $\varepsilon > 0$, there exists a constant $c > 0$, which was not computable by his techniques, such that $h_\Delta > c|\Delta|^{1/2-\epsilon}$. Tatuzawa [359][5.4.6] in 1951, showed that Siegel's theorem holds for an *effectively computable* constant $c > 0$ for all $\Delta < 0$ (except for at most one exceptional value of Δ). From Chapter Four, we know that Gauss' class number 1 problem for $\Delta_0 < 0$ was ultimately solved by Baker–Heegner–Stark, who demonstrated that the tenth field cannot exist.

Now, let's get back to the real case. Tatuzawa's result for the real side is

Theorem 5.4.1. (Tatuzawa [359].) *If $\Delta = \Delta_0 > 0$ is a fundamental discriminant with $1/2 > \varepsilon > 0$ and $\Delta \geq \max\{e^{1/\epsilon}, e^{11.2}\}$, then*

$$L(1, \chi) > .655\varepsilon\Delta^{-\epsilon}$$

with one possible exceptional value of Δ remaining, where χ is a real, non-principal, primitive character modulo Δ (see Appendix F).

We now illustrate how Theorem 5.4.1 can be used, in conjunction with the GRH, to solve some class number problems for $\Delta = \Delta_0 > 0$.

The most outstanding class number problem is that conjectured by Gauss in [91, Article 304, p. 363], namely there are infinitely many fundamental discriminants $\Delta = \Delta_0 > 0$ with $h_{\Delta_0} = 1$. This remains an open and deeply difficult problem to this day. However, there is serious heuristic evidence to support its validity. For instance, the Cohen–Lenstra heuristics elucidated in [58] say that the probability of prime discriminants $p = \Delta$ having $h_\Delta = 1$ is 75.466%. This is supported, for example, by the tables provided by Stephens and Williams [348]. Therein, they use the infrastructure to determine the regulator R, then they use the GRH to rapidly estimate $L(1, \chi)$ to the accuracy needed for determining whether or not $h_\Delta = 1$ (see Chapter Eight).

We now turn to solving certain class number one problems by using the GRH and Tatuzawa's result. We begin with a proof of the following, under the assumption of the validity of the GRH.

Conjecture 5.4.1. (Chowla's conjecture [51], 1976.) *If $p = 4m^2 + 1$ is prime and $m > 13$, then $h_p > 1$.*

We remark that by Exercise 5.2.2, if $h_\Delta = 1$ for a fundamental discriminant Δ, then by Theorem 1.3.3, Δ *must* be prime, so Chowla's conjecture has no loss of generality by assuming that $4m^2 + 1$ is prime.

Theorem 5.4.2. [5.4.7] *If the GRH is valid, then Chowla's conjecture is true.*

Proof. Let $\chi(q) = (p/q)$ be the Kronecker symbol where q is prime and set

$$T_1(y) = \prod_{q \leq y} \left[\frac{q}{q - \chi(q)}\right],$$

and

$$T_2(y) = \prod_{q > y} \left[\frac{q}{q - \chi(q)}\right].$$

[5.4.5] See Goldfeld [94] for a simple proof of Siegel's result.

[5.4.6] See Hoffstein [125] for a simple proof of Tatuzawa's result.

[5.4.7] This was proved in [247].

Thus,

$$L(1,\chi) = T_1(y)T_2(y).^{(5.4.8)}$$

Since $T_1(y) \geq \prod_{q\leq y} \frac{q}{q+1}$, then

$$\log T_1(y) \geq -\sum_{q\leq y} \frac{\log(q+1)}{q} > -\sum_{q\leq y}\frac{1}{q} > -\log y, \tag{5.4.2}$$

where the last inequality is Exercise 5.4.1(a). Furthermore, from Exercise 5.4.1(b)–(c)

$$\log T_2(y) = -\sum_{q>y} \log\left(1 - \frac{\chi(q)}{q}\right) > \left(\sum_{q>y}\frac{\chi(q)}{q}\right) - \frac{1}{y}. \tag{5.4.3}$$

By ideas of Cornell and Washington (see [247]), where the GRH is assumed, we get

$$\sum_{q>y}\frac{\chi(q)}{q} \geq -B(y)\left(\frac{4+3\log y}{\sqrt{y}}\right), \tag{5.4.4}$$

where

$$B(y) = \frac{\log p}{\pi\log y} + \frac{5.3\log p}{(\log y)^2} + \frac{4}{\log y} + \frac{1}{\pi}.$$

Now, from (5.4.3)–(5.4.4), we get

$$\log T_2(y) > -\frac{B(y)(4+3\log y)}{\sqrt{y}} - \frac{1}{y}. \tag{5.4.5}$$

Hence, from (5.4.2) and (5.4.5), we have

$$L(1,\chi) = T_1(y)T_2(y) > \frac{1}{y}\exp\left(\frac{-B(y)(4+3\log y)}{\sqrt{y}} - \frac{1}{y}\right).$$

If $y = (\log p)^2$, then $\log y = 2\log\log p$. Therefore,

$$-B(y)\frac{(4+3\log y)}{\sqrt{y}} - \frac{1}{y} =$$
$$-\left[\frac{2}{\pi\log\log p} + \frac{15}{(\log\log p)^2} + \frac{8}{(\log\log p)\log p} + \frac{4}{\pi\log p} + \frac{3}{\pi}\right.$$
$$\left. + \frac{3(5.3)}{2\log\log p} + \frac{12}{\log p} + \frac{6\log\log p}{\pi\log p} + \frac{1}{(\log p)^2}\right] = f(p),$$

say.

Hence, $L(1,\chi) > e^{f(p)}/(\log p)^2$. However, $2h_\Delta R = \sqrt{p}L(1,\chi)$, where

$$R = \log(2m+\sqrt{p}) < \log 2\sqrt{p}.$$

Therefore,

$$h_\Delta > \frac{\sqrt{p}e^{f(p)}}{2(\log p)^2(\log 2\sqrt{p})} = F(p),$$

say. Moreover, $-f(p)$ is a decreasing function of p, so $e^{f(p)}$ is an increasing function of p, forcing $F(p)$ to be one as well. Thus, if $p > 10^{13}$, then $-f(p) < 4.68$,

(5.4.8) This is called the *Euler product* for the Dirichlet L-function (see Appendix F).

$e^{f(p)} > 0.009279493$, and $\sqrt{p}/(2(\log p)^2(\log 2\sqrt{p})) \geq 112.6838154$. Hence, $F(p) > 1$ for $p > 10^{13}$.

To deal with the primes $p < 10^{13}$ when $p = 4m^2 + 1$, we note that $m < 10^{6.5}/2 \approx 1.6 \times 10^6$. For the balance, we use the fact from Exercise 5.2.2, that $h_p = 1$ is equivalent to all primes $q < m$ satisfying $(p/q) = -1$. Select some positive integer k and the first k primes $\{q_i\}_{i=1}^k$ with $q_1 = 5$. For these q_i, find (by trial) and tabulate those S_{ij} such that $0 \leq S_{ij} \leq q_i - 1$ and $((4S_{ij}^2 + 1)/q_i) \neq -1$, where $(/)$ is the Legendre symbol. There are approximately $q_i/2$ of these. If any $m \equiv S_{ij} \pmod{q_i}$ and $m > q_i$, then delete this m since either $4m^2+1 \equiv 0 \pmod{q_i}$, forcing $4m^2 + 1$ to be composite, or $((4m^2 + 1)/q_i) = 1$ and $q_i < m$ which means that $h_p > 1$. Since half of the m's are eliminated for each q_i, then a value of k such that $2^k > 1.6 \times 10^6$ should suffice to complete the task. We used $k = 30$ and a Fortran program to sieve out[(5.4.9)] as many values of $m < 1.6 \times 10^6$ as possible. In a matter of a few minutes, we found that if $m > 13$ and $p = 4m^2 + 1$ is a prime with $p < 10^{13}$, then there exists some $q_i < m$ with $(p/q_i) = 1$. □

The above shows the power of the GRH in solving some outstanding problems. However, this highly conditional result is less satisfactory than what we now demonstrate.

Chowla's conjecture is just one of many involving ERD-types. We now cite the balance of them.

Conjecture 5.4.2. (Yokoi [395], 1986). *If $\Delta = \Delta_0 = m^2 + 4$ is a fundamental radicand with odd $m > 17$, then $h_\Delta > 1$.*

Conjecture 5.4.3. (Mollin [223], 1987). *If $\Delta = \Delta_0 = m^2 - 4$ is a fundamental radicand with odd $m > 21$, then $h_\Delta > 1$.*

Conjecture 5.4.4. (Mollin–Williams [246], 1989). *If $\Delta = \Delta_0 = 4D_0 = 4(m^2 \pm 2)$ is a fundamental discriminant with $m > 20$, then $h_\Delta > 1$.*

Conjecture 5.4.5. (Mollin-Williams [246], 1989). *If $\Delta = \Delta_0 = D_0 = m^2 - p$ is a fundamental discriminant with $p \mid m$ an odd prime and $m > 42$, then $h_\Delta > 1$.*

Conjecture 5.4.6. (Mollin-Williams [246], 1989). *If $\Delta = \Delta_0 = D_0 = m^2 \pm 4r$ is a fundamental discriminant with $r > 1$ an odd integer dividing m and $m > 35$, then $h_\Delta > 1$.*

Each of these conjectures can be established via the GRH, as we have done in Theorem 5.4.2. However, we seek a stronger result. For this, we need Theorem 5.4.1. The following was proved in [251].

Theorem 5.4.3. *If $\Delta = \Delta_0 > 0$ is a fundamental discriminant of ERD-type with radicand $D = D_0$, then $h_\Delta = 1$ if and only if $D \in$ {2,3,5,6, 7, 11, 13, 14, 17, 21, 23, 29, 33, 37, 38, 47, 53, 62, 69, 77, 83, 93, 101, 141, 167, 173, 197, 213, 227, 237, 293, 398, 413, 437, 453, 573, 677, 717, 1077, 1133, 1253, 1293, 1757}, with one possible exceptional value whose existence would be a counterexample to the GRH.*

Proof. Let $\chi(n) = (\Delta/n)$ be the Kronecker symbol (see Appendix F). By Exercise

(5.4.9) Briefly put, sieving is a process which finds solutions to systems of single variable linear congruences by simply searching through all of the rational integers up to a certain prescribed bound. This is usually accomplished by eliminating possible solution candidates for the various moduli in parallel, leaving only the solutions.

5.4.3, the regulator R satisfies $R < \log 3D$. If $\varepsilon = 1/16$, then by equation (5.4.1) and Theorem 5.4.1

$$h_\Delta > \frac{0.655\Delta^{7/16}}{32\log 3D} \geq \frac{0.655D^{7/16}}{32\log 3D} > 1,$$

with one possible exceptional value. Thus, $h_\Delta > 1$ for all $\Delta > e^{16}$. For the values of $\Delta < e^{16}$, we used a sieving technique similar to that described in the proof of Theorem 5.4.2, and eliminated all possible values of D except those listed in the theorem.

Finally, we note that since the GRH can be used to establish that our list is complete, then the exceptional value arising from Tatuzawa's theorem, if it were to exist, would be a counterexample to the GRH. □

Remark 5.4.2. What Theorem 5.4.3 establishes is the remarkable fact that we have now proved that five out of the six Conjectures 5.4.1–5.4.6 are true, but we don't know which ones! Furthermore, if any one of them fails, then it fails for at most one value of Δ which would be a counterexample to the GRH! Therefore, although Conjectures 5.4.1–5.4.6 remain open, we have come as close to proving them as it is possible to do so without actually accomplishing the task. In fact, we are now, at this point in mathematical history, exactly where Gauss' class number one problem for $\Delta = \Delta_0 < 0$ was at the Heilbronn–Linfoot result in 1934, which said that there could be at most one more (a tenth) complex quadratic field, with class number one, whose existence would be a counterexample to GRH. Unfortunately, the methods of Baker–Heegner–Stark do not apply here to unconditionally solve the class number one problem for real quadratic fields of ERD-type. The existence of a regulator in the real case makes the problem much more difficult, and so it remains an open problem. It is worth noting that in [400], Zhang Ming-yao proved that if the exceptional value exists, then $\Delta > \exp(3.7 \times 10^8)$ in Conjecture 5.4.2, whereas Lu and Zhang [205] have proved that in Conjecture 5.4.1, if the exceptional Δ exists, then $\Delta > 10^{3.8\times 10^7}$.

As established in Exercise 3.2.10, if $\Delta = \Delta_0 > 0$ is of ERD-type and $h_\Delta = 1$, then $\ell(w_\Delta) \leq 4$. There is nothing in the above techniques which does not allow us to extend to larger $\ell(w_\Delta)$. In fact, we have from [256]

Theorem 5.4.4. *If $\Delta = \Delta_0 > 0$ is a fundamental discriminant with $\ell(w_\Delta) = \ell \leq 24$, then $h_\Delta = 1$ if and only if D is an entry in Table A1 (of Appendix A) with one possible exceptional value which is ruled out by GRH.*

Proof. Let χ be a real, non-principal primitive character modulo Δ (see Appendix F). Since $R < \ell \log\sqrt{\Delta}$ (see Chapter Eight, section one), then we may use the techniques of the proof of Theorem 5.4.3 to verify that if $h_\Delta = 1$, then $\ell > (.655\varepsilon\Delta^{1/2-\varepsilon})/\log\Delta$, whenever $\Delta > \max(e^{1/\varepsilon}, e^{11.2})$ with possibly one exception. Thus, if $\Delta > B > e^{11.2}$, $\varepsilon = 1/\log B$ and $f(B) = \frac{\lfloor .655B^{(1/2-1/\log B)}\rfloor}{(\log B)^2}$, then $h_\Delta = 1$ implies that $\ell > f(B)$ with one possible exception.

We choose $B = 2^{31} - 1$, for convenience sake, at the machine level. With this B, we get $f(B) > 24.1$. Therefore, for $\ell \leq 24$, then $h_\Delta > 1$ if $\Delta > B$. We now deal with the case where $1 \leq \Delta \leq B$. By Exercise 2.1.13(d), one of the equations

$$P_i = P_{i+1} \tag{5.4.6}$$

or

$$Q_i = Q_{i+1} \tag{5.4.7}$$

holds, where $\ell(w_\Delta) = 2i$ and $\ell = 2i+1$, respectively, in the simple continued fraction expansion of w_Δ. Thus we need only check the continued fraction expansion of w_Δ up to $i = 12$. We first check whether (5.4.6) or (5.4.7) occurs for Δ and discard those Δ with $\ell > 24$. We also store the values of Q_i/σ. If $p < \sqrt{\Delta}/2$ is prime with $(\Delta/p) = 1$, then $h_\Delta = 1$ implies that $p \in \Lambda^*(\Delta)$, i.e. $p = Q_i/\sigma$ for some $i \leq \ell/2$. Thus, we need only search for a prime $p < \sqrt{\Delta}/2$ such that $p \neq Q_i/\sigma$ for $i \leq n = \ell/2$ and $(\Delta/p) = 1$ in order to be assured that $h_\Delta > 1$. When this simple exclusion method was used for all numbers in excess of 50000 for which (5.4.6) or (5.4.7) held with $n \leq 12$, we found that there were no possible values of $\Delta \geq 50000$ such that $\ell \leq 24$ and $h_\Delta = 1$. The entire computational process took about 2 hours and 10 minutes on an Amdahl 5870 computer. The values of $\Delta < 50000$ such that $\ell \leq 24$ and $h_\Delta = 1$ were then identified using the standard techniques for computation of class number one, as outlined in Chapter Eight. □

Remark 5.4.3. As Table A1 of Appendix A shows, the number of Δ with $h_\Delta = 1$ tends to increase (not monotonically, however) as ℓ increases. If one could prove this, then one would have the Gauss conjecture solved. In fact, the Gauss conjecture can be strengthened as $\#\ell \to \infty$ as $\ell \to \infty$ where $\#\ell$ is the number of discriminants $\Delta = \Delta_0 > 0$ with $h_\Delta = 1$ and $\ell(w_\Delta) = \ell$.

Remark 5.4.4. The case $\Delta = \Delta_0 \equiv 1 \pmod 8$ is very special. In what follows, we are able to show that those $\Delta \equiv 1 \pmod 8$ in Table A1 are precisely those for which $h_\Delta = 1$, i.e. if the exceptional value Δ exists, then $\Delta \not\equiv 1 \pmod 8$. We emulate Hendy's ideas in Theorem 3.4.1.

Theorem 5.4.5. *Let $\Delta = \Delta_0 > 0$ be a fundamental discriminant with $\ell = \ell(w_\Delta)$ and p a prime with $(\Delta/p) = 1$. If $\Delta > 4p^{\ell+1}$, then $h_\Delta > 1$.*

Proof. Suppose $h_\Delta = 1$. By hypothesis, $\Delta > 4p^{\ell+1}$, so $p^{(\ell+1)/2} < \sqrt{\Delta}/2$. If $m = \lfloor(\ell+1)/2\rfloor$, then $p^i < \sqrt{\Delta}/2$ for $i = 1, 2, \ldots, m$. Since $(p) = \mathcal{P}\mathcal{P}'$, then $N(\mathcal{P}) = N(\mathcal{P}') = p$, so the set of ideals $S = \{\mathcal{P}^i, \mathcal{P}'^i\}_{i=1}^m$ satisfies $N(I) < \sqrt{\Delta}/2$ for all $I \in S$. Thus, S consists of distinct reduced ideals. Therefore, together with the trivial ideal $(1) = \mathcal{O}_\Delta$, we have $2m+1$ reduced ideals. By Theorem 2.1.2, $\ell \geq 2m+1$, so $m = \lfloor(\ell+1)/2\rfloor > (\ell-1)/2$, i.e. $2m+1 > \ell$, a contradiction. □

Two facts emerge from this result, the first of which is immediate.

Corollary 5.4.1. *For a given integer n, there are at most finitely many fundamental discriminants $\Delta = \Delta_0 > 0$ with $h_\Delta = 1$ and $\ell(w_\Delta) \leq n$.*

Corollary 5.4.2. *If $\Delta = \Delta_0 > 0$ is a fundamental discriminant with $\Delta \equiv 1 \pmod 8$ and $\ell(w_\Delta) = \ell \leq 24$, then $h_\Delta = 1$ if and only if D is an entry in Table A2 of Appendix A.*

Proof. By Theorem 5.4.5, $h_\Delta > 1$ when $\Delta > 2^{\ell+3}$, since $(\Delta/2) = 1$. Since we have already checked, on a computer, all values of D up to $2^{31} - 1$ as noted in the proof of Theorem 5.4.4, then the result follows since we are only concerned with $\Delta > 2^{27}$, a smaller bound. □

What Corollary 5.4.2 illustrates is that $\Delta \equiv 1 \pmod 8$ is easier to handle because of the "splitting" of 2.

Remark 5.4.5. There is another facet which is hidden in all of this which deserves comment. If $\Delta > 0$ is a discriminant, then there are h_Δ classes $\{I_i\}$, and there are $\ell(I_i) = \ell_i$ reduced ideals in each class (assuming that I_i is reduced for each i). The number $c_\Delta = \sum_{i=1}^{h_\Delta} \ell_i$ is called the *caliber* by Lauchaud in [155]. This is not new since it all emerges from the continued fraction algorithm, which tells us, in particular, that there are only finitely many fundamental discriminants $\Delta = \Delta_0 > 0$ such that $c_\Delta \leq n$ for a given positive $n \in \mathbf{Z}$. In particular, there are only finitely many fundamental discriminants $\Delta = \Delta_0 > 0$ such that $c_\Delta = 1$. In fact, we can list these under an assumption of GRH which yields that $h_\Delta R < 4.23 c_\Delta$. The list for $c_\Delta = 1$ is then $D = D_0 \in \{2, 5, 13, 29, 53, 173, 293\}$.

Remark 5.4.6. There is nothing particularly special about the choice of $\ell \leq 24$ in Theorem 5.4.4, except that for computational considerations, we stopped there. The techniques used are still valid for any period lengths $\ell(w_\Delta) \leq B$ for *any* bound B we care to choose. This, then, is limited only by a finite amount of computation,(5.4.10) albeit, in practice, a potentially difficult and time-consuming one given the current state of computational efficiency. We can take this a step (or n steps) further, since there is also nothing particularly special about the class number one problem from the perspective of using these techniques. Thus, we have from [259]

Theorem 5.4.6. *If $\Delta = \Delta_0 > 0$ is a fundamental discriminant with $\ell(w_\Delta) = \ell \leq 24$, then $h_\Delta = 2$ if and only if D is an entry in Table A3 (of Appendix A) with one GRH-ruled out exception.*

Before establishing this result, we are actually able to refine the previous techniques in order to reduce the amount of work required. The following is an improvement over the inequality $R < \ell \log \sqrt{\Delta}$ used in the proof of Theorem 5.4.4. This is, therefore, of interest in its own right.

Lemma 5.4.1. *If $\Delta = \Delta_0 > 0$ is a fundamental discriminant and $R = \log(\varepsilon_\Delta)$, then*

$$R < \lfloor 3(\ell + 1)/4 \rfloor \log \sqrt{\Delta},$$

where $\ell = \ell(w_\Delta)$.

Proof. By Theorem 2.1.3, we have

$$\varepsilon_\Delta = \prod_{i=1}^{\ell} \gamma_i \tag{5.4.8}$$

where $\gamma_i = (P_i + \sqrt{D})/Q_i$ and by Exercise 2.1.13(e), $0 < P_i < \sqrt{D}$ from the simple

(5.4.10)This is also a conclusion at which Goldfeld [95] arrived when discussing the solution of Gauss' class number problem on the complex side. Granted, their solution is effective and unconditional, whereas ours depends on GRH/Tatuzawa. Yet, the comments on the algorithms pertaining to creating the lists, remain valid nevertheless.

continued fraction expansion of w_Δ. We can write (5.4.8) as

$$\varepsilon_\Delta = \lambda \prod_{i=1}^{\lfloor \ell/2 \rfloor} \chi_i, \tag{5.4.9}$$

where $\chi_i = \gamma_i \gamma_{\ell-i+1}$, $\lambda = 1$, when ℓ is even, and $\lambda = \gamma_{(\ell+1)/2}$ otherwise. By Exercise 2.1.13(g), we have that $Q_i = Q_{\ell-i}$ and $P_{i+1} = P_{\ell-i}$ for $0 \le i < \ell$, so

$$\chi_i = ((P_i + \sqrt{D})/Q_i)((P_i + \sqrt{D})/Q_{i-1}) = (\sqrt{D} + P_i)/(\sqrt{D} - P_i),$$

from (2.1.5). Furthermore, if ℓ is odd, then by Exercise 2.1.13(d), $Q_{(\ell-1)/2} = Q_{(\ell+1)/2}$, so $D = P^2_{(\ell+1)/2} + Q^2_{(\ell+1)/2}$. Therefore,

$$\lambda = \gamma_{\frac{\ell+1}{2}} < \sqrt{\Delta},$$

since $\sigma \mid Q_i$ for all i.

If we define

$$\nu = \begin{cases} 0 & \text{when } \ell \text{ is even} \\ 1 & \text{when } \ell \text{ is odd,} \end{cases}$$

then it is an easy exercise for the reader to verify that

$$\lfloor \ell/2 \rfloor + \lfloor (\ell+2)/4 \rfloor + \nu \le \lfloor 3(\ell+1)/4 \rfloor. \tag{5.4.10}$$

If $\sigma = 2$ and $\lfloor \sqrt{D} \rfloor$ is even, then since P_i is odd for all $i \ge 0$, we cannot have $P_i = \lfloor \sqrt{D} \rfloor$. In this case, therefore, $\chi_i < 2\sqrt{D}$ and $\varepsilon_\Delta < \lambda(2\sqrt{\Delta})^{\lfloor \ell/2 \rfloor}$, so

$$\varepsilon_\Delta < 2^{\lfloor \ell/2 \rfloor}(\sqrt{\Delta})^{\lfloor \ell/2 \rfloor + \nu}. \tag{5.4.11}$$

When $\Delta > 16$, we have

$$2^{\lfloor \ell/2 \rfloor} < (\sqrt{\Delta})^{\lfloor (\ell+2)/4 \rfloor},$$

so by (5.4.10), we have the results in the case. For the remaining values of $\Delta < 16$, we see that $\sigma = 2$ and $\lfloor \sqrt{D} \rfloor$ even forces $D = 5$, for which the theorem is easily verified.

If $P_i \ne \lfloor \sqrt{D} \rfloor$ and $\lfloor \sqrt{D} \rfloor \equiv 1 \pmod{\sigma}$, then we must have $P_i \le \lfloor \sqrt{D} \rfloor - \sigma$ and $\chi_i < (2/\sigma)\sqrt{D} = \sqrt{\Delta}$. Furthermore, if j is the least positive integer such that $P_j = P_{j+1}$, we must have $\ell = 2j$. Therefore, the case in which we have the largest possible number of P_i-values equal to $\lfloor \sqrt{D} \rfloor$ can only occur when

$$P_1 = P_2 = \cdots = P_{\lfloor (\ell+2)/4 \rfloor} = \lfloor \sqrt{D} \rfloor.$$

Thus, if n is the number of values of P_i with $P_i = \lfloor \sqrt{D} \rfloor$ for $i \le \lfloor \ell/2 \rfloor$, then

$$\varepsilon_\Delta < \lambda(\sqrt{\Delta})^{\lfloor \ell/2 \rfloor - n}(\sqrt{D} + \lfloor \sqrt{D} \rfloor)/(\sqrt{D} - \lfloor \sqrt{D} \rfloor) \tag{5.4.12}$$

and $n \le \lfloor (\ell+2)/4 \rfloor$.

Since $D - \lfloor \sqrt{D} \rfloor^2 \ge \sigma^2$, we have $(\sqrt{D} + \lfloor \sqrt{D} \rfloor)/(\sqrt{D} - \lfloor \sqrt{D} \rfloor) < (2\sqrt{D}/\sigma)^2 = \Delta$. Hence,

$$\varepsilon_\Delta < \lambda(\sqrt{\Delta})^{\lfloor \ell/2 \rfloor - n}\Delta^n = (\sqrt{\Delta})^{\lfloor \ell/2 \rfloor + n + \nu}.$$

By (5.4.10), the result follows. □

This enables us to prove

Lemma 5.4.2. *If $\ell(w_\Delta) \leq 24$ and $\Delta = \Delta_0 > 6 \times 10^9$, then with at most one possible exceptional fundamental discriminant Δ (ruled out by GRH), $h_\Delta > 2$. Furthermore, if $\Delta \equiv 1 \pmod 4$, then the result holds for $\Delta > 4.75 \times 10^9$.*

Proof. By Theorem 5.4.1, we have (with at most one exception)

$$L(1, \chi) > 0.655\varepsilon\Delta^{-\varepsilon}$$

for $0 < \varepsilon < 1/2$ and $\Delta \geq \max(e^{1/\varepsilon}, e^{11.2})$, where χ is a real, non-principal, primitive character modulo Δ. By equation (5.4.1) and Lemma 5.4.1, we have

$$2h_\Delta > 2\Delta^{\frac{1}{2}-\varepsilon}(0.655\varepsilon)/(\lfloor 3(\ell+1)/4 \rfloor \log \Delta) > 4$$

when $\Delta > 6 \times 10^9$, $\varepsilon = 0.04442$ and $\ell(w_\Delta) \leq 24$.

When $\Delta \equiv 1 \pmod 4$, we can improve upon the above somewhat.

The reader can verify the case where $\lfloor \sqrt{D} \rfloor$ is even, using the methods of Lemma 5.4.1, and the above. When $\lfloor \sqrt{D} \rfloor$ is odd, then write (5.4.12) as

$$\varepsilon_\Delta < (\sqrt{\Delta})^{\lfloor(\ell+1)/2\rfloor - n}(\sqrt{D} + \lfloor \sqrt{D} \rfloor)^n \beta^{-n}$$

where $\beta = \sqrt{D} - \lfloor \sqrt{D} \rfloor$. We also note that the value n (the number of values of $P_i = \lfloor \sqrt{D} \rfloor$ for $i \leq \lfloor \ell/2 \rfloor$) cannot exceed the number of divisors of $(D - \lfloor \sqrt{D} \rfloor^2)/4$. This is a fact because each Q_i associated with one of the P_i-values must be distinct from any other, they must be even, and they must each be a divisor of $(D - \lfloor \sqrt{D} \rfloor^2)/2$.

Now,

$$\varepsilon_\Delta < (\sqrt{\Delta})^{\lfloor(\ell+1)/2\rfloor - n}(2\sqrt{\Delta})/\beta)^n = (\sqrt{\Delta})^{(\ell+1)/2}(2/\beta)^n,$$

and

$$R < \lfloor(\ell+1)/2\rfloor \log \sqrt{\Delta} + n \log(2/\beta).$$

Hence, if $\Delta > 4.75 \times 10^9$, $n \log(2/\beta) < 52.6$, $\varepsilon = 0.045$ and $\ell \leq 24$, then

$$2h_\Delta > \Delta^{\frac{1}{2}-\varepsilon}(0.655\varepsilon)/(\lfloor(\ell+1)/2\rfloor \log \sqrt{\Delta} + n \log(2/\beta)) > 4.$$

If $n \log(2/\beta) \geq 52.6$, then, since $\ell \leq \lfloor(\ell+2)/4\rfloor \leq 6$, we have $-\log \beta > 8.0$, $\beta < 0.000335$ and $D - \lfloor \sqrt{D} \rfloor^2 < 2\sqrt{D}\beta < 46.2$. It follows that in this case, $(D - \lfloor \sqrt{D} \rfloor^2)/4 < 11$. However, the maximum value of the number of divisors of m for $1 \leq m \leq 11$ is 4. Therefore, $n \geq 4$. This now means that $-\log \beta > 12.4$ and and $D - \lfloor \sqrt{D} \rfloor^2 < 1$, which is impossible. □

Now we can prove Theorem 5.4.6.

Proof of Theorem 5.4.6. Given Lemmas 5.4.1–5.4.3, to find all $\Delta = \Delta_0 > 0$ with $\ell \leq 24$ and $h_\Delta = 2$ (with at most one exceptional value remaining), we need only examine those $D < 1.5 \times 10^9$, when $D \not\equiv 1 \pmod 4$, and those $D < 4.75 \times 10^9$, when $D \equiv 1 \pmod 4$.

A computer search was run on all numbers of these forms, up to the bounds given above, to find all values of D such that $\ell \leq 24$. Once this had been done, we used the following algorithm for eliminating most of the values of Δ for which $h_\Delta > 2$. Let

$$S = \{\text{primes } p < \sqrt{\Delta}/2 : (\Delta/p) \neq -1\}.$$

By Theorem 2.1.2, if $h_\Delta = 1$, then $S \subseteq \Lambda^*(\Delta)$. Furthermore, if a prime $p \in S$, but $p \notin \Lambda^*(\Delta)$, then we may form the ideal $I = [p, (b+\sqrt{D})/\sigma]$ where $0 \le b < p$ for a suitably chosen $b \in \mathbb{Z}$. Clearly then, if $h_\Delta = 2$, $S \subseteq \Lambda_l^*(\Delta) \cup \Lambda^*(\Delta)$.

The algorithm involves the determination of the primes in $\Lambda_l^*(\Delta)$ and $\Lambda^*(\Delta)$ for $\ell \le 24$. We then attempted to find a prime $q \in S$ with $q \notin \Lambda_l^*(\Delta) \cup \Lambda^*(\Delta)$. We can eliminate such Δ since $h_\Delta \ge 3$, by the above. The few remaining numbers were then treated more carefully for the value of h_Δ. In this final step, we eliminated all values except those for which $h_\Delta = 2$. □

Remark 5.4.7. We separately tabulated all of the ERD-types with $h_\Delta = 2$ in Table A9 of Appendix A as a complement to Theorem 5.4.3, and since it motivated Theorem 5.4.6.

Remark 5.4.8. As observed in Remark 5.4.6, there is nothing, beyond computational considerations, which prevents us from using the above techniques to tabulate class numbers up to any bound we like, and we are guaranteed by the GRH and Theorem 5.4.1 that the list is complete with one possible GRH ruled-out value. We conclude that, in fact, *any* list of class numbers can be so tabulated. We have done the job up to reasonable bounds for $h_\Delta = 1$ and 2. The reader so inclined can use these techniques to compile lists of class numbers $h_\Delta = 3, 4, \ldots$ for $\ell(w_\Delta) \le B$ where B is limited only by computational considerations. Thus, we may now consider this to be a routine task.

There is another place where the GRH (but unfortunately not Theorem 5.4.1) plays an interesting and decisive role. We close this section, and hence this chapter, with a consideration of quadratic non-residues and prime-producing polynomials which links us back to Chapter Four and its main topic.

In Chapter Four, we showed how the initial consecutive prime-producing capacity of the Euler–Rabinowitsch polynomial is linked to class number problems. We also saw, via Exercises 5.2.2–5.2.3, for example, how quadratic non-residues come into play.

In what follows, $F_{\Delta,q}(x)$ is the qth Euler–Rabinowitsch polynomial (see Definition 4.1.4).

Theorem 5.4.7. [(5.4.11)] *If $\Delta > 0$ is a discriminant with $\Delta \equiv 1 \pmod 4$, and $\Delta > 17$, then the following are equivalent:*

(1) *$|F_{\Delta,1}(x)|$ is 1 or prime for all integers x with $1 \le x < (\sqrt{\Delta} - 1)/2$.*

(2) *$(\Delta/p) = -1$ for all primes $p < \sqrt{\Delta - 1}/2$ and $\Delta \equiv 5 \pmod 8$.*

Proof. Exercise 5.4.5. □

Theorem 5.4.8. *If $\Delta > 0$ is a discriminant, $\Delta \equiv 0 \pmod 8$, with radicand $D \equiv 2 \pmod 4$, and $D \ne 2p^2$ for any prime p, then the following are equivalent:*

(1) *$|F_{\Delta,2}(x)|$ is prime or 1 for all integers x with $0 \le x \le (\sqrt{D} - 1)/2$.*

[(5.4.11)] Theorems 5.4.7–5.4.10 were all taken from [250] where this problem was first discussed. The form of $F_{\Delta,i}(x)$ for $i = 1, 2$ is slightly different in [250], but equivalent nevertheless.

(2) $(D/p) = -1$ *for all odd primes* $p < \sqrt{D}/2$.

Proof. Exercise 5.4.6. □

Theorem 5.4.9. *If* $\Delta > 0$ *is a discriminant with radicand* $D \equiv 3 \pmod 4$, *and* $D \neq 2p^2 + 1$ *for any primes p, then the following are equivalent:*

(1) $|F_{\Delta,2}(x)|$ *is* 1 *or prime for all non-negative* $x \leq \sqrt{D-1}/2$.

(2) $(D/p) = -1$ *for all odd primes* $p < \sqrt{D-2}/2$.

Proof. Exercise 5.4.7. □

The following tables illustrate Theorems 5.4.8–5.4.9.

Table 5.4.1.

D	$\lvert F_{\Delta,2}(x)\rvert = \lvert 2x^2 - D/2\rvert$ for $0 \leq x < \sqrt{D}/2$
6	3, 1
10	5, 3
14	7, 5
26	13, 11, 5
38	19, 17, 11, 1
62	31, 29, 23, 13
122	61, 59, 53, 43, 29, 11
362	181, 179, 173, 163, 149, 131, 109, 83, 53, 19
398	199, 197, 191, 181, 167, 149, 127, 101, 71, 37

Table 5.4.2

D	$\lvert F_{\Delta,2}(x)\rvert = \lvert 2x^2 + 2x + (1-D)/2\rvert$ for $0 \leq x \leq (\sqrt{D}-1)/2$
3	1
7	3
11	5, 1
15	7, 3
23	11, 7
35	17, 13, 5
47	23, 19, 11
83	41, 37, 29, 17, 1
143	71, 67, 59, 47, 31, 11
167	83, 79, 71, 59, 43, 23
227	113, 109, 101, 89, 73, 53, 29, 1

Now we show that Tables 5.4.1–5.4.2 tell the whole story for Theorems 5.4.8–5.4.9. For Theorem 5.4.7, see Exercise 5.4.8.

Theorem 5.4.10. *If the GRH holds, then Tables* 5.4.1–5.4.2 *contain all of the values for which Theorems* 5.4.8–5.4.9 *hold.*

Proof. Set $b = b(D) = \lfloor\sqrt{D/2}\rfloor$ if $D \equiv 2 \pmod 4$, and $b = \lfloor\sqrt{D-1}/2\rfloor$ if $D \equiv 3 \pmod 4$. Our problem is to find all $D \equiv 2$ or $3 \pmod 4$ such that

$$(D/p) = -1 \text{ for all odd primes } p < b(D). \tag{5.4.13}$$

First, we set $S(t) = \sum_{p<t}(D/p)$ where the sum is taken over all odd primes $p < t$. In order for (5.4.13) to hold, we must have

$$|S(b)| = \pi(b) - 1 - \varepsilon$$

where, as usual, $\pi(x)$ denotes the number of primes not exceeding x and $\varepsilon = 0$, unless b is prime, in which case $\varepsilon = 1$.

If we let $\chi(p) = (\Delta/p)$, then

$$A(t) = \sum_{p<t} \chi(p),$$

we get $A(t) = S(t)$. Also, by assuming the truth of the GRH, we can use [284, Theorem 3] to get

$$|A(t)| < B(t, \Delta),$$

where

$$B(t,\Delta) = \sqrt{t}\left(\left\{\frac{1}{\pi} + \frac{5.3}{\log t}\right\}\log\Delta + 2\left\{\frac{\log t}{2\pi} + 2\right\}\right).$$

From a result of Rosser and Schoenfeld [313], we have

$$\pi(t) - 1 - \varepsilon > (t/\log t) - 1 - \varepsilon = T(t)$$

for $t > 17$. If we put $t = b(D)$, it can be shown that for all $D > 10^{11}$, we have

$$B(b, 4D) < T(b).$$

Hence, if (5.4.13) holds, and $D > 10^{11}$, then

$$|S(b)| = |A(b)| < B(b, 4D) < T(b) < |S(b)|,$$

a contradiction. It follows that if condition (5.4.13) is satisfied by some D, then $D < 10^{11}$.

We need only deal now with the values of $D < 10^{11}$. We denote by $N_i(q)$ the least positive integer N such that $N \equiv i \pmod 4$ and $(N/p) = -1$ for all odd primes $p \leq q$. To compute the values of $N_i(q)$ for various values of q, we used a sieving device.(5.4.12)

Consider now the case of $D \equiv 3 \pmod 4$. If $D > 3363$, then $\sqrt{(D-1)}/2 > 29$. The least value of D such that (5.4.13) holds for all $p \leq 29$ is 13163. But if $D \geq 13163$, then $\sqrt{(d-1)}/2 > 57$; and this means that if (5.4.13) is satisfied, then $D \geq 331427$ and $(\sqrt{D}-1)/2 > 287$. Since $N_3(167) > 10^{12}$, then (5.4.13) can hold only for values of $D \leq 3363$. Similarly, when $D \equiv 2 \pmod 4$, the condition (5.4.13) can hold only for values of $D \leq 2738$.

A direct search of all values of $D \leq 3363$ revealed that only those values given in Tables 5.4.1 and 5.4.2 satisfy condition (5.4.13). □

Exercises 5.4

1. Let y be a real number. Prove
   (a) If $y > 1$, then $\sum 1/q < \log y$ where the sum ranges over all primes $q \leq y$.
   (b) If $y > 0$, then $\sum(1/q^2) < 1/y$ where the sum ranges over all primes $q > y$.

(5.4.12)UMSU (the University of Manitoba Siefving Unit) was built by Cam Patterson (see footnote (4.3.2)). UMSU is a shift register design similar to Dick Lehmer's SRS-181 (which was mistakenly sold as scrap before it was completed). UMSU and MSSU are related in the sense that, as Cam Patterson has put it, MSSU is "UMSU on a chip." UMSU has rings corresponding to the first 32 primes (up to 131), while MSSU only has 30 rings (up to 113). The sieving rate of UMSU is 133 million integers per second, whereas the sieving rate of a single MSSU chip has a rate of 192 million integers per second. UMSU has not been used for a number of years (see footnote (6.2.3)).

(c) If $|y| \leq 1/2$, then $|\log(1-y)+y| < y^2$.

2. Prove that the Kronecker symbol (Δ/a) is a character as defined in Appendix F.

3. Let $\Delta = \Delta_0 > 0$ be a fundamental discriminant, with regulator $R = \log(\varepsilon_\Delta)$, where Δ is of ERD-type. Prove that $R < \log 3D$ where D is the associated radicand.

4. Let p be a prime and $n \in \mathbf{Z}$ positive with $\gcd(n, p) = 1$, and let g be a primitive root[5.4.13] modulo p. We call g an *nth-Fibonacci primitive root modulo p*[5.4.14] (or simply nth-FPR $\pmod{p}$) if $g^2 \equiv g + n \pmod{p}$. For the balance of the question, assume that p is an odd prime and $g, n \in \mathbf{Z}$ positive, $\gcd(n,p) = 1$, and define the sequence $\{F_i(n)\}$ recursively as follows:

$$F_0(n) = 1,\ F_1(n) = g \in \mathbf{Z} \text{ positive and for } i > 1,\ F_i(n) = F_{i-1}(n) + nF_{i-2}(n).$$

$\{F_i(n)\}$ is called the *nth-Fibonacci sequence with base g* (or nth-FS base g).

(Thus $n = 1$ yields the ordinary Fibonacci primitive roots, and if $n = g = 1$, the ordinary Fibonacci numbers.) Prove the following:

(a) If g is an nth-FPR $\pmod{p}$, then $F_i(n) \equiv gF_{i-1}(n) \equiv g^i \pmod{p}$ for all $i > 0$. (*Hint*: Use induction on i.)

(b) If g is an nth-FPR $\pmod{p}$, then $((4n+1)/p) \not\equiv -1$ where $(\ /\ )$ is the Kronecker symbol.

(c) If $(-n/p) = -1$, then there exists at most one nth-FPR $\pmod{p}$. (*Hint*: Use the quadratic formula on $x^2 - x - n = 0$.)

(d) If $n = 6$ and $p = 7$, then $g_1 = 3$ and $g_2 = 5$ are 6th-FPR's $\pmod{7}$.

(e) If two nth-FPR's $\pmod{p}$ exist, say g_1 and g_2 with $0 < g_i < p$ for $i = 1, 2$, then $g_1 + g_2 = 1 + p$.

(f) Suppose that either $n = 1$ or $p > n > 2$ and $p = 1 + 2q$ where q is prime and n has order g modulo p. If g is an nth-FPR $\pmod{p}$, then g is a primitive root modulo p if and only if $g - 1$ is a primitive root modulo p. (*Hint*: Look at $g(g-1) \equiv n \pmod{p}$.)

(g) Suppose that either $n = 1$ or $p > n > 2$ and $((4n+1)/p) = 1$ where $p = 1 + 2q$ is a prime and q is an odd prime. If either $n = 1$ or n has order q modulo p, then p has an nth-FPR. (*Hint*: Look at the primitive roots modulo $p \equiv 3 \pmod{4}$.)

(h) Suppose that either $n = 1$ or $p > n > 2$ and $p = 1 + 2q$ where $q > 2$ is prime and n has order q modulo p. If g is an nth-FPR $\pmod{p}$, then $g - 1$ and $g - (n+1)$ are primitive roots $\pmod{p}$. (*Hint*: Use part (f).)

(i) g is an nth-FPR $\pmod{p}$ if and only if the nth-FS base g satisfies $F_{i+1}(n)F_{i-1}(n) \equiv F_i^2(n) \pmod{p}$ for some $i > 1$. (*Hint*: Prove that for all $i > 0$, $F_{i+1}(n)F_{i-1}(n) - F_i^2(n) = (-n)^{i-1}(g + n - g^2)$.)

[5.4.13] In other words $g^{p-1} \equiv 1 \pmod{p}$ but $g^i \not\equiv 1 \pmod{p}$ whenever $1 \leq i < p - 1$.

[5.4.14] This was introduced in [227] as a spinoff of an investigation of Conjecture 5.4.1. The case $n = 1$ yields the *ordinary Fibonacci primitive roots* introduced by Dan Shanks.

(j) Suppose that $n = 4p + 1 = m^2 + 4$ where p is prime and $m \in \mathbf{Z}$ is positive. If $s < \sqrt{p}$ is an odd prime, then $p \equiv t \pmod{s}$ for $0 \leq t < s$. If there exists an integer $u > 0$ such that $1 + 4t \equiv (2u-1)^2 \pmod{s}$, then $f(u) = -u^2 + u + p \equiv 0 \pmod{s}$ where $0 < u < s < \sqrt{p}$. Hence, $h_\Delta > 1$. (See Exercise 5.2.3.)[(5.4.15)]

5. Prove Theorem 5.4.7. (*Hint*: See Theorem 4.2.6 and Exercises 5.2.2–5.2.3. Although these are stated only for fundamental discriminants, the reader may extrapolate.)

6. Prove Theorem 5.4.8. (*Hint*: See Chapter Four, section two.)

7. Prove Theorem 5.4.9.

8. Use the GRH to prove that Table 4.2.3 contains all the values for which Theorem 5.4.7 holds.

9. With reference to Exercise 3.2.11,

   (a) Prove that if $n_\Delta \neq 0$, then there are only finitely many such Δ with $h_\Delta = 1$. (*Hint*: Use Theorem 5.4.1.)

   (b) Prove that the following are equivalent:

      (C_1) There exist infinitely many fundamental discriminants $\Delta > 0$ with $h_\Delta = 1$ (Gauss' conjecture.)

      (C_2) There exist infinitely many fundamental discriminants $\Delta > 0$ with $n_\Delta = 0$ and $h_\Delta = 1$.

      (C_3) For a given positive $n_0 \in \mathbf{Z}$, there exists at least one fundamental discriminant Δ with $h_\Delta = 1$ and $U \geq n_0$.

   (d) Prove that if $n_\Delta \geq \sqrt{D-1}/2$ where $D \equiv 1 \pmod{4}$ is a fundamental radicand, then $h_\Delta = 1$ if and only if D is of narrow RD-type. (*Hint*: See Theorem 5.2.2.)

(5.4.15) Based on the above, the following was posed in [227]: *Conjecture*: If $n = q^2$ where $q > 13$ is prime and $4q^2 + 1$ is prime, then there is an nth-FS base g, $\{F_i(n)\}$, and some g satisfying $F_{i+1}F_{i-1} \equiv F_i^2 \pmod{p}$ for some prime p with $0 < g < p < q$, (where $F_i = F_i(n)$).

Chapter 6

Ambiguous Ideals.

We look at ambiguous cycles of ideals in arbitrary quadratic orders. There is much in the literature which deals with this issue, but some of it is confusing or simply incorrect. We will cite these instances, clarify or correct them, and give a complete overview from the point of view of continued fractions and ideal theory. We begin with the real case where we introduce a tool for classifying those ambiguous cycles of ideals which have *no* ambiguous ideals in them.

6.1 Ambiguous Cycles in Real Orders: The Palindromic Index.

In Chapter One, section five, we dealt with the thorny issue of identifying the class group of an order in two distinct ways. We maintain the *general* definition of $C_\Delta = I(\Delta)/P(\Delta)$ given in Definition 1.5.2 rather than the more restrictive identification given in Exercise 1.5.5. The reasons for this are twofold. First, we don't need such a restrictive definition, since it is used to get identification with the class group of forms or with the maximal order (e.g. see Cox [67, Chapter 2, section 7]) which we are not considering. Secondly and more importantly, the class group of Exercise 1.5.5 masks a phenomenon which we want to investigate and classify, namely that of ambiguous *cycles* without ambiguous ideals. In other words, as we shall see, it is possible to have an ambiguous *cycle* of reduced ideals without any ambiguous ideals in it, without there being an ambiguous *class* in C_Δ without any ambiguous ideals in it. At this point, it's not clear what all this means, so let's begin.

Definition 6.1.1. Let $\Delta > 0$ be a discriminant with radicand D. If I is a reduced ideal in $\mathcal{O}_\Delta$, then I is said to be in an *ambiguous cycle* if $I_j = I'$ (where $I_j = [Q_{j-1}/\sigma, (P_{j-1} + \sqrt{D})/\sigma]$) for some $j \in \mathbb{Z}$ with $2 \leq j \leq \ell + 1$ where $\ell = \ell(I)$. (See Chapter Two.)

Of course, if $I = I'$ then $j = \ell + 1$ and I is itself an ambiguous ideal. However, it is possible to have an ambiguous cycle which has no ambiguous ideals in it. To study this phenomenon the following tool was introduced in [237].[(6.1.1)]

(6.1.1) In fact all the results in this section come from any one of [233], [237] or [239].

Definition 6.1.2. Let $\Delta > 0$ be a discriminant with radicand D and let

$$I = [Q/\sigma, (P+\sqrt{D})/\sigma] = I_1 = [Q_0/\sigma, (P_0+\sqrt{D})/\sigma]$$

be a reduced ideal in $\mathcal{O}_\Delta$ with

$$0 < (\sqrt{D} - P_0)/Q_0 < 1.$$

If I is in an ambiguous cycle, then $I' = I_{p+1}$ for some $p \in \mathbf{Z}$ with $1 \le p \le \ell(I) = \ell$. We call $p = p(I)$ the *palindromic index* of I. (Observe that for a given I as above, p is unique.)

As we have done with $\ell(I)$, we will write p for $p(I)$ whenever the context is clear. It follows from Exercise 2.1.13(f) that an ambiguous cycle of reduced ideals can have at most two ambiguous ideals in it. However, we want to classify those ambiguous cycles *without any* ambiguous ideals. Toward this end, we have need of the following useful technical lemma.

Lemma 6.1.1. *If I, ℓ and p are as in Definition* 6.1.2 *then $(I_{i+1})' = I_{p+1-i}$ for $0 \le i \le p$, and*

$$a_i = a_{p-i},\ Q_i = Q_{p-i} \text{ and } P_{i+1} = P_{p-i}.$$

Moreover, if $\ell \ge p+2$, then

$$(I_{i+1})' = I_{\ell+p+1-i} \text{ for } p+1 \le i \le \ell - 1,$$

and

$$a_i = a_{\ell+p-i},\ Q_i = Q_{\ell+p-i} \text{ and } P_i = P_{\ell+p-i+1}.$$

Proof. We have that $I = I_1 = [Q_0/\sigma, (P_0+\sqrt{D})/\sigma]$ with $0 < (\sqrt{D}-P_0)/Q_0 < 1$ and $I' = I_{p+1} = [Q_p/\sigma, (P_p+\sqrt{D})/\sigma] = [Q_0/\sigma, (P_0 - \sqrt{D})/\sigma] = [Q_0/\sigma, (-P_0 + \sqrt{D})/\sigma] = [Q_p/\sigma, (-P_0+\sqrt{D})/\sigma]$. Therefore, $P_p = yQ_0 - P_0$ for some $y \in \mathbf{Z}$. However, $P_p - yQ_0 + \sqrt{D} = \sqrt{D} - P_0$ so $0 < (P_p + \sqrt{D})/Q_p - y < 1$ which implies that $y < (P_p+\sqrt{D})/Q_p < y+1$, so $a_p = y$. Thus, $P_p = a_pQ_p - P_0$, i.e. $P_0 = P_{p+1}$. By Exercise 4.2.9 we have that $I' = (I_1)' = [Q_0/\sigma, (P_1+\sqrt{D})/\sigma]$, so $P_p = zQ_0 + P_1 = zQ_0 + a_0Q_0 - P_0$ for some $z \in \mathbf{Z}$. Since $P_p - zQ_0 - a_0Q_0 + \sqrt{D} = \sqrt{D} - P_0$, then

$$0 < (P_p + \sqrt{D})/Q_p - z - a_0 < 1,$$

so $a_p = z + a_0$, as above. Thus $P_p = (a_p - a_0)Q_0 + P_1$. However, by Exercise 4.2.9 again,

$$a_p = \lfloor (P_p+\sqrt{D})/Q_p \rfloor = \lfloor (P_{p+1}+\sqrt{D})/Q_p \rfloor = \lfloor (P_0+\sqrt{D})/Q_0 \rfloor = a_0.$$

Hence $P_p = P_1$. The reader may now verify that, by induction, we have $P_{i+1} = P_{p-i}$ for $0 \le i \le p$.

Since $Q_0Q_1 = D - P_1^2 = Q_pQ_{p-1} = Q_0Q_{p-1}$, then $Q_1 = Q_{p-1}$ and

$$a_{p-1} = \lfloor (P_{p-1}+\sqrt{D})/Q_{p-1} \rfloor = \lfloor (P_1+\sqrt{D})/Q_1 \rfloor = a_1.$$

The reader may now verify by induction that we have $a_i = a_{p-i}$ and $Q_i = Q_{p-i}$ for $0 \le i \le p$ as well. Also observe that $I_{p+1-i} = [Q_{p-i}/\sigma, (P_{p-i}+\sqrt{D})/\sigma] = [Q_i/\sigma, (P_{i+1}+\sqrt{D})/\sigma] = I'_{i+1}$ by Exercise 4.2.9. The last statement may be verified by the reader as an exercise, using the above techniques. □

Now we can give the previously promised criterion.

Theorem 6.1.1. *If $\Delta > 0$ is a discriminant and I is a reduced ideal in $\mathcal{O}_\Delta$, then I is in an ambiguous cycle without an ambiguous ideal if and only if $\ell(I)$ is even and $p(I)$ is odd.*

Proof. By Lemma 6.1.1 we have that $(I_{i+1})' = I_{i+1}$ for some i with $0 \le i < \ell = \ell(I)$ if and only if either $i+1 = p+1-i$ or $i+1 = \ell+p+1-i$, i.e. $p = 2i$ or $\ell + p = 2i$, where $p = p(I)$. □

Now we illustrate Theorem 6.1.1 with an eye to verifying a comment made at the outset of this section.

Example 6.1.1. Consider the ideal $I = [9, 15+\sqrt{306}]$ of Example 1.5.1, which is not invertible. Also look at the continued fraction expansion of $\gamma = (15+\sqrt{306})/9$ given in Example 2.1.8. We see that $p(I) = 5$ and $\ell(I) = 6$, so this is in an ambiguous cycle without an ambiguous reduced ideal in it by Theorem 6.1.1, and we can see why this is the case. Although $N(I) \mid \Delta$, this does not mean that I is ambiguous (see Exercise 1.2.5) as it *would* be if such a phenomenon were to occur in a *maximal* order. In fact, from Example 2.1.8 we obtain that $I = I_1 = [9, 15+\sqrt{306}] \sim I_2 = [18, 12+\sqrt{306}] \sim I_3 = [15, 6+\sqrt{306}] \sim I_4 = [15, 9+\sqrt{306}] = I_3' \sim I_5 = [18, 6+\sqrt{306}] = I_2' \sim I_6 = [9, 12+\sqrt{306}] = I_1'$. Observe that these equivalences are not in the class group since the ideals are not invertible.[(6.1.2)] In fact, $h_\Delta = 4$ and C_Δ is cyclic (Exercise 6.1.1). Consider the ideal $J = [5, 1+\sqrt{306}]$ and look at the simple continued fraction expansion of $(1+\sqrt{306})/5$:

i	0	1	2	3	4
P_i	1	14	8	14	16
Q_i	5	22	11	10	5
a_i	3	1	2	3	6

We see that $J = J_1 \sim J_2 = [22, 14+\sqrt{306}] \sim J_3 = [11, 8+\sqrt{306}] \sim J_4 = [5, 16+\sqrt{306}]$. Moreover, the reader may verify that these ideals are invertible so we are looking at equivalences in C_Δ. Here we see the distinction quite clearly marked between the reduced (hence primitive) ideals I_j and the strictly reduced (hence strictly primitive) ideals J_i. Furthermore, the reader may verify (Exercise 6.1.1) that $C_\Delta = \langle\{J\}\rangle$. Finally, in the simple continued fraction expansion of γ, observe

(6.1.2) Recall, from Exercise 1.5.3(b), that we cannot multiply non-invertible ideals in a given order, i.e. the multiplication formulae (1.2.1)–(1.2.5) fail. The reason for this is that we cannot expect multiplication of non-invertible ideals to be *closed* in the given order, in any well-defined manner. However, if one merely multiplies them as $\mathbb{Z}$-modules (see Butts and Pall [45]), then $I^2 = [81, 45+\sqrt{2754}]$ in the order of discriminant 11016, and $[81, 45+\sqrt{2754}] = (9)[9, 5+\sqrt{34}] = (9)(5+\sqrt{34})$ in the maximal order. However, this is misleading and masks the phenomenon which we are describing in this section, namely that *in the order*, I is in an *ambiguous cycle of ideals without any ambiguous reduced ideals in it*. In fact, the ideal $J = [3, 5+\sqrt{34}]$ in the maximal order, is in an ambiguous class of C_{136} without any ambiguous reduced ideals in it. Therefore, even viewing I as (3) J does not circumvent the issue. It merely translates it to another order. Ignoring such cycles can lead to major complications. For example, see Theorems 6.1.3–6.1.5 and footnote (E.5) of Appendix E. Furthermore, as explained in Appendix E, there do not exist any ambiguous classes of forms without ambiguous forms in them. Some view this as making form theory "simpler" (e.g. see Butts and Pall [45, p. 32]). However, in this author's opinion, the existence of such ideal cycles makes ideal theory "richer", as evidenced by the pleasant connections with structure and representation of a given discriminant displayed herein. Finally, as evidenced by footnote (E.1) of Appendix E, it is a matter of taste and opinion as to what is "simpler".

the palindromy about the palindromic index $\gamma = \langle 3; \overline{1,1,1,1,3,3}\rangle$, $a_i = a_{p-i}$ as predicted by Lemma 6.1.1. Observe, as well, in Example 6.1.1, that $p(I) = \ell(I) - 1$. This is no accident, as the following shows.

Lemma 6.1.2. *If $\Delta > 0$ is a discriminant and $I = [a, (b+\sqrt{\Delta})/2]$ is a reduced ideal in an ambiguous cycle with $0 < (\sqrt{\Delta} - b)/(2a) < 1$, then $\Delta = 4a^2 + b^2$ if and only if $I' = I_{\ell-1}$, i.e. $p = \ell - 1$.*

Proof. Exercise 6.1.2. □

Now that we have brought sums of squares into the picture, we present the following generalization of a well-known result (e.g. see Cohn [57]).

Lemma 6.1.3. *Let $\Delta > 0$ be a discriminant. In $\mathcal{O}_\Delta$ there exists an ambiguous cycle of reduced ideals containing no ambiguous ideal if and only if $N(\varepsilon_\Delta) = 1$ and D is a sum of two squares.*

Proof. Exercise 6.1.3. □

Now we need a definition which will lead us into a result which basically says that if we have one ambiguous cycle of reduced ideals without ambiguous ideals, then we can generate the maximum possible number of such cycles. For maximal orders this says that if we have one such class, then we may generate the class group entirely by such classes.

Definition 6.1.3. Let $\Delta > 0$ be a discriminant with radicand D, and let s_Δ be one half the excess of the number of divisors of D of the form $4j+1$ over those divisors of the form $4j+3$, i.e. $s_\Delta = \frac{1}{2}(n_1 - n_2)$ where n_1 is the number of divisors of D of the form $4j+1$ and n_2 is the number of divisors of D of the form $4j+3$. Thus, s_Δ corresponds to the number of *distinct* sums of squares $D = a^2 + b^2$ (where distinct here means that $1 \le a \le b$ are the only possibilities counted).

Theorem 6.1.2. *If $\Delta > 0$ is a discriminant and there exists an ambiguous cycle of reduced ideals without ambiguous reduced ideals, then there are s_Δ such cycles when $D = \sigma^2\Delta/4$ is even, and there are $s_\Delta/2$ of them when D is odd.*

Proof. Exercise 6.1.4. □

Remark 6.1.1. In some places in the literature(6.1.3) it is asserted that there can be *at most* one ambiguous class without an ambiguous ideal in it when considering the maximal order. Theorem 6.1.2 shows this to be false. We believe that this confusion arises from the following. Let $C_{\Delta,1}$ be the subgroup of C_Δ consisting of classes with ambiguous reduced ideals, and as usual, let $C_{\Delta,2}$ be the elementary abelian 2-subgroup of C_Δ, then $|C_{\Delta,2}|/|C_{\Delta,1}| = 2$ whenever there exists an ambiguous class without ambiguous ideals (Exercise 6.1.5). In fact, by Lemma 6.1.3 and Theorem 6.1.2, we see that there exists an ambiguous class without ambiguous ideals in C_Δ, for a maximal order, if and only if there exist $2^{t_\Delta - 1}$ of them (t_Δ as in Definition 1.3.3). To see this, we note that by Theorem 1.3.3, $|C_{\Delta,2}| = 2^{t_\Delta}$, i.e. there are

(6.1.3)For example see Cohn [57, Exercises 9, p. 190].

2^{t_Δ} pairwise inequivalent ambiguous classes of ideals. If one of these classes has *no* ambiguous reduced ideal in it, then by Lemma 6.1.3, D is a sum of two squares. By Exercise 1.1.4(c), $t_\Delta + 1$ must be the number of distinct prime divisors of Δ. Now observe that the subgroup $C_{\Delta,1}$ of C_Δ must have $t_\Delta - 1$ generators. To see this we note that, since $N(\varepsilon_\Delta) = 1$ by Lemma 6.1.3, then $\ell(w_\Delta)$ is even (by Theorem 2.1.3) and so by Exercise 2.1.13(f), each class of $C_{\Delta,1}$ has exactly two ambiguous ideals in it. If we take an ambiguous class $\{I\}$ without ambiguous reduced ideals and form its product with each of the aforementioned $t_\Delta - 1$ generators, then these new $t_\Delta - 1$ classes, together with $\{I\}$ itself, yield t_Δ classes which generate $C_{\Delta,2}$, where the zero product means just $\{I\}$ itself times the principal class when $t_\Delta = 1$. Each of these t_Δ classes has no ambiguous reduced ideal in it (observing that the product of an ambiguous class without ambiguous reduced ideals and that of an ambiguous class with ambiguous reduced ideals yields an ambiguous class without ambiguous reduced ideals (see Exercise 6.1.5)).

The above elucidation contains a subtle point missed in [57], namely that there are either *no* ambiguous classes without ambiguous reduced ideals (in which case $C_{\Delta,2} = C_{\Delta,1}$ is of order 2^{t_Δ}), or their number *coincides* with the number of ambiguous classes *with* ambiguous reduced ideals (in which case $|C_{\Delta,2}| = 2|C_{\Delta,1}| = 2^{t_\Delta}$). Moreover, as shown above, in the latter case $C_{\Delta,2}$ is actually generated by ambiguous classes without ambiguous reduced ideals. Thus we believe that what was meant in [57] was that $|C_{\Delta,2}|/|C_{\Delta,1}| = 2$ in this instance.

Given the problems delineated in Remark 6.1.1, we clarify the situation with the following criterion.

Theorem 6.1.3. *If $\Delta > 0$ is a discriminant, then the following are equivalent:*

(1) *$D = \sigma^2\Delta/4$ is a sum of two squares and $N(\varepsilon_\Delta) = 1$.*

(2) *There are s_Δ ambiguous cycles of ideals without ambiguous reduced ideals in $\mathcal{O}_\Delta$ when D is even, and there are $s_\Delta/2$ of them when D is odd, where s_Δ is given in Definition* 6.1.3.

(3) *There is an ambiguous cycle of ideals without ambiguous reduced ideals in $\mathcal{O}_\Delta$.*

Proof. The equivalence of (1) and (3) is Lemma 6.1.3. Theorem 6.1.2 tells us that (3) implies (2), and that (2) implies (3) is obvious. □

Now we look at some illustrations of the above. As an exercise, the reader may verify assertions made in the following examples.

Example 6.1.2. Let $\Delta = 2^3 \cdot 13^2 \cdot 5^2 \cdot 17 = 574600$ with $D = 2 \cdot 13^2 \cdot 5^2 \cdot 17 = 143650$. Here $\sigma = \sigma_0 = g = 1$ and $D_0 = 34$ with $f_\Delta = 65 = 5 \cdot 13$. D is representable as a sum of squares in nine distinct ways, namely

$$\begin{aligned} D &= 379^2 + 3^2 = 363^2 + 109^2 = 333^2 + 181^2 \\ &= 267^2 + 269^2 = 195^2 + 325^2 = 377^2 + 39^2 \\ &= 375^2 + 55^2 = 143^2 + 351^2 = 305^2 + 225^2. \end{aligned}$$

However, the last five of these arise from 13 and 5 and they do not give rise to strictly primitive ideals. For instance, $I = [195, 325 + \sqrt{143650}]$ is not invertible and the simple continued fraction expansion of $(325+\sqrt{143650})/195$ is

i	0	1	2	3	4	5	6
P_i	325	260	130	195	130	260	325
Q_i	195	390	325	325	390	195	195
a_i	3	1	1	1	1	3	3

This is an ambiguous cycle with $p(I) = 5 = \ell(I) - 1$, so there are no ambiguous reduced ideals in it.

The first four representations of D as a sum of two squares do represent classes in C_Δ, namely

$$J = [3, 379 + \sqrt{D}) \sim [363, 109 + \sqrt{D}],$$

$$L = [181, 33 + \sqrt{D}] \sim [267, 269 + \sqrt{D}],$$

$$M = [379, 3 + \sqrt{D}] \sim [109, 363 + \sqrt{D}],$$

$$N = [333, 181 + \sqrt{D}] \sim [269, 267 + \sqrt{D}].$$

Moreover $C_{\Delta,2} = \langle\{J\}\rangle \times \langle\{L\}\rangle \times \langle\{M\}\rangle \times \langle\{N\}\rangle$ generated by ambiguous classes without ambiguous reduced ideals. (In fact, the reader may verify that $h_\Delta = 112$ using footnote (1.5.9).)

Example 6.1.3. Let $\Delta = D_0 = 45305 = 5 \cdot 13 \cdot 17 \cdot 41$ and let $I = [106, (19 + \sqrt{45305})/2]$. The reason for choosing this ideal is that $\Delta = 19^2 + 4 \cdot 106^2$, so I arises from this representation as a sum of two squares. The simple continued fraction expansion of $(19+\sqrt{45305})/212$ is

i	0	1	2	3	4	5	6	7	8	9	10	11	12
P_i	19	193	187	85	195	169	153	187	117	91	87	125	155
Q_i	212	38	272	140	52	322	68	152	208	178	212	140	152
a_i	1	10	1	2	7	1	5	2	1	1	1	2	2

13	14	15	16	17	18	19	20	21	22	23	24	25	26
149	155	125	87	91	117	187	153	169	195	85	187	193	19
152	140	212	178	208	152	68	322	52	140	272	38	212	212
2	2	1	1	1	2	5	1	7	2	1	10	1	1

Here we see, as predicted by Lemma 6.1.2, $p(I) = 25 = \ell(I) - 1$ so by Theorem 6.1.1 this is an ambiguous class having no ambiguous reduced ideals in it. Observe as well that I generates the representation $\Delta = 149^2 + 4 \cdot 76^2$, since $P_{13} = 149$ and $Q_{12} = Q_{13} = 152$ (as predicted in Lemma 6.1.1). A note of caution is that although $Q_{15} = 212$, this does not give rise to I' since there is no symmetry about $i = 15$, which Lemma 6.1.1 tells us would have to happen. Now we consider $J = [14, (211 + \sqrt{45305})/2]$ which arises from $\Delta = 211^2 + 4 \cdot 14^2$. The continued fraction expansion of $(211 + \sqrt{45305})/28$ is

i	0	1	2	3	4	5	6	7	8	9	10
P_i	211	209	197	139	93	65	195	197	35	155	181
Q_i	28	58	112	232	158	260	28	232	190	112	112
a_i	15	7	3	1	1	1	14	1	1	3	3

11	12	13	14	15	16	17	18	19	20
155	35	197	195	65	93	139	197	209	211
190	232	28	260	158	232	112	58	28	28
1	1	14	1	1	1	3	7	15	15

We see that J also generates $\Delta = 181^2+4\cdot 56^2$, since $P_{10} = 181$ and $Q_{10} = Q_9 = 112$. Here $p(J) = 19 = \ell(J)-1$, so J is an ambiguous class without an ambiguous reduced ideal. Note, as above, that although $Q_6 = 28$, this does not represent the conjugate.

Finally, we let $L = [62, (173+\sqrt{D})/2]$, which arises from $\Delta = 173^2 + 4\cdot 62^2$. The continued fraction expansion of $(173+\sqrt{\Delta})/124$ is

i	0	1	2	3	4	5	6	7	8	9	10	11
P_i	173	199	169	195	205	123	61	165	155	111	137	77
Q_i	124	46	364	20	164	184	226	80	266	124	214	184
a_i	3	8	1	20	2	1	1	4	1	2	1	1

12	13	14	15	16	17	18	19	20	21	22	23	24
107	77	137	111	155	165	61	123	205	195	169	199	173
184	214	124	266	80	226	184	164	20	364	46	124	124
1	1	2	1	4	1	1	2	20	1	8	3	3

Thus, $p(L) = 23 = \ell(L) - 1$ and again, as predicted, we have an ambiguous class without an ambiguous reduced ideal. Moreover, the other sum of two squares which L generates is $\Delta = 107^2 + 4\cdot 92^2$ since $P_{12} = 107$ and $Q_{12} = Q_{11} = 184$. As above, although $Q_9 = 124$, this does not represent the conjugate due to lack of palindromy.

We note that there is one more pair of representations of Δ as a sum of squares (since there are $2^{t_\Delta - 1} = 8$ such representation), namely $\Delta = 83^2 + 4\cdot 98^2 = 203^2 + 4\cdot 32^2$. The ideal which arises from this pair is $M = [98, (83+\sqrt{45305})/2]$. However, $M \sim IJK$, whereas there are no such relationships between I, J and L, so $C_\Delta = \langle\{I\}\rangle \times \langle\{J\}\rangle \times \langle\{L\}\rangle$, a class group generated by ambiguous classes without ambiguous ideals.

Given the striking symmetry properties about the palindromic index, we are led to ask what analogue of Exercise 2.1.13(f) holds for ambiguous classes in general. The answer is

Theorem 6.1.4. *If $\Delta > 0$ is a discriminant and $I = [a, (b+\sqrt{\Delta})/2]$ with $0 < (\sqrt{\Delta} - b)/(2q) < 1$ is a reduced ideal in an ambiguous cycle of $\mathcal{O}_\Delta$, then $I = I'$ if and only if one of the following holds:*

(1) *$p = \ell$ and $i = 0$ or ℓ.*

(2) *p is even and $i = p/2$.*

(3) *p and ℓ have the same parity and $i = (p+\ell)/2$.*

Proof. This is an exercise for the reader to apply Lemma 6.1.1. □

Remark 6.1.2. We reaffirm here what we said in Example 6.1.1, in light of Theorem 6.1.4, namely that $Q_i/\sigma \mid \Delta$ does not necessarily imply that Q_i/σ is the norm of an ambiguous ideal. The conductor plays havoc with that possibility, as we have seen. Furthermore, as observed earlier, some treatments of this material avoid such problems by identifying C_Δ with the group in Exercise 1.5.5 where norms are prime to the conductor. However, we now see that in doing this, we miss all of the above theory which exposes the palindromy as well as representations of cycles of reduced ideals by sums of two squares. This is the perfect complement to the theory exposed in Chapter Four where we saw how to represent the class group of a complex quadratic order by *differences of squares*.

All of the above gives rise to the necessity for

Definition 6.1.4. If $\gamma = (P + \sqrt{D})/Q$ is a reduced quadratic irrational, then γ is said to have *pure symmetric period* if $\gamma = \langle \overline{a_0; a_1, \ldots, a_{\ell-1}} \rangle$ where the word $a_0 a_1 \ldots a_{\ell-1}$ is a *palindrome*.(6.1.4)

Lemma 6.1.2 gave us a criterion for looking at the discriminant as a sum of two squares in terms of the penultimate ideal in the cycle, and Theorem 6.1.3 gave us criteria for determining when we have an ambiguous cycle without ambiguous ideals. Furthermore, as illustrated by Examples 6.1.1–6.1.3, we can always arrange for the palindromic index to be the penultimate index in the cycle, when considering ambiguous cycles without ambiguous ideals. This leads us to the following.

Lemma 6.1.4. *If $\Delta > 0$ is a discriminant and $\mathcal{C}$ denotes an ambiguous cycle of reduced ideals in $\mathcal{O}_\Delta$, then the following are equivalent:*

(1) *$\mathcal{C}$ has at most one ambiguous ideal in it.*

(2) *There exists a reduced ideal I in $\mathcal{C}$ with $I' = I_{\ell-1}$ where $\ell = \ell(\mathcal{C}) = \ell(I)$.*

Proof. If (2) holds and there are two ambiguous ideals in $\mathcal{C}$, then by Theorems 6.1.1 and 6.1.4, ℓ is forced to be even. However, $p(I) = \ell - 1$ contradicting Theorem 6.1.1.

If (1) holds and there are no ambiguous ideals in $\mathcal{C}$, then we have the result by Lemmas 6.1.2–6.1.3. If there is exactly one ambiguous ideal $I = [Q_0/\sigma, (P_0 + \sqrt{D})/\sigma]$ in $\mathcal{C}$, then ℓ must be odd by Theorems 6.1.1 and 6.1.4. Thus, in the continued fraction expansion of $(P_0 + \sqrt{D})/Q_0$, we have $Q_{\frac{\ell+1}{2}} = Q_{\frac{\ell-1}{2}}$ by Lemma 6.1.2. Thus, by (2.1.5), $D = P^2_{\frac{\ell+1}{2}} + Q^2_{\frac{\ell+1}{2}}$. The result now follows from Lemma 6.1.2. □

Now we have the means to classify those ideals with pure symmetric period.

Theorem 6.1.5. *If $\mathcal{C}$ is a cycle of reduced ideals in $\mathcal{O}_\Delta$ where $\Delta > 0$ is a discriminant, then the following are equivalent:*

(1) *There exists a reduced quadratic irrational γ with pure symmetric period such that $[\gamma] \in \mathcal{C}$.*

(2) *There exists a reduced quadratic irrational γ such that $\gamma\gamma' = -1$ and $[\gamma] \in \mathcal{C}$.*

(3) *$\mathcal{C}$ is an ambiguous class containing at most one ambiguous ideal.*

Proof. If $\gamma = \langle \overline{a_0; a_1, \ldots, a_{\ell-1}} \rangle$ then by Corollary 2.1.1, $-1/\gamma' = \langle \overline{a_{\ell-1}; a_{\ell-2}, \ldots, a_0} \rangle$ whence we have the equivalence of (1) and (2). If (2) holds, then let

$$I = [\gamma] = [Q_0/\sigma, (P_0 + \sqrt{D})/\sigma].$$

(6.1.4) A palindrome is "never odd or even", it is "a toyota". Of course the classical palindrome is "able was I ere I saw elba".

Thus,

$$D = P^2 + Q^2 = P_0^2 + Q_0^2 = P_\ell^2 + Q_\ell^2$$

in the simple continued fraction expansion of $(P+\sqrt{D})/Q$. However, $D = P_\ell^2 + Q_\ell Q_{\ell-1}$ so $Q_{\ell-1} = Q_\ell = Q_0$. Moreover, by Exercise 4.2.9,

$$a_{\ell-1} = \lfloor (P_{\ell-1}+\sqrt{D})/Q_{\ell-1} \rfloor = \lfloor (P_\ell+\sqrt{D})/Q_{\ell-1} \rfloor = \lfloor (P_0+\sqrt{D})/Q_0 \rfloor = a_0,$$

so $P_\ell = a_{\ell-1}Q_{\ell-1} - P_{\ell-1} = a_0Q_0 - P_{\ell-1} = P_0$, and so $P_1 = P_{\ell-1}$. Since Exercise 4.2.9 also tells us that $I' = [Q_0/\sigma, (P_1+\sqrt{D})/\sigma]$, then

$$I' = [Q_{\ell-1}/\sigma, (P_{\ell-1}+\sqrt{D})/\sigma] = I_{\ell-1},$$

so $I \sim I'$, and I is in an ambiguous class. By Lemma 6.1.4, we have established that (2) implies (3).

If (3) holds, then by Lemma 6.1.4 there exists a reduced ideal $I \in C$ with $I' = I_{\ell-1}$. Set $I = [Q_0/\sigma, (P_0+\sqrt{D})/\sigma]$. Thus, $I' = [Q_{\ell-1}/\sigma, (P_{\ell-1}+\sqrt{D})/\sigma]$, which implies that $I = (I')' = [Q_{\ell-1}/\sigma, (P_\ell+\sqrt{D})/\sigma]$, by Exercise 4.2.9 again. However, by Lemma 6.1.1, $P_\ell = P_0$ and $Q_{\ell-1} = Q_0$, so $D = P_0^2 + Q_0^2$ by (2.1.5). Setting $\gamma = (P_0+\sqrt{D})/Q$ yields (2). □

Remark 6.1.3. In the case where an ambiguous class contains two ambiguous reduced ideals (excluded by Theorem 6.1.4), we 'just miss' having pure symmetric period, i.e. if $[\gamma] = [Q/\sigma, (P+\sqrt{D})/\sigma]$ is in an ambiguous class having two ambiguous ideals, then $\ell(\gamma)$ is even and $\gamma = \langle \overline{a_0; a_1, \ldots, a_{\ell-1}} \rangle$ where $a_1a_2\ldots a_{\ell-1}$ is a palindrome, but $a_0a_1\ldots a_{\ell-1}$ is not. For instance, we have

Example 6.1.4. If $D = 385 = 5 \cdot 7 \cdot 11$, and $[\gamma] = [7, (7+\sqrt{385})/2]$, then $\gamma = \langle \overline{1; 1, 9, 6, 2, 3, 2, 6, 9, 1} \rangle$. Here $\ell(\gamma) = 10$ and $[\gamma]$ is ambiguous and equivalent to $J = [5, (15+\sqrt{385})/2]$, the other ambiguous ideal in the class.

Thus, we have shown that symmetry is ultimately tied to ambiguity.

As illustrations of Theorem 6.1.5, we have

Example 6.1.5. If $D = 145 = 5 \cdot 29$, then letting $[\gamma] = [4, (9+\sqrt{145})/2]$ we get $\gamma = \langle \overline{2; 1, 1, 1, 2} \rangle$ and the class of $[\gamma]$ contains exactly one ambiguous reduced ideal, namely $\left[\gamma_{\frac{\ell-1}{2}}\right] = [5, (5+\sqrt{145})/2]$. Here $\ell(\gamma) = 5$ and $[\gamma]' = [\gamma_{\ell-1}] = [4, (7+\sqrt{145})/2]$. Also, $N(\varepsilon_\Delta) = N((9+\sqrt{145})/8) = -1$.

Example 6.1.6. If $D = 221 = 13 \cdot 17$, and $[\gamma] = [5, (11+\sqrt{221})/2]$, then $\gamma = \langle \overline{2; 1, 1, 2} \rangle$ and the class of $[\gamma]$ contains no ambiguous reduced ideals. Also, note that $N((11+\sqrt{221})/10) = -1$, and compare with Exercise 1.3.7(c).

Observe that D is a sum of two integer squares if and only if there is an element $\beta \in K$ with $N(\beta) = -1$. Of course, if $N(\varepsilon_\Delta) = 1$, then $\beta \notin \mathcal{O}_\Delta$ (see footnote E.5). Thus, if $N(\varepsilon_\Delta) = 1$ but -1 is a norm from K, then each ambiguous class of C_Δ has two ideals corresponding to elements of norm -1. For instance, in Example 6.1.6, both $[\gamma]$ and $[\gamma']$ correspond to elements of norm -1, i.e. $\gamma = (11+\sqrt{221})/10$ and $\gamma' = (11-\sqrt{221})/10$. Unfortunately, this does not help us to solve the open problem: When is -1 the norm of a unit? (see footnote (1.1.6)).

Now we are in a position to give a general criterion for the description of the class group of a maximal order in terms of continued fractions. A consequence of this result will be a class number one criterion, which will bring us back to a link with Chapter Five, section two. Our device will be the palindromic index.

Definition 6.1.5. Let $\Delta = \Delta_0 > 0$ be a fundamental discriminant and set $I = [a, (b + \sqrt{\Delta})/2]$, a reduced ideal in $\mathcal{O}_\Delta$. In the following, we are considering the continued fraction expansion of $(b + \sqrt{\Delta})/(2a)$. Let

$$T(I) = \{j : 1 \le j \le \ell(I) = \ell,\ Q_j/\sigma < \sqrt{\Delta}/2 \text{ and } Q_j/\sigma \text{ is prime counted } \textit{without} \text{ multiplicity } \},$$

$$R(I) = \{j : 1 \le j \le \ell \text{ and one of } j = \ell = p(I) = p, j = p/2 \text{ or } j = (p+\ell)/2 \text{ holds}\}.$$

Remark 6.1.4. We observe that $R(I)$ is empty, unless I is an ambiguous class. Moreover, we need only concern ourselves with multiplicity in $T(I)$ when I is in an ambiguous class, via Theorem 6.1.4.

We now link back to Chapter Five, via Theorem 5.1.1, in the following result from [238]. In what follows, $T_i = T(I_i)$, $R_i = R(I_i)$ and $a_{(i,j)} = \lfloor (P_j + \sqrt{D})/Q_j \rfloor$, in the simple continued fraction expansion arising from I_i.

Theorem 6.1.6. *If $\Delta = \Delta_0 > 0$ is a fundamental discriminant, then $C_\Delta = \bigcup_{i=1}^{n} \langle \{J_i\} \rangle$ for primitive ideals J_i if and only if there exists a reduced ideal $I_i \in \{J_i\}$ for each i with $1 \le i \le n$ such that*

$$\sum_{p<\sqrt{\Delta}/2} d(p, F) = \sum_{i=1}^{n} \sum_{j \in T_i - R_i} (a_{(i,j)} + 1), \tag{6.1.1}$$

where the left hand sum ranges over all unramified primes, and

$$\sum_{p<\sqrt{\Delta}/2} d(p, F) = \sum_{i=1}^{n} \sum_{j \in R_i \cap T_i} \lfloor (a_{(i,j)} + 1)/2 \rfloor, \tag{6.1.2}$$

where the left hand sum ranges over all ramified primes.

Proof. By Theorem 5.1.1 and Theorem 6.1.4, the right hand sides of (6.1.1)–(6.1.2) actually represent $\sum d(p, F)$ where the sum ranges over only those $p < \sqrt{\Delta}/2$ which appear in $\Lambda^*_{I_i}(\Delta)$ (see Definition 5.3.1). Therefore, the equality fails to hold precisely when some prime $p < \sqrt{\Delta}/2$ does not appear in the right but *does* (as it *must* by Lemma 4.1.2) appear on the left. The result is now a consequence of Theorem 1.3.2. □

Here is an illustration.

Example 6.1.7. If $\Delta = 2^3 \cdot 11 \cdot 23 = 2024$, then $F_\Delta(x) = x^2 - 506$ and $\frac{x}{|F_\Delta(x)|}$:

1	2	3	4	5	6	7	8	9	10
$5 \cdot 101$	$2 \cdot 251$	$7 \cdot 71$	$2 \cdot 5 \cdot 7^2$	$13 \cdot 37$	$2 \cdot 5 \cdot 47$	457	$2 \cdot 13 \cdot 17$	$5^2 \cdot 17$	$2 \cdot 7 \cdot 29$

11	12	13	14	15	16	17	18	19	20	21	22
$5 \cdot 7 \cdot 11$	$2 \cdot 181$	337	$2 \cdot 5 \cdot 31$	281	$2 \cdot 5^3$	$7 \cdot 31$	$2 \cdot 7 \cdot 13$	$5 \cdot 29$	$2 \cdot 53$	$5 \cdot 13$	$2 \cdot 11$

Thus, observing that the non-inert primes less than $\sqrt{\Delta}/2$ are $2, 5, 7, 13$ and 17, we have $d(2, F) = 11$, $d(5, F) = 9$, $d(7, F) = 6$, $d(11, F) = 2$, $d(13, F) = 4$ and $d(17, F) = 2$.

Now consider the ideals $I_1 = [1, \omega_\Delta]$, $I_2 = [2, \sqrt{506}]$, $I_3 = [5, 1 + \sqrt{506}]$, and $I_4 = [7, 17 + \sqrt{506}]$.

The continued fraction expansion of ω_Δ is

i	0	1	2
P_i	0	22	22
Q_i	1	22	1
a_i	22	2	22

The continued fraction expansion of $\sqrt{506}/2$ is

i	0	1	2
P_i	0	22	22
Q_i	2	11	2
a_i	11	4	22

The continued fraction expansion of $(1 + \sqrt{506})/5$ is

i	0	1	2	3	4
P_i	1	19	10	18	21
Q_i	5	29	14	13	5
a_i	4	1	2	3	8

and, the continued fraction expansion of $(17 + \sqrt{506})/7$ is

i	0	1	2	3	4	5	6
P_i	17	18	8	9	16	14	17
Q_i	7	26	17	25	10	31	7
a_i	5	1	1	1	3	1	5

Hence, we have that $T_1 = \emptyset$, $T_2 = \{1, 2\}$, $T_3 = \{3, 4\}$ and $T_4 = \{2, 6\}$. Also, $R_1 = \{1, 2\} = R_2$, whereas $R_3 = \emptyset = R_4$. Therefore,

$$\sum_{i=1}^{4} \sum_{j \in T_i - R_i} (a_{(i,j)} + 1) = \sum_{i=3}^{4} \sum_{j \in T_i} (a_{(i,j)} + 1) = 4 + 9 + 2 + 6 = 21,$$

and

$$\sum_{i=1}^{4} \sum_{j \in R_i \cap T_i} \lfloor (a_{(i,j)} + 1)/2 \rfloor = \sum_{j \in T_2} \lfloor (a_{(2,j)} + 1)/2 \rfloor = 2 + 11 = 13.$$

On the other hand, for unramified p we have

$$\sum_{p < \sqrt{\Delta}/2} d(p, F) = d(5, F) + d(7, F) + d(13, F) + d(17, F) = 21,$$

and for ramified p we have

$$\sum_{p < \sqrt{\Delta}/2} d(p, F) = d(2, F) + d(11, F) = 13.$$

Hence, by Theorem 6.1.6 we have that $C_\Delta = \bigcup_{i=1}^{4} \langle \{I_i\} \rangle$. By Exercise 6.1.7, we have that $C_\Delta = \langle \{I_2\} \rangle \times \langle \{I_4\} \rangle = \langle \{I_3\} \rangle$ with $h_\Delta = 6$.

Observe that Theorem 5.2.6 and Exercise 5.1.3 are immediate consequences of Theorem 6.1.6.

Remark 6.1.5. If we compare Exercise 5.1.3 with Theorem 5.2.6, we see (as noted by Dubois and Levesque in [76]) that Lu's $c + \sum_{i=1}^{\ell} a_i$ is precisely the number of all those ideals $[Q, (P + \sqrt{\Delta})/2]$ such that $P^2 + 4QQ' = \Delta$, where Q appears as some $Q_i \in \Lambda^*(\Delta)$. In Exercise 5.1.3, our $\sum_{j=1}^{\lfloor \ell/2 \rfloor} \beta_j$ is precisely the number of ideals (counted without multiplicity) $[Q, (P + \sqrt{\Delta})/2]$, where $P^2 + 4QQ' = \Delta$ with $Q \in \Lambda^*(\Delta)$ being *prime*. Thus, Exercise 5.1.3 is a class number one criterion, which is easier to implement and closer to the kernel of the truth of the matter, in view of Theorem 1.3.1. For instance, if we look back at $\Delta = 4 \cdot 94$ in Example 5.2.4, and consider $I = [5, (P + \sqrt{\Delta})/2]$, we get that $P^2 + 4 \cdot 5 \cdot Q' = \Delta$ precisely when $2x = P = 4, 6, 14, 16$. Thus, there are exactly four ideals of type I, namely $[5, 2 + \sqrt{94}] = [5, 7 + \sqrt{94}] \sim [5, 8 + \sqrt{94}] = [5, 3 + \sqrt{94}]$. Moreover, these account for exactly those four values of x : 2, 3, 7, and 8 for which 4 divides $|F_\Delta(x)|$ when $1 \le x \le \lfloor w_\Delta \rfloor$, i.e. $d(5, F) = 4 = a_3 + 1$. The reader may work out the balance of this example via Exercise 5.1.3 and Theorem 5.2.6 for comparison.

Exercises 6.1

1. In Example 6.1.1, verify that $h_\Delta = 4$ and C_Δ is cyclic. (*Hint*: Use the formula in footnote (1.5.9) to get $h_\Delta = 4$. To get that C_Δ is cyclic, look at the continued fraction expansion of $\sqrt{306}/2$ and verify that $J^2 \sim H = [2, \sqrt{306}] \not\sim 1$.)

2. Prove Lemma 6.1.2. (*Hint*: If $\Delta = 4a^2 + b^2$, verify that $\sigma^2\Delta/4 = Q_\ell Q_{\ell-1} + P_\ell^2$. Conversely, if $I' = I_{\ell-1}$, then use the fact that $Q_{\ell-1} = Q_0 = Q_\ell$ and (2.1.5).)

3. Prove Lemma 6.1.3. (*Hint*: Use Exercise 2.1.13(d) to prove that D is a sum of two squares. To prove the converse, use Theorem 2.1.3, Lemma 6.1.2 and Theorem 6.1.1.)

4. Prove Theorem 6.1.2. (*Hint*: By Lemma 6.1.3, $s_\Delta > 0$, and by Lemma 6.1.2 choose $I' = I_{\ell-1}$ so $D = P_\ell^2 + Q_\ell^2 = P_{(p+1)/2}^2 + Q_{(p+1)/2}^2$. Analyze $[Q_\ell/\sigma, (P_\ell + \sqrt{D})/\sigma] \sim [Q_{(p+1)/2}/\sigma, (P_{(p+1)/2} + \sqrt{D})/\sigma]$ for each case in terms of sums of two squares.)

5. Prove that $|C_{\Delta,2}|/|C_{\Delta,1}| = 2$, whenever there is an ambiguous class without ambiguous ideals (see Remark 6.1.1). Also verify, for $\Delta > 0$, any ambiguous class containing an ambiguous ideal, contains an ambiguous reduced ideal.

6. Let $\Delta = D = 45305$. Find two inequivalent reduced ambiguous ideals. Show that there do not exist three pairwise inequivalent reduced ambiguous ideals.

7. In Example 6.1.7, verify that $C_\Delta = \langle\{I_2\}\rangle \times \langle\{I_4\}\rangle = \langle\{I_3\}\rangle$. (*Hint*: Prove that $I_3^2 \in \{I_4\}$ and $I_3^3 \in \{I_2\}$.)

8. With reference to Chapter Three, section five, we were looking at solutions to $x^2 - Dy^2 = -3$ for square-free $D > 0$. We remarked that, although we only

considered the principal cycle therein, ambiguous cycles contained signals. Prove the following: If $x^2 - Dy^2 = -3$ has $s > 0$ primitive solutions and D is divisible by exactly ν distinct primes congruent to 1 modulo 3, then:

(i) if $3 \mid D$, then $s = 1$, and there are $2^{\nu+1}$ ambiguous ideals and hence 2^{ν} cycles containing an ambiguous ideal. Moreover, there are just $2^{\nu-1}$ essentially distinct representations of D with discriminant -3, but each gives rise to signals of two kinds. Thus, there are 2^{ν} conjugate pairs of distinct signals, one conjugate pair for each cycle containing an ambiguous ideal.

(ii) Henceforth, assume that $D \not\equiv 0 \pmod 3$. If $D \equiv -1 \pmod 4$, then $s = 2$, and there are two ambiguous ideals per cycle containing an ambiguous ideal, and two conjugate pairs of signals are permitted in each such cycle. There are in fact 2^{ν} ambiguous ideals in all, and hence $2^{\nu-1}$ cycles containing an ambiguous ideal. There are just $2^{\nu-1}$ essentially distinct representations of D with discriminant -3, but each gives rise to signals of the two kinds. Thus, there are 2^{ν} conjugate pairs of distinct signals, two conjugate pairs in each cycle containing an ambiguous ideal.

(iii) If $D \equiv 1 \pmod 4$, then $s = 1$ or 2 depending on $N(\varepsilon_\Delta) = \pm 1$, and there are just $2^{\nu-1}$ ambiguous ideals of odd norm, and as many as $2^{\nu-1}$ essentially distinct representations of D with discriminant -3. Thus, there is a signal of the second kind, to wit, a signal arising from a representation $D = P^2 + PQ + Q^2$ with Q odd, for each ambiguous ideal of odd norm. There are exactly s conjugate pairs of signals of the second kind in each ideal cycle containing ambiguous ideals of odd norm. Similarly, there are exactly s conjugate pairs of signals of the first kind in each ideal cycle containing ambiguous ideals of even norm.

.2 Exponent Two.

In this section, we concentrate upon maximal quadratic orders of positive or negative discriminant whose class groups have exponent two. In the preceding section, we looked at the influence of the infrastructure on generation of ambiguous ideals in real quadratic orders. Also, in Chapter Four we classified all maximal complex quadratic orders with class groups of exponent 2. In Chapter Four we looked in detail at those fundamental discriminants $\Delta = \Delta_0$ for which $|F_{\Delta,q}(x)|$ is 1 or prime for all non-negative $x < (M_\Delta - 1)/2$ where q is a positive square-free divisor of Δ. However, one can look at a slightly different bound when $\Delta > 0$. Just as we improved upon the Minkowski bound to the Gauss bound for $\Delta < 0$ (see footnote (1.3.2)), we may also improve upon $M_\Delta = \sqrt{\Delta}/2$ to $\overline{M}_\Delta = \sqrt{\Delta/5}$ (e.g. see Cohn [57, Theorem 11, p. 141] and Hendy [118, p. 268]). Thus we may replace M_Δ by $\overline{M}_\Delta$ in Theorem 1.3.1. We may then look at the primality of $|F_{\Delta,q}(x)|$ up to $x < (\overline{M}_\Delta - 1)/2$. If q is a prime divisor of Δ with gcd $(f_\Delta, q) = 1$, and $|F_{\Delta,q}(x)|$ is 1 or prime for all non-negative $x \leq (\overline{M}_\Delta - 2)/2$, then $C_\Delta = \{1, Q\}$, where Q is the unique $\mathcal{O}_\Delta$-prime above q. This was proved by Halter-Koch in [107]. However, his result fails if q is not a prime. For example,

Example 6.2.1. If $\Delta = 285 = 3 \cdot 5 \cdot 19$ and $q = 15$ then $F_{\Delta,q}(x) = 15x^2 + 15x - 1$. Here $\lfloor \overline{M}_\Delta \rfloor = 7$ and $|F_{\Delta,q}(x)| = 1, 29, 89$ and 179 for $x = 0, 1, 2$ and 3 respectively. However, $C_\Delta \neq \{1, Q\}$ where $Q = [15, (15 + \sqrt{285})/2]$ since $Q \sim 1$ and $C_\Delta = \langle\{Q_3\}\rangle = \langle\{Q_5\}\rangle$ where Q_3 is the unique $\mathcal{O}_\Delta$-ideal over 3 and Q_5 the one above 5.

Motivated by Halter-Koch's work, let's look at what sufficient conditions might exist for the generation of C_Δ by ambiguous ideals when $\Delta = \Delta_0$ is a fundamental discriminant. What emerged from this line of investigation is the following result.

Theorem 6.2.1. [(6.2.1)] *Let $q_i \geq 1$ for $1 \leq i \leq n$ be pairwise relatively prime square-free divisors of a fundamental discriminant $\Delta = \Delta_0$. If, for each prime $p < \overline{M}_\Delta = \sqrt{\Delta/5}$ with $(\Delta/p) \neq -1$ and $p \neq q_i$ for any positive $i \leq n$, there exists a $q = \prod_{i \in S} q_i$ for some $S \subseteq \{1, 2, \ldots, n\}$ such that $|F_{\Delta,q}(x)| = p$ for some non-negative integer x, then*

$$C_\Delta = \bigcup_{i=1}^{n} \langle\{Q_i\}\rangle$$

where Q_i is the unique $\mathcal{O}_\Delta$-ideal over q_i for $i = 1, 2, \ldots, n$.

Proof. By the improvement of Theorem 1.3.1 given above, C_Δ is generated by the non-inert $\mathcal{O}_\Delta$-primes $\mathcal{P}$ with $N(\mathcal{P}) < \overline{M}_\Delta$. If $p \neq q_i$ for any non-negative $i \leq n$, then by hypothesis $|F_{\Delta,q}(x)| = p$ for q as above and $x \geq 0$. By Lemmas 4.1.2–4.1.4 $\mathcal{P} \sim Q$, and the result follows. □

Illustrations of Theorem 6.2.1 are as follows:

Example 6.2.2. Let $\Delta = 1157 = 13 \cdot 89$ and set $q = 13$. Since $\lfloor \overline{M}_\Delta \rfloor = 15$ and the only non-inert prime $p \neq 13$, $p < \overline{M}_\Delta$ is $p = 7$, then the fact that $F_{\Delta,13}(1) = 7$ implies that $C_\Delta = \langle\{Q_3\}\rangle$, by Theorem 6.2.1, where $N(Q_3) = 13$.

Example 6.2.3. Let $\Delta = 4 \cdot 195 = 4 \cdot 3 \cdot 5 \cdot 13$, and set $q_1 = 3$, $q_2 = 5$. Since $F_{\Delta,15}(1) = 2$ (which is the only non-inert prime $p < \overline{M}_\Delta$ with $p \neq 3, 5$), then $C_\Delta = \langle\{Q_3\}\rangle \times \langle\{Q_5\}\rangle$, by Theorem 6.2.1.

Example 6.2.4. Let $\Delta = 4 \cdot 1155 = 4 \cdot 3 \cdot 5 \cdot 7 \cdot 11$, and set $q_1 = 3$, $q_2 = 5$, $q_3 = 7$. The only non-inert primes $p < \overline{M}_\Delta$ with $p \neq q_i$ for $i = 1, 2, 3$, are $p = 2, 11, 17$ and 29. Since $F_{\Delta,35}(1) = 2$, $F_{\Delta,105}(0) = 11$, $F_{\Delta,15}(2) = 17$ and $F_{\Delta,21}(2) = 29$, then by Theorem 6.2.1,

$$C_\Delta = \langle\{Q_3\}\rangle \times \langle\{Q_5\}\rangle \times \langle\{Q_7\}\rangle,$$

where Q_{q_i} is the $\mathcal{O}_\Delta$-ideal above q_i.

Furthermore, we get the following results, parts of which we saw in Theorems 4.2.4–4.2.5. Also, compare with Theorems 5.4.7–5.4.9.

Corollary 6.2.1. *If $\Delta = \Delta_0 = 4(4m^2 \pm 2) > 8$ and $|F_{\Delta,2}(x)| = |2x^2 - D/2|$ is 1 or prime for all integers x with $0 \leq x < \sqrt{D}/2$, then $h_\Delta = 1$.*

Proof. By Theorem 6.2.1, $C_\Delta = \{1, Q_2\}$ where $N(Q_2) = 2$ and $Q_2 \sim 1$, since $|N((2m \pm 2 + \sqrt{\Delta})/2)| = 2$. □

[(6.2.1)] This appeared in [221].

Corollary 6.2.2. *If* $\Delta = \Delta_0 = 4((2m+1)^2 \pm 2)$ *with* $m > 0$ *and* $|F_{\Delta,2}(x)| = |2x^2 + 2x + (1-\Delta)/2|$ *is* 1 *or prime for all non-negative integers* $x < (\sqrt{D} - 1)/2$, *then* $h_\Delta = 1$.

In [107] Halter-Koch remarks: "It is not known whether there exist positive discriminants satisfying the hypothesis of Theorem 3.1 and not being of Richaud–Degert type." We now address this assertion with the following table.

Table 6.2.1. *This table lists all radicands* $0 < D < 10^6$ *such that* $|F_{\Delta,q}(x)|$ *is* 1 *or prime for all non-negative integers* $x \leq (\overline{M}_\Delta - 1)/2$ *where* q *is a positive square-free divisor of* Δ.

D	q	D	q	D	q	D	q	D	q
2	1	15	6	53	1	182	14	437	23
2	2	15	30	62	2	182	26	447	6
3	1	21	1	66	6	195	6	453	3
3	2	21	3	69	3	203	14	483	6
3	3	21	7	77	1	213	3	573	3
3	6	21	21	77	7	215	10	635	10
5	1	23	2	77	11	222	6	645	15
5	5	23	46	77	77	227	2	678	6
6	2	26	2	78	6	230	10	717	3
6	3	26	13	83	2	237	3	843	6
6	6	29	1	85	5	255	6	917	7
7	1	29	29	85	17	258	6	957	3
7	2	30	6	87	6	285	3	965	5
7	7	30	10	93	3	285	5	1077	3
7	14	30	15	95	10	285	15	1085	5
10	2	30	30	102	6	293	1	1085	7
10	5	33	3	110	10	318	6	1085	35
10	10	33	11	110	22	327	6	1133	11
11	1	35	2	122	2	341	11	1245	15
11	2	35	5	123	6	357	3	1253	7
11	11	35	7	138	6	357	7	1293	3
11	22	38	2	141	3	362	2	1685	5
13	1	38	38	143	2	365	5	1757	7
13	13	39	6	143	22	365	10	1965	15
14	2	42	6	143	26	395	10	2085	15
14	7	42	14	165	3	398	2	2373	21
14	14	42	42	165	5	402	6	2397	3
15	2	47	2	167	2	413	7	4245	15
15	3	47	94	167	334	437	1		
15	5	51	6	173	1	437	19		

Remark 6.2.1. If $\Delta = 917 = 7 \cdot 131$, then $\lfloor \overline{M}_\Delta \rfloor = 13$ and the only non-inert prime $p < 13$, $p \neq 7$ is $p = 11 = F_{\Delta,7}(2)$, so by Theorem 6.2.1, $C_\Delta = \langle \{Q_7\} \rangle$ where Q_7 lies over 7. We note that 917 is not an ERD-type. The only other non-ERD type in Table 6.2.1 is $\Delta = 341 = 11 \cdot 31$. Since the only non-inert prime $p < \overline{M}_{341}$, $p \neq 11$ is $5 = |F_{341,11}(0)|$, then $C_{341} = \langle \{Q_{11}\} \rangle$ where $N(Q_{11}) = 11$. The above computational evidence leads to

Conjecture 6.2.1. *If* $D > 4245$ *is a fundamental radicand with discriminant* Δ, *then* $|F_{\Delta,q}(x)|$ *is composite for some non-negative integer* $x < (\overline{M}_\Delta - 1)/2$.

We note that by the techniques of Chapter Five, we can show that Conjecture 6.2.1 holds with one GRH-ruled out exception (Exercise 6.2.1).

Remark 6.2.2. It is interesting to note that the only ERD-types with class number one (from Theorem 5.4.3) which do *not* appear in Table 6.2.1 are $D \in \{17, 37, 101, 197, 677\}$. Also the values which have class number bigger then one are $D \in \{10, 15, 26, 30, 35, 39, 42, 66, 78, 85, 87, 95, 102, 110, 122, 138, 143, 165, 182, 203, 215, 222, 230, 258, 285, 318, 327, 357, 362, 365, 395, 402, 447, 635, 645, 678, 957, 965, 1085, 1245, 1685, 1965, 2085, 2373, 2397, 4245\}$ and all of these have class number two. Furthermore, if we raise the bound $(\overline{M}_\Delta - 1)/2$ to $(M_\Delta - 1)/2$, then 341 and 917 are eliminated from Table 6.2.1. Also, if we raise the bounds to those in Tables 4.2.1–4.2.2, we see that the only values remaining in Table 6.2.1 are those in Tables 4.2.1–4.2.5. If we could unconditionally verify that $|F_{\Delta,q}(x)|$ being 1 or prime (up to the bounds in those tables) necessarily implies $h_\Delta \leq 2$, then the work done in Chapter Four, section two, would be considerably reduced. However, as it stands, we must rely on the techniques of Chapter Five to get the result.

On the other hand, as we saw in Chapter Four, we have the finite list of maximal complex quadratic orders with class groups of exponent two (subject to GRH of course). We have no such nice finiteness condition on the real side. However, if we restrict ourselves to ERD-types, then we do. That is our next project, namely to classify all maximal real quadratic orders with discriminants of ERD-type having class groups of exponent two. This will then complete the tasks begun in Chapters Four and Five where we determined all such orders with $h_\Delta \leq 2$.

Theorem 6.2.2. [(6.2.2)] *If* $\Delta = \Delta_0 > 0$ *is a fundamental discriminant with radicand* $D = D_0$ *of ERD-type and* C_Δ *has* $e_\Delta = 2$, *then* D *is an entry in Table A4 of Appendix A, with one possible exceptional value remaining whose existence would be a counterexample to the GRH.*

Proof. We prove two claims:

Claim 1. If I is an $\mathcal{O}_\Delta$-ideal with $N(I) \in \Lambda(\Delta)$ and $N(I)$ does not divide D, then $N(I) > \sqrt{D}/\gamma$ where

$$\gamma = \begin{cases} 2 & \text{if } D \not\equiv 1 \pmod 4 \\ 3 & \text{otherwise.} \end{cases}$$

Claim 2. If $p^2 < \sqrt{D}/\gamma$ for a prime p, then $(D/p) \neq 1$.

In this way, Claim 2 provides us with an efficient test in order to sieve discriminants up to large values, and Theorem 5.4.1 provides us with an upper bound on the discriminants Δ of ERD-type with $e_\Delta = 2$.

Let $D = m^2 + k$ where $4m \equiv 0 \pmod k$. By Exercise 6.2.2(a), in the continued fraction expansion of w_Δ we find that if either $D \not\equiv 1 \pmod 4$ or the combined conditions, $D \equiv 1 \pmod 4$ and $m \not\equiv 0 \pmod k$ hold, then

$$Q_i/\sigma > \sqrt{D}/2 \tag{6.2.1}$$

[(6.2.2)] This was established by Louboutin *et al.* in [201], and reproduced (with corrections) by permission of Oxford University Press.

for $i \neq 0$, $\ell/2$, $\ell = \ell(w_\Delta)$; and if $D \equiv 1 \pmod 4$ and $m \equiv 0 \pmod k$, then for $i \neq 0$, $\ell/2$, ℓ we have

$$Q_i/\sigma \geq (m - (k-1)/2)/2 \quad \text{when } k > 0, \tag{6.2.2}$$

and

$$Q_i/\sigma > (m - (|k|+1)/2)/2 \quad \text{when } k < 0. \tag{6.2.3}$$

Also, m being even and k being odd together imply that k divides $m/2$ forcing $|k| \leq m/2$, observing that when $D \equiv 1 \pmod 4$ and $k \mid m$, then m cannot be odd.
If (6.2.2) holds, then by Exercise 6.2.2(b)

$$Q_i/\sigma > \sqrt{D}/3 \quad (\text{for } D > 9), \tag{6.2.4}$$

and if (6.2.3) holds, then

$$Q_i/\sigma > \sqrt{D}/3 \quad (\text{ for } D > 36). \tag{6.2.5}$$

From (6.2.1) and (6.2.4)–(6.2.5) we conclude that Claim 1 is secured.
Furthermore, since $D \geq m^2 - |k|$, then $m^2 - 4m \leq D$. Thus, $m - 2 \leq \sqrt{D+4}$, and $\varepsilon_\Delta \leq (m + \sqrt{D})^2/|k|$ which is also a unit. Hence,

$$\varepsilon_\Delta \leq (\sqrt{D} + \sqrt{D+4} + 2)^2/2 < 2(\sqrt{D}+2)^2.$$

Therefore,

$$R = \log \varepsilon_\Delta < \log 2(\sqrt{D}+2)^2 \text{ for } |k| \geq 2.$$

If $|k| = 1$, then $\varepsilon_\Delta \leq m + \sqrt{D}$ and $R < \log 2(\sqrt{D}+2)$.

Case A. $D \neq x^2 + y^2$ for any $x, y \in \mathbf{Z}$.

By Theorem 5.4.1, we know that

$$L(1,\chi) \geq .655\varepsilon\Delta^{-\varepsilon} \text{ for } \Delta \geq \max\{e^{11.2}, e^{1/\varepsilon}\} \text{ with } 1/2 > \varepsilon > 0,$$

where χ is a real non-principal character modulo Δ, with one possible exceptional value of Δ remaining. Since, by (5.4.1),

$$2h_\Delta R/\sqrt{\Delta} = L(1,\chi) > .655\varepsilon\Delta^{-\varepsilon},$$

then

$$2h_\Delta R/\varepsilon\Delta^{1/2-\varepsilon} > .655.$$

Moreover, $\Delta \geq e^{1/\varepsilon}$ means $1/\varepsilon \leq \log\Delta$, so

$$.655 < 2h_\Delta R/(\varepsilon\Delta^{1/2-\varepsilon}) \leq 2h_\Delta R \log\Delta/\Delta^{1/2-\varepsilon}.$$

Let s be the number of distinct odd prime divisors of D. If $d_s = \prod_{i=1}^{s} p_i$ where $p_1 = 3$, $p_2 = 5 \ldots$ (i.e. the ordered product of the first s odd primes), then $D \geq d_s$ by Theorem 1.3.3. Set

$$f(D,\varepsilon) = \log(2\sqrt{D}+2)^2 \log(4D)/(4D)^{1/2-\varepsilon}.$$

If $s \geq 12$, then $d_s \geq 1.522 \cdot 10^{14}$. If $\varepsilon = .02938$ and $D \geq d_s$, then we get

$$2^s f(D,\varepsilon) < .51253 < .655.$$

We must have

$$2h_\Delta R\log(\Delta/\Delta^{1/2-\varepsilon}) < (2/\sigma)2^{s-1}\log(2(\sqrt{\Delta}+2)^2)\log(\Delta/\Delta^{1/2-\varepsilon}).$$

Therefore (since for fixed ε, $f(D,\varepsilon)$ is strictly decreasing),

$$2^s f(D,\varepsilon) > 2h_\Delta R\log(\Delta/\Delta^{1/2-\varepsilon}) > .655.$$

This is a contradiction. Thus, if $s \geq 12$ and $D > 1.522 \cdot 10^{14}$, then C_Δ cannot have exponent 2. On the other hand, if $D > 1.522 \cdot 10^{14}$ and $s < 12$, then we still have $2^s f(D,\varepsilon) < .655$. Hence, if C_Δ is to have $e_\Delta = 2$, we must have $D \leq 1.522 \cdot 10^{14}$ (with one possible exception).

Case B. $D = x^2 + y^2$ for some $x, y \in \mathbf{Z}$.

Let $d_s = q_1 q_2 \ldots, q_s$ where the q_i are the first s primes congruent to 1 modulo 4 ($d_{10} > 1.021 \cdot 10^{15}$). In this case, if $q > 2$ is prime dividing D, then $q \equiv 1 \pmod 4$. By Exercise 6.2.2(c), $s < 10$. If $D > 2 \cdot 10^{13}$, then

$$2^s f(D,\varepsilon) < .2985988 < .655,$$

where $\varepsilon = .03266$. Thus, in this case, if C_Δ has $e_\Delta = 2$, then $D < 2 \cdot 20^{13}$ (again with one possible exception). Hence, we are left with the remaining task from Cases A–B of finding all those values of D which are of ERD-type with $D < B = 1.53 \cdot 10^{14}$, and $e_\Delta = 2$.

For a given prime p, let $M_{p,i}$ be the least positive integer such that $M_{p,i} \equiv i \pmod 4$ and $(M_{p,i}/q) \neq 1$ for all primes q with $3 \leq q \leq p$, where $(\ /\ )$ is the Legendre symbol.

Claim 3. If D is of ERD-type with $\gamma^2 p^4 < D < M_{p,i}$ and $D \equiv i \pmod 4$, then $e_\Delta \neq 2$ for C_Δ. (Exercise 6.2.2(d).)

In Tables A5(1)–A5(3) of Appendix A we give, opposite the prime q_i, the least positive integer N (or what we called $M_{p,j}$ earlier, congruent to $j = 3, 2$ or 1 modulo 4 respectively), such that $(N/q) \neq 1$ for all q such that $3 \leq q \leq p$ where $q_i < p \leq q_{i+1}$.[(6.2.3)] If we, for example, examine Table A5, we note that $M_{223,2} > B = 1.53 \cdot 10^{14}$. If $4 \cdot 223^4 < D < M_{223,2}$ (when $D \equiv 2 \pmod 4$), then $e_\Delta \neq 2$ for C_Δ by Claim 3. However, $4 \cdot 223^4 < 9.9 \cdot 10^9 < M_{151,2}$ from Table A5. Thus, if $D > 4 \cdot 151^4$, then $e_\Delta \neq 2$ for C_Δ. Furthermore,

$$4 \cdot 151^4 < 2.08 \cdot 10^9 < M_{127,2}$$

and

$$4 \cdot 127^4 < 1.04 \cdot 10^9 < M_{113,2}.$$

Thus, if $e_\Delta = 2$ for C_Δ in the case $D \equiv 2 \pmod 4$, then $D < 4 \cdot 113^4$, i.e. $\sqrt{D} < 25538$. By a similar analysis which the reader may undertake, we get from the other tables that if $D \equiv 3 \pmod 4$, then we need $\sqrt{D} < 25538$, and if $D \equiv 1 \pmod 4$ we need $\sqrt{D} < 57963$ in order that $e_\Delta = 2$ for C_Δ.

(6.2.3) These tables were produced by OASiS (see Stephens–Williams [351]). OASiS is the acronym for "Open Architecture Sieve System". OASiS made UMSU obsolete (see footnote (5.4.12)). Despite the fact that OASiS has variable-size rings, MSSU has made OASiS obsolete, (see footnote (4.3.2), so it is now mostly idle. On the other hand, both of the MSSU sieves have been running 24 hours a day for several years now.

If D is of ERD-type, then D is one of the forms r^2k^2+k, r^2k^2+2k or r^2k^2+4k with k odd. We searched all such D values with $r|k| < 30000$ in the middle case and $r|k| < 60000$ in the other two cases. We first eliminated any D values whenever we found a prime q such that $q^2 < \sqrt{D}/\gamma$ and $(D/q) = 1$ (by Claim 2). This succeeded in eliminating almost all the possible D values. To test the remaining D values, we used the following algorithm:

(1) For each prime $p < \sqrt{\Delta}/2$ such that $(\Delta/q) = 1$, find P_0 such that $P_0^2 \equiv D \pmod{\sigma q}$ and put $Q_0 = \sigma q$.

(2) Determine the continued fraction expansion of $(P_0 + \sqrt{\Delta})/Q_0$ until we find the least k such that $Q_k = Q_0$.

(3) If for each such continued fraction we get $P_k \equiv -P_0 \pmod{Q_0}$, then $e_\Delta \leq 2$ for C_Δ.

This leaves us with only the values in Table A4. By Theorem 5.4.1, we have secured the result. □

Remark 6.2.3. Theorem 6.2.2 opens up another interesting question. Included in Tables A4 are all those ERD-types satisfying the property that $e_\Delta = 2$ and there are *no* split primes $p < M_\Delta$. In Chapter Four, section two, we saw how we could classify (via Theorems 4.2.4–4.2.6, Conjecture 4.2.1 and Theorem 4.2.7) those ERD-types with $h_\Delta \leq 2$ and no split primes $p < M_\Delta$. Thus, this completes the task, and so we obtain from Table A4 all such values of D which are listed in Table A6 of Appendix A, i.e. it contains all ERD-types having no split primes less than M_Δ. We note that we answered, similarly, the even more restrictive question, namely: Which fundamental discriminants $\Delta > 0$ have *only* inert primes less than M_Δ? The answer to this question is contained in Table 4.2.3. To be precise, if we reduce M_Δ to $A_\Delta = \sqrt{\Delta - \sigma + 1}/2$, then the entries in Table 4.2.3 are exactly those with no non-inert primes $p < A_\Delta$, and these are exactly the narrow R-D types of class number one. All of this is modulo the techniques in Chapter Five, i.e. the table is complete with one GRH-ruled out exception. One more comment on A_Δ is in order. Of course, if $\Delta \not\equiv 1 \pmod 4$, then $A_\Delta = M_\Delta$, and if $\Delta \equiv 1 \pmod 4$, then the only difference which reducing M_Δ to A_Δ makes is that we pick up $\Delta = 677$, 197, 101, 37, 17 since in each case $\sqrt{\Delta - 1}/2 = p$ where $(\Delta/p) = 1$. Thus the entries in Table 4.2.3 represent all fundamental discriminants $\Delta > 0$ with only inert primes less than A_Δ (with one possible exception). We'll have more to say about A_Δ later. Now we turn to a more general question which is natural to ask in view of the above. We would like to determine those fundamental discriminants $\Delta = \Delta_0 > 0$ which satisfy

Property P_0. There are no split primes less than M_Δ. [(6.2.4)]

In view of Theorem 1.3.1, we see that Property P_0 forces $e_\Delta \leq 2$. What we now demonstrate is that it also forces Δ to be of ERD-type. Hence, Table A6 represents a complete list of all fundamental discriminants $\Delta = \Delta_0 > 0$ satisfying Property P_0 (with one possible GRH-ruled out exception).

[(6.2.4)] M_Δ will now suffice since A_Δ was needed only to pick up those special narrow RD-types $\Delta = m^2 + 1$ with $h_\Delta = 1$.

Theorem 6.2.3. [(6.2.5)] *If $\Delta = \Delta_0 > 0$ is a fundamental discriminant with radicand $D = D_0$, then the following are equivalent:*

(a) *Δ satisfies Property P_0.*

(b) *If $F_\Delta(x) \equiv 0 \pmod p$ for some prime $p < \sqrt{\Delta}/2$ and $0 \le x < (\sqrt{\Delta}-\sigma+1)/2$, then $p \mid \Delta$.*

(c) *One of the following holds:*

(1) *$D = m^2 + 4$ is prime and $h_\Delta = 1$.*

(2) *$D = m^2 - 1$ and either $m - 1$ or $m + 1$ is prime $(m > 1)$ and $e_\Delta \le 2$.*

(3) *$D = m^2 - 4$ and either $m - 2$ or $m + 2$ is prime $(m > 2)$ and $e_\Delta \le 2$.*

(4) *$D = m^2 + m$ with either m or $m + 1$ prime $(m > 0)$ and $e_\Delta \le 2$.*

(5) *$\Delta = m^2 \pm 4r$ with $q < r < m/2$ and $r \mid m$. Moreover, C_Δ is generated by the primes with norm dividing r. Finally, if $\Delta \equiv 1 \pmod 4$, then Δ/r is prime and if $\Delta \equiv 0 \pmod 4$, then the odd part of Δ/r is prime.*

Proof. The equivalence of (a) and (b) is a special case of Exercise 4.1.9.

Now we assume (a) and prove (c). By Theorem 1.3.1, C_Δ is generated by prime ambiguous ideals $I = [N(I), b + w_\Delta]$. If $\ell(I) = 1$, then either $D = 2$ which falls into case (4) or $D = m^2 + 4$ which falls into case (1).

If $\ell(I) = 2$, then $\Delta = m^2 + 4r$ where $r \mid m$. If $r < m/2$ then we are in case (5). By Exercise 4.2.10, Δ can be written in only one such way. Therefore, the continued fraction expansion for any ramified prime $\mathcal{P}$ with $N(\mathcal{P}) < \sqrt{\Delta}/2$ yields that $p \mid r$. Thus, by Theorem 1.3.1, C_Δ is generated by the $\mathcal{O}_\Delta$-primes with norms dividing r. To complete this consideration of case (5), it suffices to show that if $\Delta \equiv 1 \pmod 4$, then Δ/r is prime since the remainder is similar. If $\Delta/r = d_1 d_2$ with $1 < d_1 < d_2$, then since $\Delta/d_1 = r d_2 > d_2 > d_1$, the ideal with norm d_1 is reduced, so $d_1 \mid r$, a contradiction.

If $\ell(I) = 2$ and we are *not* in case (5), then $r \ge m/2$, so $r = m$ or $m/2$. Thus, $\Delta = m^2 + 4m$ or $\Delta = m^2 + 2m$. If $\sigma = 1$, then either $D = (m/2)^2 + (m/2)$, which is case (4), or $D = (m/2)^2 + m = (m/2+1)^2 - 1$, which is case (2). If $\sigma = 2$, then either $D = m^2 + 4m = (m+2)^2 - 4$, which is case (3) or $D = m^2 + 2m = (m+1)^2 - 1$, which is case (2). The reader is left with the task of verifying that in each case one of the factors is always prime. Furthermore, we leave the analysis of the case $\ell(I) = 4$ with the reader since it is similar to the above. We have shown that (a)$\Rightarrow$(c). To show that (c)$\Rightarrow$(a) we prove that (5)$\Rightarrow$(a), and leave the other cases as an exercise for the reader. All we need is that there are no split primes $p < \sqrt{\Delta}/2$. This follows from Theorem 1.3.1 and Exercise 6.2.4. □

Now that Table A6 of Appendix A and Theorem 6.2.3 classify all fundamental discriminants with Property P_0, it is natural to look at the next step up. Which fundamental discriminants $\Delta = \Delta_0 > 0$ satisfy the following property?

Property P_1. There is exactly one non-inert prime less than M_Δ.

The answer is contained in

(6.2.5)This was established by Louboutin *et al.* in [200].

Theorem 6.2.4. [6.2.6] *Let $\Delta = \Delta_0 > 0$ be a fundamental discriminant with radicand D.*

(a) *If $\Delta \equiv 0 \pmod 4$, then Δ satisfies Property P_1 if and only if $h_\Delta = 1$ and $D = m^2 \pm 2 > 3$.*

(b) *If $\Delta \equiv 1 \pmod 4$ and Δ satisfies Property P_1, then one of the following must hold where p denotes the unique prime $p < \sqrt{\Delta}/2$ with $(\Delta/p) \neq -1$.*

(i) $\Delta = D = (pb)^2 + 4p \equiv 5 \pmod 8$, $h_\Delta = 1$, $b > 1$ *and* $pb^2 + 4$ *is prime with* $p \equiv pb^2 + 4 \equiv 3 \pmod 4$.

(ii) $\Delta = D = q^2 + 4qp^n \equiv 5 \pmod 8$ *for some* $n \geq 0$ *with* q *and* $q + 4p^n$ *primes and* $p^n < \sqrt{\Delta}/2 < q$.

(iii) $\Delta = D = r^2 + 4p^{2m} \equiv 5 \pmod 8$ *for some* $m \geq 0$ *where* r *is prime with* $p^m < \sqrt{\Delta}/2 < r < \sqrt{\Delta}$ *and* Δ *is a product of at most two primes both larger than* $\sqrt{\Delta}/2$.

(iv) $\Delta = D = (2a-1)^2 + 2^t$ *with* $t \geq 3$ *and* Δ *is a product of at most two primes both larger than* $\sqrt{\Delta}/2$.

Proof.

Case A. $\Delta \equiv 0 \pmod 4$.

First assume that $D = m^2 \pm 2 > 3$ and $h_\Delta = 1$. If Property P_1 fails, then there exists a prime p with $2 < p < \sqrt{D}$ and $(D/p) \neq -1$. Therefore, $p \in \Lambda^*(\Delta)$. However, if $D = m^2 + 2$, then $\ell(w_\Delta) = 2$ and $\Lambda^*(\Delta) = \{1, 2\}$, a contradiction. If $D = m^2 - 2$, then $\ell(w_\Delta) = 4$ and $\Lambda^*(\Delta) = \{1, 2, 2m - 3\}$, so $2m - 3 = p$ is forced. However, $2m - 3 \geq m > \sqrt{D}$, a contradiction. If there are no non-inert primes less than $\sqrt{D}$, then $\sqrt{D} < 2$ (given that $(D/2) \neq -1$ when $\Delta \equiv 0 \pmod 4$) i.e. we have a contradiction, so Property P_1 must hold.

Conversely, assume that Property P_1 holds. By Exercise 4.2.10, we may always write Δ uniquely as $\Delta \equiv m^2 + 4r$ where either $1 < r < m/2$ and $m = \lfloor\sqrt{\Delta}\rfloor$, or $-1 > r > -m/2$ and $m = \lfloor\sqrt{\Delta}\rfloor + 1$. If r is divisible by an odd prime q, then $(\Delta/q) \neq -1$. However, $(\Delta/2) = 0$, so if $r > 0$, then $r \geq q > \sqrt{\Delta}/2 > m/2$, and if $r < 0$, then $-r \geq q > \sqrt{\Delta}/2 > (m-1)/2$, both of which are contradictions. Thus, $|r| = 2^t$ for $t \geq 0$. If $t > 1$, then $D \equiv 0$ or $1 \pmod 4$, a contradiction. If $t = 0$, then $\Delta = m^2 \pm 4$. If $\Delta = m^2 + 4$, then $m/2$ is odd and any odd prime q dividing m satisfies $(\Delta/q) = 1$, so $m/2 \geq q > \sqrt{\Delta}/2$, a contradiction. If $\Delta = m^2 - 4 > 12$, then there exists a prime q dividing Δ with $q < \sqrt{\Delta}/2$ contradicting property P_1. Thus, we have $D = m^2 \pm 2$ and $h_\Delta = 1$, since Property P_1 prevents the existence of any generator of C_Δ other than the principal ideal over $2 \in \Lambda^*(\Delta)$. This completes the proof of (a).

Case B. $\Delta \equiv 1 \pmod 4$.

Let $p < \sqrt{\Delta}/2$ be the non-inert prime. We leave it as an exercise for the reader to verify that if p ramifies, then we get (i).

Assume p splits.

Case B(a). $\Delta \equiv 5 \pmod 8$.

If $\ell(w_\Delta)$ is even, then $\ell(w_\Delta) = 2i$ for $i \geq 1$ and $Q_i/2$ divides Δ by Exercise 2.1.13(f). Also, $Q_i/2$ must be prime since Property P_1 precludes the possibility

[6.2.6] This was established in [264], as was Theorem 6.2.5.

of any other prime divisor of $Q_i/2$. If $q_1 = Q_i/2$, then $q_1 > \sqrt{\Delta}/2$, so $a_i = 1$ by Exercise 4.2.6. Thus, $P_{i+1} = a_iQ_i - P_i = 2q_1 - P_i$, so $q = P_i = P_{i+1}$ by Lemma 6.1.1. Therefore, $\Delta = P_i^2 + Q_iQ_{i+1} = q_1^2 + 2q_1Q_{i-1}$. By Exercise 6.2.5, $Q_{i-1} < \sqrt{\Delta}/2$ so the only odd prime dividing Q_{i-1} is p. Since $\Delta \equiv 5 \pmod 8$, then $\Delta = q_1^2 + 4q_1p^n$ for some $n \geq 0$. Moreover, $\Delta/q_1 = q_2$ is prime since there would be, otherwise, a ramified prime less than $\sqrt{\Delta}/2$. Moreover, we clearly have $\sqrt{\Delta}/2 < q_1 < \sqrt{\Delta}$ and $p^n < \sqrt{\Delta}/2$. This is (ii).

If $\ell(w_\Delta) = \ell$ is odd, then by Lemma 6.1.1 we have that $Q_{\frac{\ell-1}{2}} = Q_{\frac{\ell+1}{2}}$ and $\Delta = P^2_{\frac{\ell+1}{2}} + Q^2_{\frac{\ell+1}{2}}$. Since $\Delta \equiv 5 \pmod 8$ satisfies Property P_1, then the only possible odd prime which can divide $Q_{\frac{\ell-1}{2}}$ is p. Thus, $Q_{\frac{\ell-1}{2}} = 2p^m$ for $m \geq 0$ and $\Delta = P^2_{\frac{\ell+1}{2}} + 4p^{2m}$ with $p^m < \sqrt{\Delta}/2$. If $P_{\frac{\ell+1}{2}}$ is not prime, then it has a prime divisor q which splits and for which $q < \sqrt{\Delta}/2$, a contradiction. Thus, $P_{\frac{\ell+1}{2}} = r$, a prime and $\Delta = r^2 + 4p^{2m}$ with $\sqrt{\Delta} > r > \sqrt{\Delta}/2$. Since Property P_1 is satisfied, Δ is either prime or $\Delta = q_1q_2$ for primes $q_1 > \sqrt{\Delta}/2$ and $q_2 > \sqrt{\Delta}/2$.

Case B(b). $\Delta \equiv 1 \pmod 8$.

Thus, $p = 2$. If an odd prime q divides Q_1 in the simple continued fraction expansion of w_Δ, then $q > \sqrt{\Delta}/2$, so $\Delta = (2a-1)^2 + 2^tq$ for $t \geq 3$. Thus, $\Delta > (2a-1)^2 + 4\sqrt{\Delta} > (\sqrt{\Delta}-2)^2 + \sqrt{\Delta} > \Delta$, a contradiction. Therefore, $\Delta = (2a-1)^2 + 2^t$ for $t \geq 3$. Moreover, Δ must be the product of at most two primes, both of which are greater than $\sqrt{\Delta}/2$. □

Now we use the techniques of Chapter Five to provide a list of all fundamental discriminants satisfying Property P_1 which we parametrically classified in Theorem 6.2.4.

Theorem 6.2.5. *If $\Delta = \Delta_0 > 0$ is a fundamental discriminant with radicand D and Δ satisfies Property P_1, then D is one of the following 63 values (with one possible value remaining, the existence of which would be a counterexample to the GRH).*

(a) *If $D \not\equiv 1 \pmod 4$, then* $D \in \{6, 7, 11, 14, 23, 38, 47, 62, 83, 167, 227, 398\}$.

(b) *If $D \equiv 1 \pmod 4$, then* $D \in \{17, 33, 37, 41, 61, 65, 69, 85, 89, 93, 101, 113, 133, 137, 149, 157, 197, 213, 237, 257, 269, 317, 341, 353, 377, 397, 413, 453, 461, 557, 593, 629, 677, 717, 733, 773, 853, 941, 1077, 1097, 1133, 1217, 1253, 1333, 1553, 1877, 2273, 2917, 3053, 5297, 7213\}$.

Proof. Since we already know that if the non-inert prime ramifies, Δ is of ERD-type and $h_\Delta = 1$ (see Theorem 6.2.4), then we concentrate on the split non-inert prime. Suppose that $p^{h_\Delta} < \sqrt{\Delta}$, where p splits. Since Δ satisfies Property P_1, then the only elements of $\Lambda^*(\Delta)$ are of the form p^{ih_Δ} or a prime $q > \sqrt{\Delta}/2$. Observe that if the p^{ih_Δ} case holds, then $j < \log_p(\sqrt{\Delta})/h_\Delta$.

In the simple continued fraction expansion of w_Δ, if $Q_j/\sigma = p^{ih_\Delta}$ and $Q_m/\sigma = p^{ih_\Delta}$, then $I_{j+1} = (I_{m+1})' = I_{\ell-m+1}$ by Lemma 6.1.1. Hence, at most two of the principal reduced ideals can have norm p^{ih_Δ}. It follows that there are no more than $2\log_p(\sqrt{\Delta})/h_\Delta$ reduced ideals with norms of the form p^{ih_Δ} in $\Lambda^*(\Delta)$. Thus, by Exercise 6.2.5 there is at most one ideal between any two of these with norm not of the form p^{ih_Δ}. Hence, there can be at most

$$2\log_p(\sqrt{\Delta})/h_\Delta + 2\log_p(\sqrt{\Delta})/h_\Delta + 2 = 4\log_p(\sqrt{\Delta})/h_\Delta + 2$$

reduced ideals in the principal class. Thus,

$$\ell(w_\Delta) = \ell < 4\log_p(\sqrt{\Delta})/h_\Delta + 3.$$

Now, since $R = \log \varepsilon_\Delta < \ell \log \sqrt{\Delta}$ (see Lemma 5.4.1), then

$$h_\Delta R < 4\log(\sqrt{\Delta})\log_p(\sqrt{\Delta}) + 3h_\Delta \log(\sqrt{\Delta}).$$

Since $p^{h_\Delta} < \sqrt{\Delta}$ we get that

$$h_\Delta R < 7\log(\sqrt{\Delta})\log_p(\sqrt{\Delta}) \tag{6.2.6}$$

it follows from (5.4.1) that

$$7\log(\sqrt{\Delta})\log_p(\sqrt{\Delta}) > \sqrt{\Delta}L(1,\chi)/2. \tag{6.2.7}$$

Now consider the case where $p^{h_\Delta} > \sqrt{\Delta}$, which forces $Q_1/2$ to be a prime, $q > \sqrt{\Delta}/2$ and $Q_2/2$ can only be 1. Therefore, $\Delta = q^2 + 4q$ with both q and $q+4$ primes. Since $p < \sqrt{\Delta}/2$, then $h_\Delta \geq 2$.

Put $m = \lfloor h_\Delta/2 \rfloor$, then if $p^m > \sqrt{\Delta}/2$, there exists a reduced ideal I with $N(I) < \sqrt{\Delta}/2$ and $I \sim \mathcal{P}^m$ where $(p) = \mathcal{P}\mathcal{P}'$, by Remark 1.4.1. If $N(I) = p^n$, then $n < m$. Also, either $I = \mathcal{P}^n$ or $I = \mathcal{P}'^n$. An easy check shows that $\mathcal{P}^m$ is not equivalent to $\mathcal{P}^n$. Thus, $I = \mathcal{P}'^n$ and $\mathcal{P}^m \sim \mathcal{P}'^n$, so $\mathcal{P}^{m+n} \sim 1$. Thus, $m + n \equiv 0 \pmod{h_\Delta}$, a contradiction. Hence, $p^m < \sqrt{\Delta}/2$. Assume that h_Δ is even. If p' denotes the palindromic index of $I^{h_\Delta/2}$ (see Definition 6.1.2) and is even, then by Lemma 6.1.2

$$P_{p'/2} = q,\ Q_{p'/2} = 2q \text{ and } \Delta = q^2 + Q_{p'/2}Q_{p'/2-1} = q^2 + 4q,$$

so $Q_{p'/2-1} = 2$, i.e. $I^{h_\Delta/2} \sim 1$, a contradiction. Thus, p' is odd. Therefore by Theorem 6.1.1, $I^{h_\Delta/2}$ is in an ambiguous class without an ambiguous ideal. Moreover,

$$Q_{(p'-1)/2} = Q_{(p'+1)/2} \text{ and } \Delta = P^2_{(p'+1)/2} + Q^2_{(p'+1)/2},$$

by Lemma 6.1.1. Since $Q_{(p'+1)/2} < \sqrt{\Delta}$, then $Q_{(p'+1)/2} = 2p^j$ (some $j \in \mathbf{Z}$) so $I^{h_\Delta/2} \sim I^{\pm j}$, i.e. $j \equiv h_\Delta/2 \pmod{h_\Delta}$. However, $p^{h_\Delta} > \sqrt{\Delta}$ so $j = h_\Delta/2$, i.e. $Q_{(p'+1)/2} = 2p^{h_\Delta/2}$. Yet $Q_{p'} = 2p^{h_\Delta/2}$ and is the first such Q_i with this property, a contradiction. We have shown that h_Δ must be odd. If there exists a $Q_i < \sqrt{\Delta}$ for any i with $0 < i < \ell(\mathcal{P})$, in the simple continued fraction expansion of $(b + w_\Delta)/p$, where $\mathcal{P} = [p, b + w_\Delta]$, then $Q_i = 2p^j$ for some non-negative $j \in \mathbf{Z}$, forcing $I \sim I^{\pm j}$. Therefore, $j \equiv \pm 1 \pmod{h_\Delta}$, but $p^{h_\Delta} > \sqrt{\Delta}$, forcing $j = h_\Delta - 1$ or $j = 1$. If $j = 1$, then $I \sim I'$, and if $j = h_\Delta - 1$, then $I \sim I^{h_\Delta - 1} \sim I^{-1} \sim I'$. In either case, I is in an ambiguous class. Since h_Δ is odd, then $I \sim 1$. However, $\ell(w_\Delta) = 2$ and $Q_1 = 2q \neq 2p$, a contradiction. Hence, there does not exist any integer i with $0 \leq i < \ell(\mathcal{P}) = \ell$ such that $Q_i < \sqrt{\Delta}$, in the simple continued fraction expansion of $(b + w_\Delta)/p$. Yet, $\Delta = P^2 + Q_iQ_{i-1}$ for $0 < i < \ell$ and we cannot have both Q_i and Q_{i-1} bigger than $\sqrt{\Delta}$. Thus, $\ell = 2$ with $Q_1 = 2r > \sqrt{\Delta}$, r a prime. Also, $P_1 = P_2 = a_1Q_1 - P_1$, so $P_1 = ra_1$, forcing r to divide Δ since $\Delta = P_1^2 + 4rp$. Thus, $r = q$ and so $\Delta = q^2(P_1/q)^2 + 4pq = q^2 + 4q$, a contradiction. We have shown that the case $p^{h_\Delta} > \sqrt{\Delta}$ cannot occur.[6.2.7]

[6.2.7]This use of the results of section 6.1 obviated the need for Theorem 5.4.1 in this case. The authors of the original work [264] could not eliminate the use of Theorem 5.4.1 in this case, but conjectured that it could be done. This was accomplished in [237].

By Theorem 5.4.1, we know that, with one possible exception, we must have

$$L(1,\chi) > .655\varepsilon\Delta^{-\varepsilon},$$

for $0 < \varepsilon < 1/2$ and $\Delta > \max\{e^{1/\varepsilon}, e^{11.2} < 73131\}$. Taking $\varepsilon > 1/\log\Delta$, which decreases toward $1/\log\Delta$, we get

$$L(1,\chi) \geq 0.655/(e\log\Delta) > .24/\log\Delta,$$

and so

$$\sqrt{\Delta}L(1,\chi)/2 > .12\Delta/\log\Delta,$$

for $\Delta \geq 73131$.

Hence, (6.2.7) cannot hold for $\Delta > 2\cdot 10^{11}$. A computer check for the remaining Δ yielded only those in the list. Using the results of Chapter Five, section four, we know that the above exceptional value would be a counterexample to the GRH. □

We actually proved something of separate value in the above, namely (6.2.6), which we now isolate.

Corollary 6.2.3. *If $\Delta = \Delta_0 > 0$ is a fundamental discriminant and there exists exactly one non-inert prime $p < \sqrt{\Delta}/2$ which splits, then*

$$h_\Delta R < 7\log(\Delta)\log_p(\Delta)/4.$$

Some comments on Theorem 6.2.5 are in order, since we have now strayed from the investigation of exponent two and ERD-types. For instance, in the list $h_{257} = 3 = h_{733} = h_{2917} = h_{5297}$. All others have $h_\Delta \leq 2$, but not all are ERD-types. For instance, $h_{7213} = 1$ and $\Delta = 7213$ is not an ERD-type. Nevertheless, Theorems 6.2.3–6.2.4 suggest a connection between the number of non-inert primes less than M_Δ and how $F_\Delta(x)$ factors.

We now make this explicit so that we may compare it with results in Chapters Four and Five.

Definition 6.2.1. Let $\Delta = \Delta_0 > 0$ be a fundamental discriminant and let N_Δ be the number of non-inert primes less than E_Δ where $E_\Delta \in \{M_\Delta, A_\Delta\}$. Define

$$H(\Delta) = \max\{d(|F_\Delta(x)|) : 2-\sigma \leq x < E_\Delta\}.$$

(see Definition 4.1.3).

Lemma 6.2.1. $H(\Delta) \leq N_\Delta + 1$.

Proof. For any $x \geq 0$, any prime divisor of $F_\Delta(x)$ is non-inert, by Lemma 4.1.2. In particular, if $N_\Delta + 2$ distinct primes divide $|F_\Delta(x_0)|$ for some x_0 with $2-\sigma \leq x_0 < E_\Delta$, then all of these primes are non-inert. By the definition of N_Δ, we must have at least two of these primes bigger than E_Δ. Thus, $|F_\Delta(x_0)| > E_\Delta^2 \geq (\Delta-\sigma+1)/4 \geq (\Delta-1)/4$. However, $|F_\Delta(x)| \leq (\Delta-1)/4$ for all $x \in \mathbf{Z}$ with $2-\sigma \leq x < E_\Delta$, a contradiction. □

Remark 6.2.4. A consequence of Lemma 6.2.1 is that when $N_\Delta = 0$ and $E_\Delta = A_\Delta$, then $|F_\Delta(x)|$ is 1 or prime for all $x \in \mathbf{Z}$ with $2 - \sigma \leq x < E_\Delta$. Compare this with Theorem 4.2.6 and Theorem 5.2.2. As we have seen, what emerges from this are the narrow R-D types of class number one, and using the techniques of Chapter Five, section four, the list is complete (with one possible GRH-ruled out exception).

When $N_\Delta = 1$ and $E_\Delta = M_\Delta$, this is the content of Theorems 6.2.4–6.2.5. However, the general question is very much an open one, albeit there are some obvious relationships. For instance, we have

Lemma 6.2.2. $h_\Delta \leq e_\Delta^{N_\Delta}$.

Proof. Exercise 6.2.6. □

For instance, if $e_\Delta = 2$, $E_\Delta = M_\Delta$ and $N_\Delta = 1$, then $h_\Delta \leq 2$ as in our list in Theorem 6.2.5.

It remains an open problem to determine the general relationship between $H(\Delta)$ and h_Δ. Here is an interesting question.

Query 6.2.1. For what fundamental discriminants $\Delta > 0$ does $H(\Delta) = N_\Delta + 1$?

When $N_\Delta = 0$ and $E_\Delta = A_\Delta$, we have already answered this question via Table 4.2.3 and Theorem 4.2.6. Observe that since we allow $x = 0, 1$ in $H(\Delta)$, we do not pick up all of the values $\Delta \not\equiv 5 \pmod 8$ in Table 4.2.3. However, that case is trivial since 2 is non-inert which means that $N_\Delta = 0$ implies $E_\Delta < 2$. We actually only miss $D \in \{7, 11\}$ anyway. We do pick up all of the $\Delta \equiv 5 \pmod 8$, since adding $x = 0$ only possibly *repeats* a prime factor of $F_\Delta(0)$ so that $d(|F_\Delta(x)|)$ remains unchanged, and $H(\Delta) \leq 1$ remains unchanged.

When $N_\Delta = 1$ and $E_\Delta = M_\Delta$, we answered this question in Theorem 6.2.5 where all values listed there have $H(\Delta) = 2$, except $D \in \{6, 14, 101, 197, 677\}$ for which $H(\Delta) = 1$. Again, the list is complete with one possible GRH-ruled out exception.

We also have

Query 6.2.2. For which fundamental discriminants $\Delta > 0$ do we have $h_\Delta = e_\Delta^{N_\Delta}$ when $h_\Delta > 1$?

We answered Query 6.2.2 for the case where $N_\Delta = 1$ in Theorem 6.2.5, namely if $e_\Delta = 2$, then $\Delta \in \{65, 85, 377, 629\}$; if $e_\Delta = 3$, then $\Delta \in \{257, 733, 2917, 5297\}$; and if $e_\Delta > 3$, then $\Delta \in \{\ \}$. As usual, these lists are complete with one possible GRH-ruled out exception. Other than this, however, Queries 6.2.1–6.2.2 remain open questions. One can ask similar questions for negative fundamental discriminants. With reference to Definition 4.1.1, we ask

Query 6.2.3. For which fundamental discriminants $\Delta = \Delta_0 < 0$ is $F(\Delta) = N + 1$, where $|\Delta|$ is divisible by exactly $N + 1$ ($N \geq 0$) distinct primes?

Theorem 4.1.4 tells us that in the presence of $h_\Delta = 2^{F(\Delta)-1}$ we have the answer, namely all $\Delta = \Delta_0 < 0$ where $e_\Delta \leq 2$. However, in the absence of this condition, the question is open.

We now turn to some other conjectures and open questions concerning the Minkowski bound M_Δ for fundamental discriminants $\Delta = \Delta_0 > 0$.

In [182], Leu posed

Conjecture 6.2.2. *Let $\Delta = \Delta_0 > 0$ be a discriminant with radicand D and set*

$$T_\Delta = \{r : r \text{ is prime, } r < \sqrt{\Delta}/2 \text{ and } (\Delta/r) \neq -1\}.$$

If $D = n^2 + 4$, then $h_\Delta = 2$ if and only if $D = pq$ for primes $p < q$ with $p \equiv q \equiv 1 \pmod 4$, and $1 \leq |T_\Delta| \leq 2$ such that if $r \in T_\Delta$ with $(D/r) = 1$, then $r^2 > \sqrt{\Delta}/2$.

We now prove the "easy" part of Conjecture 6.2.2, i.e. the sufficiency for $h_\Delta = 2$.

Theorem 6.2.6. [(6.2.8)] *If $D = n^2 + 4 = pq$ for primes $p < q$ and $1 \leq |T_\Delta| \leq 2$ with $r^2 > \sqrt{\Delta}/2$ whenever $(D/r) = 1$ and $r \in T_\Delta$, then $h_\Delta = 2$.*

Proof. By Theorems 1.3.1 and 3.2.1, $h_D = 1$ if and only if $S_\Delta = \emptyset$. Therefore, we may assume that $h_\Delta > 1$. Consider the reduced ideal $I = [p, (p + \sqrt{D})/2]$. In the continued fraction expansion of $\alpha = (p + \sqrt{D})/(2p)$, we must have that $\ell(\alpha) = \ell$ is odd by Theorem 2.1.3 (and $\ell > 1$ since $\ell = 1$ implies that $D = P_1^2 + 4p^2$ forcing $p^2 \mid D$, a contradiction). By Exercise 2.1.13(g), we must have that $Q_{\frac{\ell+1}{2}} = Q_{\frac{\ell-1}{2}} < \sqrt{\Delta}$. Hence, $D = P^2_{\frac{\ell+1}{2}} + Q^2_{\frac{\ell+1}{2}}$. Clearly, p cannot divide $Q_{\frac{\ell+1}{2}}$ since D is square-free. Thus, $Q_{\frac{\ell+1}{2}} = 2r_1^{s_1}r_2^{s_2}$ with $r_i \in T_\Delta$ and $(D/r_i) = 1$ for $s_i \geq 0$. If $s_i > 0$ for $i = 1, 2$, then

$$D = P^2_{\frac{\ell+1}{2}} + 4r_1^{2s_1}r_2^{2s_2} \geq P^2_{\frac{\ell+1}{2}} + 4r_1^2r_2^2 > D$$

(since $r_i^2 > \sqrt{D}/2$ by hypothesis), a contradiction. Therefore, $s_2 = 0$ say, and $Q_{\frac{\ell+1}{2}} = 2r_1^{s_1}$. If $s_1 > 1$, then $Q_{\frac{\ell+1}{2}} > 2r_1^2 > \sqrt{D}$, a contradiction, so $s_1 = 1$. Since $Q_{\frac{\ell+1}{2}} = 2r_1$, then $\mathcal{R}_1 \sim \mathcal{P}$ and so $\mathcal{R}_1^2 \sim \mathcal{P}^2 \sim 1$, where $\mathcal{R}_1$ lies over r_1 and $\mathcal{P}$ lies over p. Furthermore, $\mathcal{R}_1 \not\sim 1$ since $\ell(w_\Delta) = 1$. We have thus far shown that if $|S_\Delta| = 1$ or if $p \in T_\Delta$, then $h_\Delta = 2$, so we now assume that $T_\Delta = \{r_1, r_2\}$ with $(D/r_i) = 1$ for $i = 1, 2$. Consider $D = P_1^2 + Q_0Q_1 = P_1^2 + 2pQ_1$. Since $T_\Delta = \{r_1, r_2\}$, then $p > \sqrt{D}/2$, so $Q_1 < \sqrt{D}$. Moreover, $Q_1 \neq 2$ since $R_1 \sim 1$ as above. Thus, the only odd prime which can divide Q_1 is r_2, so $Q_1 = 2r_2^{s_2}$. If $s_2 > 1$, then $Q_1 > \sqrt{D}$ (since $r_2^2 > \sqrt{D}/2$ by hypothesis), a contradiction. Hence, $Q_1 = 2r_2$, so $\mathcal{R}_2 \sim \mathcal{P}$ and $\mathcal{R}_2^2 \sim \mathcal{P}^2 \sim 1$. Hence, $h_\Delta = 2$ and the result is secured. □

Now we look at the converse of Theorem 6.2.6.

Since $D = n^2 + 4$, then it follows from Theorem 1.3.3 that $h_\Delta = 2$ necessarily implies $D = pq$ for primes $p \equiv q \equiv 1 \pmod 4$. Suppose that $q > p$ and

$$p = a^2 + 4b^2 \text{ with } a, b > 0$$

and

$$q = s^2 + 4t^2 \text{ with } s, t > 0.$$

Since D must be a sum of 2 squares in essentially two distinct ways, we must have that

$$D = (as + 4bt)^2 + 4(bs - at)^2,$$

and

$$D = (as - 4bt)^2 + 4(bs + at)^2,$$

(6.2.8) This, together with the balance of the results of this section, appeared in [266].

from which it follows that

$$bs - at = \epsilon = \pm 1, \tag{6.2.8}$$

and

$$bs + at = c, \tag{6.2.9}$$

where c is divisible only by primes in T_Δ.

Remark 6.2.5. It is evident that $T_\Delta \neq \emptyset$. In fact, as noted in the proof of Theorem 6.2.6, $h_\Delta = 1$ if and only if $T_\Delta = \emptyset$. If D were not of the form $D = n^2 + 4$, then we would not be able to assert that $h_\Delta = 1$ implies $T_\Delta = \emptyset$, since it is possible, in general, to have $T_\Delta \neq \emptyset$ while $h_\Delta = 1$, when the primes in T_Δ have principal prime ideals above them. However, in our special case $\ell(w_\Delta) = 1$, which means that there are ***no nontrivial principal reduced ideals.*** However, we may always assert that $T_\Delta = \emptyset$ implies $h_\Delta = 1$ by Theorem 1.3.1.

It is worth noting that in earlier work, Leu showed ***unconditionally*** that if there are ***no*** inert primes less than M_Δ, and T_Δ consists ***only*** of primes p with $(\Delta/p) = 1$, then $\Delta > 0$ implies that $\Delta \in \{8, 12, 5, 13, 17, 33, 73, 97\}$, none of which satisfies our criterion. Therefore, we must have inert primes less than M_Δ. However, in Theorem 6.2.5, we were able to classify those values of D for which $|T_\Delta| = 1$, and were able to list all of them with one GRH-ruled out exception. One sub-class of that classification is, naturally, our $D = n^2 + 4$, but the only ones with $h_\Delta = 2$ for such D on that list are $D = 85$ and 269.

Now we return to an examination of the converse of Theorem 6.2.6.

As delineated earlier, we know all of the square-free $D = n^2 + 4$ with $h_\Delta = 2$, with one (GRH-ruled out) exception. We now list them here with their associated continued fraction expansions for the classes of order 2 in $\mathcal{O}_D$.

Example 6.2.5. (i) $D = 85 = 5 \cdot 17 = p \cdot q = 9^2 + 4$.

The simple continued fraction expansion of $(5 + \sqrt{D})/6$ is :

i	0	1	2	3
P_i	5	7	5	5
Q_i	6	6	10	6
a_i	2	2	1	2

(ii) $D = 365 = 5 \cdot 73 = p \cdot q = 19^2 + 4$.

The continued fraction expansion of $(15 + \sqrt{D})/14$ is:

i	0	1	2	3
P_i	15	13	15	15
Q_i	14	14	10	14
a_i	2	2	3	2

(iii) $D = 533 = 13 \cdot 41 = p \cdot q = 23^2 + 4$.

The continued fraction expansion of $(15 + \sqrt{D})/22$ is:

i	0	1	2	3	4	5
P_i	15	7	15	13	13	15
Q_i	22	22	14	26	14	22
a_i	1	1	2	1	2	1

(iv) $D = 629 = 17 \cdot 37 = p \cdot q = 25^2 + 4$.

The continued fraction expansion of $(17 + \sqrt{D})/10$ is:

i	0	1	2	3
P_i	17	23	17	17
Q_i	10	10	34	10
a_i	4	4	1	4

(v) $D = 965 = 5 \cdot 193 = p \cdot q = 31^2 + 4$.

The continued fraction expansion of $(9 + \sqrt{D})/26$ is:

i	0	1	2	3	4	5
P_i	9	17	9	25	25	9
Q_i	26	26	34	10	34	26
a_i	1	1	1	5	1	1

(vi) $D = 1685 = 5 \cdot 337 = p \cdot q = 41^2 + 4$.

The continued fraction expansion of $(11 + \sqrt{D})/34$ is:

i	0	1	2	3	4	5
P_i	11	23	11	35	35	11
Q_i	34	34	46	10	46	34
a_i	1	1	1	7	1	1

(vii) $D = 1853 = 17 \cdot 109 = p \cdot q = 43^2 + 4$.

The continued fraction expansion of $(29 + \sqrt{D})/22$ is:

i	0	1	2	3	4	5
P_i	29	37	29	17	17	29
Q_i	22	22	46	34	46	22
a_i	3	3	1	1	1	3

(viii) $D = 2813 = 29 \cdot 97 = p \cdot q = 53^2 + 4$.

The continued fraction expansion of $(39 + \sqrt{D})/38$ is:

i	0	1	2	3	4	5
P_i	39	37	39	29	29	39
Q_i	38	38	34	58	34	38
a_i	2	2	2	1	2	2

Remark 6.2.6. We observe in Example 6.2.5 that all values of p are of the form $p = k^2 + 4$ or $4k^2 + 1$. We now prove Conjecture 6.2.2 when p is of the form $4k^2 + 1$.

Theorem 6.2.7. *Conjecture 6.2.2 holds when $p = 4k^2 + 1$ for some integer $k \geq 1$.*

Proof. If $q = r^2 + 4s^2$, then $r = 2m + 1$ and

$$D = (4k^2 + 1)(r^2 + 4s^2) = (r + 4ks)^2 + 4(s - kr)^2 = (r - 4ks)^2 + 4(s + kr)^2.$$

Since D is representable as a sum of 2 squares in only 2 (essentially, i.e. up to sign and order of the summands) distinct ways, then we must have $s - kr = \varepsilon$ where $|\varepsilon| = 1$.

Since $n = r + 4ks$, then $n = r + 4k(kr + \varepsilon) = pr + 4k\varepsilon$. Therefore, $D = (pr + 4k\varepsilon)^2 + 4 = p^2r^2 + 8\varepsilon prk + 4p$. Now consider the continued fraction expansion of $(p + \sqrt{D})/(2p)$.

Case 1. $\epsilon = 1$

i	0	1	2	3
P_i	p	pr	$(4k^2 - 1)r + 4k$	pr
Q_i	$2p$	$4rk + 2$	$4rk + 2$	$2p$
a_i	$m + 1$	$2k$	$2k$	r

Case 2. $\epsilon = -1$ (in which case $r \geq 3$ since if $r = 1$, then $q = 1 + 4(k - 1)^2 < p$, a contradiction).

i	0	1	2	3	4	5
P_i	p	$pr - 2p$	$(p - 4k)r + 2$	$(p - 2)r - 4k$	$(p - 4k) + r + 2$	$pr - 2p$
Q_i	2p	$(2p - 4k)r - 2(p - 1)$	$4kr - 2$	$4kr - 2$	$(2p - 4k)r - 2(p - 1)$	2p
a_i	m	1	$2k - 1$	$2k - 1$	1	$r - 2$

Now, if we assume that $h_D = 2$, then by Theorem 5.3.4, all $Q_i/2$ in either case *must* be prime. Hence, in Case 1, $|T_\Delta| \leq 2$ clearly. In Case 2, we would have $|T_\Delta| \leq 2$ if we could show that $Q_1/2 > \sqrt{D}/2$. Suppose, to the contrary, that $Q_1/2 < \sqrt{D}/2$, then

$$Q_1/2 = (p - 2k)r - p + 1 < \sqrt{\Delta}/2,$$

which implies that

$$pr - 2kr - p + 1 \leq (pr - 4k)/2,$$

from which a calculation shows that

$$4k^2(r - 2) + 4k(1 - r) + r \leq 0,$$

or

$$(2(r - 2)k - r)(2k - 1) \leq 0.$$

Since $k \geq 1$, then we must have

$$k \leq r/(2r - 4).$$

Hence,

$$k < \begin{cases} 2 & \text{if } r = 3 \\ 1 & \text{if } r \neq 3. \end{cases}$$

Since $k \geq 1$, then $r = 3$, $k = 1$. This implies that $s = 2$ and $q = 25$, a contradiction. The converse is Theorem 6.2.6. □

We now examine the only other case for the values of p appearing in Example 6.2.5, namely $p = k^2 + 4$. From (6.2.8)–(6.2.9), we get that $b = 1$ and $a = k$. Therefore, $s = kt \pm 1$.

Case 1. $s = kt - 1$. The continued fraction expansion of $(p + \sqrt{D})/(2p)$ is:

i	0	1	2	3	4	5
P_i	p	$tp - p$	$tp - c$	$kc - tp$	$tp - c$	$tp - p$
Q_i	$2p$	$\frac{(2tp-c-p)}{2}$	$2c$	$2c$	$\frac{(2tp-c-p)}{2}$	$2p$
a_i	$\frac{t}{2}$	2	$\frac{(k-1)}{2}$	$\frac{(k-1)}{2}$	2	$t - 1$

Case 2. $s = kt + 1$. The continued fraction expansion of $(p + \sqrt{D})/(2p)$ is:

i	0	1	2	3	4	5	6	7
P_i	p	$tp - p$	$\frac{(c+p)}{2}$	$tp - c$	$kc - tp$	$tp - c$	$\frac{(c+p)}{2}$	$tp - p$
Q_i	$2p$	$\frac{(2tp+c-p)}{2}$	$\frac{(2tp-c+p)}{2}$	$2c$	$2c$	$\frac{(2tp-c+p)}{2}$	$\frac{(2tp+c-p)}{2}$	$2p$
a_i	$\frac{t}{2}$	1	1	$\frac{(k-1)}{2}$	$\frac{(k-1)}{2}$	1	1	$t - 1$

Remark 6.2.7. Again, by Theorem 5.3.4, all $Q_i/2$ in either case must be prime. However, there is a good reason why they cannot all be prime in general. For example, if $D = 87029 = 29 \cdot 3001$, then the continued fraction expansion of $(29 + \sqrt{D})/58$ has period length 7 and all $Q_i/2$ are prime. Moreover, $[29, (29 + \sqrt{D})/2]$ is ambiguous. However, $h_\Delta = 10$, and so there is (of course) another ideal, namely $[5, (3+\sqrt{D})/2]$ which has order 5. Nevertheless, in cases 1-2 above, there is no clear algebraic way to show that D is a quadratic residue modulo any integer $m < \sqrt{\Delta}/2$ where $m \neq Q_i/2$ for any i with $1 \leq i \leq l(I)$, where the Q_i appear in the simple continued fraction expansion of $(p + \sqrt{D})/(2p)$ with $I = [p, (p + \sqrt{D})/2]$ (as is the case with $D = 87029$ where $(D/5) = 1$ and $t = 10$). It is, in fact, quite frustrating that in cases 1-2 above we have $|T_\Delta| \leq 3$, and we cannot eliminate the additional prime. If this could be done, then we would have shown that the conjecture is true for p of the form either $k^2 + 4$ or $4k^2 + 1$. Thus, in order to complete the proof of the conjecture, we clearly would need only to show that if $h_\Delta = 2$, then $b = 1$. This remains open.

In his review of Leu's paper [182], S. Louboutin (see MR # 93f: 11075) says that Conjecture 6.2.2 is a "deceptively reasonable one". He goes on to say that "... it is reasonable to conjecture that ..."

Conjecture 6.2.3. *For all integers $m > 0$, there exists a prime p such that whenever $D = pq = m^2 + 4$ where $q > p$ is also prime, we have $\ell(\alpha) \geq 2m + 3$ where $\alpha = (\sqrt{D} + p)/(2p)$.*

Our earlier contention that the proof of Conjecture 6.2.2 is seriously difficult is borne out by Louboutin's last comment in his review pertaining to Conjecture 6.2.3. He says, "Hence, the author's conjecture could not be proved algebraically even if he changed $|T_\Delta| = 1$ or 2 into $1 \leq |T_\Delta| \leq \ell$ for any $\ell \geq 2$." Therefore, any advance toward the proof of Conjecture 6.2.2 should be viewed as significant progress. We now prove Conjecture 6.2.3. We begin with results for more general radicands D. In what follows, P and Q are not necessarily primes. (See Exercise 2.1.2 for the definition of $\{A_i\}$ and $\{B_i\}$.)

Theorem 6.2.8. *Let $D = PQ$ where $P = A^2 + B^2$, $\gcd(A, B) = 1$, $A > B > 0$, and $A/B = \langle q_0; q_1, \ldots, q_\ell \rangle$. If $Q = (rA_\ell + 2A_{\ell-1})^2 + (rB_\ell + 2B_{\ell-1})^2$ with $r \geq 1$ odd, then the continued fraction expansion of $(P + \sqrt{D})/(2P)$ is given by*

$$\langle (r+1)/2; \overline{a_\ell, a_{\ell-1}, \ldots, a_0, a_0, a_1, \ldots, a_\ell, r} \rangle.$$

Proof. By Exercise 2.1.1(c)

$$\langle a_\ell; a_{\ell-1}, \ldots, a_1, a_0\rangle = A_\ell/A_{\ell-1}$$

and

$$\langle a_\ell; a_{\ell-1}, \ldots, a_2, a_1\rangle = B_\ell/B_{\ell-1}.$$

Put

$$L = A_{\ell-1}A_\ell + B_\ell B_{\ell-1}, \quad \text{and}$$
$$M = A_{\ell-1}^2 + B_{\ell-1}^2.$$

We then get

$$\langle a_\ell; a_{\ell-1}, \ldots, a_1, a_0, a_0, a_1, \ldots, a_{\ell-1}, a_\ell\rangle = ((A_\ell/B_\ell)A_\ell + B_\ell)/((A_\ell/B_\ell)/A_{\ell-1} + B_{\ell-1})$$

$= P/L$ and

$$\langle a_\ell; a_{\ell-1}, \ldots, a_1, a_0, a_0, a_1, \ldots, a_{\ell-1}\rangle$$
$$= ((A_{\ell-1}/B_{\ell-1})A_\ell + B_\ell)/((A_{\ell-1}/B_{\ell-1})/A_{\ell-1} + B_{\ell-1}) = L/M.$$

Let

$$\theta = \langle \overline{a_\ell; a_{\ell-1}, \ldots, a_1, a_0, a_0, a_1, \ldots, a_{\ell-1}, q_\ell, r}\rangle.$$

Then

$$\theta = (\theta(rP + L) + P)/(\theta(rL + M) + L).$$

Suppose that $r \geq 1$ and r is odd. Set

$$\lambda = \langle (r+1)/2, \theta\rangle = (r+1)/2 + 1/\theta = \langle (r+1)/2; \overline{a_\ell, a_{\ell-1}, \ldots, a_0, a_0, \ldots, a_{\ell-1}, a_\ell, r}\rangle.$$

Now

$$\theta^2(rL + M) + \theta L = \theta r P + \theta L + P,$$

which implies that

$$\theta^2(rL + M) = \theta r P + P.$$

If $\gamma = 1/\theta$, then

$$P\gamma^2 + \gamma r P - (rL + M) = 0.$$

Thus, $\lambda = (r+1)/2 + \gamma$,

$$\gamma = (-rP + \sqrt{r^2P^2 + 4P(rL + M)})/(2P)$$

and,

$$\lambda = (P + \sqrt{r^2P^2 + 4P(rL + M)})/(2P).$$

Put $N = r^2P^2 + 4P(rL + M)$, then

$$N = P[r^2P + 4(rL + M)] = PQ = D.$$

Therefore, the continued fraction expansion of $(P + \sqrt{D})/(2P)$ is given by

$$\langle (r+1)/2; \overline{a_\ell, a_{\ell-1}, \ldots, a_0, a_0, \ldots, a_\ell, r}\rangle.$$ □

Definition 6.2.2. Let $r > 1$ be a rational number, and denote by $m(r)$ the value of s where $r = \langle a_0; a_1, a_2, \ldots, a_s\rangle$ with $a_s > 1$.

Theorem 6.2.9. *For any positive integer m, there exists an infinitude of primes p of the form $A^2 + B^2$ with $A > B$ such that $m(A/B) \geq m$.*

Proof. We make use of the ideas cited in [266] attributable to Hecke, from which we can easily deduce that there exists an infinitude of primes of the form $x^2 + y^2$ with

$$c_1 < \frac{x}{y} < c_2$$

for any given pair of positive real numbers c_1 and c_2 with $c_1 < c_2$. Consider

$$A_n/B_n = \langle a_0; a_1, a_2, \ldots, a_n \rangle$$

where $n \geq m$, and the only constraint we put on the a_i's is that they be positive integers. Now, if

$$\lambda = \langle a_0; a_1, a_2, \ldots, a_n, \theta \rangle = (\theta A_n + A_{n-1})/(\theta B_n + B_{n-1}),$$

and $\theta = b/c$ with b and c being relatively prime integers then, if $\lambda = x/y$, we get $x = bA_n + cA_{n-1}$ and $y = bB_n + cB_{n-1}$. It follows that

$$b = (xB_{n-1} - yA_{n-1})(-1)^{n-1},$$

and

$$c = (yA_n - xB_n)(-1)^{n-1}.$$

If n is odd, then $A_n/B_n > A_{n-1}/B_{n-1}$, and

$$b = xB_{n-1} - yA_{n-1}, \quad c = yA_n - xB_n.$$

Let p be a prime of the form $x^2 + y^2$ where

$$A_{n-1}/B_{n-1} < x/y < A_n/B_n.$$

In this case, we have $b, c > 0$. If n is even, then $A_{n-1}/B_{n-1} > A_n/B_n$. Let p be a prime of the form $x^2 + y^2$, where $A_n/B_n < x/y < A_{n-1}/B_{n-1}$. In this case, we also have $b, c > 0$. Thus, in either case, we see that $\theta > 0$ and that the length of the continued fraction expansion of $\lambda = x/y$ is at least $n \geq m$. □

Theorem 6.2.10. *For all integers $m > 0$, there exists a prime p such that whenever $D = pq = n^2 + 4$ where $q > p$ is also prime, we have that $\ell(\alpha) \geq 2m + 3$ where $\alpha = (\sqrt{D} + p)/(2p)$.*

Proof. If $D = pq = n^2 + 4$, we may assume without loss of generality that $q > p$. By (6.2.8)-(6.2.9) we get that

$$n = as + 4bt = (ps - 4b\epsilon)/a,$$

so that $D = (ps - 4b\epsilon)^2/a^2 + 4 = (p^2s^2 - 8\epsilon psb + 4p)/a^2$. Also, $bs \equiv \epsilon \pmod{a}$. Therefore, if $b^*b \equiv 1 \pmod{a}$, then $s \equiv b^*\epsilon \pmod{a}$. Since a is odd, we may assume without loss of generality that b^* is even. Since $s \equiv b^*\epsilon \pmod{a}$, we can write $s = b^*\epsilon + ar$. Since s is odd, b^* is even, and a is odd, we must have that r is odd. Thus,

$$\begin{aligned} D &= (p^2(b^*\epsilon + ar)^2 - 8\epsilon pb(b^*\epsilon + ar) + 4p)/a^2 \\ &= (p^2a^2r^2 + 2\epsilon(b^*p - 4b)par + p(b^{*2} - 8bb^* + 4))/a^2 \\ &= p^2r^2 + 2\epsilon(b^*p - 4b)pr/a + p(b^{*2} - 8bb^* + 4)/a^2. \end{aligned}$$

Case 1. $a > 2b$. Thus, $a/2b = \langle a_0; a_1, ..., a_\ell \rangle = A_\ell/B_\ell$ with $a_\ell > 1$. We have that $A_\ell B_{\ell-1} - B_\ell A_{\ell-1} = (-1)^{\ell-1}$. We may now assume that $\epsilon = (-1)^\ell$, for if $\epsilon \neq (-1)^\ell$, set $a_{\ell+1} = 1$ replace the values of a_ℓ by that of $a_\ell - 1$ and ℓ by $\ell + 1$. We can then use $2A_{\ell-1}\epsilon$ for the value of b^*. In this instance,

$$A_\ell = a, B_\ell = 2b, B^* = 2A_{\ell-1}\varepsilon,$$

and

$$B_{\ell-1} = (B_\ell A_{\ell-1} - \varepsilon)/A_\ell = (2b\varepsilon b^*/2 - \varepsilon)/a = \varepsilon(bb^* - 1)/a.$$

Case 2. $a < 2b$. Put $2b/a = \langle q_0; q_1, ..., q_\ell \rangle = A_\ell/B_\ell$. We now assume that $\varepsilon = (-1)^{\ell-1}$ and get

$$A_\ell = 2b, B_\ell = a, b^* = 2B_{\ell-1}\varepsilon.$$

Also,

$$A_{\ell-1} = (A_\ell B_{\ell-1} - \varepsilon)/B_\ell = \varepsilon(bb^* - 1)/a.$$

In either case, we find that

$$\varepsilon(b^*p - 4b)/(2a) = a\varepsilon b^*/2 + 2b\varepsilon(bb^* - 1)/a = A_{\ell-1}A_\ell + B_\ell B_{\ell-1},$$

and

$$(b^{*2}p - 8bb^* + 4)/(4a^2) = (b^*/2)^2 + ((bb^* - 1)/a)^2 = A_{\ell-1}^2 + B_{\ell-1}^2.$$

Since $p = A_\ell^2 + B_\ell^2$, then $D/p = q =$

$$pr^2 + 4(A_{l-1}A_l + B_l B_{l-1})r + 4(A_{l-1}^2 + B_{l-1}^2) = (rA_k + 2A_{k-1})^2 + (rB_l + 2B_{l-1})^2.$$

Also, $b^*\varepsilon + ar > 0$, since $s > 0$. Thus, if $a > 2b$, then $A_\ell r + 2A_{\ell-1} > 0$ which implies that $r \geq -1$. Also, if $a < 2b$, then $B_\ell r + 2B_{\ell-1} > 0$ implies that $r \geq -1$. If $r = -1$, then $q = (A_\ell - 2A_{\ell-1})^2 + (B_\ell - 2B_{\ell-1})^2 < A_\ell^2 + B_\ell^2 = p$, a contradiction. It follows that $r > 0$. By Theorem 6.2.8, we see that the value of $\ell(\alpha) = 2\ell + 3$. By Theorem 6.2.9, we know that there must exist, for any value of $m > 0$, some $p = A^2 + B^2$ such that $A > B$ and $m(A/B) > m$. Since $\ell \geq m$, we have $\ell(\alpha) \geq 2m + 3$ for this value of p. □

We have seen, therefore, that if D is given by the above formula, there will be only one principal reduced ideal (the trivial one), but there can be an arbitrary number of reduced ideals equivalent to the reduced ideal $I = [p, (p + \sqrt{D})/2]$, depending upon the choice of the prime p. Finally, by Theorem 5.3.4, if $h_\Delta = 2$, then all $Q_i/2$ are prime and by the above, $\ell(I) \geq 2\ell + 3$.

Another conjecture given by Leu in [182] is

Conjecture 6.2.4. *If $D = n^2 + 4$ is a radicand, then $|T_\Delta| \leq 2h_D - 1$.*

As noted by Louboutin in his aforementioned review, this conjecture is false. He notes only one counterexample. We independently established this fact in [266] and did some computation and arrived at the following list of counterexamples for $D \leq 2 \cdot 10^6$, where $D = n^2 + 4$.

Table 6.2.2: *Counterexamples to Conjecture* 6.2.4:

D	$2h_D - 1$	$\lvert S_D \rvert$	factors of D
237173	21	24	prime
316973	23	27	197, 1609
552053	29	33	prime
877973	39	42	37, 61, 5197
1585085	47	49	5, 61, 5197
1760933	59	60	373, 4721
1885133	51	56	1217, 1549

We also compiled a list of counterexamples for $D \leq 10^9$ and found 518 counterexamples, too lengthy therefore to list here.

The difficulty shown here in attempting to solve problems involving the number of inert primes less than M_Δ, and the class number for ERD-types as simple as $n^2 + 4$, points to the general difficulty of this problem. Significant work remains to be done.

We conclude this section with an open problem. In Remark 6.2.6, we noted that it is known for which values of $\Delta > 0$ there are no ramified or inert primes less than M_Δ, i.e. the only $\Delta > 0$ with *only* split primes less than M_Δ are unconditionally known. Simlarly, the values of $\Delta > 0$ with *only* ramified primes (or no primes) less than M_Δ are known. They are $\Delta \in \{8, 12, 5, 13, 24, 28, 60, 120\}$. It is natural to ask

Query 6.2.4. For which fundamental discriminants $\Delta = \Delta_0 > 0$ does $|T_\Delta| = \pi(M_\Delta)$? (See Corollary 6.2.2.)

In other words, for which $\Delta > 0$ are there no inert primes less than M_Δ? This author poses

Conjecture 6.2.5. *The answer to Query* 6.2.4 *are those discriminants* $\Delta > 0$ *with associated radicands* $D \in \{2, 3, 5, 6, 7, 10, 13, 15, 17, 19, 22, 30, 31, 33, 34, 39, 46, 51, 57, 70, 73, 79, 91, 97, 105, 106, 114, 129, 130, 145, 154, 231, 235, 246, 249, 345, 385, 399, 511, 561, 609, 1065, 3705\}$.

This author has checked up to $\Delta < 10^7$ and found only these values. Furthermore, Andrew Granville has used some deep analytic techniques to prove that if $\Delta > 10^{30}$, then there are no values satisfying the criterion in Query 6.2.4. Hugh Williams has told this author that he can sieve up to 10^{20} to test Conjecture 6.2.5, but no higher. Hence, it remains for the analytic techniques of Granville to bring down the bound to the level where Williams' computational techniques can take over to solve the problem. Currently, however, it remains open.

Exercises 6.2

1. Prove Conjecture 6.2.1 (with one GRH-ruled out exception) using the techniques of Chapter Five.

2. In the Proof of Theorem 6.2.2:

   (a) Prove that (6.2.1)–(6.2.3) hold. (*Hint*: Use Theorem 3.2.1 and examine the simple continued fraction expansion of w_Δ case by case — eight such cases will suffice.)

(b) Establish (6.2.4)–(6.2.5) from (a).

(c) In Case B, prove that $s < 10$. (*Hint*: Use the same techniques as used in Case A.)

(d) Establish Claim 3. (*Hint*: Since $D < M_{p,i}$ then there is a prime q with $3 \le q \le p$ and $(D/q) = 1$. Show that $q^2 \in \Lambda^*(\Delta)$.)

3. Assume that $\Delta = \Delta_0 > 0$ is a fundamental discriminant with radicand D such that $(D/p) \neq 1$ for all primes $p < \sqrt{\Delta}/2$. Let $I = [N(I), b + w_\Delta]$ be a reduced ambiguous ideal of $\mathcal{O}_\Delta$. Prove that:

   (a) The period length $\ell(I)$, of the simple continued fraction expansion of $(b + w_\Delta)/N(I)$, divides 4. Furthermore, if $\ell(I) = 1$, then $D = m^2 + 4$ or $D = 2$; if $\ell(I) = 2$, then $\Delta = m^2 + 4r$ with $r > 1$ and $r \mid m$; and if $\ell(I) = 4$, then $\Delta = m^2 - 4r$ with $r > 1$ and $r \mid m$. (*Hint*: Use the techniques in the proofs of Theorems 4.2.4–4.2.5 and 4.2.7, which make this an easy exercise by comparison.)

   (b) If I and J are reduced ambiguous ideals of $\mathcal{O}_\Delta$ with $I \sim J$, $N(I) = a$, and $N(J) = b$, then $\Delta = m^2 \pm 4r$ with $r = ab$. (*Hint*: This is a straightforward application of part (a).)

4. Let $\Delta = m^2 \pm 4r$ be a fundamental discriminant with $r \mid m$ and $1 < r < m/2$. Prove that if s is a square-free divisor of r and I is an $\mathcal{O}_\Delta$-ideal in the class of the ideal of norm s with $N(I) < \sqrt{\Delta}/2$, then $N(I) \mid \Delta$. (*Hint*: Look at the ideal $[\delta, (m + \sqrt{\Delta})/2]$ and the simple continued fraction expansion of $(m + \sqrt{\Delta})/(2\delta)$ in the two cases $\Delta = m^2 + 4r$ and $\Delta = m^2 - 4r$.)

5. Prove that if $\Delta > 0$ is a discriminant and Q_i appears in the simple continued fraction expansion of w_Δ with $Q_{i-1}/\sigma > \sqrt{\Delta}/2$, then $Q_i/\sigma < \sqrt{\Delta}/2$.

6. Prove Lemma 6.2.2. (*Hint*: Use Theorem 1.3.1.)

Chapter 7

Influence of the Infrastructure.

In this chapter, we examine the influence which some simple assumptions concerning the infrastructure's ordering have upon class number determination, bounding the regulator, determining formulae for the period length of the principal class, and solutions of conjectures and problems involving parameterized versions of a fundamental discriminant $\Delta > 0$.

In keeping with our general theme of exploring "quadratics" from new perspectives, we begin with a concept which was introduced in [265].

7.1 Quadratic Residue Covers.

In Exercise 2.1.17, we explored the phenomenon where *all* Q_i's in the simple continued fraction expansion of w_Δ are powers of 2 and its influence upon h_Δ when $\Delta = \Delta_0 \equiv 1 \pmod 8$ is a positive fundamental discriminant. This interest arose out of work of Shanks [327], who looked at this phenomenon and calculated up to large bounds on Δ those for which $h_\Delta = 1$. He found only five values. In [255], it was conjectured that these are *all* of the possible values. In [258], this was proved in the following elementary fashion.

Theorem 7.1.1. *If $\Delta = \Delta_0 \equiv 1 \pmod 8$ is a positive fundamental discriminant such that all Q_i's in the simple continued fraction expansion of w_Δ are powers of 2, then $h_\Delta = 1$ if and only if $\Delta \in \{17, 41, 113, 353, 1217\}$.*

Proof. First, we note that by Exercise 2.1.19, if all Q_i's are powers of 2 and $h_\Delta = 1$, then $\Delta = (2^n + 1)^2 + 2^{n+2}$ with $n > 0$ and $\ell(w_\Delta) = 1 + 2n$, and observe that $127 = 2^7 - 1$. If $n \equiv i \pmod 7$, then $D \equiv j^2 \pmod{127}$ where $(i, j) \in \{(0,32), (1,25), (2,26), (3,42), (4,37), (5,57), (6,6)\}$. Hence, $(\Delta/127) = 1$ so if $N(\mathcal{P}) = 127$ for an $\mathcal{O}_\Delta$-prime $\mathcal{P}$, then $127 \in \Lambda^*(\Delta)$ whenever $127 < \sqrt{\Delta}/2$ by Corollary 1.4.3, contradicting our hypothesis. Hence, $127 \geq \sqrt{\Delta}/2$, i.e. $\Delta \leq 64516$, for which a simple computer check yields only the values of our list. (7.1.1) □

Remark 7.1.1. By Exercise 2.1.21, if $\Delta = \Delta_0 \not\equiv 1 \pmod 4$ is a positive fun-

(7.1.1) Observe that $\Delta = (2^n + 1)^2 + 2^{n+2} = (2^n + 3)^2 - 8$, which were the forms studied by Shanks in [327]. Thus, Theorem 7.1.1 not only affirmatively settles the conjecture posed in [255], but also some queries raised by Shanks in [327].

damental discriminant and all Q_i's are powers of a prime $p > 2$, then $h_\Delta > 1$. Furthermore, if $\Delta \not\equiv 1 \pmod 4$ and all Q_i's *are* powers of 2, then from [255] we have

Conjecture 7.1.1. *If $\Delta = \Delta_0 \not\equiv 1 \pmod 4$ is a fundamental discriminant with radicand D and all Q_i's in the simple continued fraction expansion of w_Δ are powers of 2, then $h_\Delta = 1$ if and only if $D \in \{2, 3, 6, 11, 38, 83, 227\}$.*

Observe that, by Theorem 5.2.3, the shape of the D's in Conjecture 7.1.1 must be $D = a^2 \pm 2$. Hence, Conjecture 7.1.1 falls into Theorem 5.4.3, so that we know (with one GRH-ruled out exception) that these are all the values.

Remark 7.1.2. The discovery of the elementary proof of Theorem 7.1.1 was the motivation for the introduction of a more general concept. The proof could have been made to work if we had used any set of primes $\mathcal{C}$ such that for any $n \geq 0$, we have $(S_n/q) = 1$ for some $q \in \mathcal{C}$ where

$$S_n = (2^n + 3)^2 - 8,$$

is the Shanks' sequence. We used $\mathcal{C} = \{127\}$, but we could have used $\mathcal{C} = \{5, 7, 13, 17, 241\}$ since $(S_n/5) = 1$ when $n \equiv 2 \pmod 4$; $(S_n/7) = 1$ when $n \equiv 0 \pmod 3$; $(S_n/13) = 1$ when $n \equiv 1, 11 \pmod{12}$; $(S_n/17) = 1$ when $n \equiv 0, 1, 4, 7 \pmod 8$; and $(S_n/241) = 1$ when $n \equiv 5, 19 \pmod{24}$. Notice that the integers are completely covered by the various congruences modulo 4, 3, 12, 8, 24, i.e. for any integer n, one of these congruences must hold. This leads us to introduce

Definition 7.1.1. If for some function $f : \mathbf{Z} \geq 0 \to \mathbf{Z} \geq 0$ (where $\mathbf{Z} \geq 0$ denotes the non-negative integers), we have a finite set of primes $\mathcal{C}$ such that, for any $n \in \mathbf{Z} \geq 0$ we get $(f(n)/q) \neq -1$ for some $q \in \mathcal{C}$, then we call $\mathcal{C}$ a *quadratic residue cover (QRC)* for f. If $K_n = Q(\sqrt{f(n)})$ and $\mathcal{C}$ is a QRC for f, we say that $\mathcal{C}$ is a *QRC for the fields K_n $(n = 0, 1, \ldots)$*.

What we now do is generalize this sequence S_n of Shanks, and show how a computer search can be used to establish a lower bound on the growth of the class number of $\mathcal{O}_\Delta$, for fundamental positive discriminants Δ corresponding to these sequences.

By Exercise 2.1.18, if $\ell(w_\Delta) \geq 3$, all elements of $\Lambda^*(\Delta)$ are powers of a single integer $c > 1$ if and only if

$$\Delta = (bc^n + (c-1)/b)^2 + 4c^n = \Delta_n(c, b), \tag{7.1.1}$$

where $c \equiv 1 \pmod b$ and $n > 0$. Also $\ell(w_\Delta) = 2n + 1$ with

$$\begin{aligned}
P_{2j} &= \sigma(bc^n - (c-1)/b)/2 && (1 \leq j \leq n),\\
Q_{2j} &= \sigma c^j, && (1 \leq j \leq n),\\
a_{2j} &= bc^{n-j} && (1 \leq j \leq n),\\
P_{2j+1} &= \sigma(bc^n + (c-1)/b)/2 && (0 \leq j \leq n),\\
Q_{2j+1} &= \sigma c^{n-j} && (0 \leq j \leq n),\\
&\text{and}\\
a_{2j+1} &= bc^j && (0 \leq j \leq n).
\end{aligned}$$

Also, from Halter-Koch [104], the value of the fundamental unit is given by

$$\varepsilon_\Delta = \gamma\beta^n, \tag{7.1.2}$$

where

$$\gamma = (\sigma(bc^n + (c-1)/b) + 2\sqrt{\Delta})/(2\sigma),$$

and

$$\beta = (\sigma(bc^n + c + 1) + 2b\sqrt{\Delta})/(2\sigma c).$$

When Δ is actually a fundamental discriminant, we have

Definition 7.1.2. Let h_{Δ_n} denote the class number of $\mathcal{O}_\Delta$ for Δ, as given in (7.1.1), when Δ represents a fundamental discriminant, and set $K_n(c,b) = \mathbf{Q}(\sqrt{\Delta})$.

For composite $\Delta_n(c,b)$, we have

Theorem 7.1.2. *If Δ in (7.1.1) is a fundamental discriminant, then $2 \mid h_{\Delta_n}$ whenever Δ is composite.*

Proof. If n is even, then Δ is a sum of two squares. Consequently, if p is an odd prime divisor of Δ, then $p \equiv 1 \pmod 4$. If n is odd and $p \mid \Delta$, we see that $(-4c/p) = (-c/p) = 1$. We also note that

$$b^2\Delta = (b^2c^n + c + 1)^2 - 4c, \tag{7.1.3}$$

so if $p \mid \Delta$, then $(4c/p) = (c/p) = 1$. Since $(-c/p) = 1 = (c/p)$, we must have $(-1/p) = 1$ and $p \equiv 1 \pmod 4$. By Corollary 1.3.2(2), we must have h_{Δ_n} even whenever Δ is composite. □

Theorem 7.1.2 tells us that the only possibility for h_{Δ_n} to be 1 occurs when $\Delta_n(c,b)$ is a prime.

Consider the case where c is a composite and let q be any prime divisor of c such that $q^2 \leq c$. It is clear in this case that $\{q\}$ is a QRC for the fields $K_n(c,b)$. We will use the notation $\mathcal{C}(c,b) = \mathcal{S}$, where $\mathcal{S}$ is some set, to denote that $\mathcal{S}$ is a QRC for $K_n(c,b)$; no uniqueness is to be attributed then to the use of the symbol $\mathcal{C}(c,b)$, as several different sets could qualify for a $\mathcal{C}(c,b)$. We have already seen, in the case of S_n, that $\mathcal{C}(2,1) = \{127\}$ or $\{5,7,13,17,241\}$. Thus, we see that $\mathcal{C}(c,b) = \{q\}$ if $q \mid c$ and $q^2 \leq c$. If c is a prime and $c \notin \mathcal{C}(c,b)$, then because of the finiteness condition on $\mathcal{C}(c,b)$, there must exist some minimal positive $\ell \in \mathbf{Z}$ such that $c^\ell \equiv 1 \pmod q$ for all $q \in \mathcal{S}$. We call such a QRC an *ℓ-QRC*, and denote it by $\mathcal{C}_\ell(c,b)$. The following theorem shows why it is useful to try to find a $\mathcal{C}_\ell(c,b)$ with a minimal ℓ. Note that this ℓ is not to be confused with $\ell(w_\Delta)$.

Theorem 7.1.3. *Let q be any prime such that $(\Delta/q) = 1$, where $\Delta > 0$ is a fundamental discriminant. If no power of q appears in $\Lambda^*(\Delta)$, then $h_\Delta > \log\Delta/(2\log q)$.*

Proof. Since $(\Delta/q) = 1$, there must exist some prime ideal Q which divides (q) such that $N(Q) = q$. Since Q^{h_Δ} is a primitive, principal ideal and $N(Q^{h_\Delta}) = q^{h_\Delta}$, we cannot have Q^{h_Δ} as a reduced principal ideal. Hence, by Corollary 1.4.3, we have $2q^{h_\Delta} > \sqrt{\Delta}$. □

Corollary 7.1.1. *If $\mathcal{C}_\ell(c,b)$ is an ℓ-QRC for the fields $K_n(c,b)$, and no element of $\mathcal{C}_\ell(c,b)$ divides $\Delta_n(c,b)$ for any $n \geq \ell$, then $h_{\Delta_n} > (n-1)/\ell$.*

Proof. Let $h = h_{\Delta_n}$ and let $q \in \mathcal{C}_\ell(c,b)$. Since $c^\ell > q$ and $q \nmid c\Delta_n(c,b)$, we must have that

$$c^{\ell h} > q^h > \sqrt{\Delta_n(c,b)}/2 > bc^n/2 \geq c^{n-1}.$$

□

Corollary 7.1.2. [(7.1.2)] *If c is composite, then $h_{\Delta_n} > 2n-1$.*

Proof. For $C(c,b) = \{q\}$ as above, we get

$$2q^h > \sqrt{\Delta_n(c,b)} > c^n > q^{2n} \geq 2q^{2n-1};$$

which implies that $h_{\Delta_n} > 2n-1$. □

We observe that this means $h_{\Delta_n} > 1$ for $n \geq 1$, when c is composite. To illustrate further the usefulness of QRC's, we consider the case where $\ell(w_\Delta) < 3$. We need only look at $\Delta \equiv 5 \pmod 8$, since $\{2\}$ is a QRC in the other cases. If $\ell(w_\Delta) = 1$, then by Theorem 3.2.1, $\Delta = (2n+1)^2 + 4$ and if $\ell(w_\Delta) = 2$, then $\Delta = c^2(2n+1)^2 + 4c$ where $c > 1$.

We now show that we have the unfortunate fact that, if $c = 1$ or a prime, then we can *never* have a QRC for $Q(\sqrt{c^2(2n+1)^2+4c})$. Thus, Conjecture 7.1.1 cannot be proved using quadratic residue covers.

First, we need an interesting technical lemma, the proof of which is in [200]. The proof is less interesting than the application of the result.

Lemma 7.1.1.

(a) *If $f(x) = ax^2 + bx + c \not\equiv 0 \pmod 4$ where $a, b, c \in \mathbb{Z}$, $a \neq 0$ and $\gcd(a,b,c)$ is square-free, then there are infinitely many $n \in \mathbb{Z}$ such that $f(n) \in \mathbb{Z}$ is square-free.*

(b) *Let $f(x) = ax^2 + bx + c \not\equiv 0 \pmod 4$, where $a, b, c \in \mathbb{Z}$, $a > 0$ and $\gcd(a,b,c)$ is square-free. If $d = b^2 - 4ac$ and C is any finite set of odd prime integers not dividing d, then there exist infinitely many $n \in \mathbb{Z}$ such that $f(n)$ is a square-free positive integer with $(f(n)/p) = -1$ and $p \in C$ (except when $p = 3 \in C$, $a \equiv 2 \pmod 3$ and $d \equiv 1 \pmod 3$, in which case $(f(n)/3) \neq -1$, $n \in \mathbb{Z}$).*

Proposition 7.1.1. *If C is any finite set of odd primes, then*

(a) *there exist infinitely many fundamental radicands $D = m^2 + 2$ with $(D/p) = -1$ for all $p \in C$, and*

(b) *there exist infinitely many fundamental radicands $D = m^2 - 8$ with $(D/p) = -1$ for all $p \in C$.*

Proof. This follows easily from Lemma 7.1.1. □

We see that part (b) of Proposition 7.1.1 shows us that there does not exist any QRC for the family $D = m^2 - 8$, whereas Theorem 7.1.1 shows that there is one for a *sub*family of that family. For the family $D = m^2 + 2$, the ideal over 2 is principal since $N(m + \sqrt{D}) = -2$ and 2 ramifies. Therefore, any QRC may be assumed not to contain 2, as with the aforementioned family. The reason is that, for those, 2 splits and the ideals over 2 are principal since $N((m+\sqrt{D})/2) = 2$. Similarly, there can be no hope of even finding a QRC for the family $D = m^2 + 4 \equiv 5 \pmod 8$ by Lemma 7.1.1, since we may set $m = 2n+1$, $D = f(n)$ where $f(x) = 4x^2 + 4x + 5$ with $d = -2^6$, and 2 is inert for this family. Also, Theorem 7.1.1 tells us that in the case of the Shanks' sequence S_n, $h_{\Delta_n} > 1$ for $n \geq 6$ when S_n is a square-free discriminant. Thus, we have a simple parametric family of fields for which

(7.1.2) This and the above results are taken from [265].

we know all the square-free discriminants Δ with $h_\Delta = 1$. There are many cases of parametric families of discriminants for which this is very difficult to determine. Consider for instance, $\Delta_n = (2n+1)^2+4$. By Theorem 5.4.3, we know that $h_{\Delta_n} = 1$ if and only if $n = 0, 1, 2, 3, 6, 8$ with one GRH-ruled out exception. We have seen numerous results of this nature in Chapter Five.

We can now improve upon the result in Corollary 7.1.1, albeit the following conditional result depends on deep analytic results from Chapter Five.

Theorem 7.1.4. *Let $c, b \in \mathbb{Z}$ be fixed and positive. If $\Delta = \Delta_n(c,b) > B > 73131$ ($> e^{11.2}$) for $B \in \mathbb{R}$, then $h_\Delta = h_{\Delta_n} > .24\sqrt{B}/(\log B)^3$ ($n \geq 2$) with one* GRH-*ruled out exception.*

Proof. Put $\eta = 1/logB$. Since $\Delta > \max\{e^{11.2}, e^{1/\eta}\}$, then by Theorem 5.4.1 we have $L(1,\chi) > .655\sqrt{\Delta}\eta\Delta^{-\eta}$ with one possible exceptional value of Δ. Thus, by the analytic class number formula (5.4.1)

$$h_\Delta > .655\sqrt{\Delta}\eta\Delta^{-\eta}/(2R), \tag{7.1.4}$$

where R is the regulator of $K_n(a, b)$.

Also, by (7.1.2) we see that

$$R < (n+1)\log\sqrt{\Delta}. \tag{7.1.5}$$

Since $\sqrt{\Delta} \geq bc^n \geq c^n$, and $c \geq 2$, we get

$$n < \log\sqrt{\Delta}/\log c < \log\Delta - 1.$$

Hence, by (7.1.4)–(7.1.5)

$$h_\Delta > \frac{.655\eta\Delta^{1/2-\eta}}{(\log\Delta)^2} > \frac{.655\eta B^{1/2-\eta}}{(\log B)^2} = \frac{.655\sqrt{B}}{e(\log B)^3} > \frac{.24\sqrt{B}}{(\log B)^3}. \qquad \square$$

Theorem 7.1.4 is certainly better than Corollary 7.1.1, but it is conditional, whereas Corollary 7.1.1 is not conditional, given a certain $C_\ell(c,b)$, and is elementary. As an example, we point out that for S_n, we have a cover $C_7(2,1) = \{127\}$ and $h_{\Delta_n} > (n-1)/7$. Thus, if $n \geq 70001$, we know that $h_{\Delta_n} > 10000$ unconditionally. By Theorem 7.1.4, we would get $h_{\Delta_n} > 10000$ for $n \geq 32$ with one GRH-ruled out exception. Later we will show how to use the computer to narrow the gap between 32 and 70001.

Now we turn to describing a search technique for QRC's and some numerical results.

We first point out that if we are attempting to find a $C_\ell(c,b)$ for $K_n(c,b)$, we may assume that $b^2 \leq c-1$. This follows from

Theorem 7.1.5. *If C is a ℓ-QRC for the fields $K_n(c,b)$, then it is also an ℓ-QRC for the fields $K_n(c,(c-1)/b)$.*

Proof. Put $\Delta'_n(c,b) = \Delta(c,(c-1)/b)$, so that

$$c^{-2n}\Delta'_n(c,b) = (bc^{-n} + (c-1)/b)^2 + 4c^{-n}.$$

Putting $m \equiv -n \pmod{\ell}$, $m \geq 0$, we get

$$c^{-2n}\Delta'_n(c,b) \equiv \Delta_m(c,b) \pmod{q},$$

for any $q \in C$. Since $(\Delta_m(c,b)/q') \neq -1$ for some $q' \in C$, we get $(\Delta'_n/q') \neq -1$. $\square$

While we are not able to provide a proof that for some ℓ, a $\mathcal{C}_\ell(c,b)$ always exists for $K_n(c,b)$, we can provide a simple heuristic reason for believing that this is the case. Let $\mathcal{N}_k$ denote the set $\{0,1,2,3,\ldots,k-1\}$. It is a well-known result of Sylvester (see Dickson [73, p. 385]) that if T_k is the set of distinct prime factors of c^k-1, then $|T_k| \geq \tau(k)-1$, where τ is the divisor function (see Definition 4.1.2). Also, if $q \in T_k$, the distinct values for $\Delta_n(c,b)$ modulo q can occur only for $n \in \mathcal{N}_k$. Now for any prime $q \in T_k$, it seems reasonable to assume that the probability that $(\Delta_n(c,b)/q) \neq -1$ is 1/2. Thus, we would expect that if $\nu_k = |T_k|$ and 2^{ν_k} is much larger than k $(= |\mathcal{N}_k|)$, then T_k is likely to be a possibility for $\mathcal{C}_\ell(c,b)$ with $\ell = k$. Notice that if $k = 2^\mu\kappa$, where κ is an odd prime, we have $\tau(k) = 2\mu+2$. In this case, the ratio

$$2^{\nu_k}/k \geq 2^{2\mu+2-1}/2^\mu\kappa = 2^{\mu+1}/\kappa.$$

Thus, if $\kappa = 3$, say, we would not expect to have a large value for μ before we found a $\mathcal{C}_\ell(c,b)$ with $\ell = 3\cdot 2^\mu$. In fact, in a preliminary computer run, we found that for all prime values of $c \leq 200$, there exists a $\mathcal{C}_\ell(c,b)$ for each b which divides $c-1$ with $\ell = 2^\mu$ or $3\cdot 2^\mu$ and $\mu \leq 4$. Also, for these covers the maximum value of $|\mathcal{C}_\ell(c,b)|$ is 6.

Encouraged by the success of this preliminary run, we ran a second program which attempted to find covers $\mathcal{C}_\ell(c,b)$ with smaller ℓ values. For a given pair (c,b) and a value of k, the program first determined T_k and $\mathcal{N}_k$. For each prime $q \in T_k$, the values of m such that $(\Delta_m(c,b)/q) \neq -1$ were determined and deleted from those in $\mathcal{N}_k$. When a q-value caused elements to be deleted from $\mathcal{N}_k$, it was added to a set $\mathcal{C}_k$, previously initialized to $\emptyset$. If, at some point, $\mathcal{N}_k = \emptyset$, then $\mathcal{C}_k$ is a k-QRC for $K_n(c,b)$. The program attempted to find k-QRCs for $k = 1, 2^i, 2^i\kappa$, where κ is an odd prime. For values of k of the form 2^i or $2^i\kappa$, the program would initialize k to either 2 or κ and then continue to double k until either a cover was found or $|\mathcal{C}_k| > 8$. This value of 8 was chosen in order to terminate what might otherwise be a long and likely fruitless attempt to find a cover with a small value of ℓ. Our previous experience indicated that if an ℓ-QRC exists for $K_n(c,b)$, then $|\mathcal{C}_\ell(c,b)|$ tends to be small. The smallest value of k of the forms mentioned above, such that $\mathcal{C}_k$ is a cover, was recorded, and then the program tried to reduce the number of elements in $\mathcal{C}_k$ by testing every possible subset of it in order to find one with the minimal number of elements, which was still a cover. This minimal cover was used for $\mathcal{C}_\ell(c,b)$.

The program was written in MAPLE and tested on $\Delta_n(c,b)$ for all prime values of c such that $2 \leq c \leq 10000$ and all values of b such that $b \mid c-1, b \geq 1, b^2 \leq c-1$. No attempt was made to eliminate values of $\Delta_n(c,b)$ which have square factors. In under a week of run time (in background) on a SUN-4 computer, a cover was found for $K_n(c,b)$ for every possible pair (c,b) under consideration. Curiously, the largest value of ℓ which was found is 22 for

$$\mathcal{C}_{22}(7,1) = \{23, 1123, 293459, 10746341\}.$$

The following list gives the number of covers found for the various values of ℓ recorded by the program. The total number of covers found is 9544.

Table 7.1.1

ℓ	Number of Covers
1	6582
2	1988
3	403
5	226
6	187
7	56
10	51
11	12
12	28
13	3
14	2
17	2
19	1
20	2
22	1

The largest value of $|\mathcal{C}_\ell(c,b)|$ found in our run is 5 for

$$\mathcal{C}_{12}(4253,4) = \{5,7,13,31,769\}.$$

The largest element in any of our covers is

$$q = 891548343411670002940792447441$$

in $\mathcal{C}_{11}(1873,1) = \{67,89,q\}$. One of the more interesting covers is that for $\ell = 19$, which is

$$\mathcal{C}_{19}(43,6) = \{229,4219,46399,2137444528747943\}.$$

The remarkable feature of this run is that a cover was always found, and found for a relatively small value of ℓ. Furthermore, there was no tendency for the value of ℓ to increase with increasing values of c (as one would tend to expect by our heuristic). Indeed, some of the covers with large ℓ-values such as

$$\mathcal{C}_{20}(5,1), \quad \mathcal{C}_{22}(7,1), \quad \mathcal{C}_{13}(17,2), \quad \mathcal{C}_{17}(43,3), \quad \mathcal{C}_{19}(43,6)$$

occurred when c is relatively small.

In view of Corollary 7.1.1, it is also of some interest to investigate the possible existence of *strict* QRCs or SQRCs. These are quadratic residue covers $\mathcal{C}(c,b)$ such that, for any $n \geq 0$, there must exist some $p \in \mathcal{C}(c,b)$ for which $(\Delta_n(c,b)/p) = 1$. The argument used above also suggests that an ℓ-SQRC should always exist for any $K_n(c,b)$. We modified our program to search for ℓ-SQRCs for $K_n(c,b)$ for c, b in the same range as before. In under two weeks of (background) run time, we found an ℓ-SQRC for every possible pair (c,b) in our range. Once again, the ℓ-values did not get very large, the largest being 28 for $\mathcal{C}_{28}(3,1)$. These results are summarized in

Table 7.1.2

ℓ	Number of Covers
1	5628
2	2032
3	692
5	459
6	332
7	136
10	139
11	29
12	73
13	6
14	6
17	3
19	1
20	5
22	2
28	1
	total 9544

In this run, the largest value for $|C_\ell(c,b)|$ is once again 5, but it occurred for 5 different covers. Also, as observed earlier, there is no tendency for ℓ to increase with increasing values of c. Given these phenomena and our heuristic, it does not seem unreasonable to pose the conjecture below.

Conjecture 7.1.2. *For any prime value of c and any b with $c \equiv 1 \pmod{b}$ and $b \geq 1$, there always exists, for some ℓ, an ℓ-QRC for $K_n(c,b)$.*

We have already found all of the square-free elements of Shanks' sequence S_n for which $h_\Delta = 1$. We can now go somewhat further, as we can bound from below the value of h_{Δ_n}. In fact, we can easily establish the following result.

Theorem 7.1.6. *If $\Delta = \Delta_0 > 0$ is a fundamental radicand such that all elements of $\Lambda^*(\Delta)$ are powers of an odd prime p, $\ell(w_\Delta) \geq 3$ and $h_\Delta = 1$, then*

$$\begin{array}{ll} \Delta \in \{37, 61, 157, 397, 7213\} & \textit{when } p = 3, \\ \Delta \in \{101, 461, 941\} & \textit{when } p = 5, \\ \Delta \in \{197, 317, 557, 1877\} & \textit{when } p = 7, \\ \Delta \in \{773\} & \textit{when } p = 11. \end{array}$$

Proof. By using the ℓ-values determined by our program and the bound of Corollary 7.1.1, we can easily establish an upper bound on Δ for h_Δ to be 1. Most of the Δ-values below this bound can be easily eliminated by finding a small prime q such that $q \neq p$, $q < \sqrt{\Delta}/2$, and $(\Delta/q) = 0, 1$. The few remaining numbers can be tested by evaluating h_Δ. □

As pointed out earlier, the result of Corollary 7.1.1 gives us an unconditional lower bound on h_{Δ_n}, but the bound is not a very good one. We will discuss a method by which this bound can be improved. Our technique is elementary, unconditional and can easily be implemented on a computer. The proof of the following technical result is in [265].

Theorem 7.1.7. *Let $\Delta = \Delta_0 > 0$ be a fundamental discriminant $h = h_\Delta$, and let R be the regulator of $K = Q(\sqrt{\Delta})$. If $p_1, p_2, \ldots, p_k$ are distinct primes such that* *the Kronecker symbol* $(\Delta/p_i) = 1$ $(i = 1, 2, \ldots, k)$, *then*

$$h > \frac{2^k \log \tau}{k! V_k R}(B - W_k)^{k-1}(B + (k-1)W_k),$$

where

$$B = \log(\sqrt{\Delta}/2), \quad V_k = \prod_{i=1}^{k} \log p_i, \quad W_k = \sum_{i=1}^{k} \log p_i, \quad \textit{and } \tau = (1 + \sqrt{5})/2.$$

We will now apply this result to our fields $K_n(c, b)$. By (7.1.3), we get

$$\sqrt{\Delta_n(c, b)} < (b^2 c^n + c + 1)/b < c^{n+1}.$$

Thus, from (7.1.5), we find that

$$R < (n+1)^2 \log c. \qquad (7.1.6)$$

Also,

$$\log(\sqrt{\Delta}/2) > n \log c + \log b - \log 2 \geq (n-1) \log c.$$

Hence, by Theorem 7.1.7, we have

$$h_{\Delta_n} > \frac{2^k \log \tau}{k! V_k (n+1)^2 \log c}((n-1)\log c - W_k)^{k-1}((n-1)\log c + (k-1)W_k).$$

Since, for $c > x > y > 0$, we get

$$(c - y)^n (c + ny) > (c - x)^n (c + nx),$$

we see that if we have values A, B such that $W_k \leq A$, $V_k \leq B$, then

$$h_{\Delta_n} > \frac{2^k \log \tau}{k! B (n+1)^2 \log c}((n-1)\log c - A)^{k-1}((n-1)\log c + (k-1)A). \qquad (7.1.7)$$

We now show how to use (7.1.7) to improve upon the bound given in Theorem 7.1.4. As mentioned in the comments after Theorem 7.1.4, if $\Delta = \Delta_n(2, 1)$, then $h_\Delta > 10000$ for $n \geq 70001$. By a simple computer search, it is easy to establish that for each c with $0 \leq n \leq 70000$, there exist 6 primes less than or equal to 163 such that $(S_n/p) = 1$ for any of these 6 primes p. Since we know that $(S_n/2) = (S_n/127) = 1$, we may assume that $k = 6$,

$$A = \log(2 \cdot 127 \cdot 149 \cdot 151 \cdot 157 \cdot 163),$$
$$B = \log 2 \log 127 \log 149 \log 151 \log 157 \log 163.$$

Thus, by (7.1.7) we find that $h_{\Delta_n} > 10000$ for $n \geq 257$. Now for each n with $0 \leq n \leq 256$, there exist 8 distinct primes less than or equal to 131 such that $(S_n/p) = 1$ for any of these eight primes. Thus, we can put $k = 8$,

$$A = \log(2 \cdot 101 \cdot 103 \cdot 107 \cdot 109 \cdot 113 \cdot 127 \cdot 131),$$
$$B = \log 2 \log 101 \log 103 \log 107 \log 109 \log 113 \log 127 \log 131,$$

and we get from (7.1.7) that $h_{\Delta_n} > 10000$ for $n \geq 145$, a considerable improvement over the 70001 bound mentioned above. This, then, completes the discussion of quadratic residue covers.

7.2 Consecutive Powers.

The concluding section of this chapter is devoted only to remarks upon consecutive powers in the continued fraction expansion of w_Δ for a given fundamental discriminant $\Delta > 0$. To present the full theory, as developed by Mollin–Williams in [262] would be beyond the scope of this book. However, it is worth mentioning what has been done.

We began, in Exercise 2.1.17, to look at consecutive powers of norms of reduced principal ideals for positive fundamental discriminants, and continued at various places in text up to the preceding section. Basically, we have been talking about the continued fraction ordering scheme, within the infrastructure, as described throughout the text starting with its introduction in Chapter Two. Because of the importance of this continued fraction ordering scheme, we will employ it here by using terms such as "consecutive" or "in a row" with reference to this particular ordering, which amounts to looking at norms of successive Lagrange neighbours. Although the assumption that the norms of *less than three consecutive* elements of $\Lambda^*(\Delta)$ are powers of a single integer provides us with very little information concerning the set of reduced principal ideals, it is rather remarkable how the simple assumption that three consecutive norms in a row are powers of $c > 1$ allowed us in [262], to determine so completely the principal period of reduced ideals of $\mathcal{O}_\Delta$, including a simple explicit formula for $\ell = \ell(w_\Delta)$. We were also able to show that there is an upper bound on the regulator of K, when the phenomenon of at least three norms in a row occur as powers of $c > 1$. In other words, the assumption is that for a fundamental discriminant $\Delta > 0$ with $\ell = \ell(w_\Delta) > 3$, there exists a $c \in \mathbf{Z}$, $c > 1$ such that $Q_i/\sigma = c^r$, $Q_{i+1}/\sigma = c^s$ and $Q_{i+2}/\sigma = c^t$ for some positive $r, s, t \in \mathbf{Z}$ where $1 \le i \le \ell - 2$.

This phenomenon was studied by Bernstein [24]–[25], Hendy [118], Levesque [183], and Levesgue and Rhin [184], previous to our developments in [262]. In particular, we were able to show that there are only finitely many $h_\Delta = 1$ when this phenomenon occurs, and we were able to list them as follows.

Theorem 7.2.1. *Assume that, in the simple continued fraction expansion of w_Δ (for a fundamental discriminant $\Delta > 0$), there are three or more powers of a single integer $c > 1$ in a row among the Q_i/σ. Then $h_\Delta = 1$ if and only if $c = p$ is a prime, and either*

(a) $\ell = \ell(w_\Delta) = 1$ *and* $D = \{2, 5, 13, 29, 53, 173, 293\}$,

or (b) $\ell > 1$ *and*

(1) *for* $p = 2$, $D \in \{3, 6, 11, 17, 38, 41, 83, 113, 227, 353, 857, 1217\}$,

(2) *for* $p = 3$, $D \in \{13, 21, 37, 61, 93, 157, 237, 397, 453, 7213\}$,

(3) *for* $p = 5$, $D \in \{101, 461, 941\}$,

(4) *for* $p = 7$, $D \in \{77, 197, 317, 557, 1253, 1877\}$,

(5) *for* $p = 11$, $D \in \{773, 1133\}$,

(6) *for* $p = 19$, $D = 437$,

with only one possible value remaining, the existence of which would be a counterexample to the GRH.

Proof. See [200]. □

The above was made possible by the obtaining of explicit upper bounds on the regulator in [262] in the case where the hypothesis of Theorem 7.2.1 is satisfied.

We note that, in particular, Theorem 7.2.1 tells us that we must look elsewhere to solve Gauss' conjecture: There are infinitely many real quadratic fields with class number 1. In other words, if the Gauss conjecture holds, then it must hold for a family of fields for which there are *at most* two consecutive powers of the Q_i/σ. Indeed, given the difficulty of the problem, it is fair to say that the distribution of the Q_i/σ may be quite random. We showed in [262] that, if three or more elements of $\Lambda^*(\Delta)$ in a row are powers of a single integer $c > 1$, then

$$D = (\sigma(qc^n + (c^k - 1)/q)/2)^2 + \sigma^2 c^n,$$

for some q dividing $c^k - 1$, and $n, k \in \mathbf{Z}$. Such parameterizations will not lead to infinitely many $h_\Delta = 1$.

The reader will by now appreciate the interplay of the algebraic and computational techniques in the determination of class number lists.

We note that the work in [262] discussed in this section was extended by Mollin–Williams in [260], and by Williams [385], especially in terms of formulae for $\ell(w_\Delta)$ and explicit determination of ε_Δ. It is noted at the end of [385] that, for some values of Δ under consideration, it is the case that $R = O((\log \Delta)^2)$, and comments that "it would be of very great interest if an infinite family" of Δ values could be produced such that for *each* Δ, the complete continued fraction period could be predicted with $R \gg (\log \Delta)^3$. There is a family of Yamamoto [391], namely

$$\Delta = (c^n r + c - 1)^2 + 4rc^n$$

where c, r are primes $c < r$ and $R \gg (\log \Delta)^3$ *infinitely often*. However, as pointed out by Williams in his penultimate comment in [385]: "Nevertheless no one knows (beyond a certain point) how to predict its period." This is the current state of research on this topic.[7.2.1]

[7.2.1] For a discussion of complexity issues, including a description of "big O" and "$\gg$", the reader is referred to Chapter Eight.

Chapter 8

Algorithms.

8.1 Computation of the Class Number of a Real Quadratic Field.

Before we begin our discussion of the history of class number computations, we must understand some basic concepts from computer science. Every positive $n \in \mathbf{Z}$ has a representation as $n = \sum_{i=0}^{s} a_i 2^i$, where the a_i are called *bits* which mean *b*inary dig*its*. Thus, the amount of time required on a computer to perform an algorithm is measured in terms of what are called *bit operations*, by which we mean addition, subtraction, or multiplication of two binary digits, the division of a two-bit integer by a one-bit integer, or the shifting of a binary digit by one place. The number of bit operations required to perform an algorithm is called its *computational complexity*. (8.1.1) To describe the number of bit operations needed, we require what is called *big-O* notation (the order of magnitude of the complexity), which is defined as follows. If f is a positive valued function (defined on either a real variable or on $\mathbf{N}$), then $O(f(x))$ denotes a quantity whose magnitude is less than some positive constant times $f(x)$ (for sufficiently large $x \in \mathbf{R}$, or for all $x \in \mathbf{N}$), i.e. if $g(x) = O(f(x))$, then there exists a positive $c \in \mathbf{R}$ such that $g(x) < cf(x)$ for all sufficiently large x. Mathematicians use the notation $g(x) \ll f(x)$ to describe this phenomenon. (See Exercise 8.1.4.) Computer scientists refer to *arithmetic operations* which are defined to be operations on (binary) integers (of a given bitsize). For instance, it can be shown that multiplication of two n-bit integers can be performed using as many as $O(n^2)$, or as little as $O(n \log n \log \log n)$ bit operations. Typically, one sees such statements as "... $O(X)$ arithmetic operations on integers no larger than $O(Y)$". Furthermore, the *actual* amount of time required to carry out a bit operation on a computer varies, depending upon the computer and current computer technology. The big O notation for measuring complexity is independent of the particular computer being used, i.e. despite the relative differences in the speeds of the various machines, the order-of-magnitude complexity of an algorithm remains the same.

A crucial and fundamental "time estimate" is called "polynomial time" (see

(8.1.1)The reader is referred to Schneier [319] for in-depth details on this topic. We will only briefly discuss *time* complexity, although *space* complexity may also be measured.

Exercise 8.1.4(g)). An algorithm A, which performs computations on inputs $n_i \in \mathbf{Z}$, having bitlengths b_i for $1 \le i \le N$, is said to be a ***polynomial time algorithm*** (or simply, ***polynomial***), if there exist non-negative $a_i \in \mathbf{Z}$ such that the number of bit operations required for A is $O(\prod_{i=1}^{N} b_i^{a_i})$. Examples of polynomial-time algorithms are the ordinary arithmetical operations of addition, subtraction, multiplication and division. However, the computation of $n!$ is not polynomial (see Exercise 8.1.4). Thus, an algorithm is polynomial if its computational complexity is $\mathcal{O}(n^c)$ for some constant c. For instance, if $c = 0$, the algorithm is *constant*; if $c = 1$, it is linear; if $c = 2$, it is quadratic, etc. On the other hand, algorithms with computational complexity $\mathcal{O}(c^{f(n)})$, where c is constant and f is a polynomial in n, are called *exponential-time algorithms* (or simply, *exponential*). An algorithm is said to be *subexponential* if it lies between polynomial and exponential. For instance, the *number-field sieve* (see Lenstra *et al.* [177]) requires roughly

$$\exp((\log n)^{1/3}(\log\log n)^{2/3})$$

operations to factor n.[(8.1.2)] If the 1/3 and 2/3 were replaced by 0 and 1 respectively, then it would be a polynomial-time algorithm. On the other hand, if 1/3 and 2/3 were replaced by 1 and 0 respectively, then it would be exponential, i.e. it is subexponential.

A problem, such as factoring (see footnote (8.1.2)), is classified by complexity theory in terms of the algorithms used to solve them. Problems which can be solved by polynomial-time algorithms are said to be in **P**. Problems which can be solved on a nondeterministic *Turing machine*[(8.1.3)] are said to be in **NP**. The principal unsolved question in complexity theory is: Does **P**=**NP**? The answer is almost certainly no, but it remains open. A problem is said to be **NP**-*complete* if it can be proved to be as difficult as any other problem in **NP** (e.g. the travelling salesman problem, see Koblitz [147]). If the solvability of **NP**-complete problems in polynomial time were even settled, then we would have an answer to the "**P**=**NP**?" question. However, it is safe to say that this issue will not be resolved soon.

Now, we proceed with a history of class number computations.

Let $\Delta = \Delta_0 > 0$ be a fundamental discriminant with radicand D. Several methods for computing h_Δ exist, but they are all essentially variants of two main themes: counting cycles or utilizing the analytic class number formula (5.4.1).

If $I = [Q/\sigma, (P + \sqrt{D})/\sigma]$ is a reduced ideal in $\mathcal{O}_\Delta$, then by Exercise 2.1.13(e) we have that $0 < Q < 2\sqrt{D}$, $0 < P < \sqrt{D}$ and by (2.1.5), $D \equiv P^2 \pmod{\sigma Q}$. It is manifest that one simple way to compute h_Δ is to find all pairs (P, Q) which satisfy these inequalities and divisibility conditions, and then count the distinct cycles to which the corresponding ideals belong. Although it is not immediately obvious, it is possible to set up this cycle counting technique for evaluating h_Δ, such that it is of complexity $O(D^{\alpha+\varepsilon})$ for any $\varepsilon > 0$ where $\alpha = .5076$. This is because of a result of Burgess [43], who showed that the least quadratic non-residue of a prime q is less than $O(q^{\delta+\epsilon})$ with $\delta = e^{-1/2}/4$. As we will see later, it follows from this result that for such a $q < 2\sqrt{D}$, we can find a value of P such that $D \equiv P^2 \pmod{Q}$ in

---

(8.1.2)See Williams and Shallit [388] for an excellent history of factoring and primality testing. Discussion of this rich area in depth is beyond the scope of this book, but it can safely be said that the theory of computational complexity has given computational number theory more mathematical rigor, especially in terms of the integer-factoring problem.

(8.1.3)A Turing machine is a "theoretical" computer which is a finite-state machine with an infinite read-write memory tape. A nondeterministic Turing machine is one which uses polynomial-time algorithms to check solutions to a problem which it "guesses" to be true. See Schneier [319, 196ff].

$O(D^{\delta/2+\epsilon})$ operations. Thus, by sieving all the possible pairs, (P, Q) can be listed in $O(D^{1/2+\delta/2+\epsilon})$ operations.

The algorithm proceeds by the continued fraction algorithm, Theorem 2.1.2, to produce the cycle of reduced ideals which belong to the ideal corresponding to the first (P, Q) pair chosen. (This ideal may have to be reduced, but the continued fraction algorithm will take care of that for us.) The action is then repeated on the next candidate pair, and a check is made to determine whether a reduced ideal equivalent to the ideal corresponding to this pair is already in a previously computed cycle; if so, then this pair is abandoned, and if not, then a new cycle of ideals is produced. This process continues until all candidate pairs have been exhausted. The value of h_Δ is the number of distinct cycles which result. Under the assumption of the GRH, the value of α above can be reduced to 1/2. A new algorithm due to Buchmann and Lenstra can evaluate h_Δ using the cycle counting method which executes ***unconditionally*** (i.e. without GRH) in $O(D^{1/2+\epsilon})$ operations.(8.1.4)

Consider the specific example of $D = 229$. In this case $\sigma = 2$ and for values of $Q \leq 30$, we must find values of $P < Q$ such that $2Q \mid 229 - P^2$. We can only have Q even; putting $Q = 2$ and then $Q = 6$, we get the table below. We also illustrate what happens when we select $Q = 22$ and $P = 3$.

Table 8.1.1.

i	0	1	0	1	2	3	0	1	2	3	0	1	2
P_i	1	15	1	11	7	13	5	13	7	11	3	−3	13
Q_i	2	2	6	18	10	6	6	10	18	6	22	10	6
a_i	8	15	2	1	2	4	3	2	1	4	0	1	4

Notice that if we take $I = [11, (3 + \sqrt{229})/2]$, then I is not a reduced ideal but is equivalent to an ideal in the second cycle here. Indeed, there are no other reduced ideals in $\mathcal{O}_\Delta$ except those listed in the first three cycles in Table 8.1.1; hence, $h_\Delta = 3$ for $\Delta = 229$.

Most tables of class numbers of real quadratic fields have been computed in this fashion. The first published tables seem to have been those of Cayley in 1862. In Table 8.1.2 below, we summarize the various tables of class numbers of real quadratic fields which have been computed by the cycle counting algorithm. (8.1.5)

(8.1.4)The authors of [261], from which all of this section was taken, were informed of this fact by H.W. Lenstra Jr.

(8.1.5)The tables of Cayley and Gauss give values different than h_Δ, namely equation (E.1) of Appendix E. Also, see Appendix C for excerpts from the last cited table.

Table 8.1.2.

Author	Approximate Publication Date	Range
Cayley [49]	1862	All non square $d \leq 99$
Gauss [90]	1863	All non square $d < 10^3$
Schaffstein [315]	1927–28	Prime values of $D < 12000$ and some others
Ince [129]	1934	Square-free values of $D < 2025$
Kloss, Newmann and Ordman [144] (see also Kloss [143])	1965	All prime values of D ($\equiv 1 \bmod 4$) $D \leq 105269$
Kuroda [152]	1965	All prime values of D ($\equiv 1 \bmod 4$) $D \leq 2776817$
Hendy [120]	1975	All square-free values of D with $10^3 < D < 10^5$
Wada [373]	1981	All square-free $D < 10^5$
Buchmann *et al.* (unpublished)	1991	All square-free $D < 10^6$

It seems remarkable that Gauss, who had done so much to establish the theory of binary quadratic forms, and who was also a calculator of prodigious talents, did not contribute the first published table of class numbers. Gauss gave some of the reasons for this in a letter to Schumacher (see Smith [344, p. 261]). Although his remarks seem to be confined to the complex quadratic case, we quote him here.

> 'If, without having seen M. Clausen's Table, I have formed a right conjecture as to its object, I shall not be able to express an opinion in favour of its being printed. If it is a canon of the classification of binary forms for some thousand determinants, that is to say, if it is a Table of the reduced forms contained in every class, I should not attach any importance to its publication. You will see, on reference to the Disq. Arith., p. 521 (note), that in the year 1800 I had made this computation for more than four thousand determinants' [viz., for the first three and tenth thousands, for many hundreds here and there, and for many single determinants besides, chosen for special reasons]; 'I have since extended it to many others; but I have never thought it was of any use to preserve these developments, and I have only kept the final result for each determinant. For example, for the determinant $-11,921$, I have not preserved the whole system, which would certainly fill several pages, but only the statement that there are 8 genera, each containing 21 classes. Thus, all that I have kept is the simple statements viii, 21, which in my own papers is expressed even more briefly. I think it quite superfluous to preserve the system itself, and much more so to print it, because (1) anyone, after a little practice, can easily, without much expenditure of time, compute for himself a Table of any particular determinant, if he should happen to want it, especially when he has a means of verification in such a statement as viii, 21; (2) because the work has a certain charm of its own, so that it is a real pleasure to spend a quarter of an hour in doing it for one's self; and the more so, because (3) it is very seldom that there is any occasion to do it... . My own abbreviated Table of the number of genera and classes I have never published, principally because it does not proceed uninterruptedly.'

Thus, we see that the principal reason behind Gauss' failure to publish a table was that his own table was not complete; however, he also mentions a number of concerns about such computations. Many readers might find these comments as compelling today as they were when he made them (1841). Certainly, we agree wholeheartedly with reason (2). However, it should be pointed out that when fields with large radicands are being investigated, even a computer as gifted as Gauss could not possibly conduct these calculations in a mere 15 minutes. For example, when $D = 350240722763374$, we get $\ell(w_\Delta) = 70400728$ and $h_\Delta = 1$. It might be argued that for large values of D, Gauss' reason (3) would be pertinent. In fact, it is only for large radicands that we get exotic class group structures — structures that are of considerable interest to mathematicians today. (The reader is referred to Shanks [334] for further information on this topic.) Suffice it to say here that the development of modern computers and the desire to discover more detailed information concerning the class groups and their structure have done much to loosen the strength of Gauss' arguments against the production of tables of class numbers and associated information. Also, to get an idea for heuristics is a key motivation for the construction of these tables. The Tables in Appendices A–D provide ample data for such investigations.

We now look at the role of the GRH in calculating h_Δ. We discussed this at great length in Chapter Five, section four. The GRH has not been tested as vigorously as the RH, but some numerical work has been done (see, for example, Shanks [330] and [334]) and nothing untoward has been found (see Appendix F). The GRH is of value here because, under its assumption, we can improve considerably our estimates on the value of $L(1,\chi)$ (see Definition 5.4.1). By Hua [126] (without assuming the GRH (see Chapter Five, section four)), we have

$$L(1,\chi) < 1 + \log\sqrt{\Delta}. \tag{8.1.1}$$

In fact, Pinz [289] has shown that[(8.1.6)]

$$L(1,\chi) \leq \frac{3}{4}(1 - 1/\sqrt{e} + o(1))\log\Delta \approx (.295 + o(1))\log\Delta.$$

In Chapter Five, we mentioned the unconditional (but ineffective) lower bound of Siegal on $L(1,\chi)$ and the GRH-conditional result of Tatuzawa. Also, under the GRH, Littlewood [187] showed that

$$\frac{1+o(1)}{2b\log\log\Delta} < L(1,\chi) < (1+o(1))2c\log\log\Delta \tag{8.1.2}$$

where $b = 6e^\gamma/\pi^2 = 6c/\pi^2$, γ is *Euler's constant,* [(8.1.7)] and $c = e^\gamma$.

The $1+o(1)$ can be replaced by an explicit constant via Oesterle's [284] effective version of Chebotarev's density theorem. To do this, we make use of the Euler product (see footnote (5.4.8)). For the case in which we have interest, we put

$$L(1,\chi) = T_1(Q)T_2(Q)$$

where T_1 and T_2 are defined in the proof of Theorem 5.4.2. By the same methods as used in that proof, we have

$$|\log T_2(Q)| < A(Q,\Delta), \tag{8.1.3}$$

(8.1.6) If f is a function of x, and g is a positive function of x, i.e. $g(x) > 0$ for all x, then $f = o(g)$ means $\lim_{x\to\infty} f(x)/g(x) = 0$ (e.g. see Hua [127, pp. 70–71] for more details).

(8.1.7) Actually, $\gamma = \lim_{n\to\infty}\sum_{i=1}^{n}\left(\frac{1}{i}\right) - \log n \approx 0.57721566$.

where

$$A(Q,\Delta) = B(Q)(4 + 3\log Q)/\sqrt{Q} + 1/Q, \tag{8.1.4}$$

and

$$B(Q) = \log\Delta\{1/(\pi\log Q) + 5\cdot 3/(\log Q)^2\} + 4/\log Q + 1/\pi. \tag{8.1.5}$$

Putting $Q = (\log\Delta)^2$, we can effectively bound $T_1(Q)$ and $T_2(Q)$ to get constants c_1 and c_2, such that (8.1.2) holds with $1 + o(1)$ replaced by c_2 on the right. Our main interest here, however, is not to put these bounds on $L(1,\chi)$, but rather to estimate its value well enough to determine h_Δ. Later, we will show how the above formulas enter into the problem. Now, we show how some methods have been developed for evaluating h_Δ by using the analytic class number formula (5.4.1).

Let $I = [Q/\sigma, (P + \sqrt{D})/\sigma]$ be any reduced ideal in $\mathcal{O}_\Delta$ and set $\gamma_i = (P_i + \sqrt{D})/Q_i$ for $i \geq 0$ where $P = P_0$ and $Q = Q_0$. From Theorem 2.1.3, we deduce that for $\ell = \ell(\gamma)$, we have

$$R = \log\varepsilon_\Delta = \sum_{i=1}^{\ell}\log\gamma_i.$$

From Exercise 2.1.13(h), if we use the continued fraction expansion of w_Δ, then

$$R = \begin{cases} \log(Q_s/Q_0) + 2\sum_{i=1}^{s}\log\gamma_i & \text{when } P_s = P_{s+1},\ \ell = 2s \\ \log((P_{t+1} + \sqrt{D})/Q_0) + 2\sum_{i=1}^{t}\log\gamma_i & \text{when } Q_t = Q_{t+1},\ \ell = 2t+1. \end{cases}$$

Now using the facts concerning θ_i from the proof of Theorem 2.1.2 as well as Exercise 2.1.2(g)(iv) and 2.1.13(h), in conjunction with the method of Stanton *et al* in [345], we conclude that the value of ℓ must satisfy

$$2R/\log\Delta < \ell < R/\log\tau$$

where $\tau = (1 + \sqrt{5})/2$. Thus the results discussed above show that the complexity of determining[(8.1.8)] R by this technique is $O(\ell) = O(R) = O(D^{1/2+\epsilon})$. We can improve the O-constant here by using a somewhat different kind of continued fraction (the NICF, see Exercise 3.2.4)[(8.1.9)], but this does not radically improve the speed of this regulator algorithm.

Assuming now that R has been calculated, we still have the problem of estimating $L(1,\chi)$. In Williams and Broere [386], an idea[(8.1.10)] was used to convert the $L(1,\chi)$ series into the form

$$L(1,\chi) = \Delta^{-1/2}\sum_{n=1}^{\infty}(\Delta/n)E(An^2) + \sum_{n=1}^{\infty}(\Delta/n)n^{-1}erfc(n\sqrt{A}),$$

where

$$A = \pi/\Delta,\ E(x) = \int_x^\infty e^{-t}t^{-1}dt,\ erfc(x) = \frac{2}{\sqrt{\pi}}\int_x^\infty e^{-t^2}dt.$$

At first glance, this expression appears to be far more formidable than the simpler formula given by (5.4.1). However, we can discuss the convergence of this

---

(8.1.8) By determining R, we mean computing an approximation to R of $O(D^\epsilon)$ significant figures.

(8.1.9) See Williams and Buhr [387], Adams [1] and Williams [380].

(8.1.10) The idea is called the method of theta functions, which is quite old. Also, see Purdy *et al.* [296] and Barrucand [17], the latter of whom seems to have rediscovered it.

series much more easily and we do not need any Riemann hypothesis. We point out that

$$0 < erfc(x) < e^{-x^2}/(x\sqrt{\pi}), \quad 0 < E(x) < e^{-x}/x.$$

Hence, if

$$T(m) = \Delta^{-1/2} \sum_{n=m+1}^{\infty} (\Delta/n)E(An^2) + \sum_{n=m+1}^{\infty} (\Delta/n)n^{-1}erfc(n\sqrt{A}),$$

then we can show that

$$|T(m)| < \Delta^{3/2}e^{-Am^2}/(\pi^2 m^3).$$

If we put our approximation $\bar{h}_\Delta$ to h_Δ equal to

$$\bar{h}_\Delta = \frac{\sqrt{\Delta}}{2R}\left(\frac{1}{\sqrt{\Delta}}\sum_{n=1}^{m}(\Delta/n)E(An^2) + \sum_{n=1}^{m}(\Delta/n)n^{-1}erfc(n\sqrt{A})\right),$$

then $|\bar{h}_\Delta - h_\Delta| < A^2e^{-Am^3}/(2Rm^3)$. By Theorem 1.3.3, we have that $2^{t_\Delta} \mid h_\Delta$. Thus, if m is chosen such that

$$A^2e^{-Am^2}/m^3 < 2^{t_\Delta}R,$$

then $h_\Delta = 2^{t_\Delta}\lfloor \bar{h}_\Delta/2^{t_\Delta} + 1/2 \rfloor$.

In Williams and Broere [386], further details are given on the selection of the m value. Suffice it to say here that, in order to guarantee the correctness of our class number value, we must have $m = O(D^{1/2})$. Also, as each of the transcendental functions $E(x)$ and $erfc(x)$ can be evaluated to sufficient accuracy by using a power series (in practice, Chebyshev approximations are more efficient) truncated at $O(\log D)$ terms, it is easy to show that this technique computes h_Δ in $O(D^{1/2+\epsilon})$ operations. In the production of the large table [386] of class numbers for all $D < 1.5 \cdot 10^5$, it was found that it is rarely necessary to use a value of m greater than 500 when $\Delta < 6 \cdot 10^5$. This method of determining h_Δ was generalized to arbitrary number fields by Eckhardt [78]. We should also mention that in 1903, Lerch [181] developed a similar method of finding h_Δ for $\mathcal{O}_\Delta$ with radicand $D = D_0$.

The method discussed above appears to be quite good. Indeed, combined with a fast (see below) method for finding R, it seems to be the best rigorous method known for finding h_Δ. However, the determination of the values of the E and $erfc$ functions is rather time consuming, and for large values of D, the method slows down profoundly.

Another method of estimating $L(1,\chi)$ has been suggested by Neild and Shanks [282]. Their idea is simply to use the partial product $T_1(Q)$ to determine $L(1,\chi)$. It should also be pointed out that Shanks [332] used the same basic idea to evaluate class numbers for certain cubic fields. The determination of $T_1(Q)$ is simple enough, the only difficulty being the evaluation of the Legendre symbols (Δ/q) by using the Jacobi algorithm. The real difficulty lies in deciding how large Q should be. Shanks has suggested that $T_1(Q)$ be evaluated by using the first 500, 1000, and 1500, etc. primes. When the estimate $\bar{h}_\Delta = \sqrt{\Delta}T_1(Q)/(2R)$ remains within .1 of the same integer for 6 consecutive partial products, then that integer should be taken as the class number. This technique has the advantages of being very fast and of working very well in practice. The difficulty is that it is entirely empirical. There is no proof that the value of h_Δ it produces is correct, although the probability that the value of h_Δ that it does yield is wrong is very small indeed. (See the discussion in Shanks [328].)

Up to this point, we see that the best available mathematically rigorous methods for finding h_Δ are of complexity $O(D^{1/2+\epsilon})$. Even Shanks' empirical method depends on the knowledge of R and this determination is of complexity $O(D^{1/2+\epsilon})$. If a large number of class number calculations are to be made for large values of h_Δ, then it is clear that some improvements must be developed. Unfortunately, in our current state of knowledge, it will be necessary to sacrifice some rigor in order to make some of these improvements. We will have to base some of our techniques on the assumption of the truth of the GRH. However, we can derive an unconditional $O(D^{1/4+\epsilon})$ algorithm for finding R. This improvement in computing R is the subject of our following discussion.

The following algorithm for multiplying ideals in $\mathcal{O}_\Delta$ is essentially that of Shanks described in section two of Chapter One. We alter the notation and description here to suit our purposes.

Let $I = [Q/\sigma, (P+\sqrt{D})/\sigma]$ and $\overline{I} = [\overline{Q}/\sigma, (\overline{P}+\sqrt{D})/\sigma]$ be primitive ideals of $\mathcal{O}_\Delta$. Our algorithm will find U, P'' and Q'' such that

$$I\overline{I} = (U)J$$

where $P'', Q'', U \in \mathbf{Z}$ and $J = [Q''/\sigma, (P''+\sqrt{D})/\sigma]$ is a primitive ideal of $\mathcal{O}_\Delta$.

We put $G = \gcd(Q/\sigma, \overline{Q}/\sigma)$ and solve, by using the extended Euclidean algorithm, the congruence

$$(Q/\sigma)x_1 \equiv G \pmod{\overline{Q}/\sigma}$$

for x_1 modulo $\overline{Q}/\sigma$. Put

$$U = \gcd(G, (P+\overline{P})/\sigma) = \gcd(Q/\sigma, \overline{Q}/\sigma, (P+\overline{P})/\sigma)$$

and solve, again by the Euclidean Algorithm,

$$x_2(P+\overline{P})/\sigma + Gy_2 = U$$

for x_2, y_2. Thus,

$$\begin{aligned} Q'' &= Q\overline{Q}/(\sigma U^2) \\ P'' &\equiv P + XQ/(\sigma u) \pmod{Q''} \end{aligned}$$

where

$$X \equiv y_2 x_1(\overline{P} - P) + x_2(D - P^2)/Q \pmod{\overline{Q}/U}.$$

Note that when, as often occurs, $U = G$, we can put $x_2 = 0$, $y_2 = 1$.

Now let $I_1 = (1)$, and invoke Exercise 2.1.9 where we defined the *distance function* $\delta_m = \delta(I_m, I_1) = \log \Psi_m$, which is a strictly increasing function of m and $\delta_{\ell+1} = R$ (so we may always assume $0 < \delta_k < R$ for $\ell > k > 1$) where $\ell = \ell(w_\Delta)$ by Theorem 2.1.3. Also, if $\delta_m = kR + \delta_s$ for $k \in \mathbf{Z}$, then $I_m = I_s$. If I_s and I_t are any two reduced ideals found by expanding the simple continued fraction of w_Δ, let J be defined by computing the product

$$(U)J = I_s I_t$$

as described above. Clearly, $J_1 = J$ is a principal ideal of $\mathcal{O}_\Delta$. If

$$J_1 = [\overline{Q}_0/\sigma, (\overline{P}_0 + \sqrt{D}/\sigma],$$

put $\overline{\gamma}_0 = (\overline{P}_0 + \sqrt{D})/\overline{Q}_0$ and expand $\overline{\gamma}_0$ into a continued fraction. There must be a least non-negative $m \in \mathbf{Z}$ such that $0 < \overline{Q}_m \le \lfloor\sqrt{D}\rfloor$ (see Exercise 2.1.2(g)(v)–(vi)). The corresponding ideal J_{m+1} is a reduced ideal in the principal class and must therefore be I_k for some positive $k \in \mathbf{Z}$. Furthermore, by Exercise 8.1.1 we have

$$\Psi_k = \Psi_s\Psi_t\overline{\Psi}_{m+1}/U, \tag{8.1.6}$$

where $\overline{\Psi}_{m+1} = |\overline{G}_{m-1} + \overline{B}_{m-1}\sqrt{D}|/\overline{Q}_0$ (see Exercise 2.1.9(i)).

If $\kappa = \log(\overline{\Psi}_{m+1}/U)$, then by Exercise 8.1.2 we have

$$-\log 4D < -\log Q_{s-1}Q_{t-1} < \kappa < \log 2. \tag{8.1.7}$$

Thus, by the definition of distance

$$\delta_k = \delta_s + \delta_t + \kappa. \tag{8.1.8}$$

By (8.1.7), κ is small, so we have

$$\delta_k \approx \delta_s + \delta_t. \tag{8.1.9}$$

Definition 8.1.1. The notation $I_s * I_t$ will denote the pair $(I_k, \overline{\Psi}_{m+1}/U)$ discussed above.

Remark 8.1.1. The notation $I_s * I_t$ has only one possible result since, although there are infinitely many reduced ideals equal to a given I_k (viz. $I_{k+i\ell}$ for $i = 0, 1, 2, 3, \ldots$), only one of these ideals is such that its distance is given by (8.1.8) with the value of m defined above.

There is also a connection between δ_m and m furnished in the following result of Lévy [186].

Theorem 8.1.1. *Let $\alpha = \langle a_0; a_1, \ldots, a_{k-1}, \alpha_k\rangle$. For all but finitely many quadratic irrationals α, we have $\lim_{k\to\infty} \sqrt[k]{\alpha_1\alpha_2\ldots,\alpha_k} = e^{\lambda}$, where $\lambda = \pi^2/(12\log 2) \approx$ 1.186569111.*

Since

$$\Psi_m = \prod_{i=1}^{m-1} \psi_i = \frac{Q_{m-1}}{Q_0}\prod_{i=1}^{m-1}\alpha_i,$$

we expect that

$$\delta_m = \log(Q_{m-1}/Q_0) + \log\prod_{i=1}^{m-1}\alpha_i \approx m\lambda,$$

as m gets large. Indeed, by Exercise 8.1.3,

$$(m-k-1)\log\tau + \delta_k < \delta_m < (m-k)\log\sqrt{\Delta} + \delta_k. \tag{8.1.10}$$

Thus, by virtue of Lévy's Theorem and (8.1.10), we see that if we have two reduced ideals I_s and I_t in the principal cycle with $I_1 = (1)$, then we can find an ideal I_k with $k \approx s+t$ by simply multiplying the two ideals I_s and I_t and then using the continued fraction algorithm to find a reduced ideal equivalent to their product. As Shanks observed, this means that we can now search through the principal cycle of ideals without having to do it one small continued fraction step at a time, i.e.

by using a step size of $\lambda \approx 1.18$. Instead, we can now use a step size corresponding to the distance δ_s of a particular ideal I_s. We can now get to a particular ideal of distance δ from I by performing about δ/δ_s of these multiplication/reduction steps, instead of about δ/λ continued fraction steps.

We will now illustrate this idea by means of an example taken from Williams and Wunderlich [389]. Let $D = 103$, $P = 0$, $Q = 1$, and use the formulas of Exercise 2.1.2 to develop the table below.

Table 8.1.3.

j	0	1	2	3	4	5	6	7	8	9	10	11	12	13
P_j	0	10	8	5	7	2	9	9	2	7	5	8	10	10
Q_j	1	3	13	6	9	11	2	11	9	6	13	3	1	3
a_j	10	6	1	2	1	1	9	1	1	2	1	6	20	6

where

j	δ_j
0	–
1	0
2	3.003149278
3	4.803146478
4	4.959124487
5	6.006298558
6	6.306311917
7	6.860661477
8	9.119759129
9	9.219101793
10	9.863810756
11	10.78947865
12	11.12363878
13	13.02817557

Note that $\ell(w_\Delta) = \ell = 2 \cdot 6 = 12$ and $R = \delta_{\ell+1} = \delta_{13} = 13.02817557$. If we select $I_3 = [13, 8 + \sqrt{103}]$ and $I_4 = [6, 5 + \sqrt{103}] = I_3^+$, then we get $I_3 I_4 = [78, 47 + \sqrt{103}]$ by using the above algorithm. We now expand $\overline{\alpha}_0 = (47 + \sqrt{103})/78$ into a continued fraction to get some $\overline{Q}_m \leq \lfloor\sqrt{103}\rfloor = 10$. This is done in

Table 8.1.4.

j	-2	-1	0	1	2	3
$\overline{P}_j$	–	–	47	-47	20	2
$\overline{Q}_j$	–	–	78	-27	11	9
$\overline{a}_j$	–	–	0	1	2	1
$\overline{G}_j$	-47	78	-47	31	15	–
$\overline{B}_j$	1	0	1	1	3	–

With the 'bar' notation introduced above, we get $m = 3$, $\overline{\Psi}_{m+1} = \overline{\Psi}_4 = |\overline{G}_2 + \overline{B}_2\sqrt{D}|/\overline{Q}_0$, so

$$I_3 * I_4 = ([9, 2 + \sqrt{103}], (15 + 3\sqrt{103})/78).$$

Now note that
$$I_9 = [9, 2 + \sqrt{103}],$$
from Table 8.1.3, and
$$\delta_3 + \delta_4 = \delta_9 - \kappa,$$
where $\kappa = \log((15 + 3\sqrt{103})/78) \approx -.540169172$.

We point out that if we define
$$\overline{\delta}_m = \delta(I'_m, I_1) \quad (m \le \ell),$$
then (via Lemma 6.1.1), we have
$$\overline{\delta}_m = \delta(I_{\ell+2-m}, I_1) = R - \delta_m + \log(Q_{m-1}/\sigma). \tag{8.1.11}$$
Notice that $\overline{\delta}_m$ is a strictly increasing function of m (and compare with Exercise 2.1.9(ii)).

Now we derive two algorithms for computing R. They both make use of the infrastructure idea, but they differ in their complexity. The first of these is of unconditional complexity $O(D^{1/4+\epsilon})$, and the second is the complexity of $O(D^{1/5+\epsilon})$ if the GRH holds.

The basic idea behind the first algorithm goes as follows: We create a list of ideals, starting with $I_1 = (1)$ up to I_s where $s \approx D^{1/4}$. If there is enough information at this point to determine R, we do so; otherwise, we compute $(M_1, \chi_1) = I_s * I_s$ and define
$$(M_{j+1}, \chi_{j+1}) = M_j * M_1$$
$$X_{j+1} = X_j \chi_{j+1}.$$
When one of the I_j's or its complement is found to be in our precomputed list $I_1, I_2, \ldots, I_s$, we know that we have gone almost all the distance up to R. This distance can be computed from Ψ_s and the corresponding X value, and it, together with the value Ψ_i for the ideal I_i in the list which is the same as our M_j (or its complement), can be used to find R. Notice that in this algorithm, we are pushing through the cycle of reduced ideals with a step size of about $2\delta_s$, which by (8.1.9), we see is of about the same order of magnitude as $2s$ or $D^{1/4}$. Since the value of R is $O(D^{1/2+\epsilon})$, we see that the complexity of this algorithm is $O(D^{1/4+\epsilon}) + O(D^{1/2+\epsilon}/D^{1/4}) = O(D^{1/4+\epsilon})$. We now present this algorithm in full. The reader is referred to Exercises 2.1.2 and 2.1.9 for notation.

Algorithm 8.1.1. *Computing R ($D > 10^6$).*
Initialization: *Select a value for a constant $c > 1$. Put*
$$L = \lceil cD^{1/4} \rceil + 1, \qquad T = 1 + \lfloor \log 4\sqrt{D}/(2\log\tau) \rfloor.$$

Step 1: *By developing the continued fraction of ω_Δ, compute and store those ideals $I_i = [Q_{i-1}/\sigma, (P_{i-1} + \sqrt{D})/\sigma]$, where $i \le t$ and $Q_{i-1} \le \lfloor\sqrt{D}\rfloor$. These ideals are stored as pairs (Q_{i-1}, P_{i-1}) which are sorted lexicographically. Call this list of ideals $\mathcal{T}$. Here $t = s + T$, where s is selected such that $s = L$ or $L + 1$, whichever has $Q_{s-1} \le \lfloor\sqrt{D}\rfloor$. If $P_n = P_{n+1}$ for a minimal positive $n \le t - 1$, then*
$$R = \log(Q_0/Q_n) + 2\log\Psi_{n+1}$$
and we can terminate the algorithm. If $Q_m = Q_{m+1}$ for a minimal positive $m \le t - 2$, then
$$R = \log(Q_0\psi_{m+1}/Q_m) + 2\log\Psi_{m+1}$$

and we can terminate the algorithm.
Step 2: *Compute* $(M_1, \chi_1) = I_s * I_s$. *Put* $X_1 = 1$, $j = 1$.
Step 3 *(test step): If* $M_j \in \mathcal{T}$, *then*

$$R = 2j \log \Psi_s + j \log \chi_1 + \log X_j - \log \Psi_i$$

and we can terminate the algorithm. If $M_j' \in \mathcal{T}$, *then*

$$R = 2j \log \Psi_s + j \log \chi_1 + \log X_j + \log \Psi_i - \log(Q_{i-1}/Q_0)$$

and we can terminate the algorithm.
Step 4: *Put*

$$(M_{j+1}, \chi_{j+1}) = M_j * M_1$$
$$X_{j+1} = X_j \chi_{j+1}$$
$$j \leftarrow j + 1$$

and go to Step 3.

The proof of this algorithm can be deduced from (8.1.6)–(8.1.8) and (8.1.11) above. Notice that we have to be a little careful in the implementation of the basic idea given previously because our distances are not exactly additive. The range of values for κ in (8.1.7) has to be taken into consideration, and this is why our list $\mathcal{T}$ must contain a few more ideals t than the s mentioned earlier. A complete proof of this algorithm is given in Stephens and Williams in [348]. We also mention that D is assumed to be in excess of 10^6 here. This is really no hardship, as it is easier to use the conventional continued fraction method to evaluate R when $D \leq 10^6$.

The next idea for finding R involves using (5.4.1) to estimate a value E for $h_\Delta R$ and a value L such that

$$|E - h_\Delta R| < L^2.$$

As in the previous algorithm, we compute a list $\mathcal{T}$ of all reduced principal ideals whose distance function δ does not exceed $L + \log 2$. If $2^k < E < 2^{k+1}$, we can use k "doubling" steps (see [331]) to find a reduced ideal I_m with $\delta_m \approx E$. That is, we find some I_s $(= M_1)$ with $\delta_s \approx E/2^k$, and then compute $(M_{j+1}, \chi_{j+1}) = M_j * M_1$ until we get $M_k = I_m$. Starting at M_k, we can use a version of Algorithm 8.1.1 with step size about L until we find some ideal or its conjugate in $\mathcal{T}$. We can do this by putting $L_1 = J_1 = M_k$ and defining $(J_{j+1}, m\mu_{j+1}) = J_i * I_t$, and $(L_{j+1}, n\mu_{j+1}) = L_j * I_t'$, where $\delta_t < L$ and $\delta_{t+1} > L$. When some J_j or the conjugate of some L_j is in $\mathcal{T}$, we can compute a value for $H_\Delta^* R$, where H_Δ^* is some positive integer, by using the values of the $m\mu_i$'s and the $n\mu_i$'s and value of δ_m. The complexity of this step, given E and L, is

$$O(LD^\epsilon) + O((\log E)D^\epsilon) + O(L^2 D^\epsilon / L) = O(LD^\epsilon),$$

where the first of these represents the number of operations needed to produce $\mathcal{T}$, the second the number of operations needed to find M_k, and the third the number of operations to find $H_\Delta^* R$ by searching through the cycle with step size L. Notice that H_Δ^* may or may not be the class number of $\mathcal{O}_\Delta$, but it is at least a rational integer. We must now find R from $H_\Delta^* R$.

If $R < B = E/\sqrt{L}$, we can find R by using basically the same idea as that used in Algorithm 8.1.1. We simply use a step size of about L to move through the cycle until we get a distance which exceeds B. If $R < B$, then we must find R by this process, and it will be found in $O((B/L)D^{\epsilon})$ operations.

If we do not find R by this process, then $R \geq E/\sqrt{L}$. We compute a list of all primes $q_1 = 2, q_2 = 3, \ldots, q_n < \sqrt{L} + L^2\sqrt{L}/E = (E + L^2)/B$. For each of these primes, determine whether or not the ideal I_{k_i}, at a distance $H_{\Delta}^{*}R/q_i$ from I_1, is such that $I_{k_i} = I_1$. If this should be the case for some i, try again for distances $H_{\Delta}^{*}R/q_i^2, H_{\Delta}^{*}R/q_i^3, \ldots$, etc., until we find that there is no ideal which is the same as I_1 at distance $H_{\Delta}^{*}R/q_i^{\alpha_i+1}$, but there is one at distance $H_{\Delta}^{*}R/q_i^{\alpha_i}$ ($\alpha_i \geq 0$). Thus, since $I_j = I_1$ if and only if $p \mid j - 1$, we must have q^{α_i} as the highest power of q_i which divides H_{Δ}^{*}. Since for each of these determinations, we know the exact distance δ that we wish to find a reduced ideal I from I_1, we know that each can be conducted in $O((\log \delta)D^{*}\varepsilon)$ operations. (We use the same idea as that used above to find an ideal at distance E from I_1.) It follows, then, that the complexity of this step is

$$O(n(\log H_{\Delta}^{*}R)^2 D^{\epsilon}) = O(D^{\epsilon}(E + L^2)/B).$$

By the algorithm which finds $H_{\Delta}^{*}R$, we know that $H_{\Delta}^{*}R < E + L^2$ and $R \geq B$. Hence,

$$H_{\Delta}^{*} = \prod_{i=1}^{n} q_i^{\alpha_i},$$

and R can now be computed easily.

It remains to show how values of E and L should be selected to minimize the total complexity of this algorithm. By referring to the definitions of $T_1(Q)$ and $T_2(Q)$, and to the results of (8.1.3)–(8.1.5), we have

$$E = \sqrt{\Delta}T_1(Q)/2 \tag{8.1.12}$$

and

$$|E - h_{\Delta}R| \leq \sqrt{\Delta}T_1(Q)|1 - T_2(Q)|/2. \tag{8.1.13}$$

Let $Q \approx D^{\alpha}$, where α (> 0) is to be determined. Under this assumption on Q and the GRH, we see by (8.1.3)–(8.1.5) that

$$A(Q, \Delta) = O(D^{-\alpha/2+\epsilon})$$

and

$$|1 - T_2(Q)| \leq \exp(A(Q, \Delta)) - 1 = O(D^{\alpha/2+\epsilon}).$$

Now

$$\prod_{q<Q} q/(q+1) \leq T_1(Q) \leq \prod_{q<Q} q/(q-1),$$

and by Mertens' Theorem (see Hardy and Wright [110, Theorem 428, p. 351]) [8.1.11], we know that

$$\prod_{q<Q} q/(q-1) \sim e^{\gamma} \log Q;$$

where γ is Euler's constant. Hence, by (8.1.12)–(8.1.13), we have

$$D^{1/2-\epsilon} \ll E \ll D^{1/2+\epsilon}, \tag{8.1.14}$$

[8.1.11] Mertens' Theorem says that (for γ = Euler's constant) $\prod_{p \leq x}\left(1 - \frac{1}{p}\right) \sim \frac{e^{-\gamma}}{\log x}$.

and

$$\sqrt{\Delta}T_1(Q)|1 - T_2(Q)|/2 = O(D^{1/2-\alpha/2-\epsilon}).$$

If we put

$$f(D) = \frac{\sqrt{\Delta}T_1(Q)(\exp(A(Q,\Delta)) - 1)}{2D^{1/2-\alpha/2}},$$

then $f(D) = O(D^{\epsilon})$. Setting $g(D) = \max\{1, f(D)\}$ and

$$L^2 = g(D)D^{1/2-\alpha/2},$$

we get $|h_\Delta R - E| < L^2$. It follows that

$$\begin{aligned} L &= O(D^{1/4-\alpha/4+\epsilon}), \\ B/L &= E/(L\sqrt{L}) = O(D^{1/8+3\alpha/8+\epsilon}), \\ (E + L^2)/B &= \sqrt{L} + L^2\sqrt{L}/E = O(D^{1/8-\alpha/8+\epsilon}). \end{aligned}$$

Since the evaluation of a Legendre symbol is logarithmic in the argument, we see that the complexity of evaluating $T_1(Q)$ is $O(D^{\alpha+\epsilon})$. Thus, in view of our previous remarks, the complexity of the entire algorithm is

$$O(D^{\alpha+\epsilon}) + O(D^{1/4-\alpha/4+\epsilon}) + O(D^{1/8+3\alpha/8+\epsilon}).$$

To optimize this, we need $\alpha = 1/4 - \alpha/4 = 1/8 + 3\alpha/8$, i.e. $\alpha = 1/5$. Hence, under the GRH we have shown that our new algorithm will determine R with complexity $O(D^{1/5+\epsilon})$, an improvement over Algorithm 8.1.1.

We have really only sketched this algorithm here, and have omitted many implementation details because the algorithm, which is of some interest from the complexity theory point of view, is not as efficient as Algorithm 8.1.1 for most applications. The value of D must be very large for this algorithm to represent a significant improvement over Algorithm 8.1.1. Further details concerning this topic can be found in Lenstra [178] where this algorithm is first mentioned, and Schoof [321]. With these faster algorithms for evaluating R, the heuristic of Shanks discussed above can be used to compute h_Δ with greater speed. This was done by Tennenhouse and Williams [361] to produce a table of all $\mathcal{O}_\Delta$ where $p = D$ is a prime, $p \leq 10^8$, and $h_\Delta = 1$.

Now we discuss the evaluation of h_Δ.

We will divide the problem of determining h_Δ into two parts. These will be the evaluation of h_Δ when h_Δ is small ($h_\Delta \leq D^{1/8}$, say), and when h_Δ is large ($h_\Delta > D^{1/8}$). Both of the methods discussed here require the assumption of the truth of the GRH in order that the value obtained for h_Δ be correct.

We first assume that h_Δ has a known factor h_1 and put $h_2 = h_\Delta/h_1$. For a fixed fundamental discriminant $\Delta = \Delta_0 > 0$ with radicand D, define

$$\overline{h}_2(Q) = Ne(\sqrt{\Delta}T_1(Q)/(2Rh_1)),$$

and

$$\kappa(Q) = \sqrt{\Delta}T_1(Q)/(2Rh_1) - \overline{h}_2(Q), \tag{8.1.15}$$

where $Ne(x)$ is as in Definition 2.2.1. Note that

$$\overline{h}_2(Q) = T_2(Q)^{-1}h_2 - \kappa(Q), \tag{8.1.16}$$

and $|\kappa(Q)| < 1/2$. We will now determine when we can be sure that $\overline{h}_2(Q) = h_2$. In this case, we then have $h_\Delta = h_1\overline{h}_2(Q)$.

We need two simple lemmas.

Lemma 8.1.1. *If* $|\log x| < \log(k+1)/(k+|y|))$, *where* $k \geq 0$ *and* $|y| < 1/2$, *then*

$$k/(k+1+y) < x < (k+1)/(k+y).$$

Lemma 8.1.2. *If* $k \in \mathbf{Z}$ *and*

$$k/(k+1+\kappa(Q)) < T_2(Q) < (k+1)/(k+1+\kappa(Q)), \tag{8.1.17}$$

then $h_2 = k$ *if and only if* $\overline{h}_2(Q) = k$.

The first of these results is a very easy exercise for the reader. The second follows easily from (8.1.16). One need only show that if (8.1.17) holds and $\overline{h}_2(Q) = k$, then $h_2 < \overline{h}_2(Q) + 1$ and $h_2 > \overline{h}_2(Q) - 1$. By similar reasoning, it is easy to show that $\overline{h}_2(Q) = h_2$, when (8.1.17) holds with $h_2 = k$.

From these results and (8.1.3), it follows that $h_2 = \overline{h}_2(Q)$ whenever Q is sufficiently large, i.e.

$$A(Q,\Delta) < \log\{(\overline{h}_2(Q)+1)/(\overline{h}_2(Q_+|\kappa(Q)|)\}. \tag{8.1.18}$$

The question that we must now address is how large must Q be for (8.1.18) to hold.

Since

$$\log\left(\frac{1+x}{|y|+x}\right) > \log\left(1+\frac{1}{2x+1}\right) > \frac{1}{2x+2} \qquad (|y| < 1/2),$$

we see that (8.1.18) will hold if

$$A(Q,\Delta) < 1/(2\overline{h}_2(Q)+2). \tag{8.1.19}$$

Also, if $h_1 > D^\alpha/R$, from (8.1.15) and Mertens' Theorem there exists some positive constant c_1 such that

$$2\overline{h}_2(Q) + 2 < c_1 D^{1/2-\alpha} \log Q.$$

From (8.1.4)–(8.1.5), we see that there exists a positive constant c_2 such that

$$A(Q,\Delta) < (c_2 \log DQ)/Q^{1/2}.$$

We will certainly have (8.1.19) holding when

$$D^{\alpha-1/2}(c_1 \log Q) > (c_2 \log DQ)/Q^{1/2}.$$

It follows that if $\beta = 1 - 2\alpha$, then (8.1.19) holds with $Q = D^{\beta+\epsilon}$. Thus, if we know a factor h_1 of h_Δ where $h_1 > D^\alpha/R$ $(0 < \alpha < 1/2)$, then in $O(D^{\beta+\epsilon})$ operations, a value for $T_1(Q)$ can be calculated such that $h_\Delta/h_1 = \overline{h}_2(Q)$.

If we let q_i denote the ith prime and select some integral value of c, say 500, then we can define

$$F_j = \prod_{i=c(j-1)+1}^{cj} q_i/(q_i - (\Delta/q_i)).$$

Given this, we can now present

Algorithm 8.1.2. *(Determination of* h_Δ *given* R *and* h_1, *where* $h_1 \mid h_\Delta$ *and* $h_1 > D^\alpha/R$.*)*

(1) *Put* $j = 1$, $F = \sqrt{\Delta}/(2Rh_1)$

(2) $F \leftarrow FF_j$

(3) *Compute* $\overline{h}_2 = Ne(F)$, $\kappa = F - \overline{h}_2$

(4) *If* $A(q_{jc}, \Delta) \geq \log((\overline{h}_2 + 1)/(\overline{h}_2 + |\kappa|))$, *go to 2.*

(5) *Otherwise,* $h = \overline{h}_2\overline{h}_1$ *and the algorithm terminates.*

Notice that this algorithm is just a version of Shanks' heuristic, but justified on the basis of the GRH. As indicated above, this algorithm will terminate in $O(D^{1-2\alpha+\epsilon})$ operations.

Let $h < D^\gamma$. By (8.1.2), we have a positive constant c_3 such that

$$Rh_\Delta > c_3\sqrt{D}/(\log\log D).$$

Thus, even if $h_1 = 1$, we have

$$Rh_1 > c_3 D^{1/2-\gamma}(\log\log D).$$

Putting $\alpha = 1/2 - \gamma$, we see that h_2 can be evaluated by Algorithm 8.1.2 in $O(D^{2\gamma+\epsilon})$ operations. Thus, the above technique will work well when $h_\Delta < D^{1/8}$. As small class numbers tend to be much more frequent than large ones (see the heuristics in Cohen–Lenstra [58]), Algorithm 8.1.2 is applicable in most cases with $h_1 = 1$.

In dealing with larger values of h_Δ, we first point out that if

$$Rh_1 > D^{2/5},$$

then Algorithm 8.1.2 will find h_2 in $O(D^{1/5+\epsilon})$ operations. We will now discuss how a divisor h_1 of h_Δ can be found such that $h_1 > D^{2/5}/R$. To solve this problem, we employ a slight modification of the baby step-giant step strategy of Shanks [328], as utilized by Lenstra [178] and Schoof [321]. Put

$$B_2(Q) = \kappa(Q) + \sqrt{\Delta}T_1(Q)(\exp(A(Q,\Delta)) - 1)/(2R).$$

From (8.1.15), (5.4.1) and (8.1.3), we have

$$|h_\Delta - \overline{h}(Q)| < B_2(Q),$$

where $\overline{h}(Q) = Ne(\sqrt{\Delta}T_1(Q)/(2R))$. If we put $Q = D^{1/5}$, $\overline{h} = \overline{h}(D^{1/5})$ and $B_2 = B_2(D^{1/5})$, we get

$$|h_\Delta - \overline{h}| < B_2,$$

and

$$RB_2 = O(D^{2/5+\epsilon}). \tag{8.1.20}$$

It follows that there exists some $m \in \mathbf{Z}$ such that $|m| < B_2$ and $\overline{h} = h + m$. Thus, if I is any ideal of $\mathcal{O}_\Delta$, we have

$$I^{\overline{h}} \sim I^m. \tag{8.1.21}$$

Given some I, we now attempt to find a value for m such that (8.1.21) holds. We select our $I = [Q/\sigma, (b+\sqrt{D})/\sigma]$ by putting $Q/\sigma = q$, a prime such that $(D/q) = 1$, finding a solution x of

$$x^2 \equiv D \pmod{q},$$

and putting $P \equiv \sigma - 1 \pmod{\sigma}$, $P \equiv x \pmod{q}$. As noted by Lehmer [170] and Shanks [329], P can be determined in $O(\log q)$ operations when a quadratic non-residue of q is known. From Bach [13], we know that under the GRH, there exists a quadratic non-residue n of q such that $n < 2(\log q)^2$ when $q > 1000$.

Let $|m| = ki + j$ $(0 \leq j \leq k)$. By using a fast exponentiation algorithm like those described in Knuth [145], we can compute reduced ideals L and M such that

$$L \sim I^{\bar{h}} \quad \text{and} \quad N \sim I^k$$

in $O((\log \bar{h})D^\epsilon)$ and $O((\log k)D^\epsilon)$ operations, respectively. If (8.1.21) holds, then we must have either

$$N'^{i} L \sim I^j \quad \text{when} \quad m \geq 0, \tag{8.1.22a}$$

or

$$N'^{i} L' \sim I^j \quad \text{when} \quad m < 0. \tag{8.1.22b}$$

We use the following algorithm to find values of i, j such that one of (8.1.22a) or (8.1.22b) holds.

Algorithm 8.1.3. *Find m such that* (8.1.21) *holds.*

Case 1. $(B_2 \geq R)$.

(1) *Put $k = \lfloor \sqrt{B_2/R} \rfloor + 1$ and compute the cycles of reduced ideals which belong to each of (1), $I, I^2, \ldots, I^{k-1}$. The total number of ideals produced here is $O(kR)$. We can then take these ideals and sort them (on their Q, P values) into a list T.*

Altogether, by (8.1.20) *this step will execute in $O((\sqrt{RB_2}+R)D^\epsilon) = O(D^{1/5+\epsilon})$ operations.*

(2) *Put $M_0 = L$, $J_0 = L'$ and compute reduced ideals M_n and J_n by using $M_{i+1} \sim M_i N'$; $J_{i+1} \sim J_i N'$. If either of M_i or $J_i \in T$, we get $m = ik + j$ when $M_i \sim I_j$ or $m = -ik - j$ when $J_i \sim I^j$. Since $|m| < B_2$ and $j \geq 0$, we see that we must find such a value of i with $i < B_2/k = O(D^{1/5+\epsilon})$.*

Case 2. $(B_2 < R)$.

(1) *In this case, we put $k = 1$ and $j = 0$ in (8.1.22a) and (8.1.22b) and compute M_n and J_n $(n = 0, 1, 2, \ldots)$ as above. Let $S = \sqrt{RB_2}$. We now move through the cycle of principal ideals with step size S to determine, as n increases, whether M_n or J_n is principal. In the former case, we get $m = n$, and in the latter $m = -n$. Since S is fixed here, our calculation of the first $O(S)$ reduced ideals in the principal cycle requires $O(SD^\epsilon)$ operations, and need only be done once. Since a value of n must exist with $n < B$, we find that we require a total of*

$$O(SD^\epsilon) + O(B_2 R D^\epsilon / S) = O(D^{1/5+\epsilon})$$

operations to find m in this case.

Having found m, we next factor $|\bar{h}_\Delta - m| = O(D^{1/2+\epsilon})$, a task that can certainly be done in $O(D^{1/6+\epsilon})$ operations (see, for example, Lehman [168]), and then find the period e_1 of I (the least positive $e_1 \in \mathbf{Z}$ such that $I^{e_1} \sim (1)$) by trying all the factors of $|\bar{h}_\Delta - m|$. Since $e_1 \mid h_\Delta$, we can use $h_1 = e_1$ if e_1 is large enough. If this is not the case, we select another ideal M and find the order e_2 of the subgroup of the class group which contains the ideal classes of I and M. Continue this process until we find some e_n such that $e_n > D^{2/5}/R$. We omit the details of how this latter computation can be performed and refer the reader to Lenstra [178], Schoof [321] and Dueck and Williams [77], where the process is discussed in greater detail. Suffice it to say here that under certain further Riemann Hypotheses, it can be shown that we can find e_n in $O(D^{1/5+\epsilon})$ operations. This means that under these Riemann Hypotheses (or the GRH), we can compute h_Δ in $O(D^{1/5+\epsilon})$ operations. We also point out that, since most of the class groups tend to be cyclic or close to it (see, once again, Cohen and Lenstra [58]), usually e_1 is large enough to determine h_Δ.

Now we show how some of the ideas developed above have been used to test the heuristic of Cohen and Lenstra mentioned in section four of Chapter Five.

By Corollary 1.3.2, as mentioned in section four of Chapter Five, if the Cohen–Lenstra heuristic is correct, then $f(x)/\pi(x)$ should tend toward .75446 as x gets large. Here, $\pi(x)$ denotes the number of primes not exceeding x and $f(x)$ denotes the number of those primes $q \leq x$ for which $h_\Delta = 1$ when $D = q$. This heuristic was tested for $x < 10^9$ in Stephens–Williams [348]. Since many of the above techniques of this section were used to do this, we will now review how these computations were performed. Let $\pi_i(x)$ $(i = 1, 3)$ denote the number of primes $q \leq x$ such that $q \equiv i \pmod 4$, and let $f_i(x)$ represent the number of these primes such that the class numbers of the corresponding $\mathcal{O}_\Delta$ have $h_\Delta = 1$. Clearly, since $h_8 = 1$,

$$\pi(x) = \pi_1(x) + \pi_3(x) + 1,$$

$$f(x) = f_1(x) + f_3(x) + 1.$$

We examined this problem by dividing up the fields according to the congruence class of the radicand, and computed

$$r_i(x) = f_i(x)/\pi_i(x) \quad (i = 1, 3).$$

By using (8.1.15)–(8.1.19), we see that if for a given Δ, we have $\Delta < B$ and

$$A(Q, B) < log(2/(1 + |\kappa(Q)|)),$$

then $h_\Delta = 1$ if and only if $\overline{h}(Q) = 1$. In Table 8.1.5 below, we give, for selected values of t, a prime Q such that $A(Q, B) < log(2/(1 + t))$ when $B = 4 \cdot 10^9$.

Table 8.1.5.

t	prime Q	$\pi(Q)$
.001	16673	1929
.005	16843	1949
.01	17077	1969
.05	19001	2159
.1	21787	2444
.2	29101	3163
.3	39901	4196
.4	56671	5746
.5	84731	8257

Thus, if $q < 10^9$ and we wish to determine h_Δ with radicand $D = q$, we need only select a value of t and evaluate $\overline{h}(Q)$ for the corresponding Q in Table 8.1.5.

If $|\kappa(Q)| < t$ and $\bar{h}(Q) = 1$, then $h_\Delta = 1$. If $|\kappa(Q)| < t$ and $\bar{h}(Q) \neq 1$, then $h_\Delta \neq 1$. If $|\kappa(Q)| > t$, select the next t value until a corresponding $\kappa(Q)$ is obtained with $|\kappa(Q)| < t$. It is this simple idea which we used to obtain h_Δ, but, as noted below, the proper implementation of this algorithm is very important for its speed of execution.

The algorithm for computing R (Algorithm 8.1.1), and the above technique, were coded in assembly language (double precision floating point was used for the accumulation of R and $T_1(Q)$) and run on an Amdahl 5870 computer. After conducting some numerical experiments, it was decided that in Algorithm 8.1.1, values of $c = 1.5$ and 1.75 be used for $D \equiv 1$ and $3 \pmod 4$, respectively. We now mention some implementation techniques that we found useful during the course of these calculations.

As mentioned by Schoof [321], instead of actually conducting a preliminary sort in Step 1 of Algorithm 8.1.1 and then using a binary search, say, to determine whether or not $M \in T$ (this is also the case in any of the algorithms for finding whether or not a given article is in a particular list), it is much faster in practice to use hashing techniques. We found that hashing on the last byte of the Q value for the ideals was most effective.

Preliminary tests showed that in our early implementation of our algorithm, over 95% of the time was being spent on the evaluation of $T_1(Q)$. This was because the routine for evaluating the Legendre symbols was too slow, no matter what was tried. (Variants of both the Jacobi algorithm and the Euler criterion power algorithm methods were used.) This problem was solved by computing $T_1(Q)$ for all D in a fixed interval under the assumption that $|\kappa(Q)| < .01$ (i.e. $Q = 17077$). This was done by first precalculating all the quadratic residues for all the primes up to 17077, and then using this as a table to drive the next phase of the procedure which was the rapid accumulation of $T_1(Q)$ for each D in the interval. This was done by multiplying each T_1 value by $q/(q-(\Delta/q))$ with only a single array look-up being necessary for determining (Δ/q). If it was necessary to go to larger Q values ($|\kappa(17077)| > .01$), the computations were continued for those particular D values by using the Euler criterion method to compute the extra Legendre symbols. By using an interval of size 10^6, the estimate of the amount of time needed by this part of the program was cut from 225 to 38 CPU hours. Empirically, it seems that for about 90% of all values of D, h_Δ can be determined with $\kappa(Q) < .01$.

In Table 8.1.6 below, we present some of the results of the computations. In total, about 80 CPU hours were needed to compute these results.

Table 8.1.6.

$x/10^6$	$\pi_3(x)$	$f_3(x)$	$r_3(x)$	$\pi_1(x)$	$f_1(x)$	$r_1(x)$
10	332398	255697	0.7692495	332180	256346	0.7717081
50	1500681	1148210	0.7651260	1500452	1151040	1.7671288
100	2880950	2201430	0.7641334	2880504	2205113	0.7655303
150	4222411	3223457	0.7634162	4221984	3228344	0.7646509
200	5540116	4226819	0.7629477	5538820	4233706	0.7643697
250	6840343	5216929	0.7626707	6838974	5225613	0.7640931
300	8126606	6195760	0.7624044	8125718	6206614	0.7638235
350	9402353	7166342	0.7621860	9401172	7177686	0.7634884
400	10668718	8129627	0.7620060	10667607	8142331	0.7632762
450	11927101	9086081	0.7618013	11925936	9100975	0.7631246
500	13179058	10037729	0.7616424	13176808	10052888	0.7629229
550	14423312	10983002	0.7614757	14422043	11000483	0.7627548
600	15662772	11925126	0.7613675	15661930	11943522	0.7625830
650	16897400	12863448	0.7612679	16895994	12882297	0.7624468
700	18127414	13799009	0.7612233	18125516	13817870	0.7623435
750	19352799	14730095	0.7611351	19350381	14749979	0.7622578
800	20573718	15658228	0.7610792	20572560	15679417	0.7621557
850	21791649	16584780	0.7610613	21790316	16605975	0.7620805
900	23005255	17505721	0.7609444	23003959	17528770	0.7619893
950	24215752	18425396	0.7608847	24215718	18449756	0.7618918
1000	25424042	19343291	0.7608267	25423491	19368166	0.7618217

At the suggestion of Henri Cohen, we also attempted to fit the values of $r_i(x)$ ($i = 1,3$) to a curve of the form $a + bx^{-\alpha}$. This was done by using a golden ratio search technique to determine that α value which yielded the minimum error, where, by the error, we mean the sum of the squares of the vertical deviations of the data points $(x^{-\alpha}, r_i(x))$ with $x = 10^5 j$ and $j = 1,2,3\ldots,10^4$ from the least squares straight line fitted to those points. In the case of $r_1(x)$, we obtained a value $\alpha = .22972$ and a y-intercept .75665; in the case of $r_3(x)$ we obtained a value $\alpha = .24007$ and y-intercept .75657. In both cases, the y-intercept is somewhat larger than .75446. This could be the result of the naïvety of the assumption that $r_i(x)$ can be accurately described by a curve as simple as $y = a + bx^{-\alpha}$.

For given fixed interval size I, let

$$\overline{\pi}_i(x) = \pi_i(x + I) - \pi_i(x),$$

$$\overline{f}_i(x) = f_i(x + I) - f_i(x),$$

$$\overline{r}_i(x) = r_i(x + I) - r_i(x).$$

To get a further idea of how the real quadratic fields of class number one are distributed, we wrote a higher precision version of the programs described above and sampled intervals of size $I = 10^7$ at values of $x = 10^i$ ($i = 9,10,11\ldots,16$). In Table 8.1.7, we present the results of these calculations. By $t_i(x)$, we denote the time in minutes that the programs required to determine $\overline{r}_i(x)$. In spite of some fluctuations, it still appears that the overall tendency of $r_i(x)$ is to decrease. Indeed, none of the calculations presented here seem to provide any inconsistency with the belief that $r_i(x)$ tends to approach .75446 very slowly.

Table 8.1.7.

$x/10^9$	$\overline{\pi}_3(x)$	$f_3(x)$	$\overline{r}_3(2)$	$t_3(x)$	$\overline{\pi}_1(x)$	$f_1(x)$	$\overline{r}_1(x)$	$t_1(x)$
1	241505	183299	.75899	53	240944	183084	.74986	50
10	271319	164904	.75881	61	217331	164650	.75760	55
10^2	197381	149300	.75641	72	197020	149039	.75647	63
10^3	180913	136536	.75504	94	180813	136567	.75513	82
10^4	167168	126296	.75550	138	167144	126234	.75524	117
10^5	155353	117451	.75603	226	155229	117163	.75478	200
10^6	144816	109113	.75346	465	144578	109414	.75678	430
10^7	135906	102648	.75529	976	135991	102811	.75601	963

There are many other formulas besides (5.4.1) from which we can determine h_Δ. The best known of these is the (closed form) formula of Dirichlet (see Appendix F):

$$\varepsilon_\Delta^{h_\Delta} = \prod_{j=1}^{\lfloor(\Delta-1)/2\rfloor} (\sin(j\pi/\Delta))^{-(\Delta/j)}. \tag{8.1.23}$$

It is clear that from (8.1.23), we get

$$h_\Delta = -R^{-1}\left(\prod_{j=1}^{\lfloor(\Delta-1)/2\rfloor} (\Delta/j)\log\sin(j\pi/\Delta)\right).$$

A glance at the chapter on the class number of binary quadratic forms in Vol. III of Dickson's History [73] will reveal many more formulas for h_Δ (see also Appendix F); however, while many of these may be of considerable mathematical interest, most of them, such as (8.1.23), require that the calculator expend $O(D^{1+\epsilon})$ operations to determine h_Δ. Even modifications of (8.1.23), like those of Hasse [112] and Bergstrom [23], do not greatly affect this complexity measure.

We also mention the method of Slavutskii [342] for computing h_Δ. He first shows that $h_\Delta \leq \sqrt{\Delta}$, and then uses a version of (8.1.23) to derive formulas for determining the value of h_Δ modulo p^t. If p^t is large enough, h_Δ can be found unequivocally. The formulas for determining h_Δ modulo p^t are sums involving Kronecker symbols and Bernoulli numbers, but can be simplified by using a congruence of Voronoi. Unfortunately, the complexity of finding h_Δ by this ingenious method is still $O(D^{1+\epsilon})$. One could also use other p-adic results (see, for example, Washington [375]) to derive techniques for computing the value of h_Δ unconditionally, but such methods also seem to be of complexity $O(D^{1+\epsilon})$. However, it is possible they might be sufficiently simple that they could be very useful for values of D which are not too large. This has been investigated by Buchmann, Sands and Williams [36]. In producing a table of class numbers for all real quadratic fields with $D < 10^6$, they found that the latter techniques work very well for small values of D, but when D is large ($\approx 10^6$), it could take as long as 40 seconds (on a MicroVax II) to find h_Δ. Thus, the technique seems to be limited in its application to many small values of D or occasional large values of D. Finally, we point out that a very general algorithm for determining h_Δ for any Δ has been developed by Pohst and Zassenhaus [291]. Although the authors do not provide a complexity analysis of their rather intricate algorithm, it appears that it would compute h_Δ rather more slowly than many of the methods described here.

It seems that we have come a long way since the time of Gauss. With the aid of computers, tables of class numbers can be produced with very little effort, and it is even possible to compute class numbers for fields with large discriminants, (see Chapters Five and Six). However, it is rather disappointing to realize that in spite of the enormous amount of ingenuity expended on the problem of computing h_Δ, we are, for the problem of rigorously computing h_Δ, little beyond where Gauss was. It is true that we can now compute R much more quickly than could be done in the past, but the problem of finding a provable method of computing h_Δ in $O(D^{\alpha+\epsilon})$ operations with $\alpha < 1/2$ still remains. Such a technique would, of course, exist if the GRH were ever proved, but this seems to be a long way off yet. Indeed, there are several prominent mathematicians who do not believe that it is true. We should also point out here that Hafner and McCurley [103] have shown that under suitable Riemann Hypotheses, the computation of h_Δ (indeed, even the structure of the class group) for an imaginary quadratic field can be achieved in expected running time $L(D)^{c+o(1)}$, where $L(D) = \exp(\sqrt{\log D \log\log D})$ and $c = \sqrt{2}$. While this is a better complexity measure than any previously found, it is still unknown as to whether this technique is practical for large values of D. Some results of a numerical study of the algorithm, reported in Buchmann and Düllmann (see [35] and [38]), have shown that Hafner and McCurley's ideas can be extended to the real quadratic case. Thus, although there may be serious implementation problems, there is some hope that a subexponential algorithm for determining h_Δ might exist. We also can say (see [39]) that if certain Riemann hypotheses hold, then the problem of determining R and h_Δ is in **NP**; that is, a short proof of the value of R and h_Δ exists (although it might take an exponential amount of time to find it). In spite of this somewhat ambiguous state of affairs, it is hoped that the reader will by now have a good appreciation of the current state of research on this problem.

Exercises 8.1

1. Verify equation (8.1.6). (*Hint*: Use Claim 2 of the proof of Theorem 2.1.2.)
2. Establish the inequality (8.1.7). (*Hint*: Use Exercise 2.1.2(g)(vi) and equation (2.1.13).)
3. Establish inequality (8.1.10). (*Hint*: Use Exercise 2.1.2(g)(viii).)
4. Prove the following, in which $n \in \mathbf{N}$.
   (a) If $f(x) = O(g(x))$, then $cf(x) = O(g(x))$ for all positive $c \in \mathbf{R}$.
   (b) If $f_i(x) = O(g_i(x))$ for $i = 1, 2$, then both $f_1(x) + f_2(x) \in O(g_1(x) + g_2(x))$, and $f_1(x)f_2(x) \in O(g_1(x)g_2(x))$.
   (c) If $f(x) = O(g(x))$, where $g(x)$ is a polynomial, then $f^n(x) = O(g^n(x))$. (This shows that f increases approximately akin to the nth power of x. For instance, if $n = 4$ and $x \in \mathbf{N}$, then doubling x means increasing f by roughly a factor of 16.)
   (d) $n! = O(n^n)$.
   (e) $n \log n = O(\log n!)$.
   (f) $\log n = O(n^\alpha)$ for any positive $\alpha \in \mathbf{R}$.
   (g) If $f(n) = \sum_{i=1}^{r} a_i n^i$ for $a_i \in \mathbf{R}$ with $a_r > 0$, then $f(n) = O(n^r)$.

.2 Cryptology.

In our modern world, we send messages every day by various methods such as fax (facsimile), e-mail (electronic mail via computers) or what has come to be known in the vernacular as "snailmail" (or regular paper missives sent via various postal services). Since time immemorial, people have tried to send "secret" or "coded" messages for reasons ranging from the strictly personal to military considerations with world-altering consequences. The advent of modern-day computers has made secrecy a primary consideration in such matters as military, diplomatic and financial transactions. The goal is to ensure that the message which we want to encode or *encrypt*, called the *plaintext*, must be designed in such a way that the intended recipient, and no one else, will understand it. This coded message is called the *ciphertext*. The process of transforming a plaintext into ciphertext is called *enciphering* or *encryption*, whereas the reverse process is called *deciphering* or *decryption*, which is done by the intended receiver who possesses the knowledge to do so. On the other hand, a *cryptanalyst* is someone whose goal is somehow to break these systems, and so, this is called *cryptanalysis*. Any method for changing plaintext into ciphertext via some transformation is called a *cipher*, whereas the *key* determines the particular transformation used, as illustrated below. The design and implementation of these secrecy codes or systems is called *cryptography*, whereas the study of these secrecy systems is called *cryptology*. The practitioners of cryptology are called *cryptologists*.

A *message unit* consists of a specified number of *letters* or *characters* which may include not only our familiar letters from A to Z, but also any other symbol from a blank to an exclamation mark. An *enciphering transformation* is a one-to-one function f from the set $\mathcal{S}$ of all plaintext message units to a set $\mathcal{T}$ of all ciphertext message units, whereas its inverse f^{-1} is called the *deciphering transformation*. The general term *cryptosystem* refers to a family of such transformations $f : \mathcal{S} \to \mathcal{T}$ described by a given algorithm, and having parameters. A simple example is to take message units to be single letters from A to Z (together with a blank, not enciphered) for both plaintext and ciphertext, and label them using $0, 1, 2, \ldots, 25$ as their *numerical equivalents*. A sample cipher using modular arithmetic is $f(s) \equiv s + 3 \pmod{26}$ where $0 \le s \le 25$. This was ostensibly invented in ancient Rome by Julius Caesar, and is now an example of what we call a *shift transformation* (see Suetonius [356]). For instance,

ET TU BRUTE

as plaintext becomes

$$4 \quad 19 \quad 19 \quad 20 \quad 1 \quad 17 \quad 20 \quad 19 \quad 4;$$

then we use the cipher $s \to s + 3$ to get

$$7 \quad 22 \quad 22 \quad 23 \quad 4 \quad 20 \quad 23 \quad 22 \quad 7.$$

Translating back to letters, we have

$$H \quad V \quad V \quad X \quad E \quad U \quad X \quad W \quad H$$

which is the ciphertext message that we send. The intended receiver first converts the letters to numbers, then performs f^{-1} upon it, the deciphering transformation, i.e. $t \to t-3$, to get the numerical version of the plaintext which is now converted to letters.

There are more general transformations such as $f(s) = as + b \pmod{26}$ with $a, b \in \mathbf{Z}$, $0 \le f(s) \le 25$ and $\gcd(a, 26) = 1$, called *affine transformations*. Thus, the parameters a and b are the key. In general, an *affine cryptosystem* on an N-letter alphabet with parameters a and b (where $\gcd(a, N) = 1$, $0 \le b < N$, $0 < a < N$) consists of $f(s) \equiv as + b \pmod{N}$ together with $f^{-1}(t) \equiv a^{-1}(t-b) \pmod{N}$ for $s \in \mathcal{S}$ and $t \in \mathcal{T}$. For instance, if we know that

IRDAPC MEPNCL

was enciphered using the affine transformation $f(s) = 5s + 21 \pmod{26}$, then to decipher it, we first convert it to its numerical equivalents:

$$8\ 17\ 3\ 0\ 15\ 2\ 12\ 4\ 15\ 13\ 2\ 11,$$

then apply $t - 21 = t - b \pmod{26}$ followed by $21(t-21) = a^{-1}(t-b) \pmod{26}$ to get

$$13\ 20\ 12\ 1\ 4\ 17\ 19\ 7\ 4\ 14\ 17\ 24,$$

which we convert back to letters: NUMBER THEORY.

Motivated by the above illustration, we adopt the following notation. The encryption and decryption keys will be denoted by k_1 and k_2 respectively, and $f = E_{k_1}$ will be the encryption transformation while $f^{-1} = D_{k_2}$ will be the decryption transformation. Hence, $D_{k_2}(E_{k_1}(s)) = s$ for $s \in \mathcal{S}$. In the above examples, k_1 and k_2 can be calculated from one another. Such ciphers are called *symmetric ciphers*. In particular, if $k_1 = k_2$, then these are called *one key ciphers*, in order to distinguish them from the other type called *public key ciphers* which we will describe below. For example, if we consider the affine transformation $E_{k_1}(s) = f(s) = s + 13 \pmod{26}$ determined by the key $k_1 : a = 1$, and $b = 13$ in the above definition of such enciphering transformations, then the key k_2 determining the deciphering transformation is the same since $f^{-1}(t) = D_{k_2}(t) = t - 13 \pmod{26}$. Thus, $s = D_k(E_k(s))$ where $k_1 = k_2 = k$.

Such single key cryptosystems require that the key be kept secret, since the knowledge of the key would allow anyone to cryptanalyze the ciphertext. Symmetric ciphers can be further subdivided into *stream ciphers* and *block ciphers*. Stream ciphers act upon plaintext one bit at a time (generally used before the advent of computers), whereas block ciphers act upon groups of bits called "blocks". One of the most popular of these block ciphers for centuries was the *Vigenère cipher*, which is described as follows. Let b be a fixed n-tuple of characters in an N-letter alphabet. This serves as a key. Encipher via the transformation $f(s) \equiv s + b \pmod{N}$ where s is a plaintext message unit, which is an n-tuple of numerical equivalents (see Exercise 8.2.6). However, if a cryptanalyst somehow can find out the values of N and n, then a frequency analysis on the first letter of each block can be used to determine the first character of the key. By induction, this cryptosystem can be broken, so it also has its weaknesses.[(8.2.1)]

There are also methods involving enciphering matrices (see Koblitz [147]), and exponentiation ciphers (see Rosen [312]). The most widely used symmetric cipher is the *Data Encryption Standard* (DES) endorsed by the military in the U.S.A. for encrypting sensitive data, such as for bank, tax, hospital records, etc. (see Schneier [319, p. 12]).

[(8.2.1)] In fact, an excellent general source for background and general historical interest is Kahn [134].

The weakness of the symmetric cryptosystems is having to keep the key secret. To avoid this pitfall, a new type of cryptosystem has been (relatively) recently introduced.[(8.2.2)] We now describe this. "Public key cryptosystems" involve two distinct keys, k_1 (called the *public* key) used to encrypt plaintext, and k_2 (called the *private* key). Thus, anyone can encrypt a message, but not decrypt it. The concept is based upon a "trap-door" in a "one-way function", both of which we now describe. First of all, we say that a function f is a *one-way function* if $f(s)$ is easy to compute for inputs s, but s is virtually impossible to compute from $t = f(s)$, given the vast amount of computer time required,[(8.2.3)] i.e. we can easily compute f, but not f^{-1}. Unfortunately, there is no rigorous mathematical proof that one-way functions exist, or that a given function is *provably* one way. For example, $f : \mathbf{N} \times \mathbf{N} \to \mathbf{N}$, defined by $f(m, n) = mn$ is (conjectured to be) a one-way function. However, there is no information available to invert the function for all inputs because factoring is "difficult" (given our current state of knowledge). However, we cannot use one-way functions as encryption transformations *unless* we have some secret way of inverting them. The "trap-door" is the secret, i.e. a *trap-door one-way function* is a one-way function (as defined above), together with a key k_2 (called a *trap-door*) which we can use to compute $f^{-1}(t)$ easily for a given $t = f(s)$. Thus, a *public key cryptosytem* involves a method of encrypting plaintext $s \in \mathcal{S}$ via a one-way function $E_{k_1} : \mathcal{S} \to \mathcal{T}$ with (public key) k_1, and a trap-door (or private key) k_2, such that $D_{k_2}(E_{k_1}(s)) = s$ is easily computed for all $s \in \mathcal{S}$. Each user in this system then has 2 keys, one private and one public.

Public key cryptosystems therefore avoid the inherent weakness of symmetric cryptosystems, since the (public) encryption keys are published (in some database). Hence, sender A gets receiver B's public key from the published list, and encrypts the message using the key. Receiver B then uses his private key to decrypt A's message.

Although public key cryptosystems alleviate the problem of distributing the secret key involved in symmetric cryptosystems, the algorithms involved are significantly slower (by at least a factor of 10^3) than the ones used in symmetric cryptosystems. Thus, it makes most practical sense to use the more secure public key cryptography to distribute the keys for a symmetric cipher. These keys (called *session keys* if they are only used for one particular session), are then used for symmetric cryptosystems. Such systems which use public key cryptography to ensure secure key exchange, then use the faster symmetric cryptosystems for the exchange of the bulk of the message, are called *hybrid cryptosystems*.

As with any other means of communication, we need a signature of some type to verify that the message sent is actually from the wanted party. With public key cryptography, this is done as follows. Suppose $\mathcal{P} \in \mathcal{S}$ is the signature, f the enciphering transformation for the sender A, and g the enciphering transformation for the receiver B. If, at say the end of the message, A sends $gf^{-1}(P)$, then when B applies g^{-1}, $f^{-1}(P)$ remains, to which B now applies f to get P. Since only B has f^{-1}, then B knows the message came from A.

There are numerous "authentication methods" which can be used to foil cryptanalytic techniques. The reader is referred to Schneier [319] for an in-depth analysis

---

(8.2.2)See Diffie and Hellman [74], and Hellman [117].

(8.2.3)This is not a rigorous mathematical statement, nor can it be made to be so, since it is very much dependent upon the current state of computer technology such as parallel processors and ultra-fast "sieving chips". Thus, the term "impossible to compute in realistic computational time" has to be taken very much in context. What is secure "today" may not be so "tomorrow".

of them. We now turn to a discussion of key exchange algorithms.

Diffie and Hellman [74] described a scheme by which two individuals A and B could exchange a secret cryptographic key. Briefly, the idea is as follows.

(i) A and B agree on a large value of q where $\mathbf{F}_q$ is a finite field of q elements, and some fixed primitive element g of F_q where both g and q can be made public.

(ii) A selects an integer n at random(8.2.4) in $\mathbf{F}_q$ (i.e. $1 \leq n \leq q-1$), and transmits $b = g^n$ ($b \in \mathbf{F}_q$) to B, and b is made public.

(iii) B selects an integer m at random and transmits $c = g^{mn}$ which is made public ($c \in \mathbf{F}_q$) to A.

(iv) A determines $k = c^n$, B determines $k = b^m$ ($k \in \mathbf{F}_q$), and $k = g^{mn}$ is then used as the secret key which both A and B can compute.

A cryptanalyst would know g, q, b, c, but not know m or n. If the cryptanalyst could determine n from knowledge of g and b and the fact that $b = g^n$, then k could easily be computed. The problem of determining $n = \log_g b$ given b and g is called the *discrete logarithm problem*.(8.2.5) We'll say more about this later.

Recently in [214], McCurley modified the Diffie–Hellman algorithm by replacing $\mathbf{F}_q$ by the ring $\mathbf{Z}/n\mathbf{Z}$ where n is a product of two large primes. The claim is that this system is at least as difficult to cryptanalyze as it is to factor n.(8.2.6) In fact, the ancient problem of factoring n is the basis for one of the most widely used public key cryptosystems, the *RSA cryptosystem* (named after Rivest, Shamir and Adleman [311]) which was introduced in the late seventies. To describe this method we first choose two large primes p and q (where large can vary, depending upon "realistic computational considerations" (see footnote (8.2.3)) say 150 digits each) and set $n = pq$. Thus, $\phi(n) = (p-1)(q-1) = n + 1 - p - q$ is easily computed. Next, we choose at random an integer e between 1 and $\phi(n)$ with $\gcd(e, \phi(n)) = 1$. To encipher an $s \in \mathcal{S}$, we first translate the characters in s into their numerical equivalents and then group the numbers into k-digit base-N integers in $\mathcal{S}$ and ℓ-digit base-N integers in $\mathcal{T}$ with $k < \ell$ chosen so that N^k and N^ℓ are suitably large. Each prime must be chosen large enough so that $N^k < n < N^\ell$. In this fashion, each message unit $s \in \mathcal{S}$ transforms to $f(s) = t \in \mathcal{T}$ as a unique ℓ-digit base-N integer. We then set $f(s) \equiv s^e = t \pmod{n}$ so the public enciphering key is (n, e). To decipher, we need knowledge of the inverse d of e modulo $\phi(n)$, i.e. $de = r\phi(n) + 1$ for some $r \in \mathbf{Z}$. Thus, $f^{-1}(e) = t^d = s^{ed} \equiv s \pmod{n}$, and so $D_k = (d, n)$ is the deciphering key. For instance, we have

Example 8.2.1. We cannot be realistic here in terms of our choice of n because, given modern-day computational abilities, no n less than 150 digits would allow a secure transmission. However, for illustrative reasons, we now choose $p = 163$ and $q = 167$, so $n = 27221$ and $\phi(n) = 26892$. Also, in practice, p and q must not be

(8.2.4)By "at random", we mean that the number was chosen by a computer program that generates sequences of digits in a way no one could duplicate — sometimes called a "random number generator" or pseudo-random number generator.

(8.2.5)In [283], Odlyzko suggests that use of $q = 2^r$ or a large prime p yields a greater level of security.

(8.2.6)Koblitz [146] and Miller [218] have used groups of points on an elliptic curve over $\mathbf{F}_q$ to develop a secure key-exchange system. The elliptic curve method seems to hold current favour. See Koblitz [147].

chosen close to each other, or else a cryptanalyst can use continued fractions to break the system. Let $e = 1673$, $k = 2$ and $\ell = 5$. If we wish to encipher FAST, we first get its numerical equivalent which is 501819. We group the 50 together to get a $k = 2$ digit integer. Taking each plaintext $s \in \{50, 18, 19\}$, we perform $s^e \pmod{n}$ which results in $t \in \{25241, 116104, 23478\}$ respectively. The receiver then applies $t^d \pmod{n}$ where $d = 13229$ is the multiplicative inverse of e modulo $\phi(n)$ achieved via application of the Euclidean algorithm. Thus, each t^d yields $\{50, 18, 19\}$ which decrypts to FAST, as required.

The RSA system provides us with a nice example of a one-way function with a trap-door. To see this in detail, let n be the RSA modulus with $ed \equiv 1 \pmod{\varphi(n)}$. For $1 \le s \le n-1$, the function $f(s) = s^e \pmod{n}$ is the "RSA" one-way function with trap-door d, namely $f(s)^d = s \pmod{n}$. Thus, $f(s)$ is a one-way trap-door function, since s is easily computed (via d) from $f(s)$ for all inputs s; but it is computationally impossible to compute s from $f(s)$ *without* the trap-door value d. The reason is that there is no known way of obtaining s without computing d. Furthermore, if one knew d, then one could easily compute the factorization of n. Thus, it is conjectured that the only way to break RSA is by factoring the modulus.

No known cryptosystem exists which can lay claim to being ***provably*** one-way (or public key). As we shall see, however, in later descriptions of public key cryptosystems, certain authors do give evidence, mostly of a probabilistic or heuristic nature, that their particular cryptosystems are indeed secure. What would be involved in such a proof would be a determination of the minimum number of bit operations required (as discussed in the last section) in order to cryptanalyze the algorithm. Similarly, it has not been proven that the RSA cryptosystem cannot be cryptanalyzed without factoring n. Yet thus far, no method (other than factoring) of cryptanalyzing RSA has been found. As we noted, the security of the system depends upon the size of the modulus n. An n as given in Example 8.2.1 would, of course, not suffice. Primes p and q of 77 digits each (i.e. 154-digit (512-bit) modulus) would suffice today, but given computational advances, this may not hold true for long, so it would be prudent to go to 1024-bit moduli anytime since advances in factoring are occurring at a steady pace. Fundamentally, the security of the RSA system depends on our inability to factor a composite integer effectively. We have now discussed some algorithms pertaining to key-exchange and those involving the discrete log problem such as Diffie–Hellman.

There are numerous others such as Massey–Omura cryptosystem, Elgamal cryptosystem and others (see Koblitz[147], Kranakis [150], and Rosen [312]) involving discrete logs.[8.2.7] As noted above, with the Diffie–Hellman system, if a cryptanalyst can get n from knowledge of g and g^n, then the cipher would be broken, i.e. by basically solving the discrete log problem.

We now describe one such very recent ***key-exchange algorithm*** based upon reduced ideals in maximal complex quadratic orders. In Chapter Two, section two, Algorithm 2.2.1 was a precursor to the following secret key-exchange due to Buchmann and Williams [37].

Two users A and B select a value of D such that $|D|$ is large ($\approx 10^{200}$) where $D < 0$ is a square-free integer, and an ideal I in O_Δ. The value of D and the ideal can be made public.

[8.2.7] The reader may also be interested in another public key cryptosystem not covered here. It is called the "Knapsack" public key cryptosystem (see Koblitz [147, p. 107ff] and Rosen [312, p. 266ff]).

can be made public.

(i) A selects at random an integer x and computes a reduced ideal J such that $I^x \sim J$, so A sends J to B.

(ii) B selects at random an integer y and computes a reduced ideal L such that $L \sim I^y$, so B sends L to A.

(iii) A computes a reduced ideal $M_1 \sim L^x$, and B computes an ideal $M_2 \sim J^y$.

Since $M_1 \sim L^x \sim (I^y)^x = (I^x)^y \sim J^y \sim M_2$, then by Theorem 1.4.2(e), $N(M_1) = N(M_2)$ and if $M_1 = [N(M_1), \alpha_1]$, $M_2 = [N(M_2), \alpha_2]$ then $|\mathrm{Tr}(\alpha_1)| = |\mathrm{Tr}(\alpha_2)|$. Thus, A and B can use either $N(M_1) = N(M_2)$ or $|\mathrm{Tr}(\alpha_1)| = |\mathrm{Tr}(\alpha_2)|$ as their secret key. Note, however, that the values of $N(M_1)$ and $|\mathrm{Tr}(\alpha_1)|$ are not independent since, of course, $N(M_1) \mid N(\alpha_1)$.

If A wants to send a secure message m to B, then A can compute for randomly selected x, $M \sim L^x$ where M is a reduced ideal and $L \sim I^y$ is in B's public file, or has been sent to A by B. Here, as above, x is known only to A and y is known only to B. The encrypted message is sent to B as $(E + N(M), J)$ where $J \sim I^x$ and E is the first block of m with $E < N(M)$. Subsequent blocks of m would be sent in the same way. However, A must change x for each new block he sends. To find E, B must determine $N(M)$. Since B has $J \sim I^x$ and y, then $J^y \sim I^{xy} \sim L^x \sim M$.

It remains to consider the problem of finding an efficient algorithm for multiplication of ideals. The binary method of exponentiation will then provide an efficient algorithm for computing I^m for large m.

In the preceding section, we have shown how to use Shanks' method to multiply ideals. Now if we let $m = (b_0\ b_1\ b_2\ \ldots\ b_k)$ be the binary representation of m with $b_0 = 1$, then let $N_0 \sim I$ and define $N_i^2 \sim M_i$ with

$$N_{i+1} \sim \begin{cases} M_i & \text{if } b_{i+1} = 0, \\ M_i I & \text{if } b_{i+1} = 1, \end{cases}$$

where M_i and N_{i+1} are reduced $\mathcal{O}_\Delta$-ideals. It follows that $N_k \sim I^m$. If we select I such that $N(I) > \sqrt{|\Delta|/3}$ is prime, then for all multiplications $M_i I$, we would likely have $\gcd(N(I), N(M_i)) = 1$. As seen in the preceding section, this simplifies the determination of N_{i+1}. Hence, we see that computing N_k requires the performance of $O(log(m \log |D|))$ elementary operations.

Now, we describe the security of this cryptosystem. The complexity of this system is greater than that of the Diffie–Hellman scheme described above and so is the *bandwidth*, i.e. to communicate about 100 digits of key, it is necessary to exchange about 200 digits of information across the communication channel. However, what is lost in complexity is made up in added security over the Diffie–Hellman system. As mentioned earlier, this cannot be proved, but we provide evidence that this is the case as follows.

From Narkiewicz [280, p. 389], we know that

$$h_\Delta \leq \frac{2}{\pi}|\Delta|^{1/2}(1 + \log(2|\Delta|^{1/2}/\pi))$$

and by results of Chapter Five, section four,

$$h_\Delta > |\Delta|^{1/2-\epsilon}$$

for any $\varepsilon > 0$ and all sufficiently large $|\Delta|$. Indeed, under the GRH, Littlewood [187], see (8.1.2), showed that

$$(\pi(1+o(1))\sqrt{\frac{|\Delta|}{12c^{\gamma}\log\log|\Delta|}} < h_{\Delta} < (2(1+o(1))\sqrt{|\Delta|}\log\log|\Delta|)/\pi,$$

(where γ is Euler's constant. See footnote (8.1.7).)

As with (8.1.2), we can replace $1 + o(1)$ by absolute constants by using the explicit results of Oesterlé [284]. Thus, we would expect h_{Δ} to be about the same order of magnitude as $|\Delta|^{1/2}$.

Now, suppose that a cryptanalyst $\mathcal{C}$ who is trying to break this system has the value of D and the ideals I, J and L, but does not know x or y. The problem then is for $\mathcal{C}$ to determine M_1 or M_2. One way $\mathcal{C}$ might approach this is to attempt to solve the discrete log problem in C_{Δ} mentioned above, i.e. given a reduced ideal I and a reduced ideal J such that $I^x \sim J$ find x.

One simple attack (called the giant step-baby step method discussed in the preceding section) is to put $q = Ne(|D|^{1/4})$ and assume that $x = qk - r$ where $0 \le r < q$. Since we can also assume that $h_{\Delta} > x$ and we know from the above that $h_{\Delta} = O(|\Delta|^{1/2+\varepsilon})$, we see that $k = O(|\Delta|^{1/4+\epsilon})$. We first compute $F \sim I^q$. We then find all the $O(|\Delta|^{1/4+\varepsilon})$ reduced ideals equivalent to $I^j J$ for $j = 0, 1, \ldots, q-1$ and check when $F^i = I^j J$ in $O(\Delta|^{1/4+\epsilon})$ operations. When this occurs, we have $x = qi - j$. For the complexity result given above to hold, we must sort the list of reduced ideals equivalent to $I^j J$ (for $j = 0, 1, \ldots, q-1$). In practice, however, it is more efficient to use hashing techniques.

The method of Pohlig and Hellman [290] can be adapted to the solution of this problem. Indeed, this technique may be regarded as a more sophisticated version of the giant step-baby step method. If p is the largest prime factor of h_{Δ} then this algorithm will find x in $O(p^{1/2+\varepsilon})$ operations, utilizing about the same amount of storage. However, in order to employ this scheme, we must know the value of h_{Δ}. The fastest known methods of determining h_{Δ} are of complexity $O(|\Delta|^{1/5+\epsilon})$ assuming GRH, even when h_{Δ} is made up exclusively of small prime factors (see the discussion in the preceding section after Algorithm 8.1.3).

The index calculus method (see Koblitz [147] or Odlyzko [283]) has proved to be a very successful method for attacking the discrete log problem. However it appears to be difficult to apply to the above scheme. Thus, what all of the discussion seems to establish is that the problem of determining x such that $I^x \sim J$ is a very difficult one. The best known methods are of complexity $O(|\Delta|^{1/4+\epsilon})$ or possibly $O(|\Delta|^{1/5+\epsilon})$. Just how difficult the problem really is, is not known. What we can show is that, using the ideas of Shanks discussed in the preceding section, if an efficient method for solving the problem is discovered, then the method will very likely allow us to factor D. This suggests that the problem of finding x is at least as difficult as the factoring problem.

Assume that D is a composite integer and we can solve the problem of finding x for a given I, J such that $J \sim I^x$. By Corollary 1.3.2, h_{Δ} is even, so the probability that for $J = (1)$, the value of x is even is at least $1/2$. If we can find x, then for $m = x/2$, we have $(I^m)^2 \sim 1$. Thus, by Exercise 1.3.7(c), there is an ambiguous ideal $L \sim I^m$. Thus, a reduced ideal equivalent to L will, with probability $\geq 1/2$, provide us with a nontrivial factorization of D. Of course, it may be possible to cryptanalyze the above scheme without having to solve the discrete log problem in C_{Δ}. To see this, suppose that A is some algorithm that upon being given I, J and

L with $I^x \sim J$, $I^y \sim L$ produces $M = A(I, J, L)$ where $M \sim I^{xy}$. By using ideas similar to those developed by Schmuely [318] and McCurley [214], it can be shown that if a cryptanalyst $\mathcal{C}$ possesses such an algorithm, then it is very likely that $\mathcal{C}$ can easily find a nontrivial factor of D.

To see this, we assume D is composite again and let $2^m \mid h_\Delta$. By Theorem 1.3.3, the Sylow 2-subgroup $C_{\Delta,2}$ of C_Δ is of the form

$$C_{\Delta,2} = \mathcal{C}_{m_1} \times \mathcal{C}_{m_2} \times \cdots \times \mathcal{C}_{m_{t_\Delta}}$$

where t_Δ is as in Definition 1.3.3, and $\mathcal{C}_{m_i}$ is cyclic of order $m_i \mid 2^m$ with $i = 1, 2, \ldots, t_\Delta$. If $2^j = \max\{m_i\}_{i=1}^{t_\Delta} \geq 2$ and $s = h_\Delta/2^m$, then for any class C_0 of C_Δ we have

$$C_0^{2^j s} = 1. \tag{8.2.1}$$

Furthermore, there must exist an ideal class C_1 in C_Δ such that

$$C_1^{2^j} = 1 \quad \text{and} \quad C_1^{2^{j-1}} \neq 1.$$

Thus, if $C_0^{2^{j-1}s} \neq 1$, then $(C_0C_1)^{2^{j-1}s} \neq 1$ and it follows that at least half of the ideal classes in C_Δ possess the property that if C_0 is such a class, then $C_0^{2^{j-1}}$ has even order.

Now let C be any element of C_Δ and put $s = 2u - 1$. By (8.2.1),

$$C^{2^{j+1}u} = C^{2^j}. \tag{8.2.2}$$

Hence, the subgroups of C_Δ generated by C^{2^j} and $C^{2^{j+1}}$ are identical. By previous remarks, we see that if we select at random an ideal G of $\mathcal{O}_\Delta$, then with probability $\geq 1/2$ we will find that $G^{2^{j-1}v}$ is not principal whenever v is odd. For such a G, put $I \sim G^{2^{j+1}}$, select x, y at random and put $J \sim G^{2^j x}$, $L \sim G^{2^j y}$. By (8.2.2), we have

$$J \sim G^{2^j x} \sim G^{2^{j+1}xy} \sim I^{xu}$$

and

$$L \sim G^{2^j y} \sim G^{2^{j+1}yu} \sim I^{yu}.$$

Thus, we may use algorithm A to produce

$$M = A(I, J, L) \sim I^{xyu^2} \sim G^{2^{j+1}xyu^2}. \tag{8.2.3}$$

Since

$$2^{j+1}u^2 \equiv 2^{j-1}s^2 + 2^{j-1} \pmod{2^j s},$$

then by (8.2.1) and (8.2.3), we have

$$M \sim G^{2^{j-1}s^2xy}G^{2^{j-1}xy}, \tag{8.2.4}$$

and

$$M^2 \sim G^{2^j xy}.$$

If $G^{2^{j-1}xy} \sim M$, then $G^{2^{j-1}s^2xy} \sim 1$ by (8.2.4) which, by the selection of G is not possible since s^2xy is odd. Thus, $H = MG'^{2^{j-1}xy}$ is not a principal ideal, but $H^2 \sim 1$. Also, we can compute H by making a call to $A(G^{2^{s-1}}, G^{2^{s-1}x}, G^{2^{s-1}y})$ to find $G^{2^{j-1}xy}$.

What we have therefore constructed is a reduced ideal H in an ambiguous class of C_Δ. By Theorem 1.4.2(f), we can now factor D very easily. It might be argued that we need to know the value of j here, but since $j \le \log_2 h_\Delta = O(\log D \log\log D)$, there is no difficulty in attempting to guess its value. In the special cases outlined below, we might obtain much better bounds on j.

One other problem which might arise in the above scheme is that the order of the class C might be very small. However, this seems to be most unlikely in view of the following. Cohen–Lenstra [58] presented several heuristic results concerning C_Δ. Some of these are:

(i) The probability that the odd part of the class group is cyclic is 97.757%.

(ii) If m is any odd integer, the probability that m divides h_Δ is

$$\prod_{p^\alpha \| m}\left[1-\prod_{i=1}^{\infty}\left(1-\frac{1}{p^i}\right)\Big/\prod_{i=1}^{\alpha-1}\left(1-\frac{1}{p^i}\right)\right].$$

(iii) If $e > 1$ is a fixed odd integer, the average number of elements of C_Δ which are of order exactly e is 1.

(iv) If p is an odd prime and r_p the p-rank of the class group, then the probability that $r_p = a$ given a is

$$p^{-a^2}\prod_{i=1}^{\infty}\left(1-\frac{1}{p^i}\right)\Big/\prod_{i=1}^{a}\left(1-\frac{1}{p^i}\right)^2.$$

Buell [42] got extensive numerical results which tend to confirm these heuristics.

If $D = -p_1p_2 \equiv 1 \pmod 4$ where p_1, p_2 are primes such that $(p_1/p_2) = -1$, then by Exercise 8.2.7, $h_\Delta \equiv 2 \pmod 4$. Thus, we can select many values of D such that the exact power of 2 dividing h_Δ is small. For such D values, we would expect the class group to be cyclic or close to it. Since in a cyclic group of order m there exist $\phi(d)$ generators for any subgroup of order $d \mid m$ and

$$\frac{\phi(d)}{d} > \left(e^\gamma \log\log d + \frac{2.50367}{\log\log d}\right)^{-1}$$

(see Rosser–Schoenfeld [313]), the chance of selecting an ideal I such that C has small order is very small when h_Δ is large.

Buchmann and Williams also provided a key exchange system which uses the infrastructure of a real quadratic field in [40] (see also Scheidler *et al.* [316] for a detailed description of the scheme, as well as considerations for the implementation of the required algorithms).

We have only just barely touched upon the topic of algorithms in number theory here. The subject is rich, extensive and growing fast. The reader is invited to go to any of the references listed in this chapter for further reading, but also to an excellent survey article by Lenstra and Lenstra in [175] where numerous algorithms are discussed in relation to factoring integers and proving primality. In particular, a fifty-page paper [388] by Williams and Shallit gives a history of computational number theory with emphasis on factorization and primality testing. Their list of over 180 references is of interest in itself. Also of value and interest is the extensive textbook by H. Cohen [56], as well as texts by Pohst and Zassenhauss [292], and Zimmer [402] on algorithms in computational number theory. Also, there is the more recent comprehensive text by Schneier [319]. Also, see the articles by Lenstra [179]-[180].

Exercises 8.2

1. Use the Caesar cipher to encrypt BEWARE THE IDES OF MARCH.
2. Use the affine transformation $f(s) = 3s+13 \pmod{26}$ to encode THE NONES OF MARCH ARE SAFE.
3. Decipher UP MUJWJ which was enciphered, using the affine transformation $f(s) = 21s + 3 \pmod{26}$.
4. How many shift transformations are there in an N-letter alphabet?
5. Suppose that we have two affine transformations f and g on an N-letter alphabet. Prove that the composition $f \circ g$ is also an affine enciphering transformation.
6. Use the Vigenère cipher modulo $N = 26$ to encipher THE KEY IS THE NTUPLE.
7. Let $\Delta = -p_1p_2 \equiv 1 \pmod 4$ where p_1 and p_2 are primes such that $(p_1/p_2) = -1$. Prove that $h_\Delta \equiv 2 \pmod 4$. (*Hint*: Since $t_\Delta = 1$, then it suffices to show that $I = [p, (1+\sqrt{\Delta})/2]$ is not a square in C_Δ. Assume that it is, and get a contradiction to $(p_1/p_2) = -1$.)

8.3 Implications for Computational Mathematics for the Philosophy of Mathematics.

This section contains philosophical insights provided by Andrew Lazarus, which we use as a closing statement to the main text of the book, since it provides a human face to what has gone before. This is an extended version of a paper by Lazarus in [165], reprinted by permission of the American Mathematical Society, as follows:

"Electronic machinery continues to play an indispensable role in mathematical research, and there is no reason to expect this phenomenon to abate. One sometimes hears that the proof of the Four Color Theorem was the first example of a proof in which the computer made a contribution that could not have been done by humans unassisted. Recall that Haken and Appel first showed that it would suffice to establish the theorem for a certain finite set of graphs; then they used a computer to four-color each of these. Mark Kac [133] wrote of this proof, "The latter task was relegated to the computer which came through with flying colors, completing the tedious and relatively uninteresting part of the proof."[(8.3.1)]

(8.3.1) Since the writing of Lazarus' article, a new proof of the Four Colour Theorem has arisen. It is due to Seymour, Robertson, Senders and Thomas. Although it still uses a computer, it is much cleaner and shorter (in terms of CPU time) than the Appel–Haken proof.

I believe that most mathematicians who care about the ramifications of the use of a computer agree with Kac, as do I. There is, however, a school — at least a trend — amongst philosophers of mathematics to describe computer-assisted proof as part of a "paradigm shift" (cf., of course, Kuhn [151]) that makes the meaning of the term 'proof' different from what it was before. The *new* tradition in philosophy of mathematics is philosophical analysis of mathematical praxis, over against the *old* traditional multifurcated arguments of the formalists, the logicists, the intuitionists, and the constructivists. I respect the new tradition as a valuable, even overdue, endeavor, but within its framework what I am espousing — and what I think most computational mathematicians believe — is a conservative position."

A brief history of computational mathematics as heuristic is now given.

"Computer-assisted proofs have grown in both number and breadth since the Four Color Theorem, even discounting results which are calculations of the properties of individual specific numbers, such as primality testing. Besides the Four Color Theorem, we may give partial credit to machines for the disproof of Mertens' Conjecture, the nonexistence of a certain projective plane, new results on high-dimensional polyhedra, and so on. Computational mathematics has come so much to the fore that it seemed remarkable to read that Andrew Wiles did not have any computer in the room where he solved Fermat's Last Theorem, not even for word processing and e-mail.

However, the novelty of computation in pure mathematics has been greatly exaggerated. One "new philosopher" of mathematics attributed the first machine-assisted proofs to D.H. Lehmer.[8.3.2] In his Gibbs Lecture of 1965, published in [171], Lehmer himself attributed the first electronic mathematics to a harmonic analyzer developed in the 1890s. According to Lehmer, its graphical output stumped Michelson and led Gibbs to his eponymous phenomenon. Thus, original computerized proofs are some 30 years old, while computerized inspiration has been around almost a century. Inspiration from calculation has been around even longer. We have Gauss' statement that he was led to conjecture the Prime Number Theorem on the basis of his own tables. In thinking of the electronic computer as a novelty, we forget the human calculators who were on occasion hired by *theoretical* mathematicians. These pre-electronic computers were persons who did not necessarily have high-level mathematics training but were unusually efficient at mental arithmetic (see Beckman, [19, p. 104]). The history of computation as heuristic is old indeed."

Now we look at the aesthetics involved.

"Having settled that computational mathematics is not just recent mathematics, we can ask, "Is it attractive mathematics?" There is no accounting for taste, but to answer this question in the negative is to repudiate, not to affirm, the traditional standard. Gauss both used and enjoyed computations. Boring, trivial mathematics can and will be created more rapidly with computers. These results, even if published, are destined for the same obscurity as the tedious hand computations of determinants that fascinated some British mathematicians of the last century. I suggest that one adopt a holistic approach when evaluating results which include

[8.3.2]Probably Lehmer *et al* [173], which was brought to my attention by its fourth-named author, where the statement of the theorem was inspired by computer-generated data, and an iterated calculation in which the computer was programmed to recognize parameter values, indicating success was used to finish off its proof. The Lehmers made a special point of remarking that their result was a statement in full generality about an *infinite* family of primes, not a mere report on the properties of some finite set.

calculations. My own ranking would note that the Four Color Theorem required considerable human ingenuity to reduce an infinite problem to an effective algorithm, the disproof of Mertens' Conjecture involved a search not known in advance to succeed, and the nonexistence of the projective plane involved an a priori finite calculation too large to do by hand. While all three are significant accomplishments, I feel that this list is in descending order of elegance.

To denigrate the Four Color Theorem as uninteresting by reflecting onto it a judgement of a proof (e.g. see Davis and Hersh [69]) is a dangerous practice. The paradoxical implication is that if another proof of the Four Color Theorem without a computer appears, four-coloring becomes interesting again. Usually problems become *less* interesting after being solved once.

Strong criticism is voiced by Jerry King [140]: computerized proof will lead to a "future without elegance, a world of disfigured mathematics." Let us assume that King's inelegance is restricted to the computational components of the proofs, and not to the theorems, to avoid the paradox above. How does one distinguish the beauty of a computation by the identity of the computing agent? One could, theoretically, argue that computation will now be less elegant because the computer's speed will lead to a loss of interest in beautiful algorithms — for example, that our computer programs will sum arithmetic series by looping, not by formula. This undesirable outcome is exactly the opposite of the actual situation in computational mathematics: efficiency and complexity are being analyzed and quantified as never before. For a fixed algorithm, I see no meaningful aesthetic distinction between a hand computation and a machine computation. Criticism like King's should be understood as applying to both. Such anti-computational sentiments consign to aesthetic inferiority some of number theory's most fertile techniques, such as sieves. I doubt if this view would survive widely if someone presented a sieve-theoretic proof of the Riemann Hypothesis.

The most controversial aspect of computer-assisted proof is whether it truly entails new concepts of proof and of mathematics itself. The novel position is that the computer has turned mathematics into an experimental, empirical science, closer to (say) physics.

That mathematics is an experimental science I readily concede. Imre Lakatos showed in his *Proofs and Refutations* [161] that mathematical knowledge increases through a process of conjecture-formation, testing, and modification: this process describes the behavior of the mathematical community much better than the sausage-grinder model of mathematicians extruding new, securely-encased theorems.

I have some difficulty, however, with the label "empirical" and with the assertion that mathematical truth is similar to physical truth. It has been said, I think accurately, that practicing mathematicians are Platonists during the week and become formalists on Sundays. I am partial to the idea that abstract objects, such as the integers, have an existence independent of human thought, i.e. notions like '17', 'polygon', and even 'semi-simple algebra' refer to objects that do not exist in a specific space or time.[8.3.3]

I am willing to accept Lakatos' label of "quasi-empirical" not because it means

[8.3.3] For an argument against mathematical Platonism based in part on Benacerraf's Dilemma (we in space-time cannot obtain knowledge of objects outside of space-time), see Kitcher [141]. I think an even better argument against Platonism derives from independence results. How can your Platonic view of the real numbers be the same as mine if we have differing views, or none at all, on the Continuum Hypothesis?

"we won't insist that mathematics is exactly like an empirical science if you'll acknowledge it's quasi-empirical" — as alleged, perhaps in jest, by the philosopher Thomas Tymoczko [367] — but because I think that the way in which we apprehend abstract objects is through empirical manifestations or avatars — a stronger concept than 'example'.

Tymoczko stated that the proof of the Four Color Theorem relegated mathematics to the status of physics because it introduced black-box testing (although he did not use this term) into the proof [365]. He conjured up Martian mathematicians who began to consult a Martian named Simon of super-human (or super-Martian!) computational capabilities. Eventually the other Martians, who could no longer understand how Simon arrived at his results, filled their proofs with "Simon says". I find this a dubious analogy to the process of computer programming and debugging. The computer is a tool, not an oracle: the reasons a computer programmer has for believing the output of his own program, even if unjustified, are not the same the Martians have for believing Simon. Mathematical knowledge, viewed as a collection of beliefs, has long featured reliance on others. The epistemic consequences of reliance are complicated, but in my view irrelevant to a discussion of *incremental* changes introduced by the computer.

I reiterate that it is dangerous to classify the Four Color Theorem on the basis of its known *proof*, which may be supplanted. But I think it is an additional error here to classify the proof on the basis of how much time an effective algorithm takes. Had Haken and Appel the use of a billion slaves, they could have produced the same proof without a computer. I assert that the computer and slave versions of the proof are epistemologically although not morally equivalent.

The point of Tymoczko's argument was to adduce the Four Color Theorem as an example of a necessary a posteriori proposition. To the extent that he considers mathematical results as necessary, I agree with him. However, *if* the a priori/a posteriori dichotomy is coherent – a question that is by no means settled, and on which I take no position here — I think the distinction must be between statements that are known *only* a posteriori and those *potentially* known a priori, say by an agent with superhuman, but not omniscient, computational abilities (cf. Krakowski [149]). Otherwise it is possible that some of the more complicated proofs done without a computer would likewise qualify. [(8.3.4)]

An even more radical attack on the status of computer-assisted proof was launched by the logician Nicolas Goodman, who claimed that such a proof is empirical because conjoined to its explicit hypotheses is the truth of the laws of physics [97, p. 122].[(8.3.5)] Furthermore, Goodman misinterprets the technical evidence for his claim in [97] that mathematics is a natural science. Take, for example, his assertion that phlogistic chemistry and Euler's manipulation of divergent series are both "completely dead".[(8.3.6)] Not a single theoretical construct of phlogistic chemistry survived Lavoisier. The vocabulary of modern chemistry is so unsuited

---

(8.3.4)The canonical example is the complete enumeration of *finite* simple groups, whose proof, in original form, ran several thousand pages. To see the problem inherent in Tymoczko's classification, assume that the only person with enough experience in the field to survey the proof and apprehend its validity without reliance on others was D. Gorenstein. What is the status of the priority of this theorem now that Gorenstein has passed away?

(8.3.5)See Koetsier [148, p. 288], for inconsistency in Goodman's attitude toward the epistemologies of mathematics and physics.

(8.3.6)Goodman's summary of Euler's attitude towards divergent series is taken from Morris Kline [142] and is significantly inaccurate. See Kitcher [141, p. 243] and the primary sources cited there for proof that Euler understood these matters much better than Goodman asserts.

to describing phlogiston theory, and conversely, that it is used as an exemplar of incommensurability of competing paradigms, while Euler's *Algebra* is still perfectly readable and remains in print. We still manipulate infinite series as Euler did, but we have a formal concept of convergence to replace his intuitive understanding of which manipulations would yield perspicuous results and which would be problematical.(8.3.7) The legacies of Euler and Priestley are significantly different. Euler is no example for "scientific revolutions" in mathematics.(8.3.8)

If mathematics by computer is empirical because it requires machines grounded in the laws of physics, we might as well say that a proof that depends on published articles relies not only on their correctness, but also on the empirical laws of optics. This extremely skeptical approach has indeed been promulgated by Philip J. Davis [69]. In Davis' concept of mathematical truth even hand computations have no exact, truthful answer, but merely one answer much more probable than the others.(8.3.9) When I first read this, I could only envision Davis at the market:

> DAVIS: I'm afraid you've short-changed me.
> CASHIER: Oh no, professor — it's just that an event of very low probability has occurred.

What the universe of outcomes is and how probabilities can be assigned to its elements — even what 'probability' means to Davis — are not clear to me.

Finally, I suggest that mathematics differs from physics in respect to revisibility. Since Einstein it has been recognized that the truth of physics grows by both expansion and revision, i.e. our understanding about physical objects is inherently incomplete and this is reflected in the "laws of physics". Our knowledge of mathematical objects is also incomplete but this is *not* reflected in theorems. We write theorems and if we are correct, in my view their validity is permanent. It is true that later mathematicians may have to refine or re-contextualize our work, as happened when the concept of convergence was improved in the nineteenth century, but, unlike the physicist, most mathematicians do not work with the expectation that such revision is inevitable. This belief I share and find well justified."

This completes the text of Lazarus' article, whose views this author shares. As a Platonist, one must consider an individual mathematician's proof of a "new" result as a "discovery". To claim to have "created" something new, is merely misplaced ego in the eye of the Platonist. If we are fortunate in discovering a new and beautiful fact, then our role is to present it in the best possible light so that we may share that discovered beauty with others. We trust that this book has gone some distance in accomplishing that task.

(8.3.7)In fact, some of the applications of divergent series that Euler *disparaged* have been resurrected (unlike phlogiston)! In the 2-adic integers $1+2+4+\cdots$ converges and the problematical equation $-1 = 1+2+4+\cdots$, obtained from the MacLaurin expansion of $1/(1-x)$, is true. The p-adic numbers are another example of the unreasonable effectiveness of mathematics that radical epistemologies do little to explain.

(8.3.8)Euclid *pace*, Goodman, was not "refuted" by Einstein, but I have no space here for similar analysis.

(8.3.9)History has not been kind to this article, in which Davis asks why mathematicians do not seriously consider the possibility that Fermat's Last Theorem is undecidable.

Appendix A

Tables.

The tables in Appendix A are reproduced from original sources as cited. Also, these tables (with the exception of Table A5) were also independently checked by this author against computer runs made by Mike Jacobson[(A.0.1)] who programmed his algorithms in C, using techniques described in Chapter Eight, and run on an IBM RS6000/590.

Table A1. This is a complete list of all $h_\Delta = 1$ for fundamental radicands $D > 0$ and $\ell(w_\Delta) \leq 24$ with one possible GRH-ruled out exception. This first appeared in [256].

ℓ	D
1	2, 5, 13, 29, 53, 173, 293
2	3, 6, 11, 21, 38, 77, 83, 93, 227, 237, 437, 453, 1133, 1253
3	17, 37, 61, 101, 197, 317, 461, 557, 677, 773, 1877
4	7, 14, 23, 33, 47, 62, 69, 133, 141, 167, 213, 398, 413, 573, 717, 1077, 1293, 1397, 1757, 3053
5	41, 149, 157, 181, 269, 397, 941, 1013, 2477, 2693, 3533, 4253
6	19, 22, 57, 59, 107, 131, 253, 278, 309, 341, 381,749, 813, 893, 1893, 2453, 2757, 3317
7	89, 109, 113, 137, 373, 389, 509, 653, 797, 853, 997, 1493, 1997, 2309, 2621, 3797, 4973
8	31, 71, 158, 206, 383, 501, 503, 581, 743, 789, 869, 917, 983, 989, 1333, 1349, 1437, 2573, 3093, 6677, 14693
9	73, 97, 233, 277, 349, 353, 613, 821, 877, 1181, 1277, 1613, 1637, 1693, 2357, 3557, 3989, 4157, 4517, 7213, 11213
10	43, 67, 86, 118, 129, 161, 301, 517, 563, 597, 669, 827, 1238, 1357, 1389, 2253, 2901, 3101, 3437, 4413, 4613, 7061, 7653
11	541, 593, 661, 701, 857, 1061, 1109, 1217, 1237, 1709, 1733, 1949, 2333, 2957, 3677, 3701, 4373, 5237, 5309, 7013, 8693, 9533, 10853, 12437
12	46, 103, 127, 177, 209, 239, 263, 479, 734, 887, 933, 973, 1149, 1541, 1589, 1661, 1797, 1837, 2229, 2933, 3269, 3309, 3453, 4829, 6261, 6333, 6797, 7637, 10757, 12381
13	421, 757, 1021, 1097, 1117, 1301, 1553, 1973, 2069, 2237, 2273, 2789, 2861, 3373, 3461, 3517, 3917, 4133, 4397, 5573, 5717, 6221, 6317, 7253, 7517, 8741, 9173, 9437, 10181, 11597, 15797

[(A.0.1)] Mike is a graduate student of Hugh Wiliams in the computer science department of the University of Manitoba.

Table A1 continued

14	134, 179, 201, 251, 262, 307, 347, 422, 467, 497, 502, 587, 683, 713, 838, 1317, 1382, 1477, 2077, 2189, 2317, 3197, 3837, 4037, 4197, 4661, 4997, 5093, 5277, 5357, 5493, 5997, 7493, 7613, 7997, 9237, 17237
15	193, 281, 1861, 1933, 2141, 2437, 2741, 2837, 3037, 3413, 4637, 4877, 6653, 8117, 11549, 13037, 15077, 23117
16	94, 191, 217, 249, 302, 311, 329, 393, 431, 446, 537, 542, 589, 647, 878, 1319, 1487, 1909, 2157, 2351, 2413, 2517, 2733, 3149, 4109, 6013, 6117, 6533, 7629, 7773, 8717, 9037, 9917, 11693, 13853, 14253, 15221, 16397, 16557, 18053
17	521, 617, 709, 1433, 1597, 2549, 2909, 3581, 3821, 4013, 5501, 5693, 5813, 6197, 7853, 8093, 8573, 9677, 10597, 10973, 13109, 13613, 15413, 17093, 20261, 22637, 26717
18	139, 163, 283, 417, 419, 566, 633, 737, 758, 781, 787, 998, 1141, 1142, 1163, 1286, 1307, 1337, 1461, 1718, 1829, 1931, 2243, 2537, 2653, 2966, 2973, 3013, 3117, 3629, 3713, 4061, 4269, 4541, 4781, 6629, 6717, 7037, 7133, 7181, 8013, 8157, 8197, 8301, 8777, 9957, 10277, 10493, 11429, 11957, 12293, 13373, 13917, 16373, 18653, 18813, 18893, 20597, 23597, 24173, 26837, 30917
19	241, 313, 449, 829, 953, 1069, 1193, 1213, 1697, 2381, 3853, 4733, 5077, 5189, 5381, 5669, 5981, 6173, 6277, 6389, 6397, 6917, 7717, 7757, 7877, 8237, 9973, 10037, 11093, 11933, 12893, 13397, 19997, 27917
20	151, 199, 367, 622, 863, 1151, 1454, 1501, 1502, 1941, 2033, 3902, 4101, 4317, 4677, 4821, 5549, 6077, 7277, 8133, 8453, 8813, 9253, 9357, 11381, 11733, 14237, 15837, 17933, 18293, 21653, 23453, 25157, 36077, 49013
21	337, 569, 977, 1453, 1669, 1741, 2053, 2293, 4093, 4349, 5437, 5557, 8861, 9341, 10133, 10709, 11117, 12917, 14549, 15053, 16253, 18413, 18917, 19013, 19973, 20117, 20333, 25373, 28493, 29333
22	166, 489, 491, 523, 643, 662, 947 ,971, 1137, 1187, 1427, 1571, 1667, 1713, 1821, 2181, 2217, 2469, 3493, 3693, 3749, 3909, 3947, 4213, 4787, 4989, 5789, 5893, 5909, 6933, 6941, 7509, 7941, 10157, 10533, 10821, 11189, 11469, 12477, 12533, 13733, 14333, 14853, 15069, 15637, 15893, 17813, 19613, 20429, 21117, 23093, 30533, 35237, 36893
23	433, 457, 641, 881, 1381, 1913, 2393, 2749, 3389, 3733, 4421, 5653, 6701, 7349, 7949, 8669, 10253, 11813, 12413, 13709, 13757, 14717, 14813, 14957, 15749, 16229, 16453, 19037, 19421, 22613, 22853, 24317, 27653, 28517, 30197, 31253, 33893, 37397
24	271, 382, 607, 753, 911, 1103, 1262, 1438, 1473, 1838, 1982, 2063, 2078, 2558, 2661, 2687, 2893, 2903, 3986, 3113, 3167, 3377, 3669, 4237, 4333, 4533, 5293, 5533, 5753, 6509, 6621, 7197, 7269, 8153, 8189, 8213, 8413, 10637, 11157, 11573, 11589, 11893, 12677, 12797, 13453, 13541, 14117, 15693, 15917, 17133, 17309, 18677, 18933, 19797, 20053, 20373, 20837, 22757, 25709, 25973, 26213, 27317, 34997, 39077

Table A2. This is a complete list of all fundamental discriminants $D = \Delta \equiv 1 \pmod 8$ with $h_\Delta = 1$ and $\ell = \ell(w_\Delta) \leq 24$. This first appeared in [256].

ℓ	D
3	17
4	33
5	41
6	57
7	89, 113, 137
9	73, 97, 233, 353
10	129, 161
11	593, 857, 1217
12	177, 209
13	1097, 1553, 2273
14	201, 497, 713
15	193, 281
16	217, 249, 329, 393, 537
17	521, 617, 1433
18	417, 633, 737, 1337, 2537, 3713, 8777
19	241, 313, 449, 953, 1193, 1697
20	2033
21	337, 569, 977
22	489, 1137, 1713, 2217
23	433, 457, 641, 881, 1913, 2393
24	753, 1473, 3113, 3377, 5753, 8153

Table A3. This is a complete list of all $h_\Delta = 2$ for fundamental discriminants $\Delta > 0$ with radicand D for $\ell = \ell(w_\Delta) \leq 24$ with one possible GRH-ruled out exception, 1958 of them to be exact. The third column contains the number of $h_\Delta = 2$ in each period. This first appeared in [259].

ℓ	D	The number of $h_\Delta = 2$
1	10, 26, 85, 122, 362, 365, 533, 629, 965, 1685, 1853, 2813	12
2	15, 30, 35, 39, 42, 51, 66, 87, 102, 110, 123, 143, 146, 165, 182, 203, 221, 230, 258, 285, 327, 357, 402, 447, 635, 645, 678, 741, 843, 902, 957, 1085, 1245, 1298, 1517, 1533, 2037, 2045, 2085, 2397, 2613, 4245, 4277, 4773, 5645, 5957, 6573, 8333	48
3	65, 185, 458, 485, 1157, 2117, 2285, 3077, 3293, 3365, 12365	11
4	34, 55, 78, 95, 119, 138, 155, 174, 194, 205, 215, 222, 266, 287, 299, 305, 318, 335, 377, 395, 429, 482, 527, 623, 755, 782, 861, 885, 1022, 1055, 1205, 1405, 1469, 1965, 2013, 2093, 2222, 2301, 2373, 2877, 3005, 3237, 3597, 3813, 4893, 5117, 5397, 5757, 5885, 6005, 6285, 6293, 7157, 7733, 7973, 8357, 9005, 9077	58
5	74, 218, 493, 565, 1037, 1565, 1781, 2138, 2165, 2173, 3869, 5165, 5213, 5837, 6485, 8021, 10397, 14213	18
6	70, 105, 111, 114, 178, 183, 187, 267, 273, 303, 371, 374, 407, 418, 470, 498, 518, 545, 551, 590, 602, 618, 642, 803, 805, 822, 923, 1005, 1007, 1034, 1118, 1167, 1173, 1178, 1202, 1581, 1605, 1623, 1653, 1707, 1749, 1790, 2103, 2109, 2147, 2245, 2261, 2445, 2717, 2723, 2765, 2845, 3405, 3605, 3638, 3737, 3893, 4085, 4301, 4445, 4605, 5133, 5453, 7805, 10237, 10317, 10653, 11837, 12845, 13253, 13277, 13445, 14405, 14573, 15197, 19445, 21677, 23693, 25437	79
7	58, 202, 314, 538, 685, 949, 1165, 1261, 2885, 3133, 3277, 3653, 5429, 5765, 6437, 7373, 9197, 9509, 12557, 16757, 17141, 17261, 18317, 22301	24
8	91, 238, 282, 638, 695, 707, 710, 854, 866, 942, 1247, 1403, 1643, 1655, 1869, 1883, 1943, 2238, 2390, 2483, 2685, 2978, 3205, 3333, 3765, 4247, 4565, 5069, 5141, 5829, 6341, 6365, 6693, 6773, 6837, 6965, 7405, 7469, 8165, 8853, 9141, 9453, 9485, 10013, 10293, 10373, 10517, 10797, 10805, 11357, 11501, 15677, 16805, 17357, 17853, 19493, 31533, 37373, 38213	59
9	106, 698, 1073, 1189, 1285, 1385, 1418, 1865, 2581, 3233, 4469, 4553, 4709, 5597, 8885, 9365, 9773, 9893, 10229, 10685, 12053, 12077, 13565, 14285, 16733, 23285, 28757, 29957	28

Table A3 continued

10	115, 154, 159, 186, 246, 259, 286, 339, 345, 354, 403, 411, 451, 465, 494, 515, 534, 543, 561, 583, 591, 598, 665, 671, 682, 687, 703, 705, 762, 779, 830, 938, 978, 1047, 1102, 1203, 1263, 1265, 1363, 1383, 1645, 1671, 1727, 1742, 2098, 2123, 2127, 2485, 2651, 2658, 2701, 2747, 2802, 2829, 2867, 2882, 3157, 3165, 3218, 3587, 3685, 3741, 3743, 3827, 3867, 4103, 4619, 4667, 5057, 5061, 5205, 5253, 5285, 5405, 5522, 6149, 6613, 6789, 7005, 7845, 8045, 8445, 8517, 8533, 8621, 9085, 9093, 9581, 9701, 9821, 10365, 10645, 10877, 11373, 11557, 11973, 12117, 12165, 12837, 14773, 14861, 16037, 16077, 16205, 17045, 17741, 17877, 18093, 18357, 18717, 19253, 21405, 21749, 21885, 22413, 22517, 22781, 23933, 23997, 24213, 24845, 25077, 25133, 26333, 26477, 27173, 28005, 28853, 30245, 30693, 33677, 37565, 39245, 41477, 47197	135
11	265, 298, 554, 794, 1322, 1658, 2218, 2509, 3242, 4181, 4682, 4685, 11413, 11773, 13085, 14453, 15685, 16085, 18485, 20285, 20765, 25565, 28013, 28685, 31037, 39797, 40157, 43733, 46637, 51917, 56117	31
12	247, 295, 355, 366, 385, 386, 426, 535, 609, 767, 802, 815, 851, 969, 995, 1027, 1113, 1162, 1207, 1343, 1353, 1355, 1358, 1535, 1538, 1703, 1717, 1799, 1910, 1946, 2018, 2047, 2054, 2105, 2231, 2318, 2327, 2334, 2365, 2438, 2507, 2735, 2855, 2987, 3002, 3263, 3302, 3563, 3695, 4322, 4382, 4415, 4453, 4542, 4717, 4917, 5447, 6605, 6853, 6905, 7365, 7413, 7797, 9429, 10262, 11077, 12341, 12453, 12485, 12605, 12669, 13837, 15333, 16365, 18557, 18805, 22893, 23253, 24293, 24485, 25397, 25413, 25517, 27053, 27389, 27605, 29141, 29405, 29861, 30173, 32357, 36533, 40533, 44117, 44693, 45485, 45573, 47157, 52037, 59213, 59573, 75677	102
13	746, 778, 1082, 1241, 1514, 1649, 2042, 2426, 3085, 3338, 3349, 4058, 4573, 4589, 4885, 5389, 7418, 7421, 8765, 9389, 9965, 10085, 12965, 14837, 16277, 17533, 19357, 21053, 22373, 25877, 30733, 31373, 31853, 36965, 38597, 39437, 40757, 53477, 69893, 81413	40
14	190, 319, 406, 430, 471, 474, 611, 667, 670, 669, 742, 745, 806, 807, 1001, 1043, 1070, 1115, 1119, 1309, 1315, 1338, 1347, 1398, 1542, 1545, 1562, 1634, 1670, 1691, 1826, 1839, 1874, 2282, 2294, 2315, 2323, 2337, 2345, 2427, 2435, 2463, 2630, 2714, 2782, 2821, 3297, 3378, 3478, 3621, 3878, 4115, 4154, 4178, 4307, 4331, 4381, 4499, 4506, 4646, 4835, 5222, 5246, 5282, 5442, 5673, 5781, 5917, 6098, 6213, 6357, 6443, 6461, 6477, 6611, 6645, 7145, 7285, 7445, 7619, 7885, 8205, 8393, 8437, 8483, 8565, 8733, 8805, 8877, 8965, 9285, 9645, 9717, 9877, 10149, 10573, 11051, 11805, 12578, 12621, 12733, 12869, 12885, 13197, 13213, 13973, 14181, 15861, 15965, 16541, 17013, 17805, 18845, 18941, 19205, 19277, 19365, 19397, 19677, 21197, 21245, 21549, 21909, 21917, 22557, 22965, 23493, 24069, 24893, 25109, 25597, 26285, 26373, 26885, 27677, 27845, 28397, 28605, 28797, 28805, 31349, 31413, 32973, 34013, 34133, 35045, 35477, 35765, 35789, 36917, 38253, 40445, 42413, 42933, 43493, 44957, 45453, 47253, 51653, 52013, 54557, 54677, 55893, 64181, 64253, 71357, 85973, 98045	169
15	481, 1417, 1466, 2858, 3065, 3589, 3785, 3977, 4538, 5317, 5941, 6641, 6749, 7082, 11861, 12701, 12833, 13793, 14909, 16589, 17153, 18185, 18365, 18581, 20885, 24221, 27989, 29069, 32885, 33365, 44813, 47165, 51173, 66197, 67973, 70493, 78917	37

Table A3 continued

16	310, 391, 415, 654, 655, 679, 955, 1038, 1146, 1166, 1267, 1282, 1346, 1391, 1578, 1662, 1739, 1833, 1858, 1895, 1902, 2183, 2195, 2198, 2407, 2526, 2553, 2615, 3227, 3278, 3374, 3497, 3565, 3611, 3755, 3818, 3918, 4043, 4069, 4087, 4142, 4233, 4298, 4405, 4955, 4958, 5123, 5198, 5267, 5543, 5558, 5579, 5726, 5855, 6062, 6167, 6254, 6383, 6501 6527, 7322, 7337, 7355, 7898, 8029, 8078, 8207, 8378, 8421, 8493, 8718, 9107, 9309, 9373, 10509, 11303, 11405, 11517, 11917, 12878, 12957, 13802, 13943, 15213, 15365, 15573, 16685, 17673, 17997, 19298, 19389, 21965, 22029, 22173, 24101, 25885, 26133, 29685, 30413, 30581, 31493, 34989, 35861, 38309, 38405, 38517, 40685, 41741, 43053, 43253, 43805, 45773, 48954, 49565, 49685, 50973, 55013, 55173, 57293, 58253, 58373, 58973, 59237, 61277, 63557, 67133, 67205, 67997, 69621, 75413	130
17	1018, 1994, 2965, 4285, 5354, 5498, 5585, 8917, 9242, 9665, 10265, 12085, 13061, 13957, 14677, 15242, 15613, 16109, 16565, 16613, 17173, 17285, 17429, 17861, 18037, 18737, 18965, 34047, 34957, 35285, 36413, 37949, 40085, 40501, 41165, 47813, 48149, 50357, 53285, 57797, 58853, 61133, 62957, 63653, 86957, 146453	46
18	519, 562, 831, 879, 951, 1185, 1199, 1209, 1281, 1310, 1362, 1379, 1505, 1506, 1526, 1606, 1630, 1686, 1698, 1842, 1903, 1919, 1923, 1983, 1991, 2202, 2219, 2283, 2363, 2631, 2697, 2771, 2985, 3183, 3282, 3414, 3470, 3642, 3702, 3707, 3830, 3839, 4029, 4287, 4343, 4430, 4562, 4697, 4791, 4803, 5027, 5363, 5705, 5797, 5845, 5870, 6177, 6182, 6278, 6407, 6470, 6758, 6767, 6830,6842, 7358, 7485, 7802, 7869, 7958, 8582, 8589, 8697, 8843, 9119, 9269, 9381, 9383, 9445, 9470, 10245, 10461, 10502, 10643, 11397, 12093, 12162, 12278, 12549, 12722, 12749, 12909, 13019, 13237, 13593, 13605, 13682, 13821, 14053, 14309, 14605, 15485, 15933, 16845, 17197, 17381, 17733, 18173, 19229, 19245, 20253, 20541, 20645, 20677, 20917, 21533, 22085, 22181, 22245, 23309, 24357, 24365, 24477, 24837, 25445, 25485, 25781, 26781, 27485, 28461, 28589, 29309, 30317, 30957, 31733, 32021, 32669, 32773, 33789, 33845, 34405, 34685, 34853, 35189, 36669, 36813, 39893, 40613, 41789, 41837, 41973, 43581, 43589, 43797, 44645, 47333, 47549, 48245, 49805, 50133, 51557, 51845, 52853, 54197, 54845, 57845, 59637, 61477, 62405, 64277, 66549, 70133, 71213, 73253, 74973, 76037, 80013, 80117, 92477, 96701, 128117, 138773, 139277, 146333, 151373, 168773, 171797	187
19	922, 1706, 2186, 2257, 3386, 8522, 8714, 9997, 16781, 17177, 20513, 20813, 21509, 24341, 26165, 28453, 29597, 30365, 31085, 35333, 35885, 36173, 37685, 37757, 38765, 41765, 43469, 46157, 50453, 52637, 53765, 57965, 59765, 62285, 65501, 70733, 75197, 79085, 82277, 84773, 107333, 109757, 139037, 144317	44
20	511, 559, 606, 790, 1002, 1065, 1079, 1182, 1195, 1374, 1415, 1510, 1513, 1537, 1603, 1687, 1961, 2193, 2215, 2455, 2471, 2627, 2863, 3155, 3239, 3295, 3383, 3647, 3746, 3857, 4163, 4295, 4458, 4535, 4595, 4727, 4782, 4847, 4922, 5038, 5143, 5195, 5258, 5678, 5709, 5759, 5803, 5822, 5962, 6107, 6141, 6338, 6415, 6467, 6702, 6914, 6943, 7189, 7295, 7343, 7367, 7787, 7813, 7895, 8238, 8258, 8507, 8567, 8903, 9527, 9861, 9885, 10622, 11949, 13413, 13461, 14541, 14565, 15029, 15203, 16237, 16463, 16989, 18821, 19085, 19221, 19337, 19509, 19533, 20517, 20642, 21093, 21765, 21962, 23069, 23133, 25197, 25557, 26637, 26697, 28205, 28965, 29261, 29365, 32053, 32997, 34773, 35085, 36093, 36165, 36237, 36573, 38685, 39093, 40557, 41285, 42605, 42749, 43085, 44765, 45933, 46277, 48237, 48453, 49181, 50213, 52917, 54893, 57189, 58493, 58533, 60205, 61053, 61773, 61973, 62237, 62933, 63213, 65285, 65813, 67373, 69293, 70213, 71405, 74261, 76677, 77957, 80693, 80717, 81317, 81357, 81437, 82869, 88413, 93197, 96773, 106685, 109997, 113693, 115037, 240077	161

Table A3 continued

21	394, 865, 1769, 1985, 2561, 2762, 3098, 4385, 5465, 5485, 5965, 6122, 7141, 7265, 10565, 11101, 11485, 11581, 13285, 13466, 14381, 14765, 16442, 21365, 22565, 28373, 34493, 35197, 36221, 44861, 47477, 48485, 48941, 51365, 54317, 57317, 58133, 58589, 69365, 75917, 78053, 78557, 78773, 80165, 84173, 85277, 85949, 89333, 91013, 92165, 94877, 97877, 104837, 120893, 127613, 130037, 156917, 167477, 212357	59
22	466, 763, 771, 834, 1059, 1194, 1266, 1334, 1558, 1563, 1798, 1835, 1843, 1905, 1963, 1986, 2001, 2082, 2270, 2274, 2279, 2406, 2514, 2519, 2546, 2585, 2643, 2778, 2823, 2859, 2931, 2937, 2947, 3063, 3107, 3131, 3147, 3182, 3207, 3310, 3417, 3506, 3635, 3657, 3687, 3938, 4119, 4145, 4187, 4202, 4433, 4630, 4645, 4814, 4863, 4883, 4938, 4965, 5111, 5163, 5315, 5345, 5367, 5603, 5703, 5718, 5747, 5862, 5989, 6023, 6061, 6378, 6387, 6403, 6431, 6585, 6635, 6738, 6743, 7026, 7122, 7143, 7257, 7334, 7545, 7553, 7622, 7701, 7842, 7982, 8018, 8027, 8270, 8365, 8531, 8630, 8749, 8897, 8998, 9113, 9138, 9158, 9205, 9263, 9687, 9709, 10190, 10298, 10307, 11507, 11747, 12257, 12261, 12305, 12405, 12662, 12827, 13067, 13265, 13405, 13817, 14205, 14510, 14845, 15085, 15113, 15617, 15765, 16149, 16797, 17445, 17549, 18002, 20301, 21837, 23777, 24141, 24405, 25149, 25653, 25682, 26197, 27101, 28673, 29805, 30165, 31965, 32469, 33045, 33429, 33549, 33645, 33989, 34005, 34933, 36581, 37077, 37317, 37445, 37677, 39693, 40605, 42117, 44373, 45237, 45605, 46949, 48045, 48813, 48885, 49629, 49677, 50045, 50861, 51861, 52085, 52421, 54645, 56069, 56357, 56397, 57749, 57893, 58197, 58205, 59285, 60277, 62733, 63237, 63437, 63453, 64085, 65189, 66557, 67469, 68933, 71021, 71885, 72597, 76053, 76773, 77357, 77693, 78869, 80597, 83765, 85917, 86933, 89957, 92573, 94085, 94317, 95933, 98813, 99413, 101213, 101909, 104357, 106757, 107645, 112013, 116045, 116261, 116597, 120245, 121733, 126245, 130613, 135453, 136037, 136253, 138437, 145277, 145613, 148133, 148973, 149357, 149957, 150293, 156053, 157493, 165557, 168797, 204245, 248093	245
23	586, 634, 1585, 2474, 3578, 4121, 4141, 5114, 6074, 6109, 6362, 6506, 6602, 7261, 8042, 8249, 10673, 12349, 20557, 22837, 24869, 26773, 26869, 33017, 34165, 34541, 37661, 37837, 43693, 51757, 55565, 56285, 56381, 58277, 59293, 63677, 64349, 67253, 74693, 76565, 77453, 86213, 87485, 90485, 90557, 90653, 94973, 99557, 107117, 107573, 107957, 113237, 119477, 139157, 154853, 160277, 172133, 176837, 247397	59
24	826, 871, 1147, 1255, 1614, 1711, 1795, 2051, 2119, 2154, 2409, 2414, 2534, 2594, 2698, 2743, 2759, 3009, 3018, 3110, 3206, 3633, 3806, 3854, 3882, 4031, 4118, 4310, 4638, 4665, 4826, 5322, 5466, 6155, 6302, 6455, 6618, 7091, 7222, 7278, 7763, 8302, 8489, 8927, 8939, 9002, 9287, 9347, 9393, 9469, 9741, 9785, 10058, 10142, 10415, 10442, 10562, 10823, 11042, 11262, 11546, 11714, 11843, 12311, 12741, 13118, 13502, 13958, 14198, 14987, 15815, 16502, 16765, 16859, 17063, 17447, 17501, 17531, 18381, 18501, 18885, 19685, 20013, 21047, 21287, 21565, 21669, 22523, 23037, 23927, 25053, 25322, 26277, 27069, 27669, 27789, 27853, 29877, 30093, 31485, 34669, 34709, 35277, 36453, 37157, 37293, 37805, 39957, 40893, 41109, 42285, 43301, 43727, 44133, 44429, 44717, 45069, 47613, 48005, 48765, 49101, 50165, 51837, 54389, 54797, 54885, 56885, 57813, 59717, 59853, 60477, 61269, 61365, 62493, 63885, 68285, 70565, 70685, 71133, 71165, 72197, 72917, 76973, 78837, 83613, 84413, 85533, 87989, 88877, 89549, 91077, 92405, 92597, 97277, 102695, 106413, 109205, 109805, 112517, 113813, 118685, 122405, 125333, 126965, 132117, 136205, 141797, 151205, 154013, 158933, 165413, 169133, 172493, 175637, 197333, 205805	176

Tables A4. This is a complete list (with one possible GRH-ruled out exception) of $e_\Delta = 2$, for fundamental discriminants $\Delta > 0$ of ERD type, with radicands D. The table is broken down into three parts according to the shape of D. This is (an extended version of) the table in [201] reproduced by permission of Oxford University Press.

Table A4(1). $D = r^2k^2 + k$, k odd.

It is interesting to note that, in Table A4(1), all h_Δ are such that $h_\Delta \leq 8$ except for $D = 19635$, $D = 451605$ and there are a total of 98 fundamental radicands $D > 0$ in the table. Although some radicands in Table A4(1) may have the shape of the radicands in Table A4(2)–A4(3) we do not repeat them there.

D	r	k
10	3	1
15	4	−1
26	5	1
30	1	5
35	6	−1
39	2	3
42	1	−7
65	8	1
78	3	−3
95	2	−5
105	2	5
110	1	−11
122	11	1
143	12	−1
170	13	1
182	1	13
195	14	−1
203	2	7
210	1	−15
222	5	−3
230	3	5
255	16	−1
290	17	1
327	6	3
362	19	1
395	4	−5
462	1	21
483	22	−1
485	22	1
530	23	1
663	2	−13
885	2	−15
903	10	3
915	2	15
930	1	−31
962	31	1
1155	34	−1
1157	34	1
1173	2	17
1190	1	−35
1218	5	−7
1230	7	5
1295	36	−1
1370	37	1
1463	2	19
1482	1	−39
1518	13	−3
1595	8	−5
1605	8	5
1722	1	41
1767	14	3
2030	9	5
2093	2	−23
2117	46	1
2210	47	1
2301	16	−3
2618	3	17
2717	4	13
3135	56	−1
3230	3	−19
3335	2	−29
3365	58	1
3597	20	−3
3605	12	5
3813	2	−31
3990	3	21
4389	2	33
4758	23	−3
4893	10	−7
4935	2	35
5187	24	3
5330	73	1
5610	5	−15
5757	4	−19
6045	2	−39
6405	16	5
6765	2	41
7035	4	−21
7733	8	−11
7755	8	11
9030	19	5
9605	98	1
10005	20	5
12045	2	−55
12155	2	55
13485	4	29
14405	24	5
14630	11	−11
14885	122	1
19565	4	−35
19635	4	35
20165	142	1
33117	26	−7
41613	68	−3
58565	242	1
77285	278	1
108885	22	−15
451605	32	21

Table A4(2). $D = r^2k^2 + 2k$, k odd.

In Table A4(2) we have $h_\Delta \leq 8$ except for $D = 25935, 33495, 81770$. There are a total of 76 values of D in Table A4(2).

D	r	k
34	6	−1
51	7	1
66	8	1
87	3	3
102	10	1
119	11	−1
123	11	1
138	4	−3
146	12	1
194	14	-1
215	3	−5
231	5	3
258	16	1
287	17	−1
318	6	−3
330	6	3
390	4	−5
402	20	1
410	4	5
435	7	−3
447	7	3
455	3	7
482	22	−1
527	23	−1
570	8	−3
615	5	−5
623	25	−1
627	25	1
635	5	5
678	26	1
770	4	−7
782	28	−1
798	4	7
843	29	1
890	6	−5
902	30	1
1022	32	−1
1095	11	3
1235	7	5
1298	36	1
1302	12	3
1515	13	−3
1547	3	13
1610	8	5
1770	14	3
1938	44	1
1995	3	−15
2015	9	−5
2310	16	3
2387	7	−7
2415	7	7
2595	17	−3
2607	17	3
2730	4	13
2910	18	−3
3003	5	−11
3255	19	3
3570	4	−15
3723	61	1
3842	62	−1
3927	3	−21
4290	2	−33
5655	5	15
6090	26	3
6555	27	−3
7215	17	−5
10010	20	5
10370	6	−17
12558	16	7
13695	39	3
19610	28	5
21318	146	1
22490	30	−5
25935	23	7
33495	61	3
81770	22	−13

Table A4(3). $D = r^2k^2 + 4k$, k odd.

There are 81 values of D in Table A4(3), all with $h_\Delta \leq 8$. Thus Tables A4(1)–A4(3) comprise a total of 255 ERD types with $e_\Delta = 2$.

D	r	k
85	9	1
165	1	−15
205	3	−5
221	1	−17
285	1	−19
357	1	−21
365	19	1
429	7	−3
533	23	1
629	25	1
645	5	5
741	9	3
957	1	−33
965	31	1
1085	1	−35
1205	7	−5
1245	7	5
1365	1	−39
1469	3	−13
1517	1	−41
1533	13	3
1685	41	1
1853	43	1
1965	3	−15
2013	15	−3
2037	15	3
2045	9	5
2085	3	15
2373	7	−7
2397	1	−51
2405	49	1
2613	17	3
2805	1	−55
2813	53	1
3005	11	−5
3045	11	5
3237	19	−3
3485	59	1
3885	3	−21
3965	1	−65
4245	13	5
4277	4	13
4485	1	−69
4773	23	3
5565	5	−15
5645	15	5
5885	7	−11
5957	11	7
6573	27	3
7157	5	−17
7293	5	17
7565	1	−89
7685	3	29
7917	1	−91
8333	7	13
8645	1	−95
9005	19	−5
9933	3	33
10205	101	1
10965	7	−15
11165	3	35
13845	3	39
14685	11	11
15645	25	5
16133	127	1
16653	43	3
17765	7	19
18245	27	5
20405	13	−11
21045	29	5
23205	3	−51
24045	31	5
25493	3	53
26565	1	−165
30597	25	−7
31317	59	−3
32045	179	1
35805	9	21
41093	7	−29
55205	47	−5
74613	13	21

Tables A5. These three tables represent runs which illustrate the OASiS program used to complete the proof of Theorem 6.2.2. These tables originally appeared in [201] and are reproduced by permission of Oxford University Press.

Table A5(1). $N \equiv 3 \pmod 4$.

N	i	$q(i)$
3	4	7
35	5	11
63	7	17
143	10	29
447	11	31
635	12	37
1295	17	59
101283	19	67
730847	21	73
18183483	22	79
26598843	23	83

N	i	$q(i)$
28203507	25	97
305875475	26	101
400744203	27	103
545559467	29	109
2342182895	33	137
7012246247	38	163
2720673629663	39	167
3166431076347	45	197
37082770710347	47	211
163402643806023	49	227

Table A5(2). $N \equiv 2 \pmod 4$.

N	i	$q(i)$
38	4	7
62	5	11
98	6	13
150	7	17
318	9	23
398	11	31
930	14	43
1722	17	59
314522	18	61
341502	20	71
845382	22	79
5505290	23	83
53772102	25	97

N	i	$q(i)$
138614418	29	109
1164652502	30	113
2815100438	35	149
24453805998	36	151
348190770402	38	163
2295693610958	39	167
3017984461778	40	173
4645014218918	41	179
7767578138858	43	191
12760252192170	46	199
19285608519470	47	211
334664789952030	48	223

Table A5(3). $N \equiv 1 \pmod 4$.

N	i	$q(i)$
5	4	7
17	5	11
33	6	13
437	11	31
5297	12	37
6905	13	41
8393	14	43
13593	15	47
64917	16	53
149597	17	59
524457	20	71
1878245	21	73
2628093	22	79
3009173	25	97

N	i	$q(i)$
46305413	30	113
1586592293	32	131
5702566397	34	139
15933687413	35	149
25777678685	36	151
181315486677	38	163
413900743497	39	167
1241083406505	42	181
1519873554353	44	193
31209910533489	45	197
84542528922717	47	211
347478177155753	49	227
442815116635157	51	233

Table A6. This represents a complete list of all fundamental radicands, $\Delta > 0$, having no split primes $p < \sqrt{\Delta}/2$ (with one possible GRH-ruled out exception.) This table originally appeared in [201], and is reproduced by permission of Oxford University Press.

$$\textit{Define } T = \{\textit{ramified primes } < \sqrt{\Delta}/2\}.$$

D	T
2	–
3	–
5	–
6	2
7	2
11	2
13	–
14	2
15	2, 3
21	–
23	2
29	–
30	2, 3, 5
35	2, 5
38	2
42	2, 3
47	2
53	–
62	2
69	3

D	T
77	–
83	2
87	2, 3
93	3
110	2, 5
138	2, 3
143	2, 11
165	3, 5
167	2
173	–
182	2, 7, 13
195	2, 3, 5, 13
213	3
215	2, 5
227	2
237	3
255	2, 3, 5
285	3, 5
293	–
318	2, 3

D	T
357	3, 7
398	2
413	7
437	–
447	2, 3
453	3
483	2, 3, 7
635	2, 5
717	3
930	2, 3, 5
957	3, 11
1077	3
1085	5, 7
1133	11
1253	7
1295	2, 5, 7
1722	2, 3, 7, 41
1965	3, 5
2084	3, 5
2397	3, 17

Table A7. This is a complete list (with one possible GRH-ruled out exception) of all fundamental discriminants, $\Delta > 0$, with $h_\Delta = 1$ and $n_\Delta \neq 0$ (see Exercise 3.2.11). We also provide the regulator $R = \log \varepsilon_D$ for each value (to nine significant digits). This first appeared in [254].

D	$\log \varepsilon_D$
2	0.881373587
3	1.3169578969
5	1.6180339887
6	2.2924316696
7	2.7686593833
11	2.9932228461
13	1.1947632173
14	3.4000844141
17	2.0947125473
21	1.5667992370
23	3.8707667003
29	1.6472311464
33	3.8281684713
37	2.4917798526
38	4.3038824281
41	4.1591271346
47	4.5642396669
53	1.9657204716
61	3.6642184609
62	4.8362189128
69	3.2172719712
77	2.1846437916
83	5.0998292455

D	$\log \varepsilon_D$
93	3.3661046429
101	2.9982229503
133	5.1532581804
141	5.2469963702
149	4.1111425009
157	5.3613142065
167	5.8171023021
173	2.5708146781
197	3.3334775869
213	4.2902717358
227	6.1136772851
237	4.3436367167
269	5.0999036060
293	2.8366557290
317	4.4887625925
341	5.6240044731
398	6.6821070271
413	4.1106050108
437	3.0422471121
453	5.0039012599
461	5.8999048596
509	6.8297949062
557	5.4638497592

D	$\log \varepsilon_D$
573	6.6411804655
677	3.9516133361
717	5.4847797157
773	4.9345256863
797	5.9053692725
917	7.0741160992
941	7.0343887062
1013	6.8276304083
1077	5.8888702849
1133	4.6150224728
1253	5.1761178117
1293	7.4535615360
1493	7.7651450829
1613	7.9969905191
1757	6.9137363626
1877	7.3796325418
2453	8.1791997198
2477	6.4723486834
2693	8.3918567515
3053	8.1550748053
3317	8.5642675624
3533	7.7985232220

Table A8. This list comprises all fundamental discriminants $\Delta \equiv 1 \pmod 8$ of ERD-type, with $h_\Delta \leq 23$. The list is known to be unconditionally complete, by applying Theorem 5.1.1.

Δ	h_Δ	Δ	h_Δ	Δ	h_Δ	Δ	h_Δ
17	1	3585	10	15377	13	125313	19
33	1	4097	10	26241	13	138353	19
65	2	4865	10	33857	13	3601	20
105	2	5777	10	35297	13	5185	20
257	3	7049	10	58553	13	12545	20
321	3	12993	10	6097	14	19041	20
473	3	13865	10	8465	14	20737	20
145	4	18497	10	12089	14	23793	20
689	4	22505	10	63497	14	24297	20
777	4	26897	10	85337	14	32385	20
897	4	1297	11	101177	14	36105	20
905	4	6081	11	30273	15	38409	20
401	5	10401	11	2305	16	48345	20
4353	5	27473	11	10001	16	57585	20
785	6	33833	11	10817	16	81809	20
1937	6	4345	12	28905	16	86457	20
2505	6	4641	12	29585	16	87617	20
577	7	6401	12	30305	16	97305	20
1601	7	7745	12	34593	16	114257	20
1761	7	10353	12	55697	17	195377	20
2913	7	14385	12	108897	17	272513	20
8097	7	15873	12	9217	18	7057	21
12977	7	15897	12	11289	18	65537	21
1785	8	16385	12	41633	18	161537	21
2705	8	16913	12	45689	18	38417	22
5513	8	19113	12	47633	18	108905	22
8105	8	28985	12	52905	18	122505	22
23705	8	33137	12	72905	18	30977	23
28217	8	44105	12	115617	18	44097	23
3137	9	60513	12	122465	18		
22497	9	8441	13	139865	18		
3129	10	13457	13	54753	19		

Table A9. This table lists all $h_\Delta = 2$, (with one GRH ruled-out exception), where Δ is a fundamental discriminant of ERD-type with radicand $D > 0$. This first appeared in [257]. All of the values of D are in the set:

{10, 15, 26, 30, 34, 35, 39, 42, 51, 65, 66, 78, 85, 87, 95, 102, 105, 110, 119, 122, 123, 138, 143, 146, 165, 182, 194, 203, 205, 215, 221, 222, 230, 258, 285, 287, 318, 327, 357, 362, 365, 395, 402, 429, 447, 482, 485, 527, 533, 623, 629, 635, 645, 678, 741, 782, 843, 885, 902, 957, 965, 1022, 1085, 1157, 1173, 1205, 1245, 1298, 1469, 1517, 1533, 1605, 1685, 1853, 1965, 2013, 2037, 2045, 2085, 2093, 2117, 2301, 2373, 2397, 2613, 2717, 2813, 3005, 3237, 3365, 3597, 3605, 3813, 4245, 4277, 4773, 4893, 5645, 5757, 5885, 5957, 6573, 7157, 7733, 8333, 9005, 14405}.

Appendix B

Fundamental Units of Maximal Orders

This list is broken up into three parts according to congruence modulo 4 of the fundamental radicand $D > 0$, and ε_Δ will be written in the form $\varepsilon_\Delta = a + bw_\Delta$. The table has been compiled using the PAR I (version 1.37) program obtained from colleagues at the University of Bordeaux I. The list goes over the range $0 < D < 2000$ for each congruence class. They were also independently checked by this author against computer runs made by Mike Jacobson (see comments at the outset of Appendix A).

Table B1. $D \equiv 1 \pmod 4$

D	ε_Δ
5	w_Δ
13	$1 + w_\Delta$
17	$3 + 2w_\Delta$
21	$2 + w_\Delta$
29	$2 + w_\Delta$
33	$19 + 8w_\Delta$
37	$5 + 2w_\Delta$
41	$27 + 10w_\Delta$
53	$3 + w_\Delta$
57	$131 + 40w_\Delta$
61	$17 + 5w_\Delta$
65	$7 + 2w_\Delta$
69	$11 + 3w_\Delta$
73	$943 + 250w_\Delta$
77	$4 + w_\Delta$
85	$4 + w_\Delta$
89	$447 + 106w_\Delta$
93	$13 + 3w_\Delta$
97	$5035 + 1138w_\Delta$
101	$9 + 2w_\Delta$
105	$37 + 8w_\Delta$
109	$118 + 25w_\Delta$
113	$703 + 146w_\Delta$
129	$15371 + 2968w_\Delta$

D	ε_Δ
133	$79 + 15w_\Delta$
137	$1595 + 298w_\Delta$
141	$87 + 16w_\Delta$
145	$11 + 2w_\Delta$
149	$28 + 5w_\Delta$
157	$98 + 17w_\Delta$
161	$10847 + 1856w_\Delta$
165	$6 + w_\Delta$
173	$6 + w_\Delta$
177	$57731 + 9384w_\Delta$
181	$604 + 97w_\Delta$
185	$63 + 10w_\Delta$
193	$1637147 + 253970w_\Delta$
197	$13 + 2w_\Delta$
201	$478763 + 72664w_\Delta$
205	$20 + 3w_\Delta$
209	$43331 + 6440w_\Delta$
213	$34 + 5w_\Delta$
217	$3583111 + 521904w_\Delta$
221	$7 + w_\Delta$
229	$7 + w_\Delta$
233	$21639 + 3034w_\Delta$
237	$36 + 5w_\Delta$
241	$66436843 + 9148450w_\Delta$

Table B1 continued

D	ε_Δ	D	ε_Δ
249	$8011739 + 1084152w_\Delta$	469	$31 + 3w_\Delta$
253	$872 + 117w_\Delta$	473	$83 + 8w_\Delta$
257	$15 + 2w_\Delta$	481	$920179 + 87922w_\Delta$
265	$5699 + 746w_\Delta$	485	$21 + 2w_\Delta$
269	$77 + 10w_\Delta$	489	$7249279379 + 686701192w_\Delta$
273	$683 + 88w_\Delta$	493	$53 + 5w_\Delta$
277	$1228 + 157w_\Delta$	497	$1147975 + 107824w_\Delta$
281	$1000087 + 126890w_\Delta$	501	$13482 + 1261w_\Delta$
285	$8 + w_\Delta$	505	$773 + 72w_\Delta$
293	$8 + w_\Delta$	509	$442 + 41w_\Delta$
301	$10717 + 1311w_\Delta$	517	$5054 + 465w_\Delta$
305	$461 + 56w_\Delta$	521	$122752931 + 11248618w_\Delta$
309	$2379 + 287w_\Delta$	533	$11 + w_\Delta$
313	$119691683 + 14341370w_\Delta$	537	$184048971 + 16600984w_\Delta$
317	$42 + 5w_\Delta$	541	$668194 + 60037w_\Delta$
321	$203 + 24w_\Delta$	545	$1877 + 168w_\Delta$
329	$2245399 + 262032w_\Delta$	553	$598073619703 + 53124435408w_\Delta$
337	$960491695 + 110671282w_\Delta$	557	$113 + 10w_\Delta$
341	$131 + 15w_\Delta$	561	$500713 + 44144w_\Delta$
345	$6397 + 728w_\Delta$	565	$148 + 13w_\Delta$
349	$8717 + 986w_\Delta$	569	$2773504827 + 242718010w_\Delta$
353	$67471 + 7586w_\Delta$	573	$367 + 32w_\Delta$
357	$9 + w_\Delta$	577	$23 + 2w_\Delta$
365	$9 + w_\Delta$	581	$3223 + 279w_\Delta$
373	$4853 + 530w_\Delta$	589	$2089876 + 179625w_\Delta$
377	$221 + 24w_\Delta$	593	$575967 + 49330w_\Delta$
381	$963 + 104w_\Delta$	597	$4675 + 399w_\Delta$
385	$90947 + 9768w_\Delta$	601	$133779272909687 + 11378061539690w_\Delta$
389	$1217 + 130w_\Delta$	609	$581151 + 49088w_\Delta$
393	$44094699 + 4684888w_\Delta$	613	$47387 + 3989w_\Delta$
397	$1637 + 173w_\Delta$	617	$39358727 + 3301978w_\Delta$
401	$19 + 2w_\Delta$	629	$12 + w_\Delta$
409	$106387620283 + 11068353370w_\Delta$	633	$423253123 + 35038248w_\Delta$
413	$29 + 3w_\Delta$	641	$34694146323 + 2853374290w_\Delta$
417	$81144379 + 8356536w_\Delta$	645	$61 + 5w_\Delta$
421	$211627 + 21685w_\Delta$	649	$1079488334066819 + 88209784190760w_\Delta$
429	$69 + 7w_\Delta$	653	$798 + 65w_\Delta$
433	$6883177307 + 694966754w_\Delta$	661	$859967 + 69605w_\Delta$
437	$10 + w_\Delta$	665	$13187 + 1064w_\Delta$
445	$10 + w_\Delta$	669	$146741 + 11803w_\Delta$
449	$180529627 + 17883410w_\Delta$	673	$46931834869907 + 3763240848050w_\Delta$
453	$71 + 7w_\Delta$	677	$25 + 2w_\Delta$
457	$56325840235 + 5528222698w_\Delta$	681	$10331486987667 + 823358031496w_\Delta$
461	$174 + 17w_\Delta$	685	$365 + 29w_\Delta$
465	$15135 + 1472w_\Delta$	689	$101 + 8w_\Delta$

Table B1 continued

D	ε_Δ
697	$127 + 10w_\Delta$
701	$11337 + 890w_\Delta$
705	$228229 + 17864w_\Delta$
709	$17560093 + 1370434w_\Delta$
713	$5088391 + 395952w_\Delta$
717	$116 + 9w_\Delta$
721	$17938278413170303 + 1387797060244224w_\Delta$
733	$13 + w_\Delta$
737	$243656915 + 18636936w_\Delta$
741	$118 + 9w_\Delta$
745	$12301181 + 935640w_\Delta$
749	$6236 + 473w_\Delta$
753	$297282714379 + 22486626968w_\Delta$
757	$1319557 + 99538w_\Delta$
761	$771 + 58w_\Delta$
769	$15777151472704763 + 1180445209689554w_\Delta$
773	$67 + 5w_\Delta$
777	$215 + 16w_\Delta$
781	$65187059 + 4838280w_\Delta$
785	$27 + 2w_\Delta$
789	$15346 + 1133w_\Delta$
793	$4237 + 312w_\Delta$
797	$177 + 13w_\Delta$
805	$698 + 51w_\Delta$
809	$418598601107 + 30506849866w_\Delta$
813	$2091 + 152w_\Delta$
817	$331 + 24w_\Delta$
821	$7812 + 565w_\Delta$
829	$14951317 + 1075930w_\Delta$
849	$1450118056618219 + 103073312660952w_\Delta$
853	$13271 + 941w_\Delta$
857	$7841243 + 554650w_\Delta$
861	$496 + 35w_\Delta$
865	$336501019 + 23688178w_\Delta$
869	$23851 + 1675w_\Delta$
877	$233177 + 16298w_\Delta$
881	$102734288607 + 7163765650w_\Delta$
885	$115 + 8w_\Delta$
889	$1278818852939020371 1 + 88757237682690700 8w_\Delta$
893	$1112 + 77w_\Delta$
897	$579 + 40w_\Delta$
901	$29 + 2w_\Delta$
905	$349 + 24w_\Delta$
913	$498665930828843 + 34136624503128w_\Delta$
917	$571 + 39w_\Delta$

Table B1 continued

D	ε_Δ
921	$2438953085696323 + 166209254278824w_\Delta$
929	$78649159403 + 5335854130w_\Delta$
933	$72799 + 4928w_\Delta$
937	$474211686958175 + 32030016105242w_\Delta$
941	$549 + 37w_\Delta$
949	$15812 + 1061w_\Delta$
953	$2657885067 + 177959354w_\Delta$
957	$15 + w_\Delta$
965	$15 + w_\Delta$
969	$13152411 + 873080w_\Delta$
973	$874267 + 57912w_\Delta$
977	$7140745763 + 472006210w_\Delta$
985	$395 + 26w_\Delta$
989	$49981 + 3283w_\Delta$
993	$2563 + 168w_\Delta$
997	$82217 + 5378w_\Delta$
1001	$1027373 + 67064w_\Delta$
1005	$875 + 57w_\Delta$
1009	$523 + 34w_\Delta$
1013	$447 + 29w_\Delta$
1021	$41531201 + 2683493w_\Delta$
1033	$181046474550150103 + 11627786092048186w_\Delta$
1037	$78 + 5w_\Delta$
1041	$747024961984083 + 47787375886984w_\Delta$
1045	$47 + 3w_\Delta$
1049	$1488915416127 + 94870818986w_\Delta$
1057	$232266499368396415 + 14741680400265984w_\Delta$
1061	$128139 + 8117w_\Delta$
1065	$80134839 + 5066320w_\Delta$
1069	$51777638 + 3267185w_\Delta$
1073	$43983 + 2770w_\Delta$
1077	$175 + 11w_\Delta$
1081	$28429693388298711967 + 17836244437456015296w_\Delta$
1085	$16 + w_\Delta$
1093	$16 + w_\Delta$
1097	$20384655 + 1269242w_\Delta$
1101	$177 + 11w_\Delta$
1105	$27631 + 1714w_\Delta$
1109	$51828 + 3209w_\Delta$
1113	$681407 + 42112w_\Delta$
1117	$3630322 + 223945w_\Delta$
1121	$147445706806331 + 9078795117240w_\Delta$
1129	$163 + 10w_\Delta$
1133	$49 + 3w_\Delta$
1137	$631178308195 + 38581252584w_\Delta$

Table B1 continued

D	ε_Δ
1141	$618715978 + 37751109w_\Delta$
1145	$1215 + 74w_\Delta$
1149	$6994671 + 425248w_\Delta$
1153	$99457149766197583 + 6035780513750146w_\Delta$
1157	$33 + 2w_\Delta$
1165	$878 + 53w_\Delta$
1169	$178847503155871 + 10776983656128w_\Delta$
1173	$133 + 8w_\Delta$
1177	$9205626607838696891 + 552767092131556920w_\Delta$
1181	$14097 + 845w_\Delta$
1185	$40215551 + 2406400w_\Delta$
1189	$12472 + 745w_\Delta$
1193	$830238463 + 49507610w_\Delta$
1201	$2418284154713706340674447 + 1437083371218884615477 0w_\Delta$
1205	$118 + 7w_\Delta$
1209	$55389179 + 3280312w_\Delta$
1213	$68573617 + 4054234w_\Delta$
1217	$26836287 + 1583938w_\Delta$
1221	$17 + w_\Delta$
1229	$17 + w_\Delta$
1237	$628627 + 36793w_\Delta$
1241	$33030027 + 1930010w_\Delta$
1245	$120 + 7w_\Delta$
1249	$89742606340459332956 47 + 522652800566192785186w_\Delta$
1253	$86 + 5w_\Delta$
1257	$98539 + 5720w_\Delta$
1261	$38393 + 2225w_\Delta$
1265	$201179 + 11640w_\Delta$
1273	$178307794652283331 + 10283292431702952w_\Delta$
1277	$3213 + 185w_\Delta$
1281	$768247427 + 44163496w_\Delta$
1285	$12632 + 725w_\Delta$
1289	$291765666488047 + 16718827867706w_\Delta$
1293	$839 + 48w_\Delta$
1297	$35 + 2w_\Delta$
1301	$211293 + 12050w_\Delta$
1309	$56939 + 3237w_\Delta$
1313	$599 + 34w_\Delta$
1317	$205126 + 11625w_\Delta$
1321	$5198029532912465162 6507 + 294126277254392233985 0w_\Delta$
1329	$685312620437267379 + 38657671901868232w_\Delta$
1333	$42346 + 2385w_\Delta$
1337	$1330034375 + 74794544w_\Delta$
1345	$4709 + 264w_\Delta$
1349	$7771 + 435w_\Delta$

Table B1 continued

D	ε_Δ
1353	$41813897 + 2337072w_\Delta$
1357	$434189 + 24231w_\Delta$
1361	$12897280186731 + 718676917058w_\Delta$
1365	$18 + w_\Delta$
1373	$18 + w_\Delta$
1381	$36686481928 + 2029018105w_\Delta$
1385	$101295 + 5594w_\Delta$
1389	$609179 + 33592w_\Delta$
1393	$3487 + 192w_\Delta$
1397	$1546 + 85w_\Delta$
1401	$160959631651785118708 3 + 8836668726675699586 4w_\Delta$
1405	$2189 + 120w_\Delta$
1409	$34517780401307163 + 1889488038692434w_\Delta$
1417	$1596641807 + 87145658w_\Delta$
1429	$92 + 5w_\Delta$
1433	$5614976487 + 304706474w_\Delta$
1437	$28216 + 1529w_\Delta$
1441	$81885948688353541357 2299 + 44309971919066220965400w_\Delta$
1453	$1489832518 + 80274961w_\Delta$
1457	$7043667526391 + 378990620624w_\Delta$
1461	$4581578 + 246169w_\Delta$
1465	$799284263647 + 42885442570w_\Delta$
1469	$56 + 3w_\Delta$
1473	$2589425679179 + 138547228440w_\Delta$
1477	$13243150 + 707589w_\Delta$
1481	$38213187514156063 + 2038919502722810w_\Delta$
1489	$96906187 + 5156290w_\Delta$
1493	$1148 + 61w_\Delta$
1497	$21934757320195 + 1163922861864w_\Delta$
1501	$1462535281 + 77500215w_\Delta$
1505	$11659853 + 617016w_\Delta$
1509	$246 + 13w_\Delta$
1513	$21707696009 + 1145606448w_\Delta$
1517	$19 + w_\Delta$
1529	$70436783893699 + 3697233202600w_\Delta$
1533	$248 + 13w_\Delta$
1537	$73165239365 + 3830180424w_\Delta$
1541	$577487 + 30191w_\Delta$
1545	$120340445 + 6283032w_\Delta$
1549	$329861957297 + 17199418961w_\Delta$
1553	$89499871 + 4660466w_\Delta$
1561	$1670691279111727603096456 63 + 86767763963557895632141 44w_\Delta$
1565	$482 + 25w_\Delta$
1569	$358132250868874150821 9 + 18550978634070790047 2w_\Delta$
1577	$109498225987835 + 5657147932856w_\Delta$
1581	$407 + 21w_\Delta$

Table B1 continued

D	ε_Δ
1585	$388941103063 + 20042281618w_\Delta$
1589	$3085994 + 158817w_\Delta$
1597	$49063993 + 2518525w_\Delta$
1601	$39 + 2w_\Delta$
1605	$625 + 32w_\Delta$
1609	$224731554507505386703938823 + 11491592762560840864333690w_\Delta$
1613	$1449 + 74w_\Delta$
1621	$2351907622159 + 119806883557w_\Delta$
1633	$321827152971025683387551 + 16332094890275578606272w_\Delta$
1637	$2703 + 137w_\Delta$
1641	$4267 + 216w_\Delta$
1645	$12995 + 657w_\Delta$
1649	$106164687 + 5360786w_\Delta$
1653	$694 + 35w_\Delta$
1657	$10462392416665400987 + 526989504041271290w_\Delta$
1661	$1422944 + 71585w_\Delta$
1669	$630845968 + 31658329w_\Delta$
1673	$3100716960431 + 155415361952w_\Delta$
1677	$20 + w_\Delta$
1685	$20 + w_\Delta$
1689	$31382894167420287481859 + 1565331358624824925992w_\Delta$
1693	$669737 + 33365w_\Delta$
1697	$11465614547 + 570504370w_\Delta$
1705	$967 + 48w_\Delta$
1709	$2630353 + 130409w_\Delta$
1713	$37664619822531 + 1865120476712w_\Delta$
1717	$50465 + 2496w_\Delta$
1721	$3097361571938919 + 153013036428682w_\Delta$
1729	$43539038777389 + 2145771424632w_\Delta$
1733	$84123 + 4141w_\Delta$
1741	$41915649881 + 2058457913w_\Delta$
1745	$3955 + 194w_\Delta$
1749	$1939 + 95w_\Delta$
1753	$878436123717399202807 + 42988059503014651114w_\Delta$
1757	$491 + 24w_\Delta$
1761	$1147 + 56w_\Delta$
1765	$41 + 2w_\Delta$
1769	$121353967 + 5911130w_\Delta$
1777	$233751094044519242095 + 11359691894794253842w_\Delta$
1781	$103 + 5w_\Delta$
1785	$165 + 8w_\Delta$
1789	$267956798117 + 12977193281w_\Delta$
1793	$256335268460419 + 12400170182568w_\Delta$
1797	$154699 + 7475w_\Delta$
1801	$620120568149173504478881155863 + 29929905864309041680160753210w_\Delta$

Table B1 continued

D	ε_Δ
1817	$23375804294311 + 1123127150704w_\Delta$
1821	$1865785232 + 89543701w_\Delta$
1829	$39776099 + 1904675w_\Delta$
1833	$4154427 + 198712w_\Delta$
1837	$14045261 + 671055w_\Delta$
1841	$216541339705689943 + 10334406229287184w_\Delta$
1853	$21 + w_\Delta$
1857	$2329002500958706995 + 110660051291833544w_\Delta$
1861	$1837582721 + 87214658w_\Delta$
1865	$237463 + 11258w_\Delta$
1869	$12564 + 595w_\Delta$
1873	$178327928243531257358796331 + 8435934108444990409358146w_\Delta$
1877	$783 + 37w_\Delta$
1885	$509 + 24w_\Delta$
1889	$4201125415104183 + 197874020057570w_\Delta$
1893	$4931 + 232w_\Delta$
1897	$42895 + 2016w_\Delta$
1901	$213 + 10w_\Delta$
1905	$10464531213 + 490759096w_\Delta$
1909	$2326179899 + 108974712w_\Delta$
1913	$4640081093767 + 217141503434w_\Delta$
1921	$69760020739307 + 3257590519250w_\Delta$
1929	$123907539 + 5773832w_\Delta$
1933	$396987713 + 18479201w_\Delta$
1937	$43 + 2w_\Delta$
1941	$5142325047 + 238862491w_\Delta$
1945	$2572599024269 + 119372178744w_\Delta$
1949	$81333 + 3770w_\Delta$
1957	$454 + 21w_\Delta$
1961	$5308473013 + 245290344w_\Delta$
1965	$65 + 3w_\Delta$
1969	$45828407842475722320774887146451 + 2113202631220407492138882654600w_\Delta$
1973	$86533 + 3986w_\Delta$
1977	$225024082138131 + 10354630760584w_\Delta$
1981	$4817354433592 + 221444665221w_\Delta$
1985	$1864786767 + 85632322w_\Delta$
1993	$6756534626472792899627 + 309627211845100703210w_\Delta$
1997	$4478 + 205w_\Delta$

Table B2. $D \equiv 2 \pmod 4$

D	ε_Δ
2	$1 + w_\Delta$
6	$5 + 2w_\Delta$
10	$3 + w_\Delta$
14	$15 + 4w_\Delta$
22	$197 + 42w_\Delta$
26	$5 + w_\Delta$
30	$11 + 2w_\Delta$
34	$35 + 6w_\Delta$
38	$37 + 6w_\Delta$
42	$13 + 2w_\Delta$
46	$24335 + 3588w_\Delta$
58	$99 + 13w_\Delta$
62	$63 + 8w_\Delta$
66	$65 + 8w_\Delta$
70	$251 + 30w_\Delta$
74	$43 + 5w_\Delta$
78	$53 + 6w_\Delta$
82	$9 + w_\Delta$
86	$10405 + 1122w_\Delta$
94	$2143295 + 221064w_\Delta$
102	$101 + 10w_\Delta$
106	$4005 + 389w_\Delta$
110	$21 + 2w_\Delta$
114	$1025 + 96w_\Delta$
118	$306917 + 28254w_\Delta$
122	$11 + w_\Delta$
130	$57 + 5w_\Delta$
134	$145925 + 12606w_\Delta$
138	$47 + 4w_\Delta$
142	$143 + 12w_\Delta$
146	$145 + 12w_\Delta$
154	$21295 + 1716w_\Delta$
158	$7743 + 616w_\Delta$
166	$1700902565 + 132015642w_\Delta$
170	$13 + w_\Delta$
174	$1451 + 110w_\Delta$
178	$1601 + 120w_\Delta$
182	$27 + 2w_\Delta$
186	$7501 + 550w_\Delta$
190	$52021 + 3774w_\Delta$
194	$195 + 14w_\Delta$
202	$3141 + 221w_\Delta$
206	$59535 + 4148w_\Delta$
210	$29 + 2w_\Delta$
214	$695359189925 + 47533775646w_\Delta$
218	$251 + 17w_\Delta$
222	$149 + 10w_\Delta$
226	$15 + w_\Delta$
230	$91 + 6w_\Delta$
238	$11663 + 756w_\Delta$
246	$88805 + 5662w_\Delta$
254	$255 + 16w_\Delta$
258	$257 + 16w_\Delta$
262	$104980517 + 6485718w_\Delta$
266	$685 + 42w_\Delta$
274	$1407 + 85w_\Delta$
278	$2501 + 150w_\Delta$
282	$2351 + 140w_\Delta$
286	$561835 + 33222w_\Delta$
290	$17 + w_\Delta$
298	$409557 + 23725w_\Delta$
302	$4276623 + 246092w_\Delta$
310	$848719 + 48204w_\Delta$
314	$443 + 25w_\Delta$
318	$107 + 6w_\Delta$
322	$323 + 18w_\Delta$
326	$325 + 18w_\Delta$
330	$109 + 6w_\Delta$
334	$63804373719695 + 3491219999244w_\Delta$
346	$93 + 5w_\Delta$
354	$258065 + 13716w_\Delta$
358	$176579805797 + 9332532726w_\Delta$
362	$19 + w_\Delta$
366	$907925 + 47458w_\Delta$
370	$327 + 17w_\Delta$
374	$3365 + 174w_\Delta$
382	$164998439999 + 8442054600w_\Delta$
386	$111555 + 5678w_\Delta$
390	$79 + 4w_\Delta$
394	$395023035 + 19900973w_\Delta$

Table B2 continued

D	ε_Δ
398	$399 + 20w_\Delta$
402	$401 + 20w_\Delta$
406	$59468095 + 2951352w_\Delta$
410	$81 + 4w_\Delta$
418	$33857 + 1656w_\Delta$
422	$7022501 + 341850w_\Delta$
426	$88751 + 4300w_\Delta$
430	$2862251 + 138030w_\Delta$
434	$125 + 6w_\Delta$
438	$293 + 14w_\Delta$
442	$21 + w_\Delta$
446	$110166015 + 5216512w_\Delta$
454	$16916040084175685 + 793909098494766w_\Delta$
458	$107 + 5w_\Delta$
462	$43 + 2w_\Delta$
466	$938319425 + 43466808w_\Delta$
470	$1691 + 78w_\Delta$
474	$193549 + 8890w_\Delta$
478	$1617319577991743 + 73974475657896w_\Delta$
482	$483 + 22w_\Delta$
494	$73035 + 3286w_\Delta$
498	$179777 + 8056w_\Delta$
502	$3832352837 + 171046278w_\Delta$
506	$45 + 2w_\Delta$
510	$271 + 12w_\Delta$
514	$4625 + 204w_\Delta$
518	$2367 + 104w_\Delta$
526	$84056091546952933775 + 3665019757324295532w_\Delta$
530	$23 + w_\Delta$
534	$3678725 + 159194w_\Delta$
538	$69051 + 2977w_\Delta$
542	$4293183 + 184408w_\Delta$
546	$701 + 30w_\Delta$
554	$174293 + 7405w_\Delta$
562	$220938497 + 9319728w_\Delta$
566	$95609285 + 4018758w_\Delta$
570	$191 + 8w_\Delta$
574	$575 + 24w_\Delta$
582	$193 + 8w_\Delta$
586	$4115086707 + 169992665w_\Delta$
590	$5781 + 238w_\Delta$

Table B2 continued

D	ε_Δ
598	$1574351 + 64380w_\Delta$
602	$687 + 28w_\Delta$
606	$42187499 + 1713750w_\Delta$
610	$71847 + 2909w_\Delta$
614	$348291186245 + 14055888354w_\Delta$
618	$10093 + 406w_\Delta$
622	$13804370063 + 553504812w_\Delta$
626	$25 + w_\Delta$
634	$65999458125 + 2621173333w_\Delta$
638	$42283 + 1674w_\Delta$
642	$5777 + 228w_\Delta$
646	$305 + 12w_\Delta$
654	$8915765 + 348634w_\Delta$
658	$1693 + 66w_\Delta$
662	$1718102501 + 66775950w_\Delta$
670	$5791211 + 223734w_\Delta$
674	$675 + 26w_\Delta$
678	$677 + 26w_\Delta$
682	$1197901 + 45870w_\Delta$
690	$1471 + 56w_\Delta$
694	$38782105445014642382885 + 1472148590903997672114w_\Delta$
698	$5099 + 193w_\Delta$
706	$34595 + 1302w_\Delta$
710	$1279 + 48w_\Delta$
714	$4115 + 154w_\Delta$
718	$89333991830360795O3 + 333391496474140716w_\Delta$
730	$27 + w_\Delta$
734	$10394175 + 383656w_\Delta$
742	$263091151 + 9658380w_\Delta$
746	$5534843 + 202645w_\Delta$
754	$20457 + 745w_\Delta$
758	$413959717 + 15035694w_\Delta$
762	$6349 + 230w_\Delta$
766	$1459336119457446380 15 + 5272795728865625208w_\Delta$
770	$111 + 4w_\Delta$
778	$54610269 + 1957873w_\Delta$
782	$783 + 28w_\Delta$
786	$785 + 28w_\Delta$
790	$6616066879 + 235389096w_\Delta$
794	$30235 + 1073w_\Delta$
798	$113 + 4w_\Delta$

Table B2 continued

D	ε_Δ
802	$295496099 + 10434330w_\Delta$
806	$6166395 + 217202w_\Delta$
814	$4206992174549 + 147454999410w_\Delta$
818	$143 + 5w_\Delta$
822	$7397 + 258w_\Delta$
826	$222239304685 + 7732694382w_\Delta$
830	$146411 + 5082w_\Delta$
834	$6552578705 + 226897244w_\Delta$
838	$42112785797 + 1454762046w_\Delta$
842	$29 + w_\Delta$
854	$1294299 + 44290w_\Delta$
858	$703 + 24w_\Delta$
862	$358022566147312125503 + 12194296994921665128w_\Delta$
866	$42435 + 1442w_\Delta$
870	$59 + 2w_\Delta$
874	$3725 + 126w_\Delta$
878	$9314703 + 314356w_\Delta$
886	$7743524593057655851637765 + 260148796464024194850378w_\Delta$
890	$179 + 6w_\Delta$
894	$299 + 10w_\Delta$
898	$899 + 30w_\Delta$
902	$901 + 30w_\Delta$
906	$301 + 10w_\Delta$
910	$181 + 6w_\Delta$
914	$5593 + 185w_\Delta$
922	$419288307 + 13808525w_\Delta$
926	$304560297142335 + 10008472361032w_\Delta$
930	$61 + 2w_\Delta$
934	$3034565 + 99294w_\Delta$
938	$17151 + 560w_\Delta$
942	$106133 + 3458w_\Delta$
946	$45225786400145 + 1470417148788w_\Delta$
958	$16762522330425599 + 541572514048560w_\Delta$
962	$31 + w_\Delta$
966	$57499 + 1850w_\Delta$
970	$328173 + 10537w_\Delta$
974	$488825745235215 + 15662987185124w_\Delta$
978	$118337 + 3784w_\Delta$
982	$8837 + 282w_\Delta$
986	$157 + 5w_\Delta$
994	$1135 + 36w_\Delta$

Table B2 continued

D	ε_Δ
998	$984076901 + 31150410w_\Delta$
1002	$206869247 + 6535248w_\Delta$
1006	$14125267563780214605455 + 445346140255574921748w_\Delta$
1010	$1303 + 41w_\Delta$
1018	$87050499 + 2728333w_\Delta$
1022	$1023 + 32w_\Delta$
1030	$884307011291 + 27553987434w_\Delta$
1034	$15885 + 494w_\Delta$
1038	$15404267 + 478126w_\Delta$
1042	$807 + 25w_\Delta$
1046	$3987042510565 + 123277842162w_\Delta$
1054	$3355576256056895 + 103358645005416w_\Delta$
1066	$3434907 + 105205w_\Delta$
1070	$2948491 + 90138w_\Delta$
1074	$3489425 + 106476w_\Delta$
1082	$1262101 + 38369w_\Delta$
1086	$725 + 22w_\Delta$
1090	$33 + w_\Delta$
1094	$14898540581538085 + 450438047467914w_\Delta$
1102	$11382443 + 342882w_\Delta$
1106	$5055 + 152w_\Delta$
1110	$1999 + 60w_\Delta$
1114	$6111788103202293 + 183115747156745w_\Delta$
1118	$4213 + 126w_\Delta$
1122	$67 + 2w_\Delta$
1126	$91205 + 2718w_\Delta$
1130	$437 + 13w_\Delta$
1138	$348711 + 10337w_\Delta$
1142	$906612101 + 26828010w_\Delta$
1146	$3447551 + 101840w_\Delta$
1154	$1155 + 34w_\Delta$
1158	$1157 + 34w_\Delta$
1162	$22616173 + 663462w_\Delta$
1166	$660427701 + 19340870w_\Delta$
1174	$46724428792065693035913655 + 1363672091213661406747774w_\Delta$
1178	$59583 + 1736w_\Delta$
1182	$27252587 + 792682w_\Delta$
1186	$6320195 + 183522w_\Delta$
1190	$69 + 2w_\Delta$
1194	$21647468749 + 626476750w_\Delta$
1198	$42062569635023592515999 + 1215254929836463164600w_\Delta$

Table B2 continued

D	ε_Δ
1202	$10817 + 312w_\Delta$
1214	$73150798374975 + 2099470260104w_\Delta$
1218	$349 + 10w_\Delta$
1222	$7734245409307 + 221249524122w_\Delta$
1226	$35 + w_\Delta$
1230	$491 + 14w_\Delta$
1234	$586327869067265 + 16691023073856w_\Delta$
1238	$902501 + 25650w_\Delta$
1246	$443896693148735 + 12575431287288w_\Delta$
1254	$362405 + 10234w_\Delta$
1258	$114669 + 3233w_\Delta$
1262	$61855669263 + 1741204708w_\Delta$
1266	$533240465 + 14986708w_\Delta$
1270	$581837113691 + 16326743118w_\Delta$
1282	$3410326403 + 95247138w_\Delta$
1286	$25363110565 + 707264778w_\Delta$
1290	$431 + 12w_\Delta$
1294	$1295 + 36w_\Delta$
1298	$1297 + 36w_\Delta$
1302	$433 + 12w_\Delta$
1306	$311761210150515 + 8626815105389w_\Delta$
1310	$125100021 + 3456382w_\Delta$
1318	$11393611468262768176276517 + 313836679699540895794554w_\Delta$
1322	$29851 + 821w_\Delta$
1326	$2549 + 70w_\Delta$
1330	$1246589 + 34182w_\Delta$
1334	$3155972095 + 86408256w_\Delta$
1338	$17308813 + 473194w_\Delta$
1342	$1099 + 30w_\Delta$
1346	$340550115 + 9282362w_\Delta$
1354	$13154959098165 + 357503394613w_\Delta$
1358	$2999823 + 81404w_\Delta$
1362	$1370184257 + 37127048w_\Delta$
1366	$2291037052846189335142389540899525 + 61987877112128467931286485364042w_\Delta$
1370	$37 + w_\Delta$
1374	$153544325045 + 4142289074w_\Delta$
1378	$42801 + 1153w_\Delta$
1382	$13003237 + 349782w_\Delta$
1390	$2298622285432981 + 61653851607582w_\Delta$
1394	$12545 + 336w_\Delta$
1398	$4866437 + 130154w_\Delta$
1402	$930015700509 + 24837980029w_\Delta$

Table B2 continued

D	ε_Δ
1406	$75 + 2w_\Delta$
1410	$751 + 20w_\Delta$
1414	$182475063042299 + 4852648311810w_\Delta$
1418	$38899 + 1033w_\Delta$
1426	$9860975 + 261132w_\Delta$
1430	$13311 + 352w_\Delta$
1434	$176669098751 + 4665369200w_\Delta$
1438	$15960728127743 + 420894506064w_\Delta$
1442	$1443 + 38w_\Delta$
1446	$1445 + 38w_\Delta$
1454	$366707758095 + 9616961884w_\Delta$
1462	$2753 + 72w_\Delta$
1466	$153352043 + 4005185w_\Delta$
1474	$54959576050729745 + 1431510810979476w_\Delta$
1478	$119717 + 3114w_\Delta$
1482	$77 + 2w_\Delta$
1486	$18974735 + 492228w_\Delta$
1490	$193 + 5w_\Delta$
1498	$321700659967 + 8311818384w_\Delta$
1502	$5966017599 + 153939320w_\Delta$
1506	$31880816705 + 821517752w_\Delta$
1510	$84053391679 + 2163051048w_\Delta$
1514	$50750707 + 1304305w_\Delta$
1518	$1013 + 26w_\Delta$
1522	$39 + w_\Delta$
1526	$102135615 + 2614568w_\Delta$
1534	$235 + 6w_\Delta$
1538	$813603 + 20746w_\Delta$
1542	$2592101 + 66010w_\Delta$
1546	$43605 + 1109w_\Delta$
1554	$160285 + 4066w_\Delta$
1558	$17873016101 + 452808030w_\Delta$
1562	$1813197 + 45878w_\Delta$
1570	$11993673 + 302693w_\Delta$
1574	$3097184950960165 + 78066512141574w_\Delta$
1578	$442367 + 11136w_\Delta$
1582	$8593649 + 216060w_\Delta$
1586	$12943 + 325w_\Delta$
1590	$319 + 8w_\Delta$
1594	$23414622020115 + 586466207477w_\Delta$
1598	$1599 + 40w_\Delta$

Table B2 continued

D	ε_Δ
1606	$24538370918885 + 612312261798w_\Delta$
1610	$321 + 8w_\Delta$
1614	$291698879051 + 7260775270w_\Delta$
1618	$1075479 + 26737w_\Delta$
1622	$2831176412101 + 70297763190w_\Delta$
1626	$266647183921549 + 6612668183130w_\Delta$
1630	$15078465611 + 373476558w_\Delta$
1634	$24720785 + 611556w_\Delta$
1642	$37744496576829 + 931466115401w_\Delta$
1646	$80655 + 1988w_\Delta$
1654	$14885 + 366w_\Delta$
1658	$43691 + 1073w_\Delta$
1662	$64353749 + 1578550w_\Delta$
1670	$100045691 + 2448162w_\Delta$
1678	$9777409878270240143 + 238686492939208596w_\Delta$
1686	$35350768325 + 860934422w_\Delta$
1698	$4947715601 + 120070380w_\Delta$
1702	$231069000116251 + 5600952727350w_\Delta$
1706	$482163035 + 11673589w_\Delta$
1714	$207 + 5w_\Delta$
1718	$137592901 + 3319590w_\Delta$
1722	$83 + 2w_\Delta$
1726	$2689844967637183200753607543760431273535 + 64745092245111302274843632152698614232w_\Delta$
1730	$72913 + 1753w_\Delta$
1738	$2801715696553663 + 67204627338744w_\Delta$
1742	$96747 + 2318w_\Delta$
1754	$206446329557 + 4929380825w_\Delta$
1758	$587 + 14w_\Delta$
1762	$1763 + 42w_\Delta$
1766	$1765 + 42w_\Delta$
1770	$589 + 14w_\Delta$
1774	$125288009537733004908595972131215 + 2974628275728557930834554244724w_\Delta$
1778	$253 + 6w_\Delta$
1786	$78064935471648262925 + 1847205775499723178w_\Delta$
1790	$60501 + 1430w_\Delta$
1794	$12582335 + 297064w_\Delta$
1798	$2089768622363 + 49283706246w_\Delta$
1802	$849 + 20w_\Delta$
1806	$85 + 2w_\Delta$
1810	$228573297 + 5372621w_\Delta$
1814	$789064983271716805 + 18526531941823254w_\Delta$

Table B2 continued

D	ε_Δ
1822	$371368817235398339023103 + 870024030399855725899 2w_\Delta$
1826	$3069505 + 71832w_\Delta$
1830	$59291 + 1386w_\Delta$
1834	$218268829391589325 + 5096735001878706w_\Delta$
1838	$24564830389263 + 572982049244w_\Delta$
1842	$5057485457 + 117839212w_\Delta$
1846	$6289018486180795 + 146375038721562w_\Delta$
1858	$763416899 + 17710830w_\Delta$
1866	$15551 + 360w_\Delta$
1870	$19682191 + 455148w_\Delta$
1874	$21455425 + 495624w_\Delta$
1878	$16901 + 390w_\Delta$
1882	$3861 + 89w_\Delta$
1886	$92415 + 2128w_\Delta$
1894	$22630910383266902129070483206998403 4245 + 52001040967946406726789343138096090 86w_\Delta$
1898	$2309 + 53w_\Delta$
1902	$2846027 + 65258w_\Delta$
1906	$10705292457 + 245209385w_\Delta$
1910	$278479 + 6372w_\Delta$
1914	$175 + 4w_\Delta$
1918	$416753 + 9516w_\Delta$
1930	$8020227 + 182561w_\Delta$
1934	$1935 + 44w_\Delta$
1938	$1937 + 44w_\Delta$
1942	$6984160935186866597 + 158485533243485118w_\Delta$
1946	$121257485 + 2748762w_\Delta$
1954	$4691555 + 106134w_\Delta$
1958	$177 + 4w_\Delta$
1966	$98509752681070086696409398 61775 + 2221710522690929421268974583 08w_\Delta$
1970	$577 + 13w_\Delta$
1974	$1376255 + 30976w_\Delta$
1978	$60973994702573 + 1370981217774w_\Delta$
1982	$438784358463 + 9855968632w_\Delta$
1986	$13209364625 + 296409628w_\Delta$
1990	$35314095433681999 + 791628731408580w_\Delta$
1994	$84198805 + 1885573w_\Delta$

Table B3. $D \equiv 3 \pmod 4$

D	ε_Δ
3	$2 + w_\Delta$
7	$8 + 3w_\Delta$
11	$10 + 3w_\Delta$
15	$4 + w_\Delta$
19	$170 + 39w_\Delta$
23	$24 + 5w_\Delta$
31	$1520 + 273w_\Delta$
35	$6 + w_\Delta$
39	$25 + 4w_\Delta$
43	$3482 + 531w_\Delta$
47	$48 + 7w_\Delta$
51	$50 + 7w_\Delta$
55	$89 + 12w_\Delta$
59	$530 + 69w_\Delta$
67	$48842 + 5967w_\Delta$
71	$3480 + 413w_\Delta$
79	$80 + 9w_\Delta$
83	$82 + 9w_\Delta$
87	$28 + 3w_\Delta$
91	$1574 + 165w_\Delta$
95	$39 + 4w_\Delta$
103	$227528 + 22419w_\Delta$
107	$962 + 93w_\Delta$
111	$295 + 28w_\Delta$
115	$1126 + 105w_\Delta$
119	$120 + 11w_\Delta$
123	$122 + 11w_\Delta$
127	$4730624 + 419775w_\Delta$
131	$10610 + 927w_\Delta$
139	$77563250 + 6578829w_\Delta$
143	$12 + w_\Delta$
151	$1728148040 + 140634693w_\Delta$
155	$249 + 20w_\Delta$
159	$1324 + 105w_\Delta$
163	$64080026 + 5019135w_\Delta$
167	$168 + 13w_\Delta$
179	$4190210 + 313191w_\Delta$
183	$487 + 36w_\Delta$
187	$1682 + 123w_\Delta$
191	$8994000 + 650783w_\Delta$
195	$14 + w_\Delta$
199	$16266196520 + 1153080099w_\Delta$
203	$57 + 4w_\Delta$
211	$278354373650 + 19162705353w_\Delta$
215	$44 + 3w_\Delta$

D	ε_Δ
219	$74 + 5w_\Delta$
223	$224 + 15w_\Delta$
227	$226 + 15w_\Delta$
231	$76 + 5w_\Delta$
235	$46 + 3w_\Delta$
239	$6195120 + 400729w_\Delta$
247	$85292 + 5427w_\Delta$
251	$3674890 + 231957w_\Delta$
255	$16 + w_\Delta$
259	$847225 + 52644w_\Delta$
263	$139128 + 8579w_\Delta$
267	$2402 + 147w_\Delta$
271	$115974983600 + 7044978537w_\Delta$
283	$138274082 + 8219541w_\Delta$
287	$288 + 17w_\Delta$
291	$290 + 17w_\Delta$
295	$2024999 + 117900w_\Delta$
299	$415 + 24w_\Delta$
303	$2524 + 145w_\Delta$
307	$88529282 + 5052633w_\Delta$
311	$16883880 + 957397w_\Delta$
319	$12901780 + 722361w_\Delta$
323	$18 + w_\Delta$
327	$217 + 12w_\Delta$
331	$2785589801443970 + 153109862634573w_\Delta$
335	$604 + 33w_\Delta$
339	$97970 + 5321w_\Delta$
347	$641602 + 34443w_\Delta$
355	$954809 + 50676w_\Delta$
359	$360 + 19w_\Delta$
367	$19019995568 + 992835687w_\Delta$
371	$1695 + 88w_\Delta$
379	$12941197220540690 + 664744650125541w_\Delta$
383	$18768 + 959w_\Delta$
391	$7338680 + 371133w_\Delta$
395	$159 + 8w_\Delta$
399	$20 + w_\Delta$
403	$669878 + 33369w_\Delta$
407	$2663 + 132w_\Delta$
411	$49730 + 2453w_\Delta$
415	$18412804 + 903849w_\Delta$
419	$270174970 + 13198911w_\Delta$
427	$62 + 3w_\Delta$
431	$151560720 + 7300423w_\Delta$
435	$146 + 7w_\Delta$

Table B3 continued

D	ε_Δ
439	$440 + 21w_\Delta$
443	$442 + 21w_\Delta$
447	$148 + 7w_\Delta$
451	$46471490 + 2188257w_\Delta$
455	$64 + 3w_\Delta$
463	$247512720456368 + 11502891625161w_\Delta$
467	$1625626 + 75225w_\Delta$
471	$7838695 + 361188w_\Delta$
479	$2989440 + 136591w_\Delta$
483	$22 + w_\Delta$
487	$51906073840568 + 2352088722477w_\Delta$
491	$93628044170 + 4225374483w_\Delta$
499	$4490 + 201w_\Delta$
503	$24648 + 1099w_\Delta$
511	$4188548960 + 185290497w_\Delta$
515	$17406 + 767w_\Delta$
519	$14851876 + 651925w_\Delta$
523	$81810300626 + 3577314675w_\Delta$
527	$528 + 23w_\Delta$
535	$1618804 + 69987w_\Delta$
543	$669337 + 28724w_\Delta$
547	$160177601264642 + 6848699678673w_\Delta$
551	$8380 + 357w_\Delta$
555	$1814 + 77w_\Delta$
559	$506568295 + 21425556w_\Delta$
563	$68122 + 2871w_\Delta$
571	$181124355061630786130 + 7579818350628982587w_\Delta$
579	$385 + 16w_\Delta$
583	$8429543 + 349116w_\Delta$
587	$1907162 + 78717w_\Delta$
591	$165676 + 6815w_\Delta$
595	$18514 + 759w_\Delta$
599	$24686379794520 + 1008658133851w_\Delta$
607	$164076033968 + 6659640783w_\Delta$
611	$236926 + 9585w_\Delta$
615	$124 + 5w_\Delta$
619	$517213510553282930 + 20788566180548739w_\Delta$
623	$624 + 25w_\Delta$
627	$626 + 25w_\Delta$
631	$4896157531299865003556 0 + 1949129537575151036427w_\Delta$
635	$126 + 5w_\Delta$
643	$1988960193026 + 78436933185w_\Delta$
647	$120187368 + 4725053w_\Delta$
651	$1735 + 68w_\Delta$
655	$737709209 + 28824684w_\Delta$
659	$5930 + 231w_\Delta$

Table B3 continued

D	ε_Δ
663	$103 + 4w_\Delta$
667	$107119097 + 4147668w_\Delta$
671	$58620 + 2263w_\Delta$
679	$17792625320 + 682818291w_\Delta$
683	$170067682 + 6507459w_\Delta$
687	$165337 + 6308w_\Delta$
691	$31138100617500578690 + 1184549173291009383w_\Delta$
695	$33639 + 1276w_\Delta$
699	$2271050 + 85899w_\Delta$
703	$1159172 + 43719w_\Delta$
707	$2526 + 95w_\Delta$
715	$75646 + 2829w_\Delta$
719	$403480310400 + 15047276489w_\Delta$
723	$242 + 9w_\Delta$
727	$728 + 27w_\Delta$
731	$730 + 27w_\Delta$
739	$9801566107361674215389 0 + 360556437651645275867 1w_\Delta$
743	$714024 + 26195w_\Delta$
751	$72933184667948824244 18960 + 2661369706772060244 56793w_\Delta$
755	$1209 + 44w_\Delta$
759	$551 + 20w_\Delta$
763	$719724601 + 26055780w_\Delta$
767	$31212 + 1127w_\Delta$
771	$2989136930 + 107651137w_\Delta$
779	$11785490 + 422259w_\Delta$
787	$34625394242 + 1234262007w_\Delta$
791	$225 + 8w_\Delta$
795	$6626 + 235w_\Delta$
799	$424 + 15w_\Delta$
803	$7226 + 255w_\Delta$
807	$51841948 + 1824923w_\Delta$
811	$138207216357861641 0 + 4853111762292119 7w_\Delta$
815	$156644 + 5487w_\Delta$
823	$2351704749036440061 68 + 819752743049763665 1w_\Delta$
827	$900602 + 31317w_\Delta$
831	$9799705 + 339948w_\Delta$
835	$34336355806 + 1188258591w_\Delta$
839	$840 + 29w_\Delta$
843	$842 + 29w_\Delta$
851	$8418574 + 288585w_\Delta$
859	$20588447719796430601 24010 + 70246877103894937291 269w_\Delta$
863	$18524026608 + 630565199w_\Delta$
871	$19442812076 + 658794555w_\Delta$
879	$107245324 + 3617295w_\Delta$
883	$34878475759617272473442 + 117375416293635780 2169w_\Delta$

Table B3 continued

D	ε_Δ
887	$469224 + 15755w_\Delta$
895	$359 + 12w_\Delta$
899	$30 + w_\Delta$
903	$601 + 20w_\Delta$
907	$123823410343073497682 + 4111488857741309517w_\Delta$
911	$371832584927520 + 12319363142953w_\Delta$
915	$121 + 4w_\Delta$
919	$4481603010937119451551263720 + 147834442396536759781499589w_\Delta$
923	$638 + 21w_\Delta$
935	$1376 + 45w_\Delta$
939	$122695 + 4004w_\Delta$
943	$737 + 24w_\Delta$
947	$13509645362 + 439004487w_\Delta$
951	$224208076 + 7270445w_\Delta$
955	$2095256249 + 67800900w_\Delta$
959	$960 + 31w_\Delta$
967	$4649532557817485528 + 149518887194649693w_\Delta$
971	$12479806786330 + 400496058813w_\Delta$
979	$360449 + 11520w_\Delta$
983	$284088 + 9061w_\Delta$
987	$377 + 12w_\Delta$
991	$379516400906811930638014896080 + 12055735790331359447442538767w_\Delta$
995	$8835999 + 280120w_\Delta$
1003	$9026 + 285w_\Delta$
1007	$476 + 15w_\Delta$
1011	$8426 + 265w_\Delta$
1015	$352871 + 11076w_\Delta$
1019	$6089923321730 + 190776436539w_\Delta$
1023	$32 + w_\Delta$
1027	$133150393 + 4154868w_\Delta$
1031	$651737448664200 + 20297537082877w_\Delta$
1039	$493194634979076339348637520 + 15300674275156671864306921w_\Delta$
1043	$576927 + 17864w_\Delta$
1047	$136807 + 4228w_\Delta$
1051	$955382766460015065912973250 + 29469722410666558657086507w_\Delta$
1055	$1689 + 52w_\Delta$
1059	$5983486610 + 183868081w_\Delta$
1063	$1745324831097224 + 53531527412685w_\Delta$
1067	$98 + 3w_\Delta$
1079	$180576876 + 5497325w_\Delta$
1087	$1088 + 3w_\Delta$
1091	$1090 + 33w_\Delta$
1095	$364 + 11w_\Delta$
1099	$1593163815326 + 48057545715w_\Delta$
1103	$2346242745024 + 70645611145w_\Delta$

Table B3 continued

D	ε_Δ
1111	$100 + 3w_\Delta$
1115	$4560126 + 136565w_\Delta$
1119	$31657255 + 946364w_\Delta$
1123	$16506645562632482 + 492571123214799w_\Delta$
1131	$368185 + 10948w_\Delta$
1135	$10614754956124 + 315073619505w_\Delta$
1139	$135 + 4w_\Delta$
1147	$9447152318 + 278945403w_\Delta$
1151	$65905671840 + 1942607807w_\Delta$
1155	$34 + w_\Delta$
1159	$1669145551424551 + 49028938575180w_\Delta$
1163	$1260321002 + 36956541w_\Delta$
1167	$373828 + 10943w_\Delta$
1171	$52268476130 + 1527430263w_\Delta$
1187	$39095175626 + 1134743775w_\Delta$
1191	$3175 + 92w_\Delta$
1195	$4650060959 + 134516232w_\Delta$
1199	$497777820 + 14375599w_\Delta$
1203	$434282 + 12521w_\Delta$
1207	$1159928 + 33387w_\Delta$
1211	$174 + 5w_\Delta$
1219	$26811112454855326 + 767914700295135w_\Delta$
1223	$1224 + 35w_\Delta$
1227	$1226 + 35w_\Delta$
1231	$2051772574240109813487200 + 58479034523501522751177w_\Delta$
1235	$246 + 7w_\Delta$
1239	$176 + 5w_\Delta$
1243	$171874 + 4875w_\Delta$
1247	$929188 + 26313w_\Delta$
1255	$31134002572889 + 878846634708w_\Delta$
1259	$111409143832810 + 3139843281069w_\Delta$
1263	$142297 + 4004w_\Delta$
1267	$625514462 + 17573127w_\Delta$
1271	$32799 + 920w_\Delta$
1279	$2823156987094185989161256000 + 78940478828277833993575041w_\Delta$
1283	$1846514088226 + 51551261505w_\Delta$
1291	$709367533957910186877445608410 + 19742774580291798609489951267w_\Delta$
1295	$36 + w_\Delta$
1299	$865 + 24w_\Delta$
1303	$36556474232606087113228808 + 1012726314518934688771461w_\Delta$
1307	$195748082 + 5414517w_\Delta$
1311	$193820 + 5353w_\Delta$
1315	$887511646 + 24474351w_\Delta$
1319	$1213059240 + 33401011w_\Delta$
1327	$65024 + 1785w_\Delta$
1335	$45124 + 1235w_\Delta$

Table B3 continued

D	ε_Δ
1339	$150175 + 4104w_\Delta$
1343	$388128 + 10591w_\Delta$
1347	$6046682 + 164753w_\Delta$
1351	$6175 + 168w_\Delta$
1355	$1823289 + 49532w_\Delta$
1363	$3451726 + 93495w_\Delta$
1367	$1368 + 37w_\Delta$
1371	$1370 + 37w_\Delta$
1379	$2517968895 + 67806016w_\Delta$
1383	$1294948 + 34821w_\Delta$
1387	$1510442 + 40557w_\Delta$
1391	$9081305 + 243492w_\Delta$
1399	$722678664789940453096 40 + 193213014430044201 5349w_\Delta$
1403	$74726 + 1995w_\Delta$
1407	$3751 + 100w_\Delta$
1411	$444071330 + 11821953w_\Delta$
1415	$22029004 + 585621w_\Delta$
1419	$12770 + 339w_\Delta$
1423	$4800114231861476037042 08 + 127247460171556263967 59w_\Delta$
1427	$9373918762 + 248147097w_\Delta$
1435	$999046 + 26373w_\Delta$
1439	$1285798517003036496 0 + 338955373761021521w_\Delta$
1443	$38 + w_\Delta$
1447	$130410331757421848 + 3428291438910093w_\Delta$
1451	$32507827844328530 + 853403156620707w_\Delta$
1455	$23764 + 623w_\Delta$
1459	$981439563379940345032810 10 + 2569424820981242287252281w_\Delta$
1463	$153 + 4w_\Delta$
1471	$390105737347282885487560183661 60 + 10171289163972762387156715367 37w_\Delta$
1479	$4404376 + 114525w_\Delta$
1483	$158695889009423019181226 + 4120928705704463238045w_\Delta$
1487	$7924224 + 205495w_\Delta$
1491	$68230 + 1767w_\Delta$
1495	$116 + 3w_\Delta$
1499	$3002576946562695360610 + 77552058394306780299w_\Delta$
1507	$4893778 + 126063w_\Delta$
1511	$38459732591163960 + 989404834498003w_\Delta$
1515	$506 + 13w_\Delta$
1523	$1522 + 39w_\Delta$
1527	$508 + 13w_\Delta$
1531	$115707050481256970516351322245 0 + 2957142327537120481274165987 7w_\Delta$
1535	$1378124 + 35175w_\Delta$
1543	$66432151290852615686191 28 + 1691201493608500620541 41w_\Delta$
1547	$118 + 3w_\Delta$
1551	$1850201 + 46980w_\Delta$
1555	$5117847907209209 + 129784305853476w_\Delta$

Table B3 continued

D	ε_Δ
1559	$2122635541940760 + 53759147640581w_\Delta$
1563	$24054459026 + 608437685w_\Delta$
1567	$2634128 + 66543w_\Delta$
1571	$1430064397610 + 36080080863w_\Delta$
1579	$852876982639022059460736831316490 + 21463242324201893177981952219891w_\Delta$
1583	$94312707829728 + 2370444310151w_\Delta$
1591	$11540919718817324 + 289337903407635w_\Delta$
1595	$639 + 16w_\Delta$
1599	$40 + w_\Delta$
1603	$72515603726 + 1811192895w_\Delta$
1607	$137050219854024 + 3418785065885w_\Delta$
1615	$3284569 + 81732w_\Delta$
1619	$78053433117061872410 + 1939851949127419161w_\Delta$
1623	$53017 + 1316w_\Delta$
1627	$14642 + 363w_\Delta$
1631	$39376 + 975w_\Delta$
1635	$115321 + 2852w_\Delta$
1639	$1401940 + 34629w_\Delta$
1643	$8958 + 221w_\Delta$
1651	$56161343343007135 + 1382177919146568w_\Delta$
1655	$320409 + 7876w_\Delta$
1659	$747655 + 18356w_\Delta$
1663	$172548834009765974150624 + 4231222898258884355775w_\Delta$
1667	$272140632242 + 6665390367w_\Delta$
1671	$9558676 + 233835w_\Delta$
1679	$1680 + 41w_\Delta$
1687	$34655415003224 + 843749566395w_\Delta$
1691	$33501079090 + 814679883w_\Delta$
1695	$249616 + 6063w_\Delta$
1699	$2743345037851573163040031857578450 + 66555468361862588059573967039601w_\Delta$
1703	$1937753 + 46956w_\Delta$
1707	$616226 + 14915w_\Delta$
1711	$12257063847449 + 296320326540w_\Delta$
1723	$52754108801097189062 6 + 12709066495146742875w_\Delta$
1727	$25433 + 612w_\Delta$
1731	$1666375 + 40052w_\Delta$
1735	$3124 + 75w_\Delta$
1739	$34520185 + 827796w_\Delta$
1743	$167 + 4w_\Delta$
1747	$722692635222980474388116882 + 17290485040281059863874103w_\Delta$
1751	$34815 + 832w_\Delta$
1759	$1712191043087396298285630692900648926 40 + 40824352105820133467328015407343336 81w_\Delta$
1763	$42 + w_\Delta$
1767	$1177 + 28w_\Delta$
1771	$505 + 12w_\Delta$
1779	$84624555410 + 2006357991w_\Delta$

Table B3 continued

D	ε_Δ
1783	$300574618222444417557224 + 7118305658345640812235w_\Delta$
1787	$216226 + 5115w_\Delta$
1795	$2389684563359 + 56403798408w_\Delta$
1799	$3024120 + 71299w_\Delta$
1803	$203137 + 4784w_\Delta$
1807	$146471428766233 + 3445671269172w_\Delta$
1811	$146690 + 3447w_\Delta$
1819	$44077149500774570 + 1033468279123611w_\Delta$
1823	$396777789408 + 9292958959w_\Delta$
1831	$1918032733134113893633567757731959058040 + 44824127254720935022058326058303814477w_\Delta$
1835	$1097124846 + 25611677w_\Delta$
1839	$26133415 + 609404w_\Delta$
1843	$49871434404842 + 1161687167529w_\Delta$
1847	$1848 + 43w_\Delta$
1851	$1850 + 43w_\Delta$
1855	$7373836 + 171207w_\Delta$
1867	$87726791690293922 + 2030299407245043w_\Delta$
1871	$1496067034725964412580 7440 + 345871008137584484948087w_\Delta$
1879	$94345422639044420558808634792532120 + 2176493849722945361406489700835109w_\Delta$
1883	$6726 + 155w_\Delta$
1887	$8992 + 207w_\Delta$
1891	$81913626778277555 0 + 18836941567957233w_\Delta$
1895	$54849639 + 1259996w_\Delta$
1903	$875034172 + 20058831w_\Delta$
1907	$17162 + 393w_\Delta$
1915	$102715020522800140 6 + 23471971442429307w_\Delta$
1919	$77244401049 + 1763313460w_\Delta$
1923	$4367227226 + 99590095w_\Delta$
1927	$46636702417448 + 1062397346883w_\Delta$
1931	$53405823410 + 1215339123w_\Delta$
1939	$6672165474011915935 + 151522771012903896w_\Delta$
1943	$83751 + 1900w_\Delta$
1947	$353 + 8w_\Delta$
1951	$226742438218831488120936558859828640 + 5133389024333877559662473405561007w_\Delta$
1955	$2874 + 65w_\Delta$
1959	$9719730001876 + 219602349925w_\Delta$
1963	$24445564472798 + 551746923459w_\Delta$
1967	$2528 + 57w_\Delta$
1979	$196956439969690 + 4427384992341w_\Delta$
1983	$354724327 + 7965804w_\Delta$
1987	$2163842 + 48543w_\Delta$
1991	$95638951 + 2143380w_\Delta$
1995	$134 + 3w_\Delta$
1999	$4027701399389138208695911951306886478800 + 90084665203202024260494303744425250249$

Appendix C

Class Numbers of Real Quadratic Fields

This table presents class numbers of maximal order $\mathcal{O}_\Delta$ of real quadratic fields with associated radicands $D < 10^4$, together with the norm of their fundamental units. It is given in matrix form (a_{ji}) where $D = 100i + j$. If *no* * appears before the a_{ji} then $a_{ji} = N(\varepsilon_\Delta)h_\Delta$. If a * appears then D is not square-free, $a_{ji} =^* k$, $D = k^2\overline{D}$ where $\overline{D}$ is square-free. This table was taken from an unpublished table by Buchmann *et al.* which goes up to $D < 10^6$. It is reproduced with the permission of the authors. The class numbers are not dependent on the GRH as are some of the techniques outlined in Chapter Eight. A description of how the class numbers were determined via OASiS is contained in [351]. Nevertheless, Appendix C was checked by this author against runs made by Mike Jacobson who used techniques of Chapter Eight which *are* dependent on the GRH. No conflicts were found. (See comments at the outset of Appendix A.)

Table C1.

j \ i	0	1	2	3	4	5	6	7	8	9	10	11	12	13	14	15	16	17	18	19
1	*1	−1	1	1	−5	1	−1	−1	*3	−4	2	3	−1	−1	1	1	−7	*9	−1	−3
2	−1	2	−2	1	2	1	2	*3	2	2	2	2	2	4	−2	1	*3	2	4	2
3	1	1	2	2	2	1	*3	2	2	4	4	1	2	1	2	*3	2	2	4	2
4	*2	*2	*2	*4	*2	*6	*2	*8	*2	*2	*2	*4	*2	*2	*6	*4	*2	*2	*2	*4
5	−1	2	2	2	*9	4	*11	2	2	4	2	−4	2	*3	2	2	2	8	*19	2
6	1	−2	1	*3	2	6	2	4	2	6	1	4	*3	−2	6	2	2	−2	12	−4
7	1	1	*3	1	2	*13	1	2	2	1	2	*3	2	1	4	4	1	2	2	3
8	*2	*6	*4	*2	*2	*2	*4	*2	*2	*2	*12	*2	*2	*2	*8	*2	*2	*2	*4	*6
9	*3	−1	1	1	−1	−1	2	−1	−1	*3	−7	−1	2	2	−1	3	−1	−1	*3	1
10	−2	2	4	2	4	4	−4	2	*9	8	−4	4	*11	2	4	2	4	*3	−4	2
11	1	2	1	1	2	2	2	*3	1	1	4	10	6	4	4	1	*3	2	3	*7
12	*2	*4	*2	*2	*2	*16	*6	*2	*2	*4	*2	*2	*2	*4	*2	*6	*2	*4	*2	*2
13	−1	−1	1	−1	1	*3	−1	1	1	1	−1	2	−1	−4	*3	2	−1	1	*7	−1
14	1	2	1	−2	*3	4	1	4	2	−4	*13	−2	1	*3	2	−2	2	−12	1	8
15	2	2	2	*3	2	2	4	4	2	4	4	2	*9	2	2	4	4	*7	*11	2
16	*4	*2	*6	*2	*4	*2	*2	*2	*4	*2	*2	*6	*8	*2	*2	*2	*4	*2	*2	*2
17	−1	*3	1	−1	1	1	−1	1	5	1	*3	−1	−1	1	−2	2	*7	2	1	*3
18	*3	1	−2	2	2	2	2	1	−4	*3	−2	2	4	1	−2	4	−4	1	*3	4
19	1	2	4	2	1	2	1	1	*3	1	1	2	2	1	4	*7	1	*3	2	2
20	*2	*2	*2	*8	*2	*2	*2	*12	*2	*2	*2	*4	*2	*2	*2	*4	*18	*2	*2	*8
21	1	*11	2	3	−1	−1	*3	1	−1	1	−1	1	4	−1	*7	*39	−1	−1	1	−2
22	1	−2	2	4	1	*3	1	*19	2	−2	2	8	2	−2	*3	−12	1	4	1	*31
23	1	2	3	4	*3	1	2	4	1	2	8	1	3	*21	1	3	2	1	1	2
24	*2	*2	*4	*18	*2	*2	*4	*2	*2	*2	*32	*2	*6	*2	*4	*2	*2	*2	*4	*2
25	*5	*5	*15	*5	*5	*5	*25	*5	*5	*5	*5	*15	*35	*5	*5	*5	*5	*5	*5	*5

Table C1 continued

j \ i	0	1	2	3	4	5	6	7	8	9	10	11	12	13	14	15	16	17	18	19
26	−2	*3	−8	3	2	1	−4	*11	2	1	*3	5	−10	4	4	2	2	1	2	*3
27	*3	1	1	2	6	2	4	5	1	*3	2	*7	4	5	1	6	3	2	*3	2
28	*2	*8	*2	*2	*2	*4	*2	*2	*6	*4	*2	*2	*2	*4	*2	*2	*2	*24	*2	*2
29	−1	1	−3	1	2	*23	−2	*27	−1	−1	*7	−9	−3	1	−5	1	*3	2	1	3
30	2	−4	2	4	2	−4	*3	−12	2	4	2	−4	4	4	4	*3	2	−4	4	−4
31	1	1	4	1	1	*3	1	4	2	*7	1	4	1	*11	*3	1	4	4	1	1
32	*4	*2	*2	*2	*12	*2	*2	*2	*8	*2	*2	*2	*4	*6	*2	*2	*4	*2	*2	*2
33	1	1	−1	*3	−1	−2	1	−3	*7	1	−1	1	*3	1	−1	2	1	−1	2	−1
34	2	1	*3	1	4	2	−2	1	2	3	2	*9	2	2	2	14	2	*17	2	7
35	2	*3	6	2	4	2	2	*7	2	4	*3	2	4	4	4	2	4	6	2	*3
36	*6	*2	*2	*4	*2	*2	*2	*4	*2	*6	*2	*4	*2	*2	*2	*16	*2	*2	*6	*44
37	−1	−1	1	−1	1	1	*7	1	*3	−1	−2	1	−1	1	1	2	−1	*3	1	−6
38	1	2	2	*13	4	−2	2	*3	1	2	2	−4	1	2	1	2	*3	2	1	4
39	2	1	1	2	5	*7	*3	1	3	4	1	4	8	6	1	*9	6	2	2	2
40	*2	*2	*4	*2	*2	*6	*8	*2	*2	*2	*4	*2	*2	*2	*12	*2	*2	*2	*4	*2
41	−1	1	−1	1	*21	−1	−1	2	*29	−1	1	1	−2	*3	1	1	5	−1	1	1
42	2	3	*11	*3	−8	1	2	2	−6	2	−4	1	*3	6	4	2	−2	2	2	1
43	1	2	*9	*7	3	2	1	1	2	4	2	*3	4	2	8	1	2	8	2	2
44	*2	*12	*2	*2	*2	*4	*2	*2	*2	*4	*6	*2	*2	*8	*38	*2	*2	*4	*2	*18
45	*3	−4	*7	2	−4	2	2	2	*13	*3	4	−4	2	6	*17	2	2	−4	*3	2
46	1	2	2	−6	1	4	8	−2	*3	2	1	2	2	2	8	−6	3	*3	2	2
47	1	*7	2	1	2	1	1	*3	*11	1	2	2	2	2	1	4	*3	1	3	8
48	*4	*2	*2	*2	*8	*2	*18	*2	*4	*2	*2	*2	*4	*2	*2	*6	*4	*2	*2	*2
49	*7	−1	1	−1	−1	*3	1	1	1	−2	−1	1	−1	1	*3	−1	−2	2	*43	−1
50	*5	*5	*5	*5	*15	*5	*5	*5	*5	*5	*5	*5	*25	*15	*5	*5	*5	*5	*5	*5

Table C1 continued

$j \backslash i$	0	1	2	3	4	5	6	7	8	9	10	11	12	13	14	15	16	17	18	19
51	2	1	1	*3	2	2	4	1	2	2	1	1	*3	8	1	4	2	4	6	1
52	*2	*2	*6	*4	*2	*2	*2	*4	*2	*2	*2	*24	*2	*26	*22	*4	*2	*2	*2	*4
53	−1	*3	1	−1	1	1	−1	1	−1	−1	*9	−1	1	2	−1	−1	2	−1	−2	*3
54	*3	2	3	2	1	−2	2	−4	2	*3	2	4	4	−2	1	4	9	−2	*3	6
55	2	2	4	2	4	4	2	2	*3	2	2	8	2	2	4	2	2	*3	4	4
56	*2	*2	*16	*2	*2	*2	*4	*6	*2	*2	*4	*34	*2	*2	*4	*2	*6	*2	*8	*2
57	1	−1	−3	2	−1	−1	*3	−1	−1	2	1	−2	3	1	1	*3	−1	1	1	3
58	−2	1	2	1	−2	*3	4	1	4	1	*23	4	−4	2	*27	2	−2	6	2	4
59	1	2	2	3	*3	2	3	4	1	4	2	2	1	*3	1	1	4	1	*13	2
60	*2	−*4	*2	*6	*2	*4	*2	*2	*2	*8	*2	*2	*6	*4	*2	*2	*2	*4	*2	*14
61	−1	1	*3	*19	−1	2	−1	−3	2	*31	−1	*3	−2	−1	1	1	1	7	−1	2
62	1	*9	1	−2	4	2	1	2	1	−4	*3	2	1	2	4	2	2	4	*7	*3
63	*3	1	1	*11	1	1	4	2	1	*3	1	1	2	2	4	2	1	4	*9	2
64	*8	*2	*2	*2	*4	*2	*2	*2	*12	*2	*2	*2	*4	*2	*2	*2	*8	*42	*2	*2
65	−2	2	−2	−2	2	−2	2	*3	−2	−2	2	−2	2	4	−2	−2	*3	−6	−2	2
66	2	1	2	2	2	1	*3	1	2	4	−4	2	2	1	−2	*3	*7	5	6	1
67	1	1	2	1	1	*9	2	2	*17	1	4	2	2	3	*3	3	1	4	1	4
68	*2	*2	*2	*4	*6	*2	*2	*16	*2	*22	*2	*4	*2	*6	*2	*28	*2	*2	*2	*4
69	1	*13	−1	*3	3	−1	1	−1	1	2	−1	1	*3	*37	2	1	−1	−2	2	1
70	2	−4	*3	−4	2	4	2	4	8	−4	2	*3	2	−4	*7	−4	2	4	4	−4
71	1	*3	1	2	2	1	2	2	2	1	*3	3	4	4	1	1	2	12	1	*3
72	*6	*2	*4	*2	*2	*2	*4	*2	*2	*18	*4	*2	*2	*14	*8	*2	*2	*2	*12	*2
73	−1	−1	2	−1	3	1	−1	−1	*3	1	−2	2	1	−3	1	*11	1	*3	−1	−1
74	−2	2	−4	2	2	6	4	*3	6	1	2	1	*7	2	2	1	*3	1	2	4
75	*5	*5	*5	*5	*5	*5	*15	*5	*5	*5	*5	*5	*5	*5	*5	*15	*5	*5	*25	*5

Table C1 continued

j \ i	0	1	2	3	4	5	6	7	8	9	10	11	12	13	14	15	16	17	18	19
76	*2	*4	*2	*2	*2	*24	*26	*2	*2	*4	*2	*14	*2	*4	*6	*2	*2	*4	*2	*2
77	1	1	−1	2	*3	−7	−1	4	−1	−1	1	1	−1	*9	1	1	4	−1	−1	1
78	2	2	1	*3	1	*17	2	−2	1	2	*7	2	*3	−4	3	2	1	4	4	2
79	3	1	*3	1	1	4	2	2	2	4	2	*3	1	2	4	1	4	2	1	1
80	*4	*6	*2	*2	*4	*2	*2	*2	*4	*14	*6	*2	*16	*2	*2	*2	*4	*2	*2	*6
81	*9	−1	−1	1	−2	1	1	1	−1	*3	1	−1	2	−1	−1	2	*41	−2	*3	1
82	−4	2	2	1	2	4	2	2	*21	5	−2	2	2	1	4	4	*29	*9	−6	1
83	1	2	1	1	4	2	1	*3	1	1	*19	*13	1	2	1	1	*3	1	2	2
84	*2	*2	*2	*8	*22	*2	*6	*28	*2	*2	*2	*4	*2	*2	*2	*12	*2	*2	*2	*8
85	−2	−2	2	2	−2	*3	−2	−6	2	−6	2	2	−2	−2	*3	−2	−2	8	4	−2
86	1	2	2	2	*9	−2	*7	6	1	−4	6	4	1	*3	5	−4	2	2	4	2
87	2	2	2	*2	1	1	2	1	1	4	7	1	*3	4	1	*23	2	3	4	3
88	*2	*2	*12	*2	*2	*14	*4	*2	*2	*2	*8	*6	*2	*2	*4	*2	*2	*2	*4	*2
89	−1	*3	*17	−1	1	1	4	1	1	1	*33	−2	−1	1	−3	1	1	−1	−1	*3
90	*3	2	−4	4	*7	2	4	2	4	*3	−12	4	8	2	−4	8	*13	2	*3	2
91	2	1	4	2	1	2	1	4	*9	1	3	6	1	2	4	2	2	*3	2	2
92	*2	*8	*2	*14	*2	*4	*2	*6	*2	*4	*2	*2	*2	*4	*2	*2	*6	*16	*2	*2
93	1	−1	−1	1	−2	−1	*3	4	1	3	−5	−1	1	5	−1	*3	−1	1	1	−1
94	1	2	*7	−2	2	*3	1	−2	6	8	1	2	7	4	*3	−2	*11	4	1	−2
95	2	4	2	2	*3	4	2	4	6	2	4	2	4	*3	12	4	4	2	2	8
96	*4	*14	*2	*6	*4	*2	*2	*2	*8	*2	*2	*2	*36	*2	*2	*2	*4	*2	*2	*2
97	−1	−1	*3	−1	1	1	−6	−1	4	−1	−1	*3	−11	1	1	−1	−1	1	5	−1
98	*7	*3	−2	1	2	2	−2	4	6	1	*3	1	2	2	2	4	2	2	−4	*3
99	*3	1	2	8	5	1	2	8	6	*3	2	2	8	1	1	12	1	2	*3	1
100	*10	*10	*10	*20	*10	*10	*10	*20	*30	*10	*10	*20	*10	*10	*10	*40	*10	*30	*10	*20

Table C1 continued

j \ i	20	21	22	23	24	25	26	27	28	29	30	31	32	33	34	35	36	37	38	39
1	2	3	1	2	*49	−4	*51	2	−1	1	−1	1	8	−1	1	*3	−20	−1	2	1
2	4	1	2	7	−8	*3	−10	4	2	3	2	4	−8	2	*9	4	6	2	−2	1
3	1	2	1	*7	*3	1	4	12	1	1	8	2	1	*3	8	4	4	*23	3	2
4	*2	*2	*2	*48	*2	*2	*2	*52	*2	*22	*2	*4	*6	*2	*2	*4	*2	*2	*2	*8
5	4	2	*21	−16	−4	6	−8	−8	4	2	2	*3	2	6	2	2	2	4	4	4
6	4	*9	1	4	2	2	1	8	2	−2	*3	4	2	16	2	2	12	−12	2	*3
7	*3	*7	3	4	2	2	4	1	4	*3	4	2	2	1	1	4	1	2	*9	1
8	*2	*2	*4	*2	*2	*2	*4	*2	*6	*2	*8	*2	*2	*2	*4	*2	*2	*6	*4	*2
9	*7	2	*47	−1	2	−2	−1	*3	*53	−1	2	−1	−1	1	1	*11	*3	−1	−2	1
10	8	2	−8	8	8	6	*3	2	−8	4	4	2	8	2	4	*3	*19	4	4	4
11	1	1	8	1	1	*9	2	3	8	2	1	4	*13	4	*3	1	2	2	14	1
12	*2	*8	*2	*34	*6	*4	*2	*2	*2	*4	*2	*2	*2	*12	*2	*2	*2	*8	*2	*2
13	2	−1	−3	*3	1	1	2	−3	−2	7	1	1	*3	−1	−1	1	−1	1	2	2
14	2	4	*3	−4	2	2	1	2	4	12	8	*3	1	−4	2	6	2	2	5	2
15	4	*3	2	2	8	2	2	4	2	8	*3	4	2	8	2	4	8	2	4	*3
16	*12	*46	*2	*2	*4	*2	*2	*2	*16	*54	*2	*2	*4	*2	*2	*2	*4	*2	*6	*2
17	−1	−2	1	1	−1	1	−1	2	*3	−3	1	1	−1	1	2	−1	−1	*3	1	−1
18	2	6	−2	2	4	1	4	*3	2	3	2	1	2	4	−2	1	*3	*13	2	2
19	8	2	2	2	12	2	*3	1	1	8	1	1	4	1	6	*3	4	9	8	1
20	*2	*2	*2	*4	*22	*6	*2	*4	*2	*2	*2	*4	*2	*2	*6	*8	*2	*2	*2	*28
21	3	2	−1	1	*3	−1	−1	1	2	1	6	−5	−3	*9	1	1	2	*61	−1	1
22	4	−2	2	*3	2	−4	4	2	4	6	1	4	*3	2	6	2	1	−10	*7	−4
23	*17	2	*3	2	1	*29	2	2	2	2	3	*3	2	1	4	2	1	4	1	1
24	*2	*6	*4	*2	*2	*2	*8	*2	*2	*2	*12	*2	*2	*2	*4	*2	*2	*14	*4	*6
25	*45	*5	*5	*5	*5	*5	*5	*5	*5	*15	*55	*25	*5	*5	*5	*5	*5	*5	*15	*5

Table C1 continued

j \ i	20	21	22	23	24	25	26	27	28	29	30	31	32	33	34	35	36	37	38	39
26	−14	1	4	1	−2	2	−4	6	*3	4	−16	2	−2	1	2	8	*7	*9	2	2
27	5	2	2	2	2	*19	2	*3	4	1	4	2	2	2	2	1	*3	1	2	8
28	*26	*4	*2	*2	*2	*4	*6	*2	*2	*4	*2	*2	*2	*16	*2	*42	*2	*4	*2	*2
29	−7	−1	1	2	3	*3	1	−1	2	−2	−4	10	−3	−1	*3	−1	1	2	1	−1
30	4	4	6	−4	*9	4	2	8	2	−4	8	−4	4	*3	*7	−4	*11	−4	2	4
31	10	1	2	*3	4	1	2	1	6	2	2	2	*3	1	2	8	1	4	2	1
32	*4	*2	*6	*2	*8	*2	*2	*2	*4	*2	*2	*6	*4	*14	*2	*2	*4	*2	*2	*2
33	1	*3	6	−1	1	4	−1	1	−1	1	*3	−2	−2	2	−1	−1	2	−1	−1	*3
34	*3	8	−2	2	8	2	2	1	−4	*3	4	5	*7	1	−8	4	4	1	*3	4
35	8	4	4	10	2	*13	12	2	*9	2	6	8	2	4	4	4	2	*3	4	2
36	*2	*2	*2	*4	*2	*2	*2	*12	*2	*2	*2	*56	*2	*2	*2	*4	*6	*2	*2	*4
37	2	−1	−1	2	−1	1	*3	2	−1	2	−1	−9	2	1	1	*3	−1	2	1	1
38	1	−2	2	4	2	*3	3	*37	4	4	*7	6	5	−2	*3	−4	2	4	2	2
39	1	12	1	1	*3	1	4	8	2	1	2	2	2	*3	2	1	2	3	2	4
40	*2	*2	*8	*6	*2	*2	*4	*2	*2	*14	*4	*2	*18	*2	*4	*2	*2	*2	*16	*2
41	2	−1	*3	−1	−1	*11	1	−1	1	−6	−1	*3	1	−4	2	−1	1	2	1	3
42	−2	*3	2	1	4	4	−4	4	*7	1	*39	1	−2	2	6	4	2	1	4	*3
43	*3	3	1	4	2	3	2	2	1	*3	4	4	8	1	4	2	1	2	*3	1
44	*2	*4	*2	*2	*2	*4	*2	*14	*6	*8	*2	*2	*2	*4	*2	*2	*2	*12	*62	*2
45	2	4	2	2	2	−4	*23	*3	2	4	4	−4	4	2	−4	−4	*27	2	−4	2
46	4	−4	1	4	1	2	*21	−2	1	2	1	*11	10	4	1	*3	1	2	6	−2
47	2	2	4	1	1	*3	1	2	4	2	6	2	8	1	*3	1	2	4	1	1
48	*32	*2	*2	*2	*12	*14	*2	*2	*4	*2	*2	*2	*4	*6	*2	*2	*8	*2	*2	*2
49	1	1	−4	*9	1	−1	1	−1	4	1	−1	1	*57	−2	−1	*13	−2	1	1	1
50	*5	*5	*15	*5	*35	*5	*5	*5	*5	*5	*5	*15	*5	*5	*5	*5	*5	*25	*5	*5

Table C1 continued

j \ i	20	21	22	23	24	25	26	27	28	29	30	31	32	33	34	35	36	37	38	39
51	2	*3	7	1	8	1	2	8	1	2	*3	2	5	2	4	2	2	*11	1	*3
52	*6	*2	*2	*28	*2	*2	*2	*8	*2	*6	*2	*4	*2	*2	*2	*4	*2	*2	*6	*4
53	-1	-5	1	2	1	2	1	-1	*3	-1	1	1	-5	1	1	2	-2	*3	-1	1
54	2	2	*7	2	4	-6	1	*9	1	2	6	2	1	4	2	-8	*3	-2	2	2
55	12	2	4	4	2	4	*3	12	2	4	4	2	8	4	2	*3	4	2	4	12
56	*2	*14	*4	*2	*2	*6	*4	*2	*2	*2	*4	*2	*2	*2	*24	*2	*2	*2	*4	*2
57	*11	1	-2	-1	*3	-3	-1	1	-3	-1	1	2	-1	*3	-1	-1	2	*17	2	3
58	*7	2	-4	*3	-2	1	2	6	-2	4	2	3	*3	4	4	2	2	1	2	3
59	2	4	*3	4	3	2	3	2	2	2	4	*9	1	1	4	1	1	4	6	2
60	*2	*12	*2	*2	*2	*16	*2	*2	*2	*4	*6	*2	*2	*4	*2	*2	*2	*4	*2	*6
61	*3	-1	2	1	1	-2	1	1	-1	*3	-1	-4	3	-1	-1	1	1	-1	*3	-2
62	1	4	4	-10	1	4	*11	-2	*3	6	3	4	4	*41	4	-4	1	*3	1	6
63	1	4	6	2	2	2	1	*3	2	1	2	3	2	8	1	2	*3	2	1	8
64	*4	*2	*2	*2	*4	*2	*6	*2	*4	*2	*2	*2	*8	*58	*2	*18	*4	*2	*2	*2
65	2	-2	2	2	-4	*3	-4	2	2	-2	-2	2	-2	-2	*3	2	-2	2	-2	4
66	2	*19	2	*13	*3	1	6	2	2	1	4	3	4	*3	-2	1	4	2	-2	10
67	4	2	1	*3	7	4	4	1	2	4	1	1	*33	4	1	4	6	1	2	5
68	*2	*2	*18	*8	*2	*2	*2	*4	*2	*2	*2	*12	*2	*2	*34	*4	*2	*2	*2	*8
69	-1	*3	-1	1	1	1	4	2	1	-1	*3	-1	1	1	-1	3	1	-1	-2	*63
70	*3	12	2	4	4	-4	4	-4	4	*3	2	-4	4	-4	2	8	6	-8	*3	-20
71	2	2	2	1	2	2	1	2	*3	3	6	4	1	1	4	1	1	*3	*7	*19
72	*2	*2	*4	*2	*2	*2	*4	*6	*2	*2	*32	*2	*2	*2	*4	*2	*6	*2	*44	*2
73	1	-2	-1	2	-1	1	*9	1	*13	1	1	3	1	-1	1	*3	-1	*7	3	-6
74	-4	1	2	1	-2	*3	4	4	2	1	-4	*23	-2	2	*3	3	2	4	-4	1
75	*5	*5	*5	*5	*15	*5	*5	*5	*5	*5	*5	*5	*5	*15	*5	*5	*35	*5	*5	*5

Table C1 continued

j \ i	20	21	22	23	24	25	26	27	28	29	30	31	32	33	34	35	36	37	38	39
76	*2	*8	*2	*6	*2	*4	*2	*2	*2	*4	*2	*2	*6	*4	*2	*2	*2	*8	*2	*2
77	1	3	*3	−1	−1	1	−3	−3	2	−2	−2	*3	−2	1	4	*7	−1	1	−3	−2
78	1	*33	4	−4	4	2	6	2	5	2	*9	2	2	2	2	−2	2	4	2	*3
79	*3	1	2	8	2	1	4	2	1	*3	1	*17	2	10	*7	6	14	1	*3	2
80	*4	*2	*2	*2	*4	*2	*2	*2	*24	*2	*2	*2	*4	*26	*2	*2	*4	*6	*2	*2
81	−5	1	−1	−1	1	−2	1	*3	1	3	8	−5	−6	*7	*59	−1	*3	1	−1	3
82	2	1	2	2	−8	1	*3	2	2	4	2	2	2	2	−6	*3	8	10	2	2
83	1	2	2	1	2	*3	1	*11	*31	2	1	2	*7	2	*9	1	2	4	8	4
84	*2	*2	*2	*4	*6	*2	*2	*4	*2	*2	*2	*4	*2	*6	*2	*16	*2	*2	*2	*4
85	2	2	−2	*3	2	2	2	−2	−2	2	−2	*7	*3	−2	−4	10	2	−2	4	−2
86	2	−2	*3	2	4	2	2	4	4	−10	1	*3	2	−2	4	2	2	2	2	−4
87	1	*27	1	4	2	2	1	2	1	2	*21	1	6	4	2	2	2	2	*13	*3
88	*6	*2	*4	*2	*2	*2	*8	*2	*38	*6	*4	*2	*2	*22	*4	*2	*2	*2	*36	*2
89	−3	1	4	−1	1	3	−1	−1	*3	*7	−1	1	2	−1	1	−2	2	*3	−3	−1
90	4	4	−4	2	8	4	−4	*3	*17	4	4	4	4	4	−4	6	*3	2	−4	8
91	4	2	2	2	2	1	*3	1	*7	2	2	1	2	3	1	*3	1	4	4	2
92	*2	*4	*2	*2	*2	*36	*2	*2	*2	*4	*2	*2	*2	*8	*6	*2	*2	*4	*2	*2
93	2	2	−1	−1	*3	−1	−1	*7	1	6	1	1	−2	*3	1	−1	1	−1	2	*11
94	2	2	2	*3	2	2	2	2	3	2	4	−2	*3	6	1	6	1	4	8	−2
95	2	2	*3	2	6	4	*7	4	4	2	2	*3	2	4	4	14	2	8	4	4
96	*4	*6	*2	*2	*8	*2	*2	*2	*4	*2	*6	*2	*4	*2	*2	*2	*4	*2	*2	*6
97	*3	*13	−1	2	1	*7	2	−1	−1	*9	1	1	2	1	2	2	−1	−1	*3	5
98	2	2	6	2	4	4	2	1	*3	1	−2	4	8	1	4	10	*43	*3	−6	1
99	3	2	*11	5	*7	14	1	*3	2	1	4	4	1	8	1	10	*3	2	2	8
100	*10	*10	*10	*20	*50	*10	*30	*20	*10	*10	*10	*40	*10	*10	*10	*60	*10	*10	*10	*20

Table C1 continued

$j \backslash i$	40	41	42	43	44	45	46	47	48	49	50	51	52	53	54	55	56	57	58	59
1	−3	1	−1	2	*3	1	1	1	−1	*13	1	−1	1	*3	1	−1	1	−1	−1	6
2	4	2	2	*3	4	1	4	7	*49	8	8	3	*51	6	8	4	2	1	2	2
3	1	2	*3	2	4	4	1	1	2	1	1	*27	*11	3	4	3	2	2	2	3
4	*2	*6	*2	*4	*2	*2	*2	*28	*2	*2	*6	*4	*2	*2	*2	*8	*2	*2	*2	*12
5	*3	2	*29	8	2	−4	2	4	*31	*3	4	4	2	−8	2	2	4	2	*3	2
6	1	−6	2	−4	1	2	*7	−12	*3	2	1	4	4	2	4	2	1	*3	1	−4
7	1	*37	2	2	4	1	8	*3	16	6	2	5	4	4	1	1	*3	2	1	8
8	*2	*2	*4	*2	*2	*14	*48	*2	*2	*2	*4	*2	*2	*2	*52	*18	*2	*2	*44	*2
9	11	1	2	1	−9	*3	1	−2	2	−1	−1	4	−1	−1	*3	1	1	2	4	1
10	4	4	−4	2	*21	4	−4	4	16	6	4	4	−8	*3	−4	4	8	2	4	8
11	12	1	1	*3	4	6	4	4	2	2	3	2	*3	2	2	12	10	5	4	4
12	*2	*4	*18	*14	*2	*4	*2	*2	*2	*4	*2	*6	*2	*8	*2	*2	*2	*4	*2	*2
13	−1	*3	1	1	1	−1	1	1	−1	*17	*3	−1	−2	4	−1	8	3	−8	−1	*9
14	*3	*11	*7	2	1	−4	4	−2	2	*3	2	−2	4	4	1	2	8	2	*3	−2
15	4	2	12	2	2	8	4	8	*3	2	4	8	4	2	*19	2	10	*3	2	*13
16	*4	*14	*2	*2	*8	*2	*2	*6	*4	*2	*2	*2	*4	*2	*2	*2	*12	*2	*2	*2
17	4	1	−1	1	1	−1	*9	2	−1	2	2	2	4	−2	−7	*3	−2	−1	4	2
18	*7	2	4	4	*47	*3	−2	4	4	1	−4	2	2	1	*3	2	*53	2	−2	6
19	1	2	1	4	*3	1	2	*11	2	1	4	1	8	*3	1	1	12	4	*23	2
20	*2	*2	*2	*12	*2	*2	*2	*4	*2	*2	*2	*32	*6	*2	*2	*4	*2	*2	*2	*4
21	−1	−2	*3	10	−1	2	−1	−1	1	2	−1	*3	1	−2	4	−9	6	1	−3	1
22	1	*3	1	2	8	4	5	2	3	2	*9	−4	2	2	1	2	8	−6	2	*3
23	*3	4	4	8	1	1	12	1	4	*3	1	2	2	1	4	4	9	2	*3	1
24	*2	*2	*8	*2	*2	*2	*68	*2	*6	*2	*4	*2	*2	*22	*4	*2	*2	*6	*8	*2
25	*5	*5	*65	*5	*5	*5	*5	*15	*5	*5	*5	*5	*5	*5	*5	*5	*75	*5	*5	*5

Table C1 continued

j \ i	40	41	42	43	44	45	46	47	48	49	50	51	52	53	54	55	56	57	58	59
26	4	1	−8	4	−2	2	*3	2	2	2	4	2	4	1	−4	*3	−28	2	2	1
27	1	1	6	1	4	*3	2	2	4	6	2	2	1	6	*9	5	4	4	7	5
28	*2	*4	*2	*2	*6	*4	*2	*2	*2	*8	*2	*2	*2	*12	*2	*2	*2	*4	*2	*2
29	2	−1	−7	*3	1	1	1	−3	1	2	1	1	*3	*73	−2	6	−12	−2	2	*77
30	4	4	*3	−4	2	4	2	4	16	8	2	*3	2	−8	4	4	10	4	4	−12
31	2	*9	1	2	4	2	2	4	1	1	*3	6	1	12	3	1	10	4	*7	*3
32	*24	*2	*46	*38	*4	*2	*2	*26	*4	*6	*2	*2	*4	*2	*2	*2	*16	*2	*54	*2
33	−2	−1	2	1	2	1	2	−1	*3	−3	1	2	−1	−3	1	1	1	*21	1	−4
34	2	4	−4	10	6	9	2	*3	2	3	2	4	−4	4	8	1	*3	14	−2	4
35	4	2	*11	*17	2	2	*3	2	2	8	4	8	4	4	2	*3	*7	4	4	2
36	*2	*2	*2	*4	*2	*18	*2	*8	*2	*2	*2	*4	*2	*2	*6	*4	*2	*2	*2	*4
37	1	2	1	−1	*3	−2	−1	1	1	−1	4	1	−1	*3	−1	*7	3	−1	−2	1
38	4	−2	6	*3	2	−2	2	8	4	2	2	4	*3	4	1	4	1	6	4	6
39	2	7	*3	1	4	4	1	2	2	12	7	*3	*13	2	*7	2	1	2	1	1
40	*2	*6	*4	*2	*2	*2	*4	*2	*22	*2	*12	*2	*2	*2	*8	*2	*2	*2	*4	*6
41	*3	−2	−1	1	−5	1	12	1	3	*9	*71	2	5	*7	−1	1	−1	−3	*3	−2
42	2	2	4	2	−2	2	2	1	*3	2	−12	2	−2	1	2	6	4	*3	4	1
43	2	2	1	2	2	4	1	*3	10	1	*41	2	*7	4	1	2	*3	7	1	8
44	*2	*4	*2	*2	*2	*8	*6	*2	*2	*4	*2	*2	*2	*4	*2	*6	*2	*4	*2	*2
45	−4	2	2	12	2	*3	2	−4	4	2	−4	*7	−4	2	*33	−4	2	2	2	4
46	*17	2	2	−4	*3	4	2	4	1	2	*29	2	2	*9	2	2	2	*13	2	2
47	4	4	2	*3	1	1	2	2	2	4	*7	1	*3	1	2	*43	1	2	2	8
48	*4	*2	*6	*2	*4	*2	*2	*2	*4	*2	*2	*6	*8	*2	*2	*2	*4	*2	*2	*2
49	−1	*3	1	−1	1	−1	−3	3	2	*7	*3	1	8	1	−1	1	2	−1	−1	*3
50	*45	*5	*5	*5	*5	*5	*5	*5	*5	*15	*5	*5	*5	*5	*5	*5	*5	*5	*15	*5

Table C1 continued

j \ i	40	41	42	43	44	45	46	47	48	49	50	51	52	53	54	55	56	57	58	59
51	1	6	16	2	1	4	3	1	*21	1	1	4	2	1	4	4	1	*9	1	14
52	*2	*2	*2	*16	*2	*2	*2	*12	*2	*2	*2	*4	*2	*2	*2	*4	*6	*2	*2	*8
53	4	−1	−1	5	2	−2	*3	*7	3	4	1	−1	2	−6	2	*3	−1	1	3	−1
54	1	2	2	16	2	*3	2	−4	2	−18	*19	2	2	−2	*3	−4	4	4	1	−12
55	2	4	4	12	*9	10	*7	12	2	2	4	2	6	*3	2	4	8	2	2	4
56	*26	*2	*4	*66	*2	*2	*4	*2	*2	*2	*8	*2	*6	*2	*4	*2	*2	*2	*4	*2
57	−1	−1	*3	−5	−1	*7	−1	5	3	−1	2	*3	1	1	2	−1	−1	2	−1	2
58	−2	*3	4	5	2	2	8	4	2	2	*3	1	2	4	−4	2	4	1	−4	*3
59	*3	3	1	10	*7	4	2	13	2	*3	1	4	4	4	2	4	1	2	*3	2
60	*2	*8	*2	*2	*2	*4	*2	*2	*18	*4	*2	*2	*2	*4	*2	*2	*2	*24	*2	*2
61	1	8	−1	*7	1	−1	1	*69	−1	*11	2	−2	−3	1	1	1	*3	1	−1	1
62	2	2	1	6	2	2	*3	−22	4	2	1	−4	2	4	1	*3	2	2	2	2
63	8	2	*7	1	1	*39	1	4	2	2	6	2	6	2	*3	1	4	4	4	4
64	*4	*2	*2	*2	*12	*2	*2	*2	*16	*2	*2	*2	*4	*6	*2	*2	*4	*2	*2	*2
65	6	*7	−2	*3	2	2	2	−6	10	2	−2	−2	*9	−4	−2	4	2	−2	8	−2
66	2	1	*3	2	4	2	−2	1	2	6	4	*3	−12	1	2	*11	4	*31	2	2
67	*7	*3	4	6	4	1	2	8	10	1	*3	1	2	2	4	2	2	4	1	*3
68	*6	*2	*2	*4	*2	*2	*2	*4	*2	*6	*2	*4	*2	*2	*2	*8	*2	*2	*6	*4
69	2	1	1	2	−2	1	8	1	*3	−1	2	1	5	6	1	−1	−1	*3	−1	1
70	4	4	8	4	8	−4	6	*3	2	12	*13	4	4	4	2	−4	*9	−4	2	4
71	4	12	5	4	12	2	*3	6	1	8	10	1	8	4	1	*3	2	2	4	2
72	*2	*2	*4	*2	*2	*6	*8	*2	*2	*2	*4	*2	*2	*2	*12	*2	*2	*2	*4	*2
73	−1	4	−1	−1	*3	−2	−1	2	1	−1	6	1	−7	*3	−2	−1	2	1	1	4
74	4	1	−4	*27	−2	1	8	4	−2	2	2	2	*3	1	8	2	−2	5	4	14
75	*5	*5	*15	*25	*5	*5	*5	*5	*5	*5	*5	*15	*5	*5	*5	*5	*5	*5	*5	*5

Table C1 continued

$j \backslash i$	40	41	42	43	44	45	46	47	48	49	50	51	52	53	54	55	56	57	58	59
76	*2	*12	*2	*2	*2	*4	*2	*2	*2	*4	*6	*2	*2	*16	*74	*2	*2	*76	*2	*6
77	*3	−1	2	1	*11	1	1	−4	−1	*3	−1	1	1	1	−3	*13	1	−10	*3	1
78	1	2	4	14	1	4	1	−6	*3	2	1	6	4	4	8	−6	2	*3	9	*7
79	1	4	6	2	2	14	1	*9	4	2	2	1	1	8	1	2	*3	1	1	4
80	*4	*2	*2	*2	*8	*2	*6	*2	*4	*2	*2	*2	*4	*2	*2	*6	*4	*34	*14	*2
81	4	−2	3	2	−3	*3	1	1	1	4	−3	4	−3	−1	*3	−1	2	2	−1	−1
82	−4	4	−2	2	*3	10	−2	2	2	6	*11	11	2	*3	−2	1	2	*7	−4	2
83	2	4	3	*3	1	1	4	1	2	4	4	8	*3	6	1	2	3	1	4	4
84	*2	*2	*6	*4	*2	*2	*2	*4	*2	*2	*2	*72	*2	*2	*2	*4	*14	*2	*2	*4
85	2	*3	−2	−2	4	2	−2	4	−2	−2	*3	−20	2	2	−2	−2	6	−4	2	*3
86	*3	4	3	4	1	−10	4	2	6	*3	1	8	2	−2	2	*7	1	2	*3	8
87	2	2	2	8	4	4	2	1	*3	1	1	8	2	1	4	2	*11	*3	*29	1
88	*2	*2	*8	*2	*2	*2	*4	*6	*2	*2	*4	*2	*2	*2	*28	*2	*6	*2	*16	*2
89	2	1	−1	4	*67	−2	*3	−1	−5	1	3	−1	2	−2	1	*9	−1	1	2	2
90	4	2	8	2	−8	*3	4	2	8	2	−4	8	*23	*7	*3	4	4	4	4	2
91	1	4	2	1	*3	5	1	2	4	4	2	2	4	*3	*17	1	4	1	2	6
92	*2	*4	*2	*6	*2	*4	*2	*2	*2	*8	*2	*2	*42	*4	*2	*2	*2	*4	*2	*2
93	−1	3	*9	1	−3	1	*19	−1	2	−1	1	*3	1	−1	1	2	−1	1	1	−2
94	6	*3	6	*13	4	−4	1	4	1	6	*3	*7	1	4	2	−2	4	2	2	*9
95	*3	2	2	4	4	2	4	4	8	*3	6	2	4	4	4	12	4	8	*3	4
96	*64	*2	*2	*2	*4	*2	*2	*2	*12	*2	*14	*2	*4	*2	*2	*2	*8	*6	*2	*2
97	−10	1	−1	−1	1	−3	2	*3	1	1	1	−1	−3	2	3	−2	*3	2	−1	1
98	6	1	2	2	−4	*11	*9	1	4	*7	−18	2	2	1	−2	*3	4	6	6	1
99	1	8	2	2	2	*3	6	1	16	1	3	2	2	1	*3	2	4	6	2	2
100	*10	*10	*10	*20	*30	*10	*10	*40	*70	*50	*10	*20	*10	*30	*10	*20	*10	*10	*10	*20

Table C1 continued

j \ i	60	61	62	63	64	65	66	67	68	69	70	71	72	73	74	75	76	77	78	79
1	4	−1	*3	−1	−12	2	6	−1	1	3	−1	*3	1	*7	1	−2	3	2	−2	−1
2	−4	*3	10	2	8	1	−2	2	2	4	*3	2	−4	2	−2	*11	4	1	2	*3
3	*3	4	1	4	2	2	4	1	1	*3	2	1	*49	2	2	4	1	1	*51	4
4	*2	*2	*2	*4	*2	*2	*2	*4	*18	*2	*2	*8	*2	*2	*2	*4	*2	*6	*2	*4
5	2	4	−4	−4	4	2	2	*3	4	2	2	*7	4	2	2	4	*39	10	2	4
6	16	2	2	2	1	−2	*3	4	12	2	12	4	4	−4	*23	*3	1	−2	2	2
7	1	2	2	12	2	*3	1	4	6	1	*7	12	1	1	*3	1	1	4	2	1
8	*2	*2	*8	*2	*6	*2	*4	*2	*2	*2	*4	*2	*2	*6	*4	*2	*2	*2	*8	*2
9	1	−2	3	*3	−8	1	1	−1	9	*7	1	−1	*9	−1	1	1	1	−6	2	1
10	−8	8	*3	2	8	8	−4	4	8	2	−4	*3	4	12	8	14	−8	4	4	4
11	1	*3	1	3	4	4	2	2	*7	1	*3	2	1	6	1	16	4	6	2	*3
12	*6	*4	*2	*2	*2	*4	*2	*2	*2	*48	*2	*2	*2	*4	*2	*2	*2	*4	*6	*2
13	1	−5	2	1	*11	2	2	*7	*3	1	−1	1	−1	1	2	5	1	*3	2	−2
14	8	2	2	4	6	2	1	*3	1	2	12	−2	1	4	2	*17	*9	4	1	2
15	4	2	4	4	2	6	*21	4	4	4	4	2	8	16	6	*3	2	2	4	2
16	*8	*2	*2	*2	*4	*6	*2	*2	*4	*2	*2	*2	*4	*2	*6	*2	*8	*2	*2	*2
17	1	1	−1	−1	*3	*7	−2	1	−2	−1	1	3	1	*3	−1	−1	1	−1	−5	4
18	4	4	−2	*9	−4	1	2	1	4	4	*11	1	*3	1	−2	4	−4	2	2	2
19	2	2	*3	4	*7	4	1	1	6	4	3	*3	1	6	4	2	2	8	2	7
20	*2	*6	*2	*4	*2	*2	*2	*8	*2	*2	*6	*4	*38	*2	*2	*4	*2	*2	*2	*12
21	*3	−1	−1	*7	−1	−1	1	2	1	*3	2	−1	10	−1	−2	2	−1	3	*3	*89
22	1	−2	4	−4	*13	2	8	8	*3	−2	1	2	2	2	2	−8	2	*3	3	−8
23	2	8	*7	1	6	2	6	*9	1	4	2	2	4	2	6	1	*33	1	1	8
24	*2	*2	*4	*2	*2	*2	*12	*82	*2	*2	*4	*2	*2	*2	*16	*6	*2	*2	*4	*2
25	*5	*35	*5	*5	*5	*15	*5	*5	*5	*5	*5	*5	*85	*5	*15	*5	*5	*5	*5	*5

Table C1 continued

j \ i	60	61	62	63	64	65	66	67	68	69	70	71	72	73	74	75	76	77	78	79
26	6	2	2	1	*3	2	−16	8	−2	1	2	6	−18	*3	16	2	4	3	4	4
27	*7	2	2	*3	3	2	*47	*31	1	6	1	1	*3	2	2	4	22	1	2	1
28	*2	*4	*6	*2	*2	*8	*2	*58	*2	*4	*2	*18	*2	*4	*2	*2	*2	*4	*2	*2
29	−1	*3	−1	−1	1	−1	1	1	−1	*13	*3	−1	−5	2	2	−1	1	1	−1	*3
30	*3	−4	4	4	2	−4	8	−4	2	*3	4	4	8	−20	2	4	4	−4	*3	−8
31	2	1	4	2	2	4	18	2	*3	2	4	4	4	5	2	4	2	*3	2	4
32	*4	*2	*2	*2	*4	*2	*2	*6	*4	*2	*2	*2	*8	*2	*2	*2	*12	*2	*2	*2
33	1	−3	1	1	1	1	*3	−1	−1	1	−2	1	1	−1	−1	*9	−2	2	8	−1
34	4	1	2	3	4	*33	2	4	12	1	−2	4	6	2	*3	1	2	2	−2	1
35	4	8	4	4	*3	2	2	4	6	4	8	2	6	*3	2	4	4	8	2	*23
36	*2	*2	*2	*24	*2	*2	*2	*4	*2	*34	*2	*4	*6	*2	*26	*4	*2	*2	*2	*16
37	−1	*19	*9	−1	−2	1	−3	−1	2	1	1	*3	−1	2	4	−3	1	1	−4	−1
38	1	*3	3	2	4	2	1	2	2	−2	*3	2	4	2	1	−4	8	−4	1	*63
39	*3	2	10	4	2	2	6	2	4	*3	1	*11	4	10	10	4	3	2	*3	4
40	*2	*2	*4	*2	*2	*2	*4	*2	*6	*2	*8	*2	*2	*2	*4	*2	*2	*6	*28	*2
41	1	2	*79	2	2	1	−2	*3	−1	1	1	−2	−2	1	3	−1	*3	−1	−1	1
42	8	2	−8	4	−2	1	*9	1	2	4	16	1	12	1	*61	*3	−2	*7	2	*19
43	1	1	6	1	2	*3	4	2	4	2	1	2	1	2	*3	2	3	12	4	*13
44	*2	*32	*2	*2	*6	*4	*2	*2	*2	*4	*2	*2	*2	*12	*2	*2	*14	*88	*2	*2
45	4	−4	−4	*3	−4	4	2	4	*37	10	−4	2	*3	−4	2	2	4	−12	2	2
46	1	8	*3	2	2	2	1	−2	4	8	2	*3	1	2	4	*7	1	10	3	−4
47	1	*3	1	4	8	1	*17	4	2	1	*9	2	1	4	2	1	2	2	8	*3
18	*12	*2	*2	*46	*4	*2	*2	*2	*8	*6	*2	*2	*4	*2	*14	*2	*4	*2	*6	*2
49	1	2	1	1	−1	6	−4	−2	*3	−5	10	1	3	−1	8	−1	−1	*3	1	−1
50	*55	*5	*25	*5	*5	*5	*5	*15	*5	*5	*5	*5	*5	*35	*5	*5	*15	*5	*5	*5

Table C1 continued

j \ i	60	61	62	63	64	65	66	67	68	69	70	71	72	73	74	75	76	77	78	79
51	4	7	4	4	1	1	*3	6	4	4	8	1	2	1	1	*3	2	2	4	1
52	*2	*2	*2	*4	*2	*6	*2	*4	*2	*2	*2	*4	*14	*2	*18	*8	*2	*2	*2	*4
53	−3	6	*13	−1	*3	−1	−1	1	2	−8	3	1	−1	*3	6	2	1	−3	−1	2
54	4	−4	2	*3	2	−4	2	2	10	4	17	*7	*3	−2	1	2	2	−2	8	12
55	4	2	*3	8	2	8	*11	4	4	4	12	*3	2	2	8	2	6	8	2	4
56	*2	*18	*4	*2	*2	*2	*16	*2	*2	*2	*84	*2	*2	*2	*4	*2	*2	*2	*4	*6
57	*3	5	−1	2	1	3	2	−8	−1	*3	−21	2	2	1	−1	4	2	−1	*9	4
58	12	1	4	*17	−2	6	2	2	*3	*7	6	2	2	2	4	1	6	*3	−12	2
59	2	2	2	1	2	18	1	*3	*19	1	8	1	12	16	3	1	*3	1	2	4
60	*2	*4	*2	*2	*2	*4	*6	*26	*14	*4	*2	*2	*22	*8	*2	*6	*2	*4	*2	*2
61	2	−2	1	−1	2	*81	−1	−1	1	−1	1	4	−2	2	*3	−1	1	4	5	1
62	2	12	2	−2	*3	−16	1	*7	2	*59	8	−2	3	*3	4	2	2	−4	5	2
63	4	1	1	*3	2	5	2	1	1	4	16	4	*3	2	2	8	4	2	2	1
64	*4	*2	*6	*2	*8	*2	*14	*2	*4	*2	*2	*6	*4	*2	*2	*2	*4	*2	*2	*2
65	−2	*3	2	2	2	−8	2	4	−2	2	*3	−2	−2	2	−18	4	12	−2	*11	*3
66	*3	1	−4	6	−4	*7	8	6	−4	*9	−2	1	4	2	−2	12	14	8	*3	16
67	1	2	4	1	2	8	2	2	*3	1	2	2	*13	2	4	12	4	*3	1	4
68	*2	*2	*2	*4	*14	*2	*2	*12	*2	*2	*2	*32	*2	*2	*2	*4	*6	*2	*2	*4
69	*17	1	−1	2	−1	−1	*3	1	−1	2	−1	1	1	−1	2	*87	−1	2	2	−2
70	2	−4	8	*7	2	*3	12	−12	4	−8	8	4	2	4	*3	−20	4	8	6	−4
71	10	*11	1	2	*3	3	8	8	1	1	2	2	2	*9	6	10	6	2	4	2
72	*2	*2	*56	*6	*2	*2	*4	*2	*2	*2	*4	*2	*6	*2	*4	*2	*2	*2	*8	*2
73	−1	−1	*3	−1	−1	2	−1	2	2	1	1	*3	3	−2	6	−9	−3	1	−9	2
74	−2	*21	−8	1	4	10	4	8	2	2	*3	2	−2	6	−4	2	10	*13	4	*3
75	*45	*5	*5	*5	*5	*5	*5	*5	*25	*15	*5	*5	*5	*5	*5	*5	*5	*5	*15	*5

Table C1 continued

j \ i	60	61	62	63	64	65	66	67	68	69	70	71	72	73	74	75	76	77	78	79
76	*14	*4	*2	*2	*2	*4	*2	*22	*6	*8	*2	*2	*2	*4	*2	*2	*2	*36	*2	*2
77	1	2	−1	1	2	−1	1	*3	*23	−1	4	−1	1	1	−1	−1	*3	2	−1	1
78	10	−16	2	2	2	4	*3	−2	2	2	1	−4	2	8	1	*3	2	2	4	−2
79	1	2	8	1	4	*3	1	5	2	2	1	2	6	2	*3	4	2	8	3	2
80	*8	*2	*2	*2	*36	*2	*2	*2	*4	*2	*2	*2	*4	*6	*2	*2	*16	*2	*2	*2
81	11	1	1	*3	−5	−1	6	−1	1	4	2	1	*3	*11	−3	*19	−1	1	12	1
82	6	2	*3	1	4	2	−8	3	4	1	−2	*3	10	1	8	8	4	4	2	2
83	8	*3	2	2	4	2	8	8	1	1	*3	2	1	4	2	1	4	2	3	*3
84	*78	*2	*2	*4	*2	*2	*2	*8	*2	*6	*2	*4	*2	*2	*2	*4	*2	*2	*6	*4
85	−10	−6	2	−2	−2	2	6	2	*9	2	−4	2	2	2	2	−8	4	*3	2	−2
86	8	2	4	4	16	−4	3	*3	4	2	2	−4	1	2	2	2	*3	16	1	*11
87	10	2	1	2	2	2	*3	4	6	4	2	1	4	6	1	*3	1	2	4	*7
88	*2	*2	*4	*2	*2	*6	*4	*2	*2	*2	*4	*2	*2	*2	*24	*2	*62	*2	*4	*2
89	−1	1	3	−1	*3	1	−1	2	*83	−4	2	2	4	*3	−1	−1	2	−1	*7	1
90	8	2	8	*3	4	2	4	4	−16	4	−4	2	*27	2	4	8	12	4	4	4
91	1	2	*3	4	1	*13	1	3	10	1	2	*3	2	2	8	5	1	*7	18	2
92	*2	*12	*22	*2	*2	*8	*2	*2	*2	*4	*6	*2	*2	*4	*2	*2	*2	*4	*2	*6
93	*3	1	2	1	1	1	2	−1	−4	*3	−4	−1	4	−1	1	1	*7	−1	*3	−1
94	2	2	10	2	4	12	1	2	*3	−20	1	4	2	10	4	−2	3	*3	1	6
95	4	8	2	10	4	2	4	*3	4	6	8	2	2	16	2	*7	*9	2	2	8
96	*4	*2	*2	*2	*4	*2	*6	*2	*4	*2	*2	*2	*8	*86	*2	*6	*4	*2	*2	*2
97	14	−1	1	−1	4	*3	2	1	*11	−3	1	1	−1	−4	*21	1	1	1	−4	1
98	2	4	2	4	*57	7	−4	12	2	1	*13	2	4	*3	2	6	2	2	2	4
99	8	1	1	*9	6	1	8	2	1	2	2	2	*3	*7	1	4	1	2	2	6
100	*10	*10	*30	*80	*10	*10	*10	*20	*10	*10	*10	*60	*10	*10	*50	*20	*10	*10	*10	*40

Table C1 continued

j \ i	80	81	82	83	84	85	86	87	88	89	90	91	92	93	94	95	96	97	98	99
1	*3	−13	1	1	1	−5	4	4	4	*3	−1	7	1	3	2	1	−1	2	*99	−1
2	4	5	2	2	2	4	4	2	*3	1	2	4	2	1	2	1	4	*21	*13	3
3	2	8	2	*19	6	6	6	*3	5	2	4	1	1	4	1	8	*3	2	5	2
4	*2	*2	*2	*4	*2	*2	*6	*16	*2	*2	*2	*4	*2	*2	*2	*12	*98	*2	*2	*4
5	−8	8	2	4	*41	*9	4	4	2	−12	2	2	2	−8	*3	14	−4	2	−12	6
6	1	8	2	6	*3	−2	2	2	12	−8	20	−4	1	*3	1	*7	18	2	1	4
7	4	*11	2	*3	4	2	4	5	1	2	1	2	*3	2	4	12	2	2	8	3
8	*2	*2	*12	*2	*2	*2	*4	*2	*2	*2	*4	*6	*2	*2	*56	*2	*2	*2	*4	*2
9	−1	*3	−1	5	1	1	−1	1	1	3	*3	−1	−1	2	*97	−2	1	2	2	*3
10	*3	22	−4	4	*29	8	8	16	4	*9	−8	2	8	*7	−20	8	*31	6	*3	2
11	1	1	8	7	2	2	2	12	*3	4	1	2	2	1	10	1	10	*3	1	4
12	*2	*52	*2	*2	*2	*8	*2	*66	*2	*4	*2	*2	*14	*4	*2	*2	*6	*4	*2	*2
13	1	6	1	2	1	−1	*3	−3	1	1	−1	2	4	1	−3	*3	−1	1	3	1
14	1	2	*37	−2	2	*3	2	−2	4	2	1	*7	20	4	*3	2	4	2	2	−2
15	4	12	4	2	*3	8	2	8	16	2	8	2	12	*9	4	4	4	4	4	4
16	*4	*2	*2	*6	*4	*2	*2	*2	*4	*2	*14	*2	*96	*2	*2	*2	*4	*2	*2	*2
17	−3	−1	*3	−1	1	2	1	1	1	−2	1	*3	−18	*11	2	3	1	2	−1	1
18	2	*3	2	1	12	3	14	2	2	*7	*3	4	6	2	−4	1	12	6	−2	*3
19	*27	4	1	4	1	6	*13	1	1	*3	2	2	12	1	3	4	1	1	*3	4
20	*2	*2	*2	*8	*2	*2	*2	*4	*42	*2	*2	*4	*2	*2	*2	*4	*2	*18	*2	*8
21	−2	5	−1	−10	2	−1	2	*3	−1	1	8	1	−1	10	−1	−1	*3	−1	2	1
22	4	2	1	8	1	−2	*3	*7	4	2	2	2	8	10	4	*69	2	−10	2	*11
23	10	1	2	4	1	*3	3	4	4	1	8	2	2	3	*3	4	1	4	16	1
24	*2	*2	*4	*2	*18	*2	*28	*2	*2	*2	*8	*2	*2	*6	*4	*2	*2	*2	*4	*2
25	*5	*25	*5	*15	*5	*5	*5	*5	*5	*5	*95	*5	*15	*5	*5	*5	*5	*5	*5	*5

Table C1 continued

j \ i	80	81	82	83	84	85	86	87	88	89	90	91	92	93	94	95	96	97	98	99
26	−2	4	*3	2	6	*7	2	1	2	1	−16	*39	2	7	2	4	−2	2	*17	6
27	2	*3	6	2	*53	1	1	2	4	2	*3	1	1	2	2	2	2	2	2	*3
28	*6	*8	*22	*2	*14	*4	*2	*2	*2	*12	*2	*2	*2	*4	*2	*2	*2	*16	*6	*2
29	2	1	8	−1	−1	1	−1	2	*9	−1	−7	2	5	1	2	−2	−1	*3	−5	−1
30	4	12	2	*7	4	−12	2	*3	2	12	8	4	4	4	16	−4	*3	4	6	4
31	2	2	1	2	1	2	*3	3	1	8	2	2	4	20	5	*3	1	10	4	1
32	*4	*2	*14	*2	*4	*6	*2	*2	*8	*2	*2	*2	*4	*2	*6	*2	*4	*2	*2	*2
33	2	1	−1	2	*3	2	2	2	*11	−1	1	−3	1	*3	−1	−1	*13	−1	−3	4
34	4	*7	2	*3	2	4	2	2	8	4	−6	1	*9	2	−8	4	−24	2	16	1
35	2	2	*3	6	4	4	8	6	16	2	8	*3	10	2	8	2	4	16	16	2
36	*14	*6	*2	*4	*2	*2	*2	*4	*94	*2	*6	*4	*2	*2	*2	*8	*2	*2	*2	*12
37	*3	1	−1	2	2	−1	3	−1	−3	*3	1	−1	1	−1	−1	*17	1	2	*3	3
38	1	4	2	2	3	6	4	8	*3	−4	1	2	2	8	*11	2	2	*3	1	−4
39	1	4	4	6	4	5	2	*3	1	2	4	4	1	4	1	7	*9	1	1	4
40	*2	*2	*4	*2	*2	*2	*24	*2	*2	*2	*4	*2	*2	*2	*4	*6	*2	*2	*4	*2
41	2	1	2	1	13	*3	−1	−1	4	−1	−1	2	−1	−1	*3	2	1	2	4	−1
42	−2	8	−4	4	*3	1	−4	12	−2	4	4	2	−2	*3	−8	2	2	5	4	4
43	4	2	1	*9	1	3	8	2	2	4	1	8	*3	1	4	2	1	1	4	2
44	*2	*4	*6	*2	*2	*4	*2	*2	*2	*4	*2	*6	*2	*8	*2	*2	*2	*4	*2	*2
45	2	*3	−4	−4	2	−12	4	12	4	−4	*3	4	*43	8	2	2	2	6	4	*3
46	*3	−4	4	4	4	4	16	−2	1	*3	1	−4	4	2	1	4	4	2	*3	−2
47	2	1	2	4	3	8	13	1	*3	2	2	4	2	2	4	1	2	*57	10	*7
48	*4	*2	*2	*2	*16	*2	*2	*54	*4	*2	*2	*2	*68	*2	*2	*2	*12	*2	*2	*2
49	5	4	−2	*11	2	1	*93	2	−1	4	−7	3	1	−1	1	*3	−1	−3	*7	−1
50	*5	*5	*5	*5	*65	*15	*5	*25	*5	*5	*5	*5	*5	*5	*15	*5	*5	*5	*5	*5

Table C1 continued

j \ i	80	81	82	83	84	85	86	87	88	89	90	91	92	93	94	95	96	97	98	99
51	6	8	2	2	*3	2	10	2	2	1	4	3	*29	*3	2	3	4	*7	1	4
52	*2	*2	*2	*12	*2	*2	*2	*4	*2	*2	*2	*8	*6	*2	*2	*4	*2	*2	*2	*4
53	−1	1	*3	−1	1	1	−8	−1	2	1	1	*9	1	1	2	−10	*7	1	1	−8
54	1	*3	1	−8	2	4	1	2	12	*11	*3	4	2	2	2	8	4	−2	2	*3
55	*3	4	12	4	4	8	12	4	8	*3	2	6	4	2	4	*7	2	2	*3	8
56	*2	*2	*8	*2	*2	*2	*4	*2	*6	*2	*4	*2	*2	*2	*4	*2	*2	*6	*8	*2
57	3	1	1	2	1	1	1	*3	−2	*13	1	−1	−1	1	*7	1	*3	5	−1	1
58	4	1	2	4	−2	2	*3	2	2	2	4	2	2	1	−4	*9	2	4	4	2
59	11	2	2	2	2	*3	2	2	4	*17	1	4	2	*7	*3	*11	6	2	1	2
60	*2	*4	*2	*2	*6	*4	*2	*2	*2	*16	*2	*2	*2	*12	*2	*2	*2	*4	*2	*2
61	1	−1	1	*3	−1	1	7	−27	−1	2	−4	−1	*21	2	−1	1	−1	1	2	1
62	2	4	*9	−4	7	2	2	4	4	8	6	*3	2	4	4	2	1	2	1	4
63	2	*3	1	3	16	1	1	4	3	1	*3	*7	2	4	1	2	2	2	4	*9
64	*24	*2	*2	*2	*92	*2	*38	*2	*4	*6	*2	*2	*4	*2	*26	*2	*8	*2	*6	*2
65	−2	2	4	2	−14	2	−2	−2	*3	2	*7	4	−4	−2	2	−6	−2	*3	−2	−2
66	−4	2	−6	4	12	1	2	*3	4	1	2	5	8	4	−2	1	*3	6	−2	4
67	4	1	6	2	1	2	*9	2	1	*7	1	2	2	4	3	*3	2	1	16	1
68	*2	*2	*2	*4	*2	*6	*2	*8	*2	*2	*2	*4	*2	*2	*6	*4	*2	*2	*2	*4
69	−3	2	−1	−1	*3	10	−1	6	*7	−1	1	−2	2	*3	2	1	10	−1	3	1
70	8	4	2	*3	*11	−4	*17	−4	2	8	2	4	*3	−4	2	8	10	−4	8	−4
71	2	1	*3	2	2	4	8	*7	2	1	2	*3	4	5	8	4	2	2	15	*13
72	*2	*6	*4	*2	*2	*2	*4	*2	*2	*2	*36	*2	*2	*2	*16	*2	*2	*2	*4	*6
73	*3	3	−1	3	8	−1	*7	7	1	*3	3	−1	2	2	−1	1	−4	−2	*3	−1
74	10	2	12	6	2	2	2	4	*3	8	−4	4	−2	2	10	1	2	*3	2	1
75	*5	*5	*5	*5	*5	*35	*5	*15	*5	*5	*55	*5	*5	*25	*5	*5	*15	*5	*5	*5

Table C1 continued

j \ i	80	81	82	83	84	85	86	87	88	89	90	91	92	93	94	95	96	97	98	99
76	*2	*4	*2	*2	*2	*8	*6	*2	*2	*4	*2	*2	*2	*4	*2	*6	*2	*4	*2	*2
77	−4	−4	6	−1	*7	*3	−1	1	2	1	2	4	−1	−1	*27	−2	−1	1	2	1
78	2	4	1	2	*3	4	1	8	4	*67	4	−4	3	*3	2	−6	2	−4	4	2
79	2	1	6	*21	10	2	4	1	2	8	4	2	*3	12	3	4	1	16	4	4
80	*4	*2	*6	*2	*4	*2	*2	*2	*4	*2	*2	*6	*8	*2	*2	*2	*44	*2	*2	*2
81	−1	*9	*91	*17	2	−3	−1	1	1	1	*3	−5	−3	2	1	2	2	−1	−4	*3
82	*3	1	−12	4	2	2	2	3	−4	*3	2	1	8	3	2	2	4	4	*9	4
83	8	*7	8	2	2	2	8	1	*3	2	2	2	1	2	4	*37	2	*3	1	2
84	*2	*2	*2	*8	*2	*2	*2	*12	*2	*2	*2	*4	*2	*2	*2	*4	*6	*2	*2	*16
85	*7	−14	−6	4	−2	4	*3	2	−2	2	2	2	2	−2	2	*3	−4	2	2	−2
86	2	−2	2	8	1	*9	2	4	2	−2	12	2	1	*19	*3	−12	2	4	15	4
87	1	2	1	1	*3	14	4	4	1	8	12	1	2	*3	2	1	2	1	1	2
88	*2	*2	*4	*6	*2	*2	*4	*26	*2	*2	*8	*2	*6	*2	*4	*2	*2	*2	*4	*2
89	−1	1	*3	−1	2	2	−5	6	1	−2	−4	*3	7	−2	1	1	−1	4	2	1
90	8	*3	−4	2	8	2	4	8	12	4	*3	2	−8	4	16	12	8	24	4	*3
91	*3	1	1	2	2	*11	2	2	4	*9	1	4	8	1	1	4	4	1	*3	8
92	*34	*64	*2	*2	*2	*4	*2	*2	*6	*4	*2	*2	*2	*4	*2	*2	*2	*24	*2	*2
93	−1	1	−1	2	2	−2	−1	*3	−1	*23	2	−2	−3	2	1	−2	*3	1	−2	1
94	16	8	8	2	2	2	*3	−10	1	2	3	−2	6	12	2	*3	2	2	4	2
95	14	4	16	4	2	*3	4	2	4	4	4	4	*13	2	*3	8	4	12	10	6
96	*4	*2	*2	*2	*12	*2	*2	*2	*8	*2	*2	*22	*4	*18	*2	*2	*4	*2	*2	*14
97	7	1	−1	*3	−2	−3	2	1	2	1	1	−2	*3	−1	−1	4	−1	4	3	−2
98	12	1	*3	8	4	2	−2	2	2	2	−2	*3	−4	2	2	1	−4	8	*7	9
99	20	*3	8	6	4	1	1	8	18	3	*3	1	2	4	4	2	4	18	2	*3
100	*90	*10	*10	*20	*10	*10	*10	*20	*10	*30	*10	*20	*10	*10	*10	*40	*10	*70	*30	*100

Appendix D

This table consists of all fundamental radicands $D < 0$ with $|D| < 2 \cdot 10^3$ together with h_Δ and C_Δ as a product of cyclic groups. For example, $(2, 5)$ means $C_\Delta \cong C_2 \times C_5$, in the column labelled C_Δ. This was compiled using PARI. Some errors in the literature are corrected (such as $h_{-485} = 20$ whereas Borevich–Shafarevich [30, Table 5, p. 426] list it as $h_{-485} = 16$).

Table D1.

D	h_Δ	C_Δ
−1	1	
−2	1	
−3	1	
−5	2	(2)
−6	2	(2)
−7	1	
−10	2	(2)
−11	1	
−13	2	(2)
−14	4	(4)
−15	2	(2)
−17	4	(4)
−19	1	
−21	4	(2,2)
−22	2	(2)
−23	3	(3)
−26	6	(2,3)
−29	6	(2,3)
−30	4	(2,2)
−31	3	(3)
−33	4	(2,2)

D	h_Δ	C_Δ
−34	4	(4)
−35	2	(2)
−37	2	(2)
−38	6	(2,3)
−39	4	(4)
−41	8	(8)
−42	4	(2,2)
−43	1	
−46	4	(4)
−47	5	(5)
−51	2	(2)
−53	6	(2,3)
−55	4	(4)
−57	4	(2,2)
−58	2	(2)
−59	3	(3)
−61	6	(2,3)
−62	8	(8)
−65	8	(2,4)
−66	8	(2,4)
−67	1	

D	h_Δ	C_Δ
−69	8	(2,4)
−70	4	(2,2)
−71	7	(7)
−73	4	(4)
−74	10	(2,5)
−77	8	(2,4)
−78	4	(2,2)
−79	5	(5)
−82	4	(4)
−83	3	(3)
−85	4	(2,2)
−86	10	(2,5)
−87	6	(2,3)
−89	12	(4,3)
−91	2	(2)
−93	4	(2,2)
−94	8	(8)
−95	8	(8)
−97	4	(4)
−101	14	(2,7)
−102	4	(2,2)

D	h_Δ	C_Δ
−103	5	(5)
−105	8	(2,2,2)
−106	6	(2,3)
−107	3	(3)
−109	6	(2,3)
−110	12	(2,2,3)
−111	8	(8)
−113	8	(8)
−114	8	(2,4)
−115	2	(2)
−118	6	(2,3)
−119	10	(2,5)
−122	10	(2,5)
−123	2	(2)
−127	5	(5)
−129	12	(2,2,3)
−130	4	(2,2)
−131	5	(5)
−133	4	(2,2)
−134	14	(2,7)
−137	8	(8)
−138	8	(2,4)
−139	3	(3)
−141	8	(2,4)
−142	4	(4)
−143	10	(2,5)
−145	8	(2,4)
−146	16	(16)
−149	14	(2,7)
−151	7	(7)
−154	8	(2,4)
−155	4	(4)
−157	6	(2,3)
−158	8	(8)
−159	10	(2,5)
−161	16	(2,8)
−163	1	
−165	8	(2,2,2)
−166	10	(2,5)
−167	11	(11)
−170	12	(2,2,3)
−173	14	(2,7)
−174	12	(2,2,3)
−177	4	(2,2)
−178	8	(8)
−179	5	(5)
−181	10	(2,5)
−182	12	(2,2,3)
−183	8	(8)
−185	16	(2,8)
−186	12	(2,2,3)
−187	2	(2)
−190	4	(2,2)
−191	13	(13)
−193	4	(4)
−194	20	(4,5)
−195	4	(2,2)
−197	10	(2,5)
−199	9	(9)
−201	12	(2,2,3)
−202	6	(2,3)
−203	4	(4)
−205	8	(2,4)
−206	20	(4,5)
−209	20	(2,2,5)
−210	8	(2,2,2)
−211	3	(3)
−213	8	(2,4)
−214	6	(2,3)
−215	14	(2,7)
−217	8	(2,4)
−218	10	(2,5)
−219	4	(4)
−221	16	(2,8)
−222	12	(2,2,3)
−223	7	(7)
−226	8	(8)
−227	5	(5)
−229	10	(2,5)
−230	20	(2,2,5)
−231	12	(2,2,3)
−233	12	(4,3)
−235	2	(2)
−237	12	(2,2,3)
−238	8	(2,4)
−239	15	(3,5)
−241	12	(4,3)
−246	12	(2,2,3)
−247	6	(2,3)
−249	12	(2,2,3)
−251	7	(7)
−253	4	(2,2)
−254	16	(16)
−255	12	(2,2,3)
−257	16	(16)
−258	8	(2,4)
−259	4	(4)
−262	6	(2,3)
−263	13	(13)
−265	8	(2,4)
−266	20	(2,2,5)
−267	2	(2)
−269	22	(2,11)
−271	11	(11)
−273	8	(2,2,2)
−274	12	(4,3)
−277	6	(2,3)
−278	14	(2,7)
−281	20	(4,5)
−282	8	(2,4)
−283	3	(3)
−285	16	(2,2,4)
−286	12	(2,2,3)
−287	14	(2,7)
−290	20	(2,2,5)
−291	4	(4)
−293	18	(2,9)
−295	8	(8)
−298	6	(2,3)
−299	8	(8)
−301	8	(2,4)
−302	12	(4,3)
−303	10	(2,5)

D	h_Δ	C_Δ
−305	16	(2,8)
−307	3	(3)
−309	12	(2,2,3)
−310	8	(2,4)
−311	19	(19)
−313	8	(8)
−314	26	(2,13)
−317	10	(2,5)
−318	12	(2,2,3)
−319	10	(2,5)
−321	20	(2,2,5)
−322	8	(2,4)
−323	4	(4)
−326	22	(2,11)
−327	12	(4,3)
−329	24	(2,4,3)
−330	8	(2,2,2)
−331	3	(3)
−334	12	(4,3)
−335	18	(2,9)
−337	8	(8)
−339	6	(2,3)
−341	28	(2,2,7)
−345	8	(2,2,2)
−346	10	(2,5)
−347	5	(5)
−349	14	(2,7)
−353	16	(16)
−354	16	(2,8)
−355	4	(4)
−357	8	(2,2,2)
−358	6	(2,3)
−359	19	(19)
−362	18	(2,9)
−365	20	(2,2,5)
−366	12	(2,2,3)
−367	9	(9)
−370	12	(2,2,3)
−371	8	(8)
−373	10	(2,5)
−374	28	(2,2,7)

D	h_Δ	C_Δ
−377	16	(2,8)
−379	3	(3)
−381	20	(2,2,5)
−382	8	(8)
−383	17	(17)
−385	8	(2,2,2)
−386	20	(4,5)
−389	22	(2,11)
−390	16	(2,2,4)
−391	14	(2,7)
−393	12	(2,2,3)
−394	10	(2,5)
−395	8	(8)
−397	6	(2,3)
−398	20	(4,5)
−399	16	(2,8)
−401	20	(4,5)
−402	16	(2,8)
−403	2	(2)
−406	16	(2,8)
−407	16	(16)
−409	16	(16)
−410	16	(2,8)
−411	6	(2,3)
−413	20	(2,2,5)
−415	10	(2,5)
−417	12	(2,2,3)
−418	8	(2,4)
−419	9	(9)
−421	10	(2,5)
−422	10	(2,5)
−426	24	(2,4,3)
−427	2	(2)
−429	16	(2,2,4)
−430	12	(2,2,3)
−431	21	(3,7)
−433	12	(4,3)
−434	24	(2,4,3)
−435	4	(2,2)
−437	20	(2,2,5)
−438	8	(2,4)

D	h_Δ	C_Δ
−439	15	(3,5)
−442	8	(2,4)
−443	5	(5)
−445	8	(2,4)
−446	32	(32)
−447	14	(2,7)
−449	20	(4,5)
−451	6	(2,3)
−453	12	(2,2,3)
−454	14	(2,7)
−455	20	(2,2,5)
−457	8	(8)
−458	26	(2,13)
−461	30	(2,3,5)
−462	8	(2,2,2)
−463	7	(7)
−465	16	(2,2,4)
−466	8	(8)
−467	7	(7)
−469	16	(2,8)
−470	20	(2,2,5)
−471	16	(16)
−473	12	(2,2,3)
−474	20	(2,2,5)
−478	8	(8)
−479	25	(25)
−481	16	(2,8)
−482	20	(4,5)
−483	4	(2,2)
−485	20	(2,2,5)
−487	7	(7)
−489	20	(2,2,5)
−491	9	(9)
−493	12	(2,2,3)
−494	28	(2,2,7)
−497	24	(2,4,3)
−498	8	(2,4)
−499	3	(3)
−501	16	(2,8)
−502	14	(2,7)
−503	21	(3,7)

D	h_Δ	C_Δ
−505	8	(2,4)
−506	28	(2,2,7)
−509	30	(2,3,5)
−510	16	(2,2,4)
−511	14	(2,7)
−514	16	(16)
−515	6	(2,3)
−517	12	(2,2,3)
−518	16	(2,8)
−519	18	(2,9)
−521	32	(32)
−523	5	(5)
−526	12	(4,3)
−527	18	(2,9)
−530	28	(2,2,7)
−533	12	(2,2,3)
−534	20	(2,2,5)
−535	14	(2,7)
−537	12	(2,2,3)
−538	10	(2,5)
−541	10	(2,5)
−542	24	(8,3)
−543	12	(4,3)
−545	32	(2,16)
−546	24	(2,2,2,3)
−547	3	(3)
−551	26	(2,13)
−553	8	(2,4)
−554	22	(2,11)
−555	4	(2,2)
−557	18	(2,9)
−559	16	(16)
−561	16	(2,2,4)
−562	8	(8)
−563	9	(9)
−565	12	(2,2,3)
−566	30	(2,3,5)
−569	32	(32)
−570	16	(2,2,4)
−571	5	(5)
−573	16	(2,8)

D	h_Δ	C_Δ
−574	16	(2,8)
−577	8	(8)
−579	8	(8)
−581	28	(2,2,7)
−582	16	(2,8)
−583	8	(8)
−586	18	(2,9)
−587	7	(7)
−589	16	(2,8)
−590	20	(2,2,5)
−591	22	(2,11)
−593	24	(8,3)
−595	4	(2,2)
−597	12	(2,2,3)
−598	8	(2,4)
−599	25	(25)
−601	20	(4,5)
−602	24	(2,4,3)
−606	12	(2,2,3)
−607	13	(13)
−609	16	(2,2,4)
−610	12	(2,2,3)
−611	10	(2,5)
−613	10	(2,5)
−614	34	(2,17)
−615	20	(2,2,5)
−617	12	(4,3)
−618	12	(2,2,3)
−619	5	(5)
−622	12	(4,3)
−623	22	(2,11)
−626	36	(4,9)
−627	4	(2,2)
−629	36	(2,2,9)
−631	13	(13)
−633	20	(2,2,5)
−634	14	(2,7)
−635	10	(2,5)
−638	20	(2,2,5)
−641	28	(4,7)
−642	16	(2,8)

D	h_Δ	C_Δ
−643	3	(3)
−645	16	(2,2,4)
−646	16	(2,8)
−647	23	(23)
−649	20	(2,2,5)
−651	8	(2,4)
−653	14	(2,7)
−654	28	(2,2,7)
−655	12	(4,3)
−658	8	(2,4)
−659	11	(11)
−661	18	(2,9)
−662	22	(2,11)
−663	16	(2,8)
−665	24	(2,2,2,3)
−667	4	(4)
−669	12	(2,2,3)
−670	12	(2,2,3)
−671	30	(2,3,5)
−673	12	(4,3)
−674	24	(8,3)
−677	30	(2,3,5)
−678	20	(2,2,5)
−679	18	(2,9)
−681	20	(2,2,5)
−682	12	(2,2,3)
−683	5	(5)
−685	12	(2,2,3)
−687	12	(4,3)
−689	40	(2,4,5)
−690	16	(2,2,4)
−691	5	(5)
−694	10	(2,5)
−695	24	(8,3)
−697	8	(2,4)
−698	26	(2,13)
−699	10	(2,5)
−701	34	(2,17)
−703	14	(2,7)
−705	24	(2,2,2,3)
−706	24	(8,3)

D	h_Δ	C_Δ
−707	6	(2,3)
−709	10	(2,5)
−710	32	(2,16)
−713	24	(2,4,3)
−714	24	(2,2,2,3)
−715	4	(2,2)
−717	16	(2,8)
−718	12	(4,3)
−719	31	(31)
−721	16	(2,8)
−723	4	(4)
−727	13	(13)
−730	12	(2,2,3)
−731	12	(4,3)
−733	14	(2,7)
−734	40	(8,5)
−737	20	(2,2,5)
−739	5	(5)
−741	24	(2,2,2,3)
−742	8	(2,4)
−743	21	(3,7)
−745	16	(2,8)
−746	26	(2,13)
−749	32	(2,16)
−751	15	(3,5)
−753	12	(2,2,3)
−754	20	(2,2,5)
−755	12	(4,3)
−757	10	(2,5)
−758	22	(2,11)
−759	24	(2,4,3)
−761	40	(8,5)
−762	12	(2,2,3)
−763	4	(4)
−766	24	(8,3)
−767	22	(2,11)
−769	20	(4,5)
−770	32	(2,2,8)
−771	6	(2,3)
−773	26	(2,13)
−777	16	(2,2,4)

D	h_Δ	C_Δ
−778	14	(2,7)
−779	10	(2,5)
−781	20	(2,2,5)
−782	24	(2,4,3)
−785	16	(2,8)
−786	16	(2,8)
−787	5	(5)
−789	32	(2,16)
−790	16	(2,8)
−791	32	(32)
−793	8	(2,4)
−794	42	(2,3,7)
−795	4	(2,2)
−797	30	(2,3,5)
−798	16	(2,2,4)
−799	16	(16)
−802	12	(4,3)
−803	10	(2,5)
−805	16	(2,2,4)
−806	28	(2,2,7)
−807	14	(2,7)
−809	32	(32)
−811	7	(7)
−813	12	(2,2,3)
−814	12	(2,2,3)
−815	30	(2,3,5)
−817	12	(2,2,3)
−818	28	(4,7)
−821	30	(2,3,5)
−822	20	(2,2,5)
−823	9	(9)
−826	12	(2,2,3)
−827	7	(7)
−829	22	(2,11)
−830	20	(2,2,5)
−831	28	(4,7)
−834	16	(2,8)
−835	6	(2,3)
−838	14	(2,7)
−839	33	(3,11)
−842	26	(2,13)

D	h_Δ	C_Δ
−843	6	(2,3)
−849	28	(2,2,7)
−851	10	(2,5)
−853	10	(2,5)
−854	44	(2,2,11)
−857	32	(32)
−858	16	(2,2,4)
−859	7	(7)
−861	24	(2,2,2,3)
−862	8	(8)
−863	21	(3,7)
−865	16	(2,8)
−866	44	(4,11)
−869	32	(2,16)
−870	16	(2,2,4)
−871	22	(2,11)
−874	20	(2,2,5)
−877	10	(2,5)
−878	20	(4,5)
−879	22	(2,11)
−881	40	(8,5)
−883	3	(3)
−885	24	(2,2,2,3)
−886	18	(2,9)
−887	29	(29)
−889	16	(2,8)
−890	24	(2,4,3)
−893	28	(2,2,7)
−894	28	(2,2,7)
−895	16	(16)
−897	16	(2,2,4)
−898	12	(4,3)
−899	14	(2,7)
−901	24	(2,4,3)
−902	28	(2,2,7)
−903	16	(2,8)
−905	24	(2,4,3)
−906	28	(2,2,7)
−907	3	(3)
−910	16	(2,2,4)
−911	31	(31)

D	h_Δ	C_Δ
−913	12	(2,2,3)
−914	36	(4,9)
−915	8	(2,4)
−917	20	(2,2,5)
−919	19	(19)
−921	20	(2,2,5)
−922	18	(2,9)
−923	10	(2,5)
−926	40	(8,5)
−929	36	(4,9)
−930	24	(2,2,2,3)
−933	16	(2,8)
−934	26	(2,13)
−935	28	(2,2,7)
−937	20	(4,5)
−938	16	(2,8)
−939	8	(8)
−941	46	(2,23)
−942	12	(2,2,3)
−943	16	(16)
−946	16	(2,8)
−947	5	(5)
−949	12	(2,2,3)
−951	26	(2,13)
−953	32	(32)
−955	4	(4)
−957	16	(2,2,4)
−958	16	(16)
−959	36	(4,9)
−962	28	(2,2,7)
−965	44	(2,2,11)
−966	24	(2,2,2,3)
−967	11	(11)
−969	24	(2,2,2,3)
−970	12	(2,2,3)
−971	15	(3,5)
−973	12	(2,2,3)
−974	36	(4,3,3)
−977	20	(4,5)
−978	24	(2,4,3)
−979	8	(8)

D	h_Δ	C_Δ
−982	10	(2,5)
−983	27	(27)
−985	24	(2,4,3)
−986	44	(2,2,11)
−987	8	(2,4)
−989	36	(2,2,9)
−991	17	(17)
−993	12	(2,2,3)
−994	16	(2,8)
−995	8	(8)
−997	14	(2,7)
−988	26	(2,13)
−1001	40	(2,2,2,5)
−1002	16	(2,8)
−1003	4	(4)
−1005	16	(2,2,4)
−1006	20	(4,5)
−1007	30	(2,3,5)
−1009	20	(4,5)
−1010	28	(2,2,7)
−1011	12	(4,3)
−1013	26	(2,13)
−1015	16	(2,8)
−1018	18	(2,9)
−1019	13	(13)
−1021	22	(2,11)
−1022	32	(2,16)
−1023	16	(2,8)
−1027	4	(4)
−1030	12	(2,2,3)
−1031	35	(5,7)
−1033	12	(4,3)
−1034	44	(2,2,11)
−1037	20	(2,2,5)
−1038	12	(2,2,3)
−1039	23	(23)
−1041	36	(2,2,9)
−1042	12	(4,3)
−1043	8	(8)
−1045	16	(2,2,4)
−1046	42	(2,3,7)

D	h_Δ	C_Δ
−1047	16	(16)
−1049	44	(4,11)
−1051	5	(5)
−1054	16	(2,8)
−1055	36	(4,9)
−1057	16	(2,8)
−1059	6	(2,3)
−1061	26	(2,13)
−1063	19	(19)
−1065	16	(2,2,4)
−1066	20	(2,2,5)
−1067	12	(4,3)
−1069	30	(2,3,5)
−1070	36	(2,2,9)
−1073	24	(2,4,3)
−1074	32	(2,16)
−1077	24	(2,4,3)
−1079	34	(2,17)
−1081	16	(2,8)
−1082	22	(2,11)
−1085	32	(2,2,8)
−1086	28	(2,2,7)
−1087	9	(9)
−1090	12	(2,2,3)
−1091	17	(17)
−1093	10	(2,5)
−1094	26	(2,13)
−1095	28	(2,2,7)
−1097	36	(4,9)
−1099	6	(2,3)
−1101	28	(2,2,7)
−1102	20	(2,2,5)
−1103	23	(23)
−1105	16	(2,2,4)
−1106	48	(2,8,3)
−1109	50	(2,25)
−1110	16	(2,2,4)
−1111	22	(2,11)
−1113	16	(2,2,4)
−1114	22	(2,11)
−1115	10	(2,5)

D	h_Δ	C_Δ
−1117	14	(2,7)
−1118	36	(2,2,9)
−1119	32	(32)
−1121	44	(2,2,11)
−1122	16	(2,2,4)
−1123	5	(5)
−1126	22	(2,11)
−1129	16	(16)
−1130	44	(2,2,11)
−1131	8	(2,4)
−1133	28	(2,2,7)
−1135	18	(2,9)
−1137	20	(2,2,5)
−1138	12	(4,3)
−1139	16	(16)
−1141	24	(2,4,3)
−1142	18	(2,9)
−1145	24	(2,4,3)
−1146	32	(2,16)
−1147	6	(2,3)
−1149	16	(2,8)
−1151	41	(41)
−1153	16	(16)
−1154	56	(8,7)
−1155	8	(2,2,2)
−1157	28	(2,2,7)
−1158	24	(2,4,3)
−1159	16	(16)
−1162	12	(2,2,3)
−1163	7	(7)
−1165	20	(2,2,5)
−1166	36	(2,2,9)
−1167	22	(2,11)
−1169	48	(2,8,3)
−1171	7	(7)
−1173	24	(2,2,2,3)
−1174	30	(2,3,5)
−1177	12	(2,2,3)
−1178	16	(2,8)
−1181	46	(2,23)
−1182	20	(2,2,5)
−1185	16	(2,2,4)
−1186	24	(8,3)
−1187	9	(9)
−1189	20	(2,2,5)
−1190	40	(2,2,2,5)
−1191	24	(8,3)
−1193	36	(4,9)
−1194	28	(2,2,7)
−1195	8	(8)
−1198	12	(4,3)
−1199	38	(2,19)
−1201	16	(16)
−1202	24	(8,3)
−1203	6	(2,3)
−1205	40	(2,4,5)
−1207	18	(2,9)
−1209	40	(2,2,2,5)
−1211	14	(2,7)
−1213	10	(2,5)
−1214	40	(8,5)
−1217	32	(32)
−1218	24	(2,2,2,3)
−1219	6	(2,3)
−1221	32	(2,2,8)
−1222	12	(2,2,3)
−1223	35	(5,7)
−1226	42	(2,3,7)
−1227	4	(4)
−1229	38	(2,19)
−1230	24	(2,2,2,3)
−1231	27	(27)
−1234	24	(8,3)
−1235	12	(2,2,3)
−1237	14	(2,7)
−1238	42	(2,3,7)
−1239	32	(2,16)
−1241	32	(2,16)
−1243	4	(4)
−1245	32	(2,2,8)
−1246	16	(2,8)
−1247	26	(2,13)
−1249	32	(32)
−1253	32	(2,16)
−1254	24	(2,2,2,3)
−1255	12	(4,3)
−1257	20	(2,2,5)
−1258	12	(2,2,3)
−1259	15	(3,5)
−1261	20	(2,2,5)
−1262	36	(4,9)
−1263	20	(4,5)
−1265	40	(2,2,2,5)
−1266	32	(2,16)
−1267	6	(2,3)
−1270	20	(2,2,5)
−1271	40	(8,5)
−1273	20	(2,2,5)
−1277	34	(2,17)
−1279	23	(23)
−1281	24	(2,2,2,3)
−1282	12	(4,3)
−1283	11	(11)
−1285	12	(2,2,3)
−1286	58	(2,29)
−1289	36	(4,9)
−1290	16	(2,2,4)
−1291	9	(9)
−1293	24	(2,4,3)
−1294	28	(4,7)
−1295	36	(2,2,9)
−1297	12	(4,3)
−1298	32	(2,16)
−1299	8	(8)
−1301	50	(2,25)
−1302	16	(2,2,4)
−1303	11	(11)
−1306	18	(2,9)
−1307	11	(11)
−1309	24	(2,2,2,3)
−1310	36	(2,2,9)
−1311	28	(2,2,7)
−1313	24	(2,4,3)

D	h_Δ	C_Δ
−1315	6	(2,3)
−1317	20	(2,2,5)
−1318	10	(2,5)
−1319	45	(9,5)
−1321	24	(8,3)
−1322	42	(2,3,7)
−1326	40	(2,2,2,5)
−1327	15	(3,5)
−1329	36	(2,2,9)
−1330	24	(2,2,2,3)
−1333	20	(2,2,5)
−1334	32	(2,16)
−1335	28	(2,2,7)
−1337	24	(2,4,3)
−1338	20	(2,2,5)
−1339	8	(8)
−1342	20	(2,2,5)
−1343	34	(2,17)
−1345	16	(2,8)
−1346	28	(4,7)
−1347	6	(2,3)
−1349	56	(2,4,7)
−1351	24	(8,3)
−1353	16	(2,2,4)
−1354	22	(2,11)
−1355	12	(4,3)
−1357	16	(2,8)
−1358	24	(2,4,3)
−1361	60	(4,3,5)
−1362	24	(2,4,3)
−1363	6	(2,3)
−1365	16	(2,2,2,2)
−1366	18	(2,9)
−1367	25	(25)
−1370	44	(2,2,11)
−1371	12	(4,3)
−1373	18	(2,9)
−1374	28	(2,2,7)
−1378	20	(2,2,5)
−1379	16	(16)
−1381	26	(2,13)

D	h_Δ	C_Δ
−1382	38	(2,19)
−1383	18	(2,9)
−1385	48	(2,8,3)
−1387	4	(4)
−1389	28	(2,2,7)
−1390	20	(2,2,5)
−1391	44	(4,11)
−1393	16	(2,8)
−1394	48	(2,8,3)
−1397	24	(2,4,3)
−1398	20	(2,2,5)
−1399	27	(27)
−1401	28	(2,2,7)
−1402	14	(2,7)
−1403	14	(2,7)
−1405	24	(2,4,3)
−1406	44	(2,2,11)
−1407	24	(2,4,3)
−1409	36	(4,9)
−1410	32	(2,2,8)
−1411	4	(4)
−1414	28	(2,2,7)
−1415	34	(2,17)
−1417	16	(2,8)
−1418	34	(2,17)
−1419	12	(2,2,3)
−1423	9	(9)
−1426	32	(2,16)
−1427	15	(3,5)
−1429	22	(2,11)
−1430	32	(2,2,8)
−1433	36	(4,9)
−1434	32	(2,16)
−1435	4	(2,2)
−1437	24	(2,4,3)
−1438	16	(16)
−1439	39	(3,13)
−1441	28	(2,2,7)
−1442	24	(2,4,3)
−1443	8	(2,4)
−1446	32	(2,16)

D	h_Δ	C_Δ
−1447	23	(23)
−1451	13	(13)
−1453	14	(2,7)
−1454	60	(4,3,5)
−1455	28	(2,2,7)
−1457	24	(2,4,3)
−1459	11	(11)
−1461	28	(2,2,7)
−1462	16	(2,8)
−1463	32	(2,16)
−1465	16	(2,8)
−1466	58	(2,29)
−1469	56	(2,4,7)
−1471	23	(23)
−1473	28	(2,2,7)
−1474	16	(2,8)
−1477	16	(2,8)
−1478	30	(2,3,5)
−1479	28	(2,2,7)
−1481	52	(4,13)
−1482	24	(2,2,2,3)
−1483	7	(7)
−1486	20	(4,5)
−1487	37	(37)
−1489	20	(4,5)
−1490	36	(2,2,9)
−1491	12	(2,2,3)
−1493	22	(2,11)
−1495	20	(2,2,5)
−1497	20	(2,2,5)
−1498	16	(2,8)
−1499	13	(13)
−1501	24	(2,4,3)
−1502	24	(8,3)
−1505	40	(2,2,2,5)
−1506	24	(2,4,3)
−1507	4	(4)
−1509	40	(2,4,5)
−1510	16	(2,8)
−1511	49	(49)
−1513	16	(4,4)

D	h_Δ	C_Δ
−1514	50	(2,25)
−1515	12	(2,2,3)
−1517	48	(2,8,3)
−1518	24	(2,2,2,3)
−1522	20	(4,5)
−1523	7	(7)
−1526	48	(2,8,3)
−1527	14	(2,7)
−1529	52	(2,2,13)
−1531	11	(11)
−1533	24	(2,2,2,3)
−1534	20	(2,2,5)
−1535	38	(2,19)
−1537	16	(2,8)
−1538	44	(4,11)
−1541	36	(2,2,9)
−1542	20	(2,2,5)
−1543	19	(19)
−1545	24	(2,2,2,3)
−1546	34	(2,17)
−1547	12	(2,2,3)
−1549	18	(2,9)
−1551	32	(2,16)
−1553	40	(8,5)
−1554	24	(2,2,2,3)
−1555	4	(4)
−1558	12	(2,2,3)
−1559	51	(3,17)
−1561	32	(2,16)
−1562	28	(2,2,7)
−1563	6	(2,3)
−1565	28	(2,2,7)
−1567	15	(3,5)
−1569	28	(2,2,7)
−1570	20	(2,2,5)
−1571	17	(17)
−1574	54	(2,27)
−1577	28	(2,2,7)
−1578	16	(2,8)
−1579	9	(9)
−1581	40	(2,2,2,5)

D	h_Δ	C_Δ
−1582	16	(4,4)
−1583	33	(3,11)
−1585	24	(2,4,3)
−1586	44	(2,2,11)
−1589	52	(2,2,13)
−1590	32	(2,4,4)
−1591	22	(2,11)
−1594	34	(2,17)
−1595	16	(2,8)
−1597	14	(2,7)
−1598	32	(4,8)
−1599	36	(2,2,9)
−1601	56	(8,7)
−1603	6	(2,3)
−1605	16	(2,2,4)
−1606	28	(2,2,7)
−1607	27	(27)
−1609	28	(4,7)
−1610	32	(2,2,8)
−1613	42	(2,3,7)
−1614	28	(2,2,7)
−1615	24	(2,4,3)
−1618	12	(4,3)
−1619	15	(3,5)
−1621	18	(2,9)
−1622	30	(2,3,5)
−1623	28	(4,7)
−1626	20	(2,2,5)
−1627	7	(7)
−1630	28	(2,2,7)
−1631	44	(4,11)
−1633	16	(2,8)
−1634	64	(2,32)
−1635	8	(2,4)
−1637	38	(2,19)
−1639	22	(2,11)
−1641	44	(2,2,11)
−1642	14	(2,7)
−1643	10	(2,5)
−1645	16	(2,2,4)
−1646	44	(4,11)

D	h_Δ	C_Δ
−1649	48	(2,8,3)
−1651	8	(8)
−1653	16	(2,2,4)
−1654	22	(2,11)
−1655	44	(4,11)
−1657	16	(16)
−1658	42	(2,3,7)
−1659	8	(2,4)
−1661	48	(2,8,3)
−1662	20	(2,2,5)
−1663	17	(17)
−1667	13	(13)
−1669	26	(2,13)
−1670	28	(2,2,7)
−1671	38	(2,19)
−1673	32	(2,16)
−1677	16	(2,2,4)
−1678	20	(4,5)
−1679	52	(4,13)
−1685	52	(2,2,13)
−1686	44	(2,2,11)
−1687	18	(2,9)
−1689	36	(2,2,9)
−1691	18	(2,9)
−1693	22	(2,11)
−1695	20	(2,2,5)
−1697	28	(4,7)
−1698	16	(2,8)
−1699	11	(11)
−1702	20	(2,2,5)
−1703	28	(4,7)
−1705	16	(2,2,4)
−1706	58	(2,29)
−1707	10	(2,5)
−1709	42	(2,3,7)
−1711	28	(4,7)
−1713	36	(2,2,9)
−1714	20	(4,5)
−1717	16	(2,8)
−1718	46	(2,23)
−1721	52	(4,13)

D	h_Δ	C_Δ
−1722	24	(2,2,2,3)
−1723	5	(5)
−1726	24	(8,3)
−1727	36	(4,9)
−1729	24	(2,2,2,3)
−1730	36	(2,2,9)
−1731	8	(8)
−1733	34	(2,17)
−1735	26	(2,13)
−1738	16	(2,8)
−1739	20	(4,5)
−1741	26	(2,13)
−1742	44	(2,2,11)
−1743	24	(2,4,3)
−1745	40	(2,4,5)
−1747	5	(5)
−1749	40	(2,2,2,5)
−1751	48	(16,3)
−1753	20	(4,5)
−1754	38	(2,19)
−1757	28	(2,2,7)
−1758	20	(2,2,5)
−1759	27	(27)
−1761	20	(2,2,5)
−1762	24	(8,3)
−1763	12	(4,3)
−1765	20	(2,2,5)
−1766	50	(2,25)
−1767	32	(2,16)
−1769	64	(2,32)
−1770	40	(2,2,2,5)
−1771	8	(2,4)
−1774	20	(4,5)
−1777	24	(8,3)
−1778	40	(2,4,5)
−1779	10	(2,5)
−1781	68	(2,2,17)
−1783	17	(17)
−1785	32	(2,2,2,4)
−1786	20	(2,2,5)
−1787	7	(7)
−1789	26	(2,13)
−1790	52	(2,2,13)
−1793	36	(2,2,9)
−1794	32	(2,2,8)
−1795	8	(8)
−1797	24	(2,4,3)
−1798	20	(2,2,5)
−1799	50	(2,25)
−1801	28	(4,7)
−1802	32	(2,16)
−1803	8	(8)
−1806	40	(2,2,2,5)
−1807	12	(4,3)
−1810	20	(2,2,5)
−1811	23	(23)
−1814	46	(2,23)
−1817	40	(2,4,5)
−1819	10	(2,5)
−1821	36	(2,2,9)
−1822	16	(16)
−1823	45	(9,5)
−1826	56	(2,4,7)
−1829	40	(2,4,5)
−1830	24	(2,2,2,3)
−1831	19	(19)
−1833	24	(2,2,2,3)
−1834	36	(2,2,9)
−1835	10	(2,5)
−1837	24	(2,4,3)
−1838	28	(4,7)
−1839	40	(8,5)
−1841	40	(2,4,5)
−1842	16	(2,8)
−1843	6	(2,3)
−1846	28	(2,2,7)
−1847	43	(43)
−1851	14	(2,7)
−1853	36	(2,2,9)
−1855	28	(2,2,7)
−1857	20	(2,2,5)
−1858	20	(4,5)
−1861	38	(2,19)
−1865	48	(2,8,3)
−1866	32	(2,16)
−1867	5	(5)
−1869	40	(2,2,2,5)
−1870	16	(2,2,4)
−1871	45	(9,5)
−1873	12	(4,3)
−1874	56	(8,7)
−1877	34	(2,17)
−1878	24	(2,4,3)
−1879	27	(27)
−1882	18	(2,9)
−1883	14	(2,7)
−1885	16	(2,2,4)
−1886	64	(4,16)
−1887	20	(2,2,5)
−1889	72	(8,9)
−1891	10	(2,5)
−1893	20	(2,2,5)
−1894	30	(2,3,5)
−1895	48	(16,3)
−1897	16	(2,8)
−1898	28	(2,2,7)
−1901	42	(2,3,7)
−1902	36	(2,2,9)
−1903	22	(2,11)
−1905	24	(2,2,2,3)
−1906	20	(4,5)
−1907	13	(13)
−1909	28	(2,2,7)
−1910	56	(2,4,7)
−1913	36	(4,9)
−1914	48	(2,2,4,3)
−1915	6	(2,3)
−1918	16	(4,4)
−1919	44	(4,11)
−1921	40	(2,4,5)
−1923	10	(2,5)
−1927	18	(2,9)
−1929	28	(2,2,7)

D	h_Δ	C_Δ
−1930	20	(2,2,5)
−1931	21	(3,7)
−1933	18	(2,9)
−1934	52	(4,13)
−1937	48	(2,8,3)
−1938	32	(2,2,8)
−1939	8	(8)
−1941	24	(2,4,3)
−1942	22	(2,11)
−1943	32	(32)
−1945	16	(2,8)
−1946	68	(2,2,17)
−1947	8	(2,4)
−1949	70	(2,5,7)
−1951	33	(3,11)

D	h_Δ	C_Δ
−1954	28	(4,7)
−1955	12	(2,2,3)
−1957	16	(2,8)
−1958	32	(2,16)
−1959	42	(2,3,7)
−1961	32	(2,16)
−1963	6	(2,3)
−1965	40	(2,2,2,5)
−1966	36	(4,9)
−1967	36	(4,9)
−1969	20	(2,2,5)
−1970	52	(2,2,13)
−1973	42	(2,3,7)
−1974	32	(2,2,8)
−1977	36	(2,2,9)

D	h_Δ	C_Δ
−1978	12	(2,2,3)
−1979	23	(23)
−1981	20	(2,2,5)
−1982	24	(8,3)
−1983	16	(16)
−1985	40	(2,4,5)
−1986	48	(2,8,3)
−1987	7	(7)
−1990	24	(2,4,3)
−1991	56	(8,7)
−1993	24	(8,3)
−1994	54	(2,27)
−1995	8	(2,2,2)
−1997	42	(2,3,7)
−1999	27	(27)

Appendix E

A Gazetteer of Forms.

In this book, we have shown how the theory of ideals, in conjunction with the theory of continued fractions, can be used to present the theory of quadratics in its various guises. One aspect of the guises is that some of what has been done has a thread going back 200 years, together with newer material such as the infrastructure, as well as new results, conjectures and advances presented throughout the text, for example, on class number problems. Most mathematicians know (vaguely) about forms, ideals, composition, and genera. The fact that the form theory of Gauss was maintained by Dirichlet, Dedekind, Landau and Mathews, whereas modern number theory requires Eisenstein's approach, only adds to the confusion. What we have shown is that Dedekind's ideal-theoretic approach is the superior, modern edifice into which the new approaches, via the infrastructure, may be plugged up to and including crytographic techniques described in Chapter Eight. An end goal is to have subsumed the older material by updating concepts and introducing notation and meaning which are consistent with modern algebra and number theory, while, at the same time, eliminating confusing conflicts. Thus, for the sake of balance and completeness, from a historical perspective, we now present a collection of facts concerning the interrelationships between the theory of binary quadratic forms and ideal theory.[(E.1)] Furthermore, this is an opportunity to clear up some confusion in the literature, which arises in the translation from form theory to ideal theory, both in terms of language, and in terms of concepts involved.

A *binary quadratic form* may be denoted by

$$f(x,y) = ax^2 + bxy + cy^2$$

or,

$$f = (a,b,c),$$

where a is called the *leading* (or first) *coefficient*, b is called the *middle* (or second) *coefficient*, and c is called the *last* (or third) *coefficient*, with $a, b, c \in \mathbf{Z}$.

Definition E.1. We say that a form f is *strictly primitive* if $\gcd(a,b,c) = 1$, and call it *imprimitive* otherwise.

We caution the reader that, in the literature, this is merely called "primitive", but we introduced this new term here to provide language which does not lead

[(E.1)] As noted by Cohn [57, p. 195], the theory of forms "... preceded ideal theory by at least 50 years... Yet ideal theory is conceptually simpler."

to confusion. As we saw in Definition 1.5.2, we needed to make this distinction for ideals in order to avoid the confusion arising from the use, in contemporary literature, of "primitive" to mean two different things when referring to ideals. The literature on forms is consistent, however, in its use of "primitive" to mean what we called strictly primitive in Definition E.1. We adopt the new term for forms here, since we will be making the link with ideals soon, and so we need uniformization of meaning. Thus, when we make the correspondence with ideals, the term *strictly primitive* will correspond. However, Gauss' treatment further confuses the issue. Gauss wrote $f = (a, b, c)$ for $f(x, y) = ax^2 + 2bxy + cy^2$, thereby always assuming that the middle coefficient is even (see [91, Article 153, p. 108]).

Gauss called his forms, $f(x, y) = ax^2 + 2bxy + cy^2$, *properly primitive* if $\gcd(a, b, c) = \gcd(a, 2b, c) = 1$; whereas he called them *improperly primitive* if $\gcd(a, b, c) = 1$, and $\gcd(a, 2b, c) = 2$. Furthermore, he called $b^2 - ac$ the "determinant" of f [91, Article 154, p. 109], or what has come to be known as the *Gauss discriminant* (see footnotes (4.1.4) and (4.1.15)). Gauss organized forms into *orders* according to the common divisors of the coefficients, and this is the origin of the term "order", maintained by Dedekind to mean what we now call an integral domain. Thus, Gauss considered his properly primitive forms to be in one order, and the improperly primitive ones in another order. Gauss' treatment is not used in modern number theory, since the correspondence between the theory of Gauss forms, and the theory of ideals, cannot readily be made. In order to make this correspondence possible, we must turn to *Eisenstein forms*. Thus, by a *form* we will mean a binary quadratic form $k_\Delta(x, y) = ax^2 + bxy + cy^2$ with *discriminant* $\Delta = b^2 - 4ac$. If $\Delta > 0$, then k_Δ can represent both positive and negative integers, so we call k_Δ an *indefinite form*. If $\Delta < 0$, then a and c must have the same sign, and k_Δ represents only positive or only negative numbers, so we call k_Δ a *definite form* in this case. When $\Delta < 0$, and a is positive, then k_Δ is called *positive definite*, and *negative definite* otherwise. The classical approach is to concentrate upon the positive definite case, for it is the simplest with the most elegant consequences. We do so here.

The fundamental classical problem involving binary quadratic forms is that of representation, i.e. which numbers are represented by a given binary quadratic form? This problem is more complicated than the simple question appears on the surface, for we may have many forms representing the same numbers. If $n \in \mathbf{Z}$, then we say that *n is represented by the form $k_\Delta(x, y)$* if $k_\Delta(x, y) = n$ for some $x, y \in \mathbf{Z}$. If $\gcd(x, y) = 1$, then we say that *k_Δ properly represents n.*

The following provides a criterion for an $n \in \mathbf{Z}$ to be (properly) represented by a form k_Δ, which is an easy exercise for the reader.

Lemma E.1. *If Δ is a discriminant, and $n \in \mathbf{Z}$ with $\gcd(n, 2\Delta) = 1$, then n is properly represented by a strictly primitive form k_Δ if and only if Δ is a quadratic residue modulo n.*

We now need a notion of equivalence which provides us with a method of *not* distinguishing between forms which represent the same numbers. To do this, we ask: which substitutions

$$x = uX + vY, \quad y = wX + zY$$

into a given form $k_\Delta(X, Y)$ will yield a form $\ell_\Delta(x, y)$ which represents the same numbers? The expression $uz - vw$ is called the *determinant of the substitution*. If the determinant of the substitution is 1, then the substitution is called *unimodular*. Gauss called forms related by a unimodular substitution *properly* equivalent, whereas those related by a substitution of determinant -1 are called *improperly* equivalent. [(E.2)]

The reader may verify that proper equivalence is an equivalence relation, but that improper equivalence is not. Hence, improper equivalence provides unnecessary complications, and actually stands in the way of producing a group structure via multiplication (or composition) of forms, so it is rarely used. In fact, the usual notion is

Definition E.2. Two forms k and ℓ are called *equivalent*, denoted $k \sim \ell$, provided that they are properly equivalent, i.e. equivalent by a unimodular transformation.

Since equivalence provides us with equivalence classes, then there is a means of composing or multiplying these classes to get a group. The notion of composition of forms, as given by Gauss, is much longer and more difficult than the method of Dirichlet (e.g. see Buell [41, p. 55], and Cox [67, p. 47]). Moreover, Dirichlet's method provides a clearer link with the ideal theory. This is done via, what Dirichlet called, "united forms".

Definition E.3. If $k_\Delta = (a_1, b_1, c_1)$ and $\ell_\Delta = (a_2, b_2, c_2)$ are forms of discriminant Δ, then they are called *united* if $\gcd(a_1, a_2, (b_1 + b_2)/2) = 1$.

The following may be found in Buell [41, Prop. 4.5, p. 55] and Cox [67, Lemma 3.2, p. 48]. We maintain the notation for k_Δ and ℓ_Δ given in Definition E.3.

Theorem E.1. *If k_Δ and ℓ_Δ are united forms, then there exists a unique integer B modulo $2a_1a_2$ such that $B \equiv b_i \pmod{2a_i}$, for $i = 1, 2$, and $B^2 \equiv \Delta \pmod{4a_1a_2}$.*

The reader may compare Theorem E.1 with formulae (1.2.1)–(1.2.5), in order to get a feeling for where we are headed.

Thus, we may formulate

Definition E.4. The form compounded (or *composed*) of strictly primitive, united forms k_Δ and ℓ_Δ is (a_1a_2, B, C), where B is as in Theorem E.1 and $C = (B^2 - \Delta)/(4a_1a_2)$, denoted $k_\Delta \circ \ell_\Delta$. We refer to $k_\Delta \circ \ell_\Delta$ as the *composition* of k_Δ and ℓ_Δ.

Furthermore, as proved by Dirichlet,

Theorem E.2. *If f_1 and f_2 are united forms, then there exist united forms f_3 and f_4 for which $f_1 \sim f_3$, $f_2 \sim f_4$, and $f_1 \circ f_2 \sim f_3 \circ f_4$.*

What Theorem E.2 establishes is that the class determined by the composition of two forms depends on the classes of the individual forms rather than the forms themselves. Hence, we have

Theorem E.3. *Under composition, the classes of strictly primitive forms of a dis-*

[(E.2)] Thus, proper equivalence occurs if $\begin{pmatrix} u & w \\ v & z \end{pmatrix} \in SL(2, \mathbb{Z})$, the *special linear group* (e.g. see Suzuki [357, p. 74]). Although equivalent forms represent the same numbers, it is not necessarily the case that forms which represent the same numbers are *properly* equivalent. In [50], Chowla conjectured that they are, but in [317] Schinzel gave the counterexample: $f(x, y) = x^2 + 3y^2$ and $g(x, y) = x^2 + xy + y^2$.

criminant Δ, *form a finite abelian group* C_Δ^1 *whose order* h_Δ^1 *is the number of classes of equivalent, strictly primitive forms, called the* class number *of the* form class group C_Δ^1. *(The identity in* C_Δ^1 *is the* principal form $(1, 0, \Delta)$ *or* $(1, 1, (1-\Delta)/4)$, *depending on the parity of* Δ.*)*

We also need the concept of "reduction".

Definition E.5. A strictly primitive form $k_\Delta = (a, b, c)$ is called *strictly reduced* if

(i) when $\Delta < 0$, then $|b| \le a \le c$, and $b \ge 0$, if either $|b| = a$ or $a = c$ and,

(ii) when $\Delta > 0$, then $0 < b < \sqrt{\Delta}$ and $\sqrt{\Delta} - b \le 2|a| \le \sqrt{\Delta} + b$.

(Note that what we call "strictly reduced" is called merely "reduced" in the literature. We make the distinction here because of the upcoming association with ideals.)

Notice that if $\Delta < 0$ and $k_\Delta(x, y) = ax^2 + bxy + cy^2$ is a strictly reduced form, then $b^2 \le a^2$ and $a \le c$, so $-\Delta = 4ac - b^2 \ge 4a^2 - a^2 = 3a^2$. Thus, $a \le \sqrt{-\Delta/3}$. (Compare this with Theorem 1.4.2(c).) If we fix $\Delta < 0$, then since $\Delta = b^2 - 4ac$, there can only be a finite number of choices for a, b, c under the assumption that $|b| \le a \le c$. Therefore, there are an only finite number of reduced forms. In fact,

Theorem E.4. *If* $\Delta < 0$ *is a discriminant, then* h_Δ^1 *is equal to the number of strictly reduced forms of discriminant* Δ.

Proof. See Cox [67, Theorem 2.13, p. 29]. □

The case where we are dealing with indefinite forms is more complicated, as we shall see.

Now we are ready for the

Correspondence between ideals and forms.

Let Δ be a discriminant, and let $I = [a, (b + \sqrt{\Delta})/2]$ be an $\mathcal{O}_\Delta$-ideal which is primitive, with $c = (b^2 - \Delta)/(4a)$. Furthermore, we always assume that $a = N(I) > 0$. We will use the notation $\langle a, b, c\rangle$ for I, as a slight abuse of notation, to keep track of the c value.

At this juncture, the correspondence virtually leaps out at the reader. We have the correspondence

$$k_\Delta = (a, b, c) \leftrightarrow I = \langle a, b, c\rangle.$$

Making this "correspondence" have precise meaning in the general case is beyond the scope of what we are illustrating here, so we refer the reader to Cox [67] for the case of the non-maximal order. This involves certain identifications via ideals with norms prime to the conductor (see Exercise 1.5.5).[E.3] We therefore demonstrate the process for only the maximal order (see Cohn [57, pp. 200–207]).

Let $I = [\alpha, \beta]$ be an ideal in $\mathcal{O}_\Delta$ with fixed basis elements $\alpha, \beta \in \mathcal{O}_\Delta$. We say that α and β are *ordered basis elements*, if $(\alpha\beta' - \beta\alpha')/\sqrt{\Delta} > 0$. It can be shown that two bases of an ideal are ordered if and only if they are equivalent under a

[E.3] Class field theory, which is necessarily stated in terms of the maximal order, does not allow a direct identification of class groups of forms with class groups of ideals (without extreme difficulty) unless one restricts to ideals with norms prime to the conductor.

unimodular transformation. If I has ordered basis elements, then we will refer to I as simply *ordered.*

If I is ordered, and strictly primitive in $\mathcal{O}_\Delta$, then

$$k_\Delta(x, y) = N(\alpha x + \beta y)/N(I)$$

is a strictly primitive form of discriminant Δ. We say that k_Δ *belongs to* I and write $I \to k_\Delta$.

Conversely, suppose that

$$\ell_\Delta = Ax^2 + Bxy + C^2 = g(ax^2 + bxy + cy^2),$$

where $\pm g = \gcd(A, B, C)$, and assume $\Delta = \Delta_0 = b^2 - 4ac$. If $B^2 - 4AC > 0$, then we let $g > 0$, and if $B^2 - 4AC < 0$, we choose g so that $a > 0$. If

$$I = [\alpha, \beta] = \begin{cases} [a, (b - \sqrt{\Delta})/2] & \text{if } a > 0, \\ [a, (b - \sqrt{\Delta})/2]\sqrt{\Delta} & \text{if } a < 0 \text{ and } \Delta > 0, \end{cases}$$

then I is an ordered $\mathcal{O}_\Delta$-ideal. Also, when $a > 0$, I is strictly primitive, whereas $I/\sqrt{\Delta}$ is strictly primitive when $a < 0$. Thus, to every form ℓ_Δ, there corresponds an ideal I to which ℓ_Δ belongs, and we write $\ell_\Delta \to I$.

Now, if we perform $\ell_\Delta \to I \to k_\Delta$, we get $\ell_\Delta = k_\Delta$. However, if we perform $I \to \ell_\Delta \to J$, we do not necessarily get that $I = J$. The explanation for this process requires

Definition E.6. Let Δ be a fundamental discriminant. Two $\mathcal{O}_\Delta$ ideals which satisfy the property that $(\alpha)I = (\beta)J$, for some $\alpha, \beta \in \mathcal{O}_\Delta$ with $N(\alpha\beta) > 0$, are called *narrowly equivalent* or sometimes called *strictly equivalent*, denoted $I \approx J$. The class group formed via narrow equivalence is called the *narrow ideal class group*, denoted C_Δ^+, and its order h_Δ^+, is called the *narrow class number*. We call C_Δ the *wide ideal class group* and h_Δ the *wide class number*.

Now we may describe the process $I \to \ell_\Delta \to J$ which results in $I \approx J$, i.e. the best we can expect is narrow equivalence rather than the equality when applying the process, $\ell_\Delta \to I \to k_\Delta$.

To see why the above is so important, we provide

Theorem E.5. *If Δ is a fundamental discriminant, then either*

(i) *if $\Delta < 0$, or if $\Delta > 0$ and $N(\varepsilon_\Delta) = -1$, then $h_\Delta = h_\Delta^+$, or,*

(ii) *if $\Delta > 0$ and $N(\varepsilon_\Delta) = 1$, then $h_\Delta^+ = 2h_\Delta$.*

Proof. See Janusz [132, Theorem 3.2, p. 204]. □

Thus, when (i) occurs, we have $C_\Delta^+ = C_\Delta$.

The relationship between C_Δ^+ and C_Δ^1 is given by

Theorem E.6. *If Δ is a fundamental discriminant, then $C_\Delta^+ \cong C_\Delta^1$.*

Proof. See Buell [41], Cohn [57], or Cox [67]. □

We have shown how the theory of forms relates to the theory of ideals. For instance, we now see that strictly primitive ideals correspond to strictly primitive forms. Furthermore, Dirichlet's notion of united forms (Definition E.3) corresponds to $g = 1$ in our ideal multiplication formulae (1.2.1)–(1.2.5). Also, strictly reduced forms correspond to strictly reduced ideals. Moreover, Lagrange neighbours introduced in Chapter Two have their analogue in form theory. Let $\ell_\Delta = (a, b, c)$ be a form, then ℓ_Δ gets taken to $\ell_\Delta^+ = (a\beta^2 + b\beta + c, -b - 2a\beta, a)$ where $\beta = \lfloor (b + \sqrt{\Delta})/(2a) \rfloor$ via the substitution induced by $\begin{pmatrix} -\beta & 1 \\ -1 & 0 \end{pmatrix}$, so ℓ_Δ^+ gets taken to ℓ_Δ via $\begin{pmatrix} 0 & -1 \\ 1 & -\beta \end{pmatrix}$. This is the analogue for forms.

Now, we compare the notions of reduction and cycles between forms and ideals. First, we need

Definition E.7. Two strictly reduced forms (a, b, a') and (a', b', c') are called *adjacent* if $b + b' \equiv 0 \pmod{2a'}$.

Since the number of strictly reduced forms is finite for a given discriminant Δ, and adjacent forms are equivalent via the transformation $\begin{pmatrix} 0 & -1 \\ 1 & (b+b')/(2a) \end{pmatrix}$, then the list of successively adjacent forms must return to the original form and they are all equivalent. In fact, two strictly reduced forms are equivalent if and only if they are in the same *cycle* as described above, and the number of forms in a given cycle (called the *period* of the cycle) is always *even*. If $\delta = (b + b')/(2a)$ as above and $\Delta > 0$, then the values of $|\delta|$ are the partial quotients in the simple continued fraction expansion of $(b + \sqrt{\Delta})/(2a)$. If $N(\varepsilon_\Delta) = -1$, then (a, b, c) and $(-a, b, -c)$ lie in the same cycle " halfway apart", and getting from one to the other is essentially multiplying by ε_Δ. Hence, the form cycle is twice the length of the ideal cycle. If $N(\varepsilon_\Delta) = 1$, then (a, b, c) and $(-a, b, -c)$ are in different cycles, and the ideal and form cycles have the same length. However, there are twice as many form cycles as ideal cycles, since *signs matter* in the form cycles.

Definition E.8. A strictly primitive form (k, kn, c) is called an *ambiguous form*, and its class is called an *ambiguous class*.

Caution must be exercised when dealing with ambiguous cycles vs ambiguous ideals.[(E.4)] In fact, if $N(\varepsilon_\Delta) = 1$, then the cycles beginning with $(1, b, *)$ and $(-1, b, *)$ are both ambiguous and distinct. The $(x, y, -x)$ class squares to $(-1, b, *)$, so the $(x, y, -x)$ classes are all of order four in the form class group. However, when we form the correspondence with the wide ideal class group, then $(-1, b, c)$ goes to the identity, so $(x, y, -x)$ becomes of order 2. These are the ambiguous (wide) classes without ambiguous ideals, which we studied in detail in Chapter Six. The related result for forms is contained in the following result which has implications for Pell's equation (see Buell [41, Th. 9.17, p. 179]).

Theorem E.7. *If Δ is an odd discriminant, then $(1, 1, (1 - \Delta)/4)$ represents -1 if and only if there exists a strictly primitive form $(-a, b, a)$ equivalent to $(1, 1, (1 - \Delta)/4)$. In other words, the principal form of discriminant Δ represents -1 if and only if the principal form of discriminant -16, $(1, 0, 4)$, represents $\Delta = b^2 + 4a^2$ in such a way that both a and b are represented by the principal form of discriminant Δ, and the related form $(-a, b, a)$ is in the principal class. This necessarily implies that $(a/p) = (b/p) = (-1/p) = 1$ for all primes p dividing Δ.*

[(E.4)] Duncan Buell has informed this author that his type 20 cycles [41, p. 28] which are listed as ambiguous, are in fact *not* ambiguous. This was a typo, about which we inform the reader to avoid any confusion when comparing our presentation with his.

Thus, the pieces of the historical puzzle come together nicely. There are some other not so well-known pieces of the puzzle which deserve mention. We begin with Theorem E.5 for the case where $\Delta > 0$. The result can be reformulated as follows for positive fundamental discriminants,

$$h_\Delta^+ = 2^{1-e_1(\Delta)}h_\Delta$$

where $e_1(\Delta) = 1$ when $x^2 - Dy^2 = -1$ is solvable, and $e_1(\Delta) = 0$ otherwise, where D is the radicand associated with Δ. This is the order of the form class group C_Δ^1. However, as noted earlier, Gauss considered only forms with even middle coefficient. Hence, Gauss' notion of a class group of forms is different, yet again. Gauss studied the number of classes of properly equivalent, properly primitive forms, which we will call h_Δ^G for the order of Gauss' group G, resulting from such equivalences. The relationship between h_Δ, h_Δ^G and h_Δ^+ is given by

$$h_\Delta^G = 3^{1-e_2(\Delta)}2^{1-e_1(\Delta)}h_\Delta = 3^{1-e_2(\Delta)}h_\Delta^+ \tag{E.1}$$

where $e_2(\Delta) = 1$ if $x^2 - Dy^2 = 4$ has a solution with $\gcd(x, y) = 1$ and $e_2(\Delta) = 0$ otherwise. This is related to a problem of Eisenstein (see Exercise 2.1.14–2.1.15 and footnote (2.1.10)).

The above facts are all buried at various depths throughout the literature. What we have done is to highlight the important features as they pertain to the interrelationships between forms and ideals, and attempt to standardize notation and terminology to avoid the confusion which often arises in this translation.

We conclude with some remarks on the genus theory of forms. We concentrate upon fundamental discriminants $\Delta < 0$, for simplicity sake. The basic idea comes from Lagrange [159], who gave methods for determining the congruence classes represented in $(\mathbf{Z}/\Delta\mathbf{Z})^*$ (the multipliciative group of non-zero elements of the ring of rational integers modulo Δ) by a single form. The classical problem of determining representations of primes by forms requires a method of separating reduced forms of the same discriminant. The fundamental idea behind genus theory was given by Lagrange when he grouped together forms which represent the same class.

Definition E.9. Two strictly primitive, positive definite forms of a fundamental discriminant $\Delta < 0$ are in the same *genus* if they represent the same values in $(\mathbf{Z}/\Delta\mathbf{Z})^*$.

Each genus consists of a finite number of classes of forms, since equivalent forms are in the same genus. We also observe that, as defined in Theorem E.3, the *principal form* is $F_\Delta(x, y) = x^2 + (\sigma - 1)xy + (\sigma - 1 - \Delta)y^2/4$, which we may compare with Definition 4.1.1.

To determine when primes p are represented by forms $g_\Delta(x, y)$ (such as the result of Fermat: $p = x^2 + y^2$ if and only if $p \equiv 1 \pmod 4$), ideal theory and its interplay, via Theorem E.5, can come into play. If $h_\Delta^+ = 1$, and $\gcd(p, \Delta) = 1$, then $F_\Delta(x, y) = p$ if and only if $(\Delta/p) = 1$. Compare this with Lemma 4.1.4.

Gauss called the genus containing the principal form, the *principal genus*. For example,

Proposition E.1. *If $n \in \mathbf{Z}$ is positive and $p > 2$ is a prime not dividing n, then p is represented by a strictly primitive form of discriminant $\Delta = -4n$ in the principal genus if and only if, for some $m \in \mathbf{Z}$, either $p \equiv m^2 \pmod{4n}$ or $p \equiv m^2 + n \pmod{4n}$.*

Proof. See Cox [67, Corollary 2.27, p. 36]. □

In particular, when the principal genus consists of one class, then $p = x^2 + ny^2$ can be uniquely characterized by congruence conditions. For example, $p = x^2+21y^2$ if and only if $p \equiv 1, 25, 37 \pmod{84}$. The more extreme case where the maximal order has only one class per genus is known under the assumption of GRH (see footnote (4.1.19)).

As proved by Gauss in 1801, we have

Theorem E.8. *Let $\Delta = \Delta_0$ be a fundamental discriminant divisible by exactly $N+1$ $(N \geq 0)$ distinct primes. The number, h_Δ^1, of equivalent, strictly primitive forms, can be subdivided equally into 2^N genera of $h_\Delta^1/2^N$ forms which form a subgroup of C_Δ^1 under composition. Furthermore, the principal genus contains precisely those form classes which are squares of some form class under composition. (See Cohn [57, pp. 224ff] for details.)*

This author finds the connection between genus theory, form theory, ideal theory and class field theory for quadratic fields, in terms of the "genus field", a particularly striking one. However, this topic is beyond the scope of this book, so we refer the reader to Janusz [132, pp. 203–211] for a particularly pleasant rendering of this topic.[(E.5)]

The results contained in this appendix were set down in order to give balance, breadth, and to attach an historical thread to the main text by showing how over 200 years of number theory have brought us to the threshold of new and exciting directions, some of which are on the interface with modern computer technology described in the closing Chapter Eight.

[(E.5)]The reader is cautioned, however, that Janusz' [132, Theorem 3.4, p. 206] is false. A counterexample to the theorem is given on page 205 of Janusz's book where he discusses the radicand $D = 34$ saying that: "... the element $(5 + \sqrt{34})/3$ has norm -1 but no unit has norm -1." The reasons for the latter fact are explained in Chapter Six. In any case, for $D = 34$ (in Janusz's notation), $(K^{(+)} : K^{(1)}) = 2$ and $(E^{(+)} : E) = 1$, contradicting the claim that $(K^{(+)} : K^{(1)}) = (E^{(+)} : E)$, in general. We point out the above only so that the reader will not be led astray. Otherwise, the material presented by Janusz is extremely well-elucidated. The reader needs only keep in mind that there may be problems when $N(\varepsilon_\Delta) = 1$, but $N(u) = -1$ for some unit in K. (This was pointed out in a private communication from Professor Janusz to this author, in which acquienscence was given for this footnote. He also informed this author that a corrected second edition of his book will appear in 1996.)

Appendix F

Analytic Considerations.

This appendix provides some background analytic number theory to aid the reader unfamiliar with the related concepts elucidated throughout the text, particularly in Chapter Five, section four. We assume that the reader is familiar with the basic terms, such as convergent and divergent series, as well as analytic continuation. The proofs of the facts provided herein may be found in a variety of number theory texts such as Hua [127], Janusz [132], Borevich–Shafarevich [30], and Narkiewicz [281].

A *Dirichlet series* is a function of the form

$$f(s) = \sum_{n=1}^{\infty} a_n n^{-s} \tag{F.1}$$

where $a_n \in \mathbf{C}$ and s is a complex variable. A special case is the Riemann zeta function of Definition 5.4.2.

In what follows, $\mathrm{Re}\,(s)$ refers to the *real part* of s, and $\arg(s)$ refers to the *argument* of s.

Theorem F.1.

(a) *If the series (F.1) is convergent at a point* s_0, *then it is also convergent in the half-plane* $\mathrm{Re}\,(s) > \mathrm{Re}\,(s_0)$, *and the convergence is uniform in the domain* $\mathrm{Re}\,(s) > \mathrm{Re}\,(s_0)$, $\arg(s - s_0) \leq \theta < \pi/2$.

(b) *If* $s(x) = \sum_{n \leq x} a_n$ *and there exist positive constants* a, b *such that* $|s(x)| \leq ax^b$ *for all* $x \geq 1$, *then* $f(s)$ *is analytic in the half-plane* $\mathrm{Re}\,(s) > b$.

(c) *If* $\lim_{x \to \infty} s(x)/x = a_0$, *then* $\lim_{s \to 1}(s - 1)f(s) = a_0$ *where* $\mathrm{Re}\,(s) \geq 1 + \delta$, $|\arg(s - 1)| \leq \pi/2 - \varepsilon$.

The *Riemann zeta function over* K, $\zeta_K(s) = \sum N(I)^{-s}$, given after Definition 5.4.2, may also be written as $\zeta_K(s) = \sum_{n=1}^{\infty} a_K(n) n^{-s}$ where $a_K(n)$ is the number of integral ideals of K with norm n. We have

Theorem F.2. *The function* $\zeta(s)$ *is a holomorphic function in the half-plane* $\mathrm{Re}\,(s) >$ 1, *and can be analytically continued to a meromorphic function in the whole plane. Its unique singularity is at the point* $s = 1$, *at which it has a simple pole with residue* 1. *Furthermore, for* $\mathrm{Re}\,(s) > 1$, *the equality*

$$\zeta(s) = \prod (1 - p^{-s})^{-1}$$

holds, where the product runs over all primes p.

Proof. See Narkiewicz [281, pp. 104–106]. □

Finally, for any complex variable s we have,

$$\zeta(s) = 2^s \pi^{s-1} \sin\left(\frac{\pi s}{2}\right) \Gamma(1-s)\zeta(1-s). \tag{F.2}$$

(F.2) is called the *fundamental functional equation of the zeta function*. The *Gamma function* is

$$\Gamma(s) = s^{-1} e^{-\gamma s} \prod_{n=1}^{\infty} \left(1 + \frac{s}{n}\right)^{-1} e^{s/n}$$

where γ is Euler's constant (see footnote (8.1.7)). For $Re\,(s) > 0$, we may write

$$\Gamma(s) = \int_0^{\infty} x^{s-1} e^{-x}\, dx.$$

The Gamma function generalizes the notion of a factorial, i.e.

$$\Gamma(n) = (n-1)! \quad \text{for positive} n \in \mathbf{Z},$$

which arises from the functional equation $\Gamma(s+1) = s\Gamma(s)$. Furthermore, it can be shown that

$$\frac{\Gamma'(s)}{\Gamma(s)} = \log s + O(|s|^{-1}),$$

as $|s| \to \infty$.

Riemann [310] proved Theorem F.2 in 1859. The result shows that the zeros of the zeta function (other than $-2, -4, \ldots$) lie symmetrically with respect to $Re\,(s) = 1/2$. Nobody, however, has proved the R.H.: All non-trivial zeros of the zeta function lie in $Re\,(s) = 1/2$. Yet, in 1896, Hadamard [102] proved

Theorem F.3. *In the closed half-plane* $\mathrm{Re}\,(s) \geq 1$, $\zeta(s) \neq 0$.

Recently, however, there has been progress on the RH. Deninger [71], has given some plausible conjectures from which RH would follow.

Now we define

Definition F.1. A *Dirichlet character* χ *modulo* positive $k \in \mathbf{Z}$ means a function χ satisfying:

(i) $\chi(mn) = \chi(m)\chi(n)$.

(ii) $\chi(n) = 0$ if $\gcd(n, k) > 1$.

(iii) $\chi(n) = \chi(m)$ if $n \equiv m \pmod{k}$.

(iv) $\chi(1) \neq 0$.

χ is called *principal* if $\chi(n) = 1$ for each n prime to k, and χ is called *primitive* if, for every divisor k_0 of k, there exists an integer a satisfying $a \equiv 1 \pmod{k_0}$, $\gcd(a, k) = 1$ and $\chi(a) \neq 1$. Also, a character takes values in $\mathbf{C}$.

The Kronecker symbol of Exercise 1.1.4 is an example of a *Dirichlet character modulo* Δ. In fact, Dirichlet proved that any real character can be expressed in this form.

The Dirichlet L-function

$$L(s,\chi) = \sum_{n=1}^{\infty} \chi(n) n^{-s} \tag{F.3}$$

of Definition 5.4.1 satisfies

Theorem F.4.

(i) *If χ is not the principal character, then the series in (F.3) is convergent for* $\mathrm{Re}\,(s) > 0$, *and defines a holomorphic function there.*

(ii) *If χ is any Dirichlet character, then for* $\mathrm{Re}\,(s) > 1$, *we have*

$$L(s,\chi) = \prod (1 - \chi(p) p^{-s})^{-1}.$$

Immediately from this is

Corollary F.1.

(i) $L(s,\chi) \neq 0$ *for* $\mathrm{Re}\,(s) > 1$.

(ii) *If χ_0 is the principal character modulo k, then for* $\mathrm{Re}\,(s) > 1$, *we have*

$$L(s,\chi_0) = \zeta(s) \prod_{p|k} (1 - p^{-s}).$$

(iii) *If χ_0 is the principal character modulo k, then $L(s,\chi_0)$ is a meromorphic function in the whole plane, and its sole singularity is a simple pole at $s = 1$.*

(iv) $\zeta_K(s) = \prod (1 - 1/N(\mathcal{P})^s)^{-1}$ *where the product ranges over all prime $\mathcal{O}_\Delta$-ideals $\mathcal{P}$.*

We remark here that much of the above is used in the proof of the formula for h_Δ given in footnote (1.5.9) (e.g. see Cohn [57, pp. 217–219]).

We conclude with the following celebrated result of Dirichlet, which follows from the analytic class number formula (5.4.1) (e.g. see Narkiewicz [280, Th'm 8.4, p. 383]).

Theorem F.5. *Let Δ be a fundamental discriminant and let r be the number of roots of unity in $\mathcal{O}_\Delta$ when $\Delta < 0$, and $R = \log(\varepsilon_\Delta)$ is the regulator of K, when $\Delta > 0$, then*

$$h_\Delta = \begin{cases} \dfrac{-r}{2|\Delta|} \displaystyle\sum_{x=1}^{|\Delta|} \left(\frac{\Delta}{x}\right) x & \text{for } \Delta < 0 \\ -R^{-1} \displaystyle\sum_{0<x<\Delta/2} \left(\frac{\Delta}{x}\right) \log \sin(\pi x/\Delta) & \text{if } \Delta > 0. \end{cases}$$

References

[1] Adams, W.W., *On the relationship between the convergents of the nearest integer and regular continued fractions*, Math. Comp. **33** (1979), 1321–1331.

[2] Adleman, L., *A subexponential algorithm for the discrete logarithm problem with applications to cryptography*, in **Proc. Twentieth Annual IEEE Symposium on the Foundations of Computer Science** (1979), 55–60.

[3] Adleman, L.M., Rivest, R.L., and Shamir, A., *A method for obtaining digital signatures and public-key cryptosystems*, Comm. of ACM **21** (1978), 120–126.

[4] Alter, R. and Kubota, K., *The diophantine equation $x^2 + D = p^n$*, Pacific J. Math. **46** (1973), 11–16.

[5] Alter, R. and Kubota, K., *The diophantine equation $x^2 + 11 = 3^n$ and a related sequence*, J. Number Theory **7** (1975), 5–10.

[6] Ankeny, N.C., Artin, E., and Chowla, S., *The class number of real quadratic number fields*, Ann. of Math. **56** (1952), 479–493.

[7] Apéry, R., *Sur une equation diophantienne*, C.R. Acad. Sci. Paris Sér. **A 251** (1960), 1263–1264, 1451–1452.

[8] Arno, S., *The imaginary quadratic fields of class number* 4, Acta. Arith. LX (1992), 321–334.

[9] Arno, S., Robinson, M.L., and Wheeler, F.S., *Imaginary quadratic fields with small odd class number*, (preprint).

[10] Ayoub, R.G., and Chowla, S., *On Euler's polynomial*, J. Number Theory **13** (1981), 443–445.

[11] Azuhata, T., *On the fundamental units and class numbers of real quadratic fields*, Nagoya Math. J. **95** (1984), 125–135.

[12] Azuhata, T. *On the fundamental units and the class numbers of real quadratic fields II*, Tokyo J. Math. **10** (1987), 259–270.

[13] Bach, E., **Analytic methods in the analysis and design of number-theoretic algorithms**, MIT Press, Cambridge, Mass. (1985).

[14] Baker, A., *Linear forms in the logarithms of algebraic numbers*, Mathematika **13** (1966), 204–216.

[15] Baker, A., *Imaginary quadratic fields with class number two*, Ann. of Math. **94** (1971), 139–152.

[16] Baker, A. and Wüstholz, A., *Logarithmic forms and group varieties*, J. Reine Angew. Math. **442** (1993), 19–62.

[17] Barrucand, P., *Sur certaines séries de Dirichlet*, C.R. Acad. Sci. Paris. Ser. A–B, **269** (1969), A294–296.

[18] Beach, B.D., Williams, H.C., and Zarnke, C.R., *Some computer results on units in quadratic and cubic fields*, Proc. 25th Summer meeting Can. Math. Congress., Lakehead Univ. (1971), 609–648.

[19] Beckmann, P., **A history of *pi***, Dorset, New York (1989).

[20] Beegner, N.G.W.H., *Report on some calculations of prime numbers*, Nieuw. Arch. Wisk. **20** (1939), 48–50.

[21] Bell, E.T., **Men of Mathematics**, Simon and Schuster, New York (1965).

[22] Bennett, C.D., Blass, J., Glass, A.M.W., Meronk, D.B., and Steiner, R.P., *Linear forms in the logarithms of three algebraic numbers* (preprint).

[23] Bergstrom, H., *Die Klassenzahlformal für reele quadratische Zaklkörper mit zusammengesetzter Diskriminante als Produkt verallgemeinerter Gausscher Summen*, J. Reine Angew. Math. **186** (1944), 91–115.

[24] Bernstein, L., *Fundamental units and cycles in the period of real quadratic number fields, II*, Pacific J. Math. **63** (1976), 63–78.

[25] Bernstein, L., *Fundamental units and cycles I*, J. Number Theory **8** (1976), 446–491.

[26] Beukers, F., **On the generalized Ramanujan–Nagell equation**, Ph.D. thesis, Mathematisch Institute, Der Rijks Universiteit Te Leiden (1979).

[27] Beukers, F., *On the generalized Ramanujan–Nagell equation I*, Acta. Arith. XXXVIII (1981), 389–410.

[28] Biermann, Kurt R., **Die Mathematik und ihre Dozenten an der Berliner Universität 1810–1920**, Akademie Verlag Berlin, 1970.

[29] Biermann, Kurt R., **Johann Peter Gustav Lejeune Dirichlet, Dokumente für sein Leben und Wirken**, Akademie Verlag Berlin (1959).

[30] Borevich, Z.I., and Shafarevich, I.R., **Number Theory**, in Pure and Applied Math **20**, Academic Press, New York, London (1966).

[31] Boston, N., and Greenwood, M.L., *Quadratics representing primes*, (to appear: American Math. Monthly).

[32] Bouniakowsky, W., *Sur les diviseurs numériques invariables des fonctions rationelles entiers*, Acad. Sci. St. Pétersbourg Mém. Sci. Math. et. Phy. **6** (1857), 305–329.

[33] Boyer, C.B., **A history of Mathematics**, Princeton University Press, New Jersey (1985).

[34] Browkin, J. and Schinzel, A., *On the equation* $2^n - D = y^2$, Bull. Acad. Polon. Sci. Sér. Sci. Math. Astronom. Phys. **8** (1960), 311–318.

[35] Buchmann, J., and Düllmann, S., *A probability class group and regulator algorithm and its implementation*, in **Computational Number Theory**, (A. Pethő et al. eds.), Walter de Gruyter, Berlin (1991), 53–72.

[36] Buchmann, J., Sands, J., and Williams, H.C., *P-adic computation of real quadratic class numbers*, Math. Comp. **54** (1990), 855–868.

[37] Buchmann, J., and Williams, H.C., *A key-exchange system based on imaginary quadratic fields*, J. Cryptology **1** (1988), 107–118.

[38] Buchmann, J., and Williams, H.C., *Quadratic fields and cryptography*, in **Number Theory and Cryptography**, (J.H. Loxton, ed.), London Math. Soc. Lecture Note Ser. **154** (1990), 2–25.

[39] Buchmann, J., and Williams, H.C., *On the existence of a short proof for the value of the class number and regulator of a real quadratic field*, in **Number Theory and Applications**, Proc. NATO ASI, **C265** (R.A. Mollin (ed.)) Kluwer Academic Publishers (1989), 481–496.

[40] Buchmann, J., and Williams, H.C., *A key exchange system based on real quadratic fields*, in **Proc. Crypto. 89**, LNCS, **435**, Springer (1990), 335–343.

[41] Buell, D.A., **Binary Quadratic Forms, Classical Theory and Modern Computations**, Springer-Verlag, New York (1989).

[42] Buell, D., *The expectation of success using a Monte-Carlo factoring method — some statistics on quadratic class numbers*, Math. Comp. **43** (1984), 313–327.

[43] Burgess, D.A., *The distribution of quadratic residues and non-residues*, Mathematika **4** (1957), 106–112.

[44] Butts, H.S., and Pall, G., *Ideals not prime to the conductor in quadratic orders*, Acta. Arith. XXI (1972), 261–270.

[45] Butts, H.S., and Pall, G., *Modules and binary quadratic forms*, Acta. Arith. XV (1968), 23–44.

[46] Carlitz, L., *A characterization of algebraic number fields with class number two*, Proc. A.M.S. **11** (1960), 391–392.

[47] Cassels, J.W.S., *On the equation* $a^x - b^y = 1$, *II*, Proc. Cambridge Phil. Soc. **56** (1961), 97–103.

[48] Catalan, E., *Quelques théorèmes emparıques*, (Mélanges Mathématiques XV), Mém. Soc. Royale Sci. de Liège, Sér 2, **12** (1885), 42–43.

[49] Cayley, A., *Tables des formes quadratiques binaries pour les déterminants négatifs depuis* $D = -1$ *jusqu'à* $D = -100$, *pour les déterminants positifs non carrés depuis* $D = 2$ *jusqu'à* $D = 99$ *et pour les treize déterminants négatifs irréguliers qui se trouvent dans les premier millier*, J. Reine Angew. Math. **60** (1862), 357–372.

[50] Chowla, S., *Some problems in elementary number theory*, J. Reine Angew. Math. **222** (1966), 979–981.

[51] Chowla, S., and Friedlander, J., *Class numbers and quadratic residues*, Glasgow Math. J. **17** (1976), 47–52.

[52] Cohen, E.L., *Sur l'équation diophantienne $x^2+11=3^k$*, C.R. Acad. Sci. Paris Sér. A, t. **275** (1972), 5–7.

[53] Cohen, E.L., *On diophantine equations of the form $x^2+D=p^k$*, L'Enseignment Math. **20** (1974), 235–241.

[54] Cohen, E.L., *The diophantine equation $x^2+11=3^k$ and related questions*, Math. Scand. **38** (1976), 240–246.

[55] Cohen, E.L., *On the Ramanujan-Nagell equation and its generalizations*, in **Number Theory** (R.A. Mollin (ed.)), Walter de Gruyter, New York/Berlin (1990), 81–92.

[56] Cohen, H., **A Course in Computational Algebraic Number Theory**, Springer–Verlag, Berlin, New York, Graduate Texts in Math. **138** (1993).

[57] Cohn, H., **A Second Course in Number Theory**, John Wiley and Sons Inc., New York, London (1962).

[58] Cohen, H., and Lenstra, H.W. Jr., *Heuristics on class groups of number fields*, in **Number Theory**, (Noordwijkerhout, 1983), Lecture Notes in Math. **1068**, Springer–Verlag, Berlin (1984), 33–62.

[59] Cohn, J.H.E., *On square Fibonacci numbers*, J. London Math. Soc. **39** (1964), 537–540.

[60] Cohn, J.H.E., *Lucas and Fibonacci numbers and some Diophantine equations*, Proc. Glasgow Math. Assoc. **7** (1965), 24–28.

[61] Cohn, J.H.E., *Eight Diophantine equations*, Proc. London Math. Soc. **16** (1966), 153–166.

[62] Cohn, J.H.E., *The diophantine equation $x^2+C=y^n$*, Acta. Arith. LXV (1993), 367–381.

[63] Cohn, J.H.E., *The diophantine equation $x^2+3=y^n$*, Glasgow Math. J. **35** (1993), 203–206.

[64] Conrey, J.B., *Two-fifths of the zeros of the Riemann zeta function are on the real line* (unpublished manuscript 1987).

[65] Coppersmith, D., *Fast evaluation of discrete logarithms in fields of characteristic two*, IEEE Trans. Inform. Theory **30** (1984), 587–594.

[66] Cowles, M.J., *On the divisibility of the class number of imaginary quadratic fields*, J. Number Theory **12** (1980), 113–115.

[67] Cox, D.A., **Primes of the form x^2+ny^2**, J. Wiley and Sons, New York (1989).

[68] Davis, P.J., *Fidelity in mathematical discourse: Is one and one really two?*, American Mathematical Monthly **79** (1972), 252–253.

[69] Davis, P.J., and Hersh, R., **The mathematical experience**, Birkhäuser, Boston (1981).

[70] Degert, G., *Uber die Bestimmung der Grundeinheit gewisser reell-quadratischer Zahlkörper*, Abh. Math. Sem. Univ. Hamburg **22** (1958), 92–97.

[71] Deninger, C., *L-functions of mixed motives*, A.M.S. Proc. Symp. Pure Math. **55** (1994), 517–525.

[72] Deuring, M., *Imaginär-quadratische Zahlkörper mit der Klassenzahl I*, Math. Z. **37** (1933), 405–415.

[73] Dickson, L.E., *History of the Theory of Numbers*, Vol. III, Chelsea, New York (1952).

[74] Diffie, W., and Hellmann, M.E., *New directions in cryptography*, IEEE Transactions in Information Theory IT–22 (1976), 644–654.

[75] Dodd, F.W., **Number Theory in the Quadratic Field with Golden Section Unit**, Polygonal Pub. House, Passaic, N.J. (1983).

[76] Dubois, E., and Levesque, C., *On determining certain real quadratic fields with class number one and relating this property to continued fractions and primality properties*, Nagoya Math. J. **124** (1991), 157–180.

[77] Dueck, G., and Williams, H.C., *Computation of the class number and class group of a complex cubic field*, Math. Comp. **45** (1985), 223–231. (Corrigendum, ibid **50** (1988).)

[78] Eckhardt, C., *Computation of class numbers by an analytic method*, J. Symbolic Computation **4** (1987), 41–52.

[79] Edgar, H.M., Mollin, R.A., and Petersen, B,L., *Class groups, totally positive units and squares*, Proc. Amer. Math. Soc. **98** (1986), 33–37.

[80] Edwards, H.M., **Riemann's Zeta Function**, Academic Press, New York (1974).

[81] Edwards, H.M., *The background of Kummer's proof of Fermat's last theorem for regular primes*, Arch. Hist-Exact. Sci. **14** (1975), 219–236.

[82] Edwards, H.M., *Postscript to "The Background of Kummer's proof..."*, Arch. Hist. Exact Sci. **17** (1977), 381–394.

[83] Erdös, P., and Szekeres, G., *Über die Anzahl der Abelschen Gruppen gegebener Ordnung und uber ein verwandtes zahlentheoretisches Problem*, Acta. Litt. Sci. Szeged. (1934), 95–102.

[84] Escott, E.B., Réponses 1133, *Formule d'Euler $x^2 + x + 41$ et formules analogues*, L'intermédiaire des Math. 6 (1899), 10–11.

[85] Euler, L., Extrait d'une lettre de M. Euler le père à M. Bernoulli concernant le memoire imprimé parmi ceux de 1771, p. 381, Nouveaux mémoirs de l'Académie des Sciences de Berlin 1772, (1774) Histoire, 35–36 Opera Omnia, I_3, Commentationes Arithmeticae, II, 335–337, Teubner, Lipsiae et Berolini 1917.

[86] Frénicle, de Bessy, *Solutio duorum problemarum circa numeros cubos et quadratos* (1657), Bibliotheque Nationale de Paris.

[87] Frei, G., *Leonhard Euler's convenient numbers*, Math. Intell. **1** (1985), 55–58, 64.

[88] Frobenius, F.G., *Über quadratische Formen die viele Primzahlen darstellen*, Sitzungsber.d.Kgl. Preuss. Akad. Wiss. Berlin (1912), 966–980.

[89] Fung, G.W., and Williams, H.C., *Quadratic polynomials which have a high density of prime values*, Math. Comp. **55** (1990), 345–353.

[90] Gauss, C.F., **Werke**, Bd II, Göttingen (1863), 475–476.

[91] Gauss, C.F., **Disquisitiones Arithmeticae**, Springer–Verlag (English Edition), 1986.

[92] Glass, A., Meronk, D., Okada, T., and Steiner, R., *A small contribution to Catalan's equation*, J. Number Theory **47** (1994), 131–137.

[93] Glass, A., O'Neil, T., and Steiner, R., *Improved upper bounds on the exponents in Catalan's equation*, (preprint).

[94] Goldfeld, D., *A simple proof of Siegel's theorem*, Proc. Nat. Acad. Sci. U.S.A. **71** (1974), 1055.

[95] Goldfeld, D., *Gauss' class number problem for imaginary quadratic fields*, Bull. A.M.S. **13** (1985), 23–37.

[96] Golomb, S.W., *Powerful numbers*, Amer. Math. Monthly **77** (1970), 848–857.

[97] Goodman, N.D., *Modernizing the philosophy of mathematics*, Synthese **88** (1991), 119–126.

[98] Gordon, D., *Discrete logarithms in $GF(p)$ using the number field sieve*, (preprint).

[99] Granville, A., *Some conjectures related to Fermat's last theorem*, in **Number Theory** (R.A. Mollin (ed.)), Walter de Gruyter, Berlin, New York (1990).

[100] Gross, B.H. and Rohrlich, D.E., *Some results on the Mordell–Weil group of the Jacobian of the Fermat curve*, Invent. Math. **44** (1978), 201–224.

[101] Guy, R.K., **Unsolved problems in number theory**, (Second Edition) Springer–Verlag, Berlin (1994).

[102] Hadamard, J., *Sur la distribution des zéros de la fonction $\zeta(s)$ et ses conséquenses arithmétiques*, Bull. Soc. Math. France **24** (1896), 199–200.

[103] Hafner, J.L., and McCurley, K.S., *A rigorous subexponential algorithm for computation of class groups*, IBM research report, San José, Ca. (1989).

[104] Halter-Koch, F., *Einige periodische Kettenbruchentwicklungen und Grundeinheiten quadratischer Ordnungen*, Abh. Math. Sem. Univ. Hamburg **59** (1989), 157–169.

[105] Halter-Koch, F., *Quadratische Ordnungen mit grosser Klassenzahl*, J. Number Theory **34** (1990), 82–94.

[106] Halter-Koch, F., *On a class of insoluble binary quadratic diophantine equations*, Nagoya Math. J. **123** (1991), 141–151.

[107] Halter-Koch, F., *Prime-producing quadratic polynomials and class numbers of quadratic orders* in **Computational Number Theory** (A. Petho, et al. (eds.)), Walter de Gruyter, Berlin (1991), 73–82.

[108] Halter-Koch, F., *Quadratische Ordnungen mit grosser Klassenzahl II*, J. Number Theory **44** (1993), 166–171.

[109] Hardy, G.H., and Littlewood, J.E., *Some problems of 'Partitio Numerorum'; III: On the expression of a number as a sum of primes*, Acta. Math. **44** (1923), 1–70.

[110] Hardy, G.H., and Wright, E.M., **An introduction to the theory of numbers**, fourth edition, Oxford at the Clarendon Press, London (1960).

[111] Hasse, H., *Uber mehrklassige aber eingeschlechtige reell-quadratische Zahlkörper*, Elem. Math. **20** (1965), 49–59.

[112] Hasse, H., *Produktformeln für verallgemeinerte Gaussche Summen und ihre Anwendung auf die Klassenzahlformel für reele quadratische Zahlkörper*, Math. Z. **46** (1940), 303–314.

[113] Heath–Brown, D.R., *Ternary quadratic forms and sums of three square-full numbers*, Séminaire de Théorie des Nombres, Paris 1986–87, Birkhauser, Boston (1988), 137–163.

[114] Heegner, K., *Diophantische Analysis und Modulfunktionen*, Math. Z. **56** (1952), 227–253.

[115] Heilbronn, H., *On the class number in imaginary quadratic fields*, Quart. J. Math., Oxford Ser. 2, **5** (1934), 150–160.

[116] Heilbronn, H., and Linfoot, E.H., *On the imaginary quadratic corpora of class number one*, Quart. J. Math. Oxford Ser. 2, **5** (1934), 293–301.

[117] Hellmann, M.E., *The mathematics of public key cryptography*, Scientific American **241** (1979), 146–157.

[118] Hendy, M.D., *Applications of a continued fraction algorithm to some class number problems*, Math. Comp. **28** (1974), 267–277.

[119] Hendy, M.D., *Prime quadratics associated with complex quadratic fields of class number two*, Proc. Amer. Math. Soc. **43** (1974), 253–260.

[120] Hendy, M.D., *The distribution of ideal class numbers of real quadratic fields*, Math. Comp. **29** (1975), 1129–1134. (Corrigendum, ibid **30** (1976), 679.)

[121] Hendy, M.D., *Class number divisors for some real quadratic fields*, Occ. Pub. Math. **5** (1977), 1-3.

[122] Hensley, D., and Richards, I., *On the incompatibility of two conjectures concerning primes*, Proc. Symp. in Pure Math. Analytic Number Theory (AMS) **24** (1973), 123–127.

[123] Higgins, O., *Another long string of primes*, J. Rec. Math. **14** (1982), 185.

[124] Hirzebruch, F., *Hilbert modular surfaces*, L'Enseignment Math. **19** (1973), 183–281.

[125] Hoffstein, J., *On the Siegel-Tatuzawa theorem*, Acta. Arith. **38** (1980), 167–174.

[126] Hua, L.-K., *On the least solution of Pell's equation*, Bull. Amer. Math. Soc. **48** (1942), 731-735.

[127] Hua, L.-K., **Introduction to Number Theory**, Springer–Verlag, Berlin (1982).

[128] Hughes, I., and Mollin, R.A., *Totally positive units and squares*, Proc. Amer. Math. Soc. **87** (1983), 613–616.

[129] Ince, E.L., *Cycles of reduced ideals in quadratic fields* in **Mathematical Tables**, IV British Assoc. for Advance. Sci. London (1934).

[130] Inkeri, K., *On Catalan's problem*, Acta. Arith. **9** (1964), 285–290.

[131] Inkeri, K. and Hyyrö, S., *On the congruence* $3^{p-1} \equiv 1 \pmod{p^2}$ *and the diophantine equation* $x^2 - 1 = y^p$, Ann. Univ. Turku, Ser. A1, **50** (1961), 2 pages.

[132] Janusz, G.J., **Algebraic Number Fields**, Academic Press, New York, London (1973).

[133] Kac, M., Rota, G-C., and Schwartz, J.T., **Discrete thoughts**, Birkhäuser, Boston (1986).

[134] Kahn, D., **The Codebreakers, the Study of Secret Writing**, Macmillan, New York (1967).

[135] Kaplan, P., *Idéaux k-réduits des ordres des corps quadratiques réels*, (to appear: J. Math. Soc. of Japan).

[136] Kaplan, P., and Williams, K.S., *Pell's equations* $x^2 - my^2 = -1, -4$ *and continued fractions*, J. Number Theory **23** (1986), 169–182.

[137] Kaplan, P., and Williams, K.S., *The distance between ideals in orders of a real quadratic field*, L'Énseignment Math. **36** (1990), 321–358.

[138] Karst, E., *The congruence* $2^{p-1} \equiv 1 \pmod{p^2}$ *and quadratic forms with high density of primes*, Elem. Math. **22** (1967), 85–88.

[139] Karst, E., *New quadratic forms with high density of primes*, Elem. Math. **28** (1973), 116–118.

[140] King, J.P., **The art of mathematics**, Plenum Press, New York and London (1992).

[141] Kitcher, P., **The nature of mathematical knowledge**, Oxford University Press, New York and Oxford (1983).

[142] Kline, M., **Mathematics: The loss of certainty**, Oxford University Press, New York (1980).

[143] Kloss, K.E., *Some number theoretic calculations*, J. Res. Nat. Bur. Standards Ser. B **69B** (1965), 335–336.

[144] Kloss, K.E., Newman, M. and Ordman, E. *Class numbers of primes of the form* $4n+1$, Math. Comp. **23** (1965), 213–214.

[145] Knuth, D.E., **The Art of Computer Programming Vol II: Seminumerical Algorithms**, (2nd ed.) Addison–Wesley, Reading, Mass. (1981).

[146] Koblitz, N., *Elliptic curve cryptosystems*, Math. Comp. **48** (1987), 203–209.

[147] Koblitz, N., **A Course in Number Theory and Cryptography**, Springer–Verlag, New York, Graduate Texts in Math. **114** (1988).

[148] Koetsier, T., *Lakatos' philosophy of mathematics*, in **Studies in the history of philosophy of mathematics**, vol. 3, North–Holland, Amsterdam (1991).

[149] Krakowski, I., *The Four-Color problem revisited*, Philosophical Studies **38** (1980), 91–96.

[150] Kranakis, E., **Primality and Cryptography**, John Wiley and Sons (1986).

[151] Kuhn, T.S., **The structure of scientific revolutions**, second revised ed., University of Chicago, Chicago (1970).

[152] Kuroda, S., *Table of class numbers* $h(p) > 1$ *for quadratic fields* $\mathbf{Q}(\sqrt{p})$, $p \leq 2776817$, Math. Comp. **29** (1975), 335–336, UMT File.

[153] Kutsuna, M., *On the fundamental units of real quadratic fields*, Proc. Japan Acad. **50** (1974), 580–583.

[154] Kutsuna, M., *On a criterion for the class number of a quadratic number field to be one*, Nagoya Math. J. **79** (1980), 123–129.

[155] Lachaud, G., *On real quadratic fields*, Bull. Amer. Math. Soc. **17** (1987), 307–311.

[156] Lagarias, J.C. and Weisser, D.P., *Fibonacci and Lucas cubes*, Fib. Quart. **19** (1981), 39–43.

[157] Lagrange, J.-L., *Solution d'une problème d'arithmètique (1766)*, Oeuvres de Lagrange **1**, 671–731, (Gauthier–Villars (1867)).

[158] Lagrange, J.-L., *Sur la solution des problèmes indéterminatés du second degré* (1769), Oeuvres de Lagrange **2**, 377–535 (Gauthiers–Villars (1868)).

[159] Lagrange, J.-L., Oeuvres **3**, Gauthier–Villars, Paris (1869).

[160] Lagrange, J.-L., Recherches d'arithmètiques, Nouv. Mém. Acad. Berlin (1773), 265-312.

[161] Lakatos, I., **Proofs and refutations**, Cambridge University Press, Cambridge (1977), reprinted with corrections.

[162] Landau, E., *Über die Klassenzahl der binären quadratischen Formen von negativer Discriminante*, Math. Ann. **56** (1902), 671–676.

[163] Landau, E., Über die Klassenzahl imaginär-quadratischer Zahlkörper, Göttingen Nachr. (1918), 285-295.

[164] Lang, S.D., *Note on the class number of the maximal real subfield of a cyclotomic field*, J. Reine Angew. Math. **290** (1977), 70–72.

[165] Lazarus, A., *Implications of Computational Mathematics for the Philosophy of Mathematics* (to appear: Math. Comp.).

[166] Lebesgue, V.A., *Sur l'mpossibilité en nombres entiers de l'équation* $x^m = y^2 + 1$, Nouv. Ann. de Math. **9** (1850), 178–181.

[167] Legendre, A.M., **Théorie les nombres**, Libraire Scientifique, A. Hermann, Paris (1798), 69–76, second ed. (1808), 61–67, third ed. (1830), 72–80.

[168] Lehman, R.S., *Factoring large integers*, Math. Comp. **28** (1974), 637–646.

[169] Lehmer, D.H., *On the function* $x^2 + x + A$, Sphinx **6** (1936), 212–214, and **7** (1937), 40.

[170] Lehmer, D.H., *Computer technology applied to the theory of numbers*, Studies in Number Theory, MAA Studies in Math. **6** (1969), 117–151.

[171] Lehmer, D.H., *Mechanized mathematics*, Bulletin of the American Mathematical Society **72** (1966), 739–750.

[172] Lehmer, D.H., **Selected Papers of D.H. Lehmer**, Volumes I–III, (D. McCarthy (ed.)), The Charles Babbage Research Centre, St. Pierre, Canada (1981).

[173] Lehmer, D.H., Lehmer, E., Mills, W.H., and Selfridge, J.L., *Machine proof of a theorem on cubic residues*, Math. Comp. **16** (1962), 407–415.

[174] Lehmer, D.H., Lehmer, E., and Shanks, D., *Integer sequences having prescribed quadratic character*, Math. Comp. **24** (1970), 433–451.

[175] Lenstra, A.K., and Lenstra Jr., H.W., *Algorithms in number theory*, The University of Chicago Technical Report 87-008 (May 1987).

[176] Lenstra, A.K., and Lenstra Jr., H.W., *The development of the number field sieve*, Lecture Notes in Math., Springer–Verlag, Berlin, **1554** (1994).

[177] Lenstra, A.K., Lenstra Jr., H.W., Manasse, M.S., and Pollard, J.M., *The number field sieve*, in **Proc. 22nd Annual ACM Symp. on Theory of Computing**, Baltimore (1990), 564–572.

[178] Lenstra Jr., H.W., *On the calculation of regulators and class numbers of quadratic fields*, London Math. Soc. Lec. Note. Ser. **56** (1982), 123–150.

[179] Lenstra Jr., H.W., *"Primality Testing," Computational Methods in Number Theory*, Math. Centrum Tracts **154** (1983), 55–77.

[180] Lenstra Jr., H.W., *Integer programming with a fixed number of variables*, Math. Oper. Res. **8** (1983), 538–548.

[181] Lerch, M., *Sur le nombre des classes de formes quadratiques binaires d'un discriminant positif fondamental*, J. de Math. **9** (1903), 377–401.

[182] Leu, M.G., *On a criterion for the quadratic fields* $Q(\sqrt{n^2+4})$ *to be of class number two*, Bull. London Math. Soc. **24** (1992), 309–312.

[183] Lévesque, C., *Continued fraction expansions and fundamental units*, J. Math. Phys. Sci. **22** (1988), 11–14.

[184] Lévesque, C., and Rhin, G., *A few classes of periodic continued fractions*, Utilitas Math. **30** (1986), 79–107.

[185] Lévy, A., *Sur les nombres premiers dérivés de trinomes du second degré*, Sphinx–Oedipe **9** (1914), 6–7.

[186] Lévy, P., *Sur le développement en fraction continue d'une nombre choisi au hasard*, Compositio Math. **3** (1936), 286–303.

[187] Littlewood, J.E., *On the class number of the corpus* $P(\sqrt{-k})$, Proc. London Math. Soc. **27** (1928), 358-372.

[188] Ljunggren, W., *New theorems concerning the diophantine equation* $Cx^2+D=y^n$, Norsk. Vid. Selsk, Forh. Trondlyem **29** (1956), 1–4.

[189] Ljunggren, W., *On the diophantine equation* $Cx^2+D=2y^n$, Math. Scand. **18** (1966), 69–86.

[190] Ljunggren, W., *On the diophantine equation* $Cx^2+D=y^n$, Pacific J. Math. **14** (1964), 585–596.

[191] Ljunggren, W., *Einige Bemerkungen über die Darstellung ganzer Zahlen durch binäre kubische Formeln mit positiver Diskriminante*, Acta. Math. **75** (1942), 1-21.

[192] Ljunngren, W., *Some theorems on the indeterminante equation* $\frac{x^n-1}{x-1}=y^q$, Norsk. Mat. Tidsskrift **25** (1943), 17–20.

[193] London, H., and Finkelstein, R., *On Fibonacci and Lucas numbers which are perfect powers*, Fib. Quart. **7** (1969), 476–481.

[194] Louboutin, S., *Arithmétique des corps quadratiques réels et fractions continues*, Thèse de doctorat, Univ. Paril **7** (1987).

[195] Louboutin, S., *Continued fractions and real quadratic fields*, J. Number Theory **30** (1988), 167–176.

[196] Louboutin, S., *Minorationes (sous l'hypothèse de Riemann généralisée) des nombres des classes les corps quadratiques imaginaires*, C.R. Acad. Sci. Paris, t. **310**, Série I (1990), 499–513.

[197] Louboutin, S., *Prime producing quadratic polynomials and class numbers of real quadratic fields*, Canad. J. Math. XLII (1990), 315–341.

[198] Louboutin, S., *Extensions du théorème de Frobenius-Rabinowitsch*, C.R. Acad. Sci. Paris t. **312**, Série I (1991), 711–714.

[199] Louboutin, S., and Mollin, R.A., *Bounds for class numbers of quadratic orders*, (in **Computational Algebra and Number Theory** (W. Bosma and A.J. van der Poorten (eds.)), Kluwer Academic Publishers (1995), 153–158.

[200] Louboutin, S., Mollin, R.A., and Williams, H.C., *Class numbers of real quadratic fields, continued fractions, reduced ideals, prime-producing quadratic polynomials and quadratic residue covers*, Canad. J. Math. **44** (1992), 824–842.

[201] Louboutin, S., Mollin, R.A., and Williams, H.C., *Class groups of exponent two in real quadratic fields*, in **Advances in Number Theory** (F. Gouvêa and N. Yui (eds.)) Clarendon Press, Oxford (1993), 499–513.

[202] Lu, H. *On the class number of real quadratic fields*, Sci. Sinica, Special Issue **2** (1979), 118–130.

[203] Lu, H., *Divisibility for class numbers of a kind of quadratic fields*, Acta. Math. Sinica **28** (1985), 756.

[204] Lu, H., *Pellian equation conjecture and absolutely nonsingular projective varieties — Hecke operator and Pellian equation conjecture IV*, Contemporary Math. **168** (1994), 217–225.

[205] Lu, H., and Zhang, M., *S. Chowla's conjecture on a class of real quadratic fields* (preprint 1988).

[206] Lucas, E., *The theory of simply periodic numerical functions*, (translated from the French by Sidney Kravitz; edited by Douglas Lind), Fibonacci Assoc. (1969).

[207] Lukes, R.F., Patterson, C.D., and Williams, H.C., *Numerical Sieving Devices: their history and some applications* (to appear: Nieuw Archief voor Wiskunde).

[208] Mahler, K., *On the greatest prime factor of* $ax^m + by^m$, Nieuw. Arch. Wisk. **1** (1953), 113–122.

[209] Maohua, Le, *On the diophantine equation* $x^2 + D = 4p^n$, J. Number Theory **41** (1992), 87–97.

[210] Marcus, D.A., **Number Fields**, Springer–Verlag, Berlin, New York (1977).

[211] Masley, J.M., *Where are number fields with small class number?* in **Number Theory** (Proc. Southern Illinois Conf., Southern Illinois Univ., Carbondale, Ill., 1979), Lecture Notes in Math **751**, Springer, Berlin (1979), 221–242.

[212] Masser, D.W., **Open problems**, Proc. Symp. Analytic Number Theory, W.W.L. Chen (ed.), London: Imperial Coll. (1985).

[213] Mathews, G.B., **Theory of Numbers**, (1892) reprinted, Chelsea, New York, (1961).

[214] McCurley, K.S., *A key distribution system equivalent to factoring*, (preprint).

[215] McDaniel, W.L., *Representations of every integer as the difference of powerful numbers*, Fibonacci Quart. **20** (1982), 85–87.

[216] Mead, D.G., *The equation of Ramanujan–Nagell and* $\lfloor y^2 \rfloor$, Proc. Amer. Math. Soc. **4** (1973), 333–341.

[217] Mignotte, M., *Un critère élementaire pour l'équation de Catalan*, C.R. Math. Rep. Acad. Sci. Canada, **15** (1993), 199–200.

[218] Miller, V., *Use of elliptic curves in cryptosystems* in **Advances in Cryptology**, Proc. of Crypto '85, Lecture Notes in Comp. Sci., **218**, Springer–Verlag, New York, (1986), 417–426.

[219] Mitsuhiro, T., Nakahara, T., and Uehara, T., *The class number formula of a real quadratic field and an estimate of the value of a unit*, Canad. Math. Bull. **38** (1995), 98–103.

[220] Mohanty, S.P., *A system of cubic diophantine equations*, J. Number Theory **9** (1977), 153–159.

[221] Mollin, R.A., *Ambiguous classes in quadratic fields*, Math. Comp. **61** (1993), 355–360.

[222] Mollin, R.A., *Applications of a new class number two criterion for real quadratic fields*, in **Computational Number Theory**, (A. Petho et al. (eds.)), Walter de Gruyter, Berlin (1991), 83–94.

[223] Mollin, R.A., *Class number one criteria for real quadratic fields I–II*, Proc. Japan Acad. **63**, Ser. A (1987), 121–125, 162–164.

[224] Mollin, R.A., *Class numbers of quadratic fields determined by solvability of diophantine equations*, Math. Comp. **48** (1987), 233–242.

[225] Mollin, R.A., *Counting norm divisors of the Euler–Rabinowitsch polynomial to determine class numbers of real quadratic fields*, Utilitas Math. **47** (1995), 247–254.

[226] Mollin, R.A., *Diophantine equations and class numbers*, J. Number Theory **24** (1986), 7–19.

[227] Mollin, R.A., *Generalized Fibonacci primitive roots and class numbers of real quadratic fields*, Fibonacci Quart. **26** (1988), 46–53.

[228] Mollin, R.A., *Lower bounds for class numbers of real quadratic and biquadratic fields*, Proc. Amer. Math. Soc. **101** (1987), 439–444.

[229] Mollin, R.A., *Necessary and sufficient conditions for the class number of a real quadratic field to be one, and a conjecture of S. Chowla*, Proc. Amer. Math. Soc. **102** (1988), 17–21.

[230] Mollin, R.A., *On the divisor function and class numbers of real quadratic fields I-IV*, Proc. Japan Acad. **66** Ser. A (1990), 109–111; 274–277; **67**, Ser. A (1991) 338–342, and **68** Ser. A (1992), 15–17.

[231] Mollin, R.A., *On the insolubility of a class of diophantine equations and the nontriviality of the class numbers of related real quadratic fields of Richaud-Degert type*, Nagoya Math. J. **105** (1985), 39–47.

[232] Mollin, R.A., *Orders in quadratic fields III*, Proc. Japan Acad. **70**, Ser. A (1994), 176–181.

[233] Mollin, R.A., *Quadratic irrationals, ambiguous classes and symmetry in real quadratic fields*, Proc. Japan Acad. **70**, Ser. A (1994), 218–222.

[234] Mollin, R.A., *Quadratic polynomials producing consecutive, distinct primes, and class groups of complex quadratic fields*, (to appear: Acta. Arith.).

[235] Mollin, R.A., *Powers in continued fractions and class numbers of real quadratic fields*, Utilitas Math. **42** (1992), 25–30.

[236] Mollin, R.A., *Solutions of diophantine equations and divisibility of class numbers of complex quadratic fields*, (to appear: Glasgow Math. J.)

[237] Mollin, R.A., *The palindromic index — a measure of ambiguous cycles of reduced ideals without any ambiguous ideals in real quadratic orders*, (to appear: Séminaire de Théorie des nombres de Bordeaux).

[238] Mollin, R.A., *Use of the Rabinowitsch polynomial to determine the class groups of a real quadratic field*, Bull. Austral. Math. Soc. **50** (1994), 435–443.

[239] Mollin, R.A., and van der Poorten, A.J., *A note on symmetry and ambiguity*, Bull. Austral. Math. Soc. **51** (1995), 215–233.

[240] Mollin, R.A., van der Poorten, A.J., and Williams, H.C., *Halfway to a solution of* $x^2 - Dy^2 = -3$, Journal de Théorie des Nombres, Bordeaux **6** (1994) 421–459.

[241] Mollin, R.A., Small, C., Varadarajan, K., and Walsh, P.G., *On unit solutions of the equation* $xyz = x + y + z$ *in the ring of integers of a quadratic field*, Acta. Arith. XLVIII (1987), 341–345.

[242] Mollin, R.A., and Walsh, P.G., *On powerful numbers*, Internat. J. Math. and Math. Sci. **9** (1986), 801–806.

[243] Mollin, R.A., and Walsh, P.G., *A note on powerful numbers, quadratic fields and the Pellian*, C.R. Rep. Acad. Sci. Canada **8** (1986), 109–114.

[244] Mollin, R.A., and Walsh, P.G., *Proper differences of non-square powerful numbers*, C.R. Math. Rep. Acad. Sci Canada **10** (1988), 71–76.

[245] Mollin, R.A., and Williams, H.C., *Period four and real quadratic fields of class number one*, Proc. Japan Acad., Ser. A **65** (1989), 89–93.

[246] Mollin, R.A., and Williams, H.C., *Prime producing quadratic polynomials and real quadratic fields of class number one*, in **Number Theory** (J.-M. DeKoninck and C. Levesque (eds.)), Walter de Gruyter, Berlin (1989), 654–663.

[247] Mollin, R.A., and Williams, H.C., *A conjecture of S. Chowla via the generalized Riemann hypothesis*, Proc. Amer. Math. Soc. **102** (1988), 794–796 (Corrigenda, Proc. Amer. Math. Soc. (ibid, **123**, p. 975, 1994)).

[248] Mollin, R.A., and Williams, H.C., *On prime valued polynomials and class numbers of real quadratic fields*, Nagoya Math. J. **112** (1988), 143–151.

[249] Mollin, R.A., and Williams, H.C., *Class number one for real quadratic fields, continued fractions and reduced ideals* in **Number Theory and Applications** (R.A. Mollin (ed.)), (NATO ASI) Kluwer Academic Publishers **C265** (1989), 481–496.

[250] Mollin, R.A., and Williams, H.C., *Quadratic non-residues and prime-producing polynomials*, Canad. Math. Bull. **32** (1989), 474–478.

[251] Mollin, R.A. and Williams, H.C., *Solution of the class number one problem for real quadratic fields of extended Richaud–Degert type (with one possible exception)*, in **Number Theory** (R.A. Mollin (ed.)) Walter de Gruyter, Berlin (1990), 417–425.

[252] Mollin, R.A., and Williams, H.C., *Continued fractions of period five and real quadratic fields of class number one*, Acta. Arith. LVI (1990), 55–63.

[253] Mollin, R.A., and Williams, H.C., *Class number problems for real quadratic fields*, in **Number Theory and Cryptography**, (J.H. Loxton, ed.), London Math. Soc. Lecture Note Series **154** (1990), 177–195.

[254] Mollin, R.A., and Williams, H.C., *Solution of a problem of Yokoi*, Proc. Japan Acad. **66**, Ser. A (1990), 141–145. (Corringenda, ibid **67**(1991), 253.)

[255] Mollin, R.A., and Williams, H.C., *Powers of two, continued fractions and the class number one problem for real quadratic fields* $Q(\sqrt{d})$ *with* $d \equiv 1$ (mod 8), in **The Mathematical Heritage of C.F. Gauss**, (G.M. Rassias (ed.)), World Scientific Pub., Singapore (1991), 505–516.

[256] Mollin, R.A., and Williams, H.C., *On a determination of real quadratic fields of class number one and related continued fraction period length less than* 25, Proc. Japan Acad. **67**, Ser. A (1991), 20–25.

[257] Mollin, R.A., and Williams, H.C., *On a solution of a class number two problem for a family of real quadratic fields*, in **Computational Number Theory** (A. Petho et. al. (ed.)), Walter de Gruyter, Berlin, New York (1991), 95–101.

[258] Mollin, R.A., and Williams, H.C., *Affirmative solution of a conjecture related to a sequence of Shanks*, Proc. Japan Acad. **67**, Ser. A (1991), 70–72.

[259] Mollin, R.A., and Williams, H.C., *On real quadratic fields of class number two*, Math. Comp. **59** (1992), 625–632.

[260] Mollin, R.A., and Williams, H.C., *On the period length of some special continued fractions*, Séminaire de Théorie des Nombres, Bordeaux **4** (1992), 19–42.

[261] Mollin, R.A., and Williams, H.C., *Computation of the class number of a real quadratic field*, Utilitas Math. **41** (1992), 259–308.

[262] Mollin, R.A., and Williams, H.C., *Consecutive powers in continued fractions*, Acta. Arith. LXL3 (1992), 233-264.

[263] Mollin, R.A., and Williams, H.C., *A complete generalization of Yokoi's p-invariants*, Colloq. Math. LXIII (1992), 285–294.

[264] Mollin, R.A., and Williams, H.C., *Classification and enumeration of real quadratic fields having exactly one non-inert prime less than a Minkowski bound*, Canad. Math. Bull. **36** (1993), 108–115.

[265] Mollin, R.A., and Williams, H.C., *Quadratic residue covers for certain real quadratic fields*, Math. Comp. **62** (1994), 885–897.

[266] Mollin, R.A., and Williams, H.C., *Proof, disproof and advances concerning certain conjectures on real quadratic fields* $Q(\sqrt{n^2+4})$, (to appear: Canad. J. Math.).

[267] Mollin, R.A., and Zhang, L.-C., *Orders in quadratic fields II*, Proc. Japan Acad. **69**, Ser. A (1993), 368–371.

[268] Mollin, R.A., and Zhang, L.-C., *Reduced ideals, the divisor function, continued fractions and class numbers of real quadratic fields*, Publ. Math. Debrecen **43** (1993), 315–328.

[269] Mollin, R.A., and Zhang, L.-C., *A new criterion for the determination of class numbers of real quadratic fields*, (to appear: Proc. of the Fourth Conf. of the Canadian Number Theory Association 1994).

[270] Mollin, R.A., Zhang, L.-C., and Kemp, P., *A lower bound for the class number of a real quadratic field of* ERD-*type*, Canad. Math. Bull. **37** (1994), 90–96.

[271] Mordell, L.J., *A norm ideal bound for a class of biquadratic fields*, Kgl. Norske Vid. Selsk. Forh. **42** (1969), 53–55.

[272] Mordell, L.J., *On the Riemann hypothesis and imaginary quadratic fields with a given class number*, J. London Math. Soc. **9** (1934), 289–298.

[273] Mordell, L.J., **Diophantine Equations**, Academic Press, London, New York (1969).

[274] Mordell, L.J., *Reminiscenses of an octogenarian mathematician*, Amer. Math. Monthly **78** (1971), 952–961.

[275] Mordell, L.J., *On the pellian equation conjecture*, Acta. Arith. **6** (1960), 137–144.

[276] Nagell, T., *Über die Klassenzahl imaginäre-quadratischer Zahlkörper*, Abh. Math. Seminar Univ. Hamburg **1** (1922), 140–150.

[277] Nagell, T., *The diophantine equation* $x^2 + 7 = 2^n$, Norsk. Mat. Tidsskr. **30** (1948), 62–64.

[278] Nakahara, T., *On real quadratic fields whose ideal class groups have a cyclic p-subgroup*, Fac. Sci. Eng., Saga Univ. **6** (1978), 91–102.

[279] Nakahara, T. *The structure of 3-class groups in the real quadratic fields* $Q(\sqrt{D})$ *for D less than* 120000 *and for a few values of D between* 2000000 *and* 4033723, Reports Fac. Sci. and Eng., Saga University **23** (1995), 9–90.

[280] Narkiewicz, W., **Elementary and Analytic Theory of Algebraic Numbers**, PWN-Polish Scientific Publishers, Warszawa (1974).

[281] Narkiewicz, W., **Number Theory**, World Scientific Publishers, Singapore (1983).

[282] Neild, C., and Shanks, D., *On the 3-rank of quadratic fields and the Euler product*, Math. Comp. **28** (1974), 279–291.

[283] Odlyzko, A.M., *Discrete logarithms in finite fields and their cryptographic significance* in **Advances in Cryptology**, Proc. of Eurocrypt **84**, Springer (1985), 224–314.

[284] Oesterlé, J., *Versions effectives du théorème de Chebotarev sous L'Hypothèse de Riemann Généralisé*, Soc. Math. France Astérisque **61** (1979), 165–167.

[285] Patterson, C., *A* 538 *billion integer per second sieve device*, in **Proc. 1991 Canad. Conf. on Electrical and Computer Eng.** (1991), 13.1.1–13.1.4.

[286] Patterson, C., *The Derivation of a High Speed Sieve Device*, Ph.D. thesis, Dept. of Comp. Sci., University of Calgary, Calgary, Canada (1991).

[287] Patterson, C., and Williams, H.C., *A report on the University of Manitoba Sieve Unit*, Congresses Numerantium **37** (1983), 85–98.

[288] Perron, O., **Die Lehre von den Kettenbrüchen**, Stuttgart (1977), Reprint Chelsea New York.

[289] Pinz, J., *Elementary methods in the theory of L-functions, VII; upper bound for* $L(1, \chi)$, Acta Arith. **32** (1977), 397–406. (Corrigendum ibid **33** (1977), 293–295.)

[290] Pohlig, S., and Hellmann, M., *An improved algorithm for computing discrete logarithms over* $GF(p)$ *and its cryptographic significance*, IEEE Trans. Inform. Theory **24** (1978), 106–110.

[291] Pohst, M., and Zassenhaus, H., *Über die Berechnung von Klassenzahlen und Klassengruppen algebraischer Zahlkörper*, J. Reine Angew. Math. **361** (1985), 50–72.

[292] Pohst, M., and Zassenhaus, H., **Algorithmic algebraic number theory**, Cambridge University Press, New York (1990).

[293] Poletti, L., *Au sujet de la décomposition des termes de la séries* $z = x^2 + x +$ 146452961, Sphinx, **9** (1939), 83–85.

[294] Pollard, H., and Diamond, H.G., **The theory of algebraic numbers**, Second Edition, MAA Carus Monograph **9** (1975).

[295] Pomerance, C., *The quadratic sieve factoring algorithm*, in **Advances in Crypt**, Proc. Eurocrypt 84, LNCS **209** (1985), 169–182.

[296] Purdy, G., Terras, R. and A., and Williams, H.C., *Graphing L-functions of Kronecker symbols in the real part of the critical strip*, Math. Student **47** (1979), 101–131.

[297] Queen, C.S., *A simple characterization of prinicpal ideal domains*, Acta. Arith. **64** (1993), 125–128.

[298] Rabinowitsch, G., *Eindeutigkeit der Zerlegung in Primzahlfaktoren in quadratischen Zahlkörpern*, Proc. Fifth Internat. Congress Math., Cambridge (1913), 418–421.

[299] Rabinowitsch, G., *Eindeutigkeit der Zerlegung in Primzahlfactoren in quadratischen Zahlkörpern*, J. Reine Angew. Math. **142** (1913), 153–164.

[300] Ramanujan, S., **Collected papers**, Chelsea publ. Co., New York (1962).

[301] Ramasamy, A.M.S., *Ramanujan's equation*, J. Ramanujan Math. Soc. **7** (1992), 133–153.

[302] Redei, L., *Über die Klassenzahl des imaginären quadratischen Zahlkörpers*, J. Reine Angew. Math. **159** (1928), 210–219.

[303] Reiner, I., **Maximal Orders**, Academic Press, London, New York (1975).

[304] Ribenboim, P., *Euler's famous prime generating polynomial and the class number of imaginary quadratic fields*, L'Enseignment Math. t. **34** (1988), 23–42.

[305] Ribenboim, P., *Remarks on exponential congruences and powerful numbers*, J. Number Theory **29** (1988), 251–263.

[306] Ribenboim, P., **The book of prime number records**, Springer–Verlag, New York (1988).

[307] Ribenboim, P., *Consecutive Powers*, Expositiones Math. **2** (1984), 193–221.

[308] Richaud, C., *Sur la résolution des équations* $x^2 - Ay^2 = \pm 1$, Atti Accad. pontif. Nuovi Lincei (1866), 177–182.

[309] Richert, H.E., *Selberg's sieve with weights*, Mathematika **16** (1969), 1–22.

[310] Riemann, B. *Uber die Anzahl der Primzahlen unter einer gegehener Grösse*, Monatsberichte Akad. Berlin, November (1859); Collected Works, Dover Pub. Inc., N.Y. (1953), 145–153.

[311] Rivest, R., Shamir, A., and Adleman, L., *A method for obtaining digital signatures and public-key cryptosystems*, Comm. of the ACM **21** (1978), 120–126.

[312] Rosen, K.H., **Elementary Number Theory and its Applications**, (Third Edition), Addison–Wesley Pub. Co., Reading, Mass. (1993).

[313] Rosser, J.B., and Schoenfeld, L., *Approximate formulas for some functions of prime numbers*, Illinois J. Math. **6** (1962), 64–94.

[314] Sasaki, R., *On a lower bound for the class number of an imaginary quadratic field*, Proc. Japan Acad. Ser. A **62** (1986), 37–39.

[315] Schaffstein, K., *Tafel der Klassenzahlen der reellen quadratischen Zahlkörper mit Primzahl-Diskriminante unter* 12,000 *und zwischen* 100,000 – 101,000 *and* 1,000,000 – 1,001,000, Math. Ann. **98** (1928), 745–748.

[316] Scheidler, R., Buchmann, J., and Williams, H.C., *A key exchange protocal using real quadratic fields*, J. Cryptology (1994), 171–199.

[317] Schinzel, A., *On the relation between two conjectures on polynomials*, Acta. Arith. **38** (1980), 285–322.

[318] Schmuely, Z., *Composite Diffie-Hellmann public key generating schemes are hard to break*, Technical report **356**, Computer Sci. Dept., Technion–Israel Institute of Technology (Feb. 1985).

[319] Schneier, B., **Applied Cryptography**, Wiley, New York (1994).

[320] Scholtz, A., *Über die Lösbarkeit der Gleichung* $t^2 - Du^2 = -4$, Math. Z. **39** (1935), 95–111.

[321] Schoof, R.J., *Quadratic fields and factorization* in **Computational Methods in Number Theory**, (H.W. Lenstra Jr. and R. Tijdeman (eds.)), Math. Centrum Tracts, Number 155, Part II, Amsterdam (1983), 235–286.

[322] Schur, I., *Einige Bemerkungen zu der vorstehenden Arbeit des Herrn Polya Über die Verteilung der quadratischen Reste und Nichtreste*, Nachr. Ges. Wiss. Göttingen (1918), 30–36.

[323] Selenius, C.-O., *Konstruktion and Theorie halbregelmässiger Kettenbrüche mit idealer relativer Approximation*, Acta. Acad. Abo. Math. Phys. **22** (1960), 77.

[324] Sentance, W.A., *Occurrences of consecutive odd powerful numbers*, Amer. Math. Monthly **88** (1981), 272-274.

[325] Shanks, D., *On the conjecture of Hardy and Littlewood concerning the number of primes of the form* $n^2 + a$, Math. Comp. **14** (1960), 320–332.

[326] Shanks, D., *Supplementary data and remarks concerning a Hardy-Littlewood conjecture*, Math. Comp. **17** (1963), 188–193.

[327] Shanks, D., *On Gauss' class number one problem*, Math. Comp. **23** (1969), 151–163.

[328] Shanks, D., *Class number, a theory of factorization and genera*, Proc. Symp. Pure Math. **20** AMS (1971), 415–440.

[329] Shanks, D., *Five number theoretic algorithms*, Proc. of the Second Manitoba Conf. on Numerical Math. (Univ. Man., Winnipeg) (1972), 51–70.

[330] Shanks, D., *Systematic examination of Littlewood's bounds in* $L(1,\chi)$, Proc. Sympos. Pure Math., Amer. Math. Soc., Providence, R.I. (1973), 267–283.

[331] Shanks, D., *The infrastructure of real quadratic fields and its application*, Proc. 1972 Number Theory Conf., Boulder, Co. (1973), 217–224.

[332] Shanks, D., *The simplest cubic fields*, Math. Comp. **28** (1974), 1137–1152.

[333] Shanks, D., *Calculation and applications of Epstein zeta functions*, Math. Comp. **29** (1975), 271–287.

[334] Shanks, D., *A survey of quadratic, cubic and quartic algebraic number fields (from a computational viewpoint)*, Congressus Numerantium **17** (1976), 15–40.

[335] Shanks, D, and Weinberger, P., *A quadratic field of prime discriminant requiring three generators for its class group, and related theory*, Acta. Arith. **21** (1972), 71–87.

[336] Shapiro, H.S. and Slotnik, D.L., *On the mathematical theory of error correcting codes*, IBM J. Res. Develop. **3** (1959), 25–34.

[337] Shorey, T.N., and Stewart, C.L., *On the Diophantine equation* $ax^{2t}+bx^ty+cy^2=d$ *and pure powers in recurrence sequences*, Math. Scand. **52** (1983), 24–36.

[338] Siegel, C.L., *Über die Klassenzahl quadratischer Zahlkörper*, Acta Arith. **1** (1935), 83–86.

[339] Sierpiński, W., **Elementary Theory of Numbers**, PWN-Polish Scientific Publishers, Warszawa (1964).

[340] Skinner, C., *The diophantine equation* $x^2=4q^n-4q+1$, Pacific J. Math. **139** (1989), 303–309.

[341] Skolem, T., Chowla, S., and Lewis, D.J., *The diophantine equation* $2^{n+2}-7=x^2$ *and related problems*, Proc. Amer. Math. Soc. **10** (1959), 663–669.

[342] Slavutskiĭ, I.Š. (Slavutsky, I.Sh.), *Upper bounds and numerical calculation of the number of ideal classes of real quadratic fields*, Amer. Math. Soc. Trans. **82** (1969), 67–71.

[343] Slavutskiĭ, I.Š. (Slavutsky, I.Sh.), *On Mordell's theorem*, Acta. Arith. **11** (1965), 57–66.

[344] Smith, H.J.S., **Report in the Theory of Numbers**, Chelsea, N.Y. (1965).

[345] Stanton, R.G., Sudler Jr., G., and Williams, H.C., *An upper bound for the period of the simple continued fractions for* $\sqrt{D}$, Pacific J. Math. **67** (1976), 525–536.

[346] Stark, H.M., *A complete determination of the complex quadratic fields of class number one*, Michigan Math. J. **14** (1967), 1–27.

[347] Stark, H.M., *A transcendence theorem for class number problems*, Ann. of Math. **94** (1971), 153–173.

[348] Stephens, A.J., and Williams, H.C., *Computation of real quadratic fields with class number one*, Math. Comp. **51** (1988), 809–824.

[349] Stephens, A.J., and Williams, H.C., *Some computational results on a problem concerning powerful numbers*, Math. Comp. **50** (1988), 619–632.

[350] Stephens, A.J., and Williams, H.C., *Some computational results on a problem of Eisenstein*, in **Number Theory**, (J-M De Koninck and C. Levesque (eds.)), Walter de Gruyter, Berlin, New York (1989), 869–886.

[351] Stephens, A.J., and Williams, H.C., *An open architecture number sieve* in **Number Theory and Cryptography** (J.J. Loxton (ed.)), London Math. Soc. Lecture Note Series **154** (1990), 38–75.

[352] Stewart, I.N.,and Tall, D.O., **Algebraic Number Theory**, Chapman–Hall Mathematics Series, John Wiley, New York (1979).

[353] Stormer, C., *Solution complète en nombres de l'équation* $m \arctan g \frac{1}{x} + n \arctan g \frac{1}{y} = k\frac{\pi}{4}$, Bull. Soc. Math. France **27** (1899), 160–170.

[354] Stormer, C., *Quelques théorèmes sur l'équation de Pell* $x^2 - Dy^2 = \pm 1$ *et leurs applications*, Christiania Videnskabens Selskabs Skrifter, Math. Nat. Kl. 1897, No. 2, 48 pages.

[355] Stroeker, R.J., *How to solve a diophantine equation*, Amer. Math. Monthly **91** (1984), 385-392.

[356] Suetonius, G. **The Lives of the Twelve Caesars**, Heritage Press, New York (1965).

[357] Suzuki, M., **Group Theory I**, Springer–Verlag, Berlin (1982).

[358] Szekeres, G., *On the number of divisors of* $x^2 + x + A$, J. Number Theory **6** (1974), 434–442.

[359] Tatuzawa, T., *On a theorem of Siegel*, Japan. J. Math. **21** (1951), 163–178.

[360] Taylor, R., and Wiles, A., *Ring theoretic properties of certain Hecke algebras* (preprint).

[361] Tennenhouse, M., and Williams, H.C., *A note on class number one in certain real quadratic and pure cubic fields*, Math. Comp. **46** (1986), 333–336.

[362] Thue, A., *Über Annäherungswerte algebraischer Zahlen*, J. Reine Angew. Math., **135** (1909), 284–305.

[363] Tijdeman, R. *On the equation of Catalan*, Acta. Arith. **29** (1976), 197–209.

[364] Trotter, H.F., *On the norms of units in quadratic fields*, Proc. Amer. Math. Soc. (1969), 198–201.

[365] Tymoczko, T., *The Four-Color problem and its philosophical significance*, Journal of Philosophy **76** (1979), 57–83.

[366] Tymoczko, T. (ed.), **New directions in the philosophy of mathematics**, Birkhäuser, Boston (1985).

[367] Tymoczko, T., *Mathematics, science and ontology*, Synthese **88** (1991), 201–228.

[368] Tzanakis, N., *On the diophantine equation $y^2 - D = 2^k$*, J. Number Theory **17** (1983), 144–164.

[369] Tzanakis, N., and Wolfskill, J., *On the diophantine equation $y^2 = 4q^n + 4q + 1$*, J. Number Theory **23** (1986), 219–237.

[370] Tzanakis, N., and Wolfskill, J., *The diophantine equation $x = 4q^{a/2} + 4q + 1$ with an application to coding theory*, J. Number Theory **26** (1987), 96–116.

[371] van de Lune, J., te Riele, H.J.J., and Winter, D.T., *On the zeros of the Riemann zeta function in the critical strip IV*, Math. Comp. **46** (1986), 667–681.

[372] Van der Pol, B., and Speziali, P., *The primes in $k(\zeta)$*, Indag. Math. **13** (1951), 9–15.

[373] Wada, H., *A table of ideal class numbers of real quadratic fields*, Kôkyûroku in Math **10** (1981), Sophia University, Tokyo.

[374] Walsh, P.G., *The Pell equation and powerful numbers*, Master's thesis, University of Calgary, Canada (1988).

[375] Washington, L.C., **Introduction to Cyclotomic Fields**, Springer–Verlag, New York (1982).

[376] Washington, L.C., and Zhang, X., *Cyclic subgroups in class groups of real quadratic fields*, Internat. Centre for Theoretical Physics, Preprint IC/92/393.

[377] Weinberger, P., *Real quadratic fields with class numbers divisible by n*, J. Number Theory **5** (1973), 237–241.

[378] Weinberger, P. *Exponents of the class groups of complex quadratic fields*, Acta. Arith. **22** (1973), 117–124.

[379] Whitford, E.E., *The Pell Equation*, Ph.D. thesis, Columbia University, Press of the New Era Printing Co., Lancaster, PA, U.S.A. (1912).

[380] Williams, H.C., *Some results concerning the nearest integer continued fraction expansion of $\sqrt{D}$*, J. für die reine und Angew. Math. **315** (1980), 1–15.

[381] Williams, H.C., *A note on the Fibonacci quotient $F_{p-\epsilon}/p$*, Canad. Math. Bull. **25** (1982), 366–370.

[382] Williams, H.C., *A note on the period length of the continued fraction expansion of certain $\sqrt{D}$*, Utilitas Math. **28** (1985), 201–209.

[383] Williams, H.C., *Continued fractions and number theoretic considerations*, Rocky Mtn. J. Math. **15** (1985), 621–655.

[384] Williams, H.C., *Eisenstein's problem and continued fractions*, Utilitas Math. **37** (1990), 145–158.

[385] Williams, H.C., *Some generalizations of the S_n sequence of Shanks*, Acta Arith. **69** (1995), 199–215.

[386] Williams H.C., and Broere, J., *A computational technique for evaluating $L(1,\chi)$ and the class number of a real quadratic field*, Math. Comp. **30** (1976), 887–893.

[387] Williams, H.C., and Buhr, P., *Calculation of the regulator of* $\mathbf{Q}(\sqrt{D})$ *by the use of the nearest integer continued fraction algorithm*, Math. Comp. **33** (1979), 369–381.

[388] Williams, H.C., and Shallit, J.O., *Factoring integers before computers*, Proc. Symp. App. Math. **48** (1994), 481–531.

[389] Williams, H.C., and Wunderlich, M.C., *On the parallel generation of the residues of the continued fraction factoring algorithm*, Math. Comp. **48** (1987), 405–423.

[390] Williams, K.S., and Buck, N., *Comparisons of the lengths of the continued fractions of $\sqrt{D}$ and $\frac{1}{2}(1+\sqrt{D})$*, Proc. Amer. Math. Soc. **120** (1994), 995–1002.

[391] Yamamoto, Y., *Real quadratic fields with large fundamental units*, Osaka J. Math. **8** (1971), 261–270.

[392] Yamamoto, Y., *On unramified Galois extensions of quadratic number fields*, Osaka Math. J. **7** (1970), 57–76.

[393] Yokoi, H., *On real quadratic fields containing units with norm* -1, Nagoya Math. J. **33** (1968), 139–152.

[394] Yokoi, H., *On real quadratic fields containing units with norm* 1, J. Number Theory **2** (1970), 106–115.

[395] Yokoi, H., *Class number one problem for certain kind of real quadratic fields*, in **Proc. Internat. Conf. on Class Numbers and Fundamental Units of Algebraic Number Fields** (Kattata Japan 1986).

[396] Zagier, D., *A Kronecker limit formula for real quadratic fields*, Math. Ann. **213** (1975), 153–184.

[397] Zagier, D., **Zetafunktionen und Quadratische Korper**, Springer–Verlag, Berlin, New York (1981).

[398] Zhang, L.-C., and Gordon, J., *On unit solutions of the equation $xyz = x+y+z$ in a number field with unit group of rank* 1, Acta. Arith. LVII (1991), 155–158.

[399] Zhang, L.-C., and Gordon, J., *On unit solutions of the equation $xyz = x+y+z$ in not totally real cubic fields*, Canad. Math. Bull. **34** (1991), 141–144.

[400] Zhang, M.-Y., *On Yokoi's Conjecture*, (to appear: Math. Comp.).

[401] Zhenfu, C., *On the Diophantine equation $x^{2n} - Dy^2 = 1$*, Proc. Amer. Math. Soc. **98** (1986), 11–16.

[402] Zimmer, H.G., **Computational problems, methods and results in algebraic number theory**, Springer Lecture Notes in Math. **262** (1972).

Index